BIOREACTORS FOR TISSUE ENGINEERING

Bioreactors for Tissue Engineering

Principles, Design and Operation

Edited by

JULIAN CHAUDHURI
University of Bath, U.K.

and

MOHAMED AL-RUBEAI
University College Dublin, Ireland

Springer

A C.I.P. Catalogue record for this book is available from the Library of Congress.

ISBN-10 1-4020-3740-6 (HB)
ISBN-13 978-1-4020-3740-5 (HB)
ISBN-10 1-4020-3741-4 (e-book)
ISBN-13 978-1-4020-3741-2 (e-book)

Published by Springer,
P.O. Box 17, 3300 AA Dordrecht, The Netherlands.

www.springeronline.com

Legend for front cover photographs:
Upper left: Fibroblast cells growing on a collagen matrix
Upper right: Schematic of the pulsatile tissue engineered heart valve bioreactor
Lower left: Minibioreactor perfusion chamber
Lower right: Scanning electron micrograph of a poly(lactic acid-co-glycolic acid) hollow fiber

Printed on acid-free paper

Printed in the Netherlands.

TABLE OF CONTENTS

PREFACE

The emerging field of tissue engineering represents the integration of concepts and ideas from the biological sciences, engineering disciplines, material science and clinical procedures. Therapies are being developed and tested in clinical trials, with the potential to help patients suffering from the loss of tissue or its function. Engineering tissues, for example bone, cartilage, blood vessels and liver, involves three key factors. Firstly, the cell source is of paramount importance. Although there has been much interest in the use of autologous cells to treat individual patients, the field has recently refocused on the potential inherent in the use of progenitor or stem cells, and their application in many forms of regenerative medicine including tissue engineering. A second factor is the design of appropriate scaffolds to mimic the role of the extracellular matrix. This is a burgeoning field where both natural and synthetic materials are being developed with enhanced cell adhesion and proliferation properties. Finally, tissue engineering relies on appropriate cell and tissue cultivation methods – at the core of which lies the bioreactor.

To date, engineered tissues have been made possible through innovations in cultivation technology and bioreactor engineering. These new therapies will need to utilise the achievements made in the last few decades in developing high performance bioreactors for the production of pharmaceutical products, particularly those derived from mammalian cell culture, which in turn demands understanding of the hydrodynamic and biochemical factors in the cell and tissue environment. In tissue engineering, the situation is made more complex by the need to generate the correct three-dimensional architecture of the tissue in order to restore function.

Bioreactor design, control and operation are well established in engineering disciplines, but they are relatively new to biology and medicine. Hence, an engineering approach to the development of functional tissue equivalents will require a new breed of interdisciplinary researchers well trained at the biology/engineering interface. While many of the obstacles to the development of 3D functional tissue construct appear formidable, none are insurmountable provided an appropriate approach to the design, characterisation and operation of the bioreactor environment is used.

This book brings together the knowledge and experience of researchers who are applying engineering principles of bioreactor cultivation of cells/tissues to the in vitro generation of functional tissue constructs. The aim of this book is to provide the reader with a number of contributions that map out the breadth of topics that are being considered with respect to providing advanced culture systems for regenerating tissue. Individual chapters discuss concepts that will be important in the development of reactors that can be use for production of clinical scale tissue. Some chapters are general in that they provide state of the art reviews of specific reactors that can be applied to several tissue types. Other chapters focus on individual tissues and describe innovations being made towards growing functional tissue. A key feature of tissue bioreactors that is represented in this volume is the use of mechanical forces during tissue culture to condition the nascent tissue, thereby giving rise to in vivo-like function. Finally, the breadth of the bioreactor field is illustrated through the use of MRI techniques for evaluating tissue during growth, and also the cryopreservation of the cultured tissue so it can be stored and transported for clinical use.

The material is addressed primarily to those interested in tissue engineering, but many topics will also be of interest to cell culture technologists and biochemical engineers, in both academia and industry. It is hoped that this book will stimulate the development of new ideas emerging from this exciting interface between engineering and the life sciences. Finally, the editors are grateful for the support of all the contributors and the publisher who have made this book possible.

Julian B. Chaudhuri
Mohamed Al-Rubeai

CHAPTER 1

BIOREACTOR SYSTEMS FOR TISSUE ENGINEERING: A FOUR-DIMENSIONAL CHALLENGE

M. ELLIS, M. JARMAN-SMITH AND J.B. CHAUDHURI

Department of Chemical Engineering, University of Bath, Bath, UK

1. INTRODUCTION

The aim of tissue engineering is to regenerate or recreate human tissue to give functional tissue in order to repair damage or disease. It is a cell-based therapy that will enable the restoration of function to a variety of tissues and organs (Freed and Vunjak-Novakovic 1997). The key components of tissue engineering are: selection of the appropriate cell types (autologous or allogenic) to give rise to the required tissue; development of a suitable scaffold or matrix to support cell attachment; and provision of a controlled environment to allow the cells to proliferate and differentiate to tissue structures. Figure 1 shows an outline process of tissue engineering that, in addition to the steps mentioned above, indicates some of the downstream issues to be considered. This chapter is primarily concerned with the provision of the correct culture environment for tissue growth. For further information on cells and scaffolds for tissue engineering there are many recent review papers on these subjects (Bock and Goode 2003).

To date, engineered tissues have generally been simple 2D structures using a single cell type, for example dermal replacements for ulcer or burns treatment. One of the key challenges in tissue engineering is the ability to grow 3D tissue structures of relevant clinical sizes. In human tissues there are considerable mechanical and biochemical interactions between cells of the same type, between different cells and also with the extracellular matrix and its components. It has been documented that the behaviour of cells in 2D can be very different in 3D culture (Abbott 2003) and this has important implications for the in vitro generation of functional human tissues. A second challenge will be the growth and 3D assembly of multiple cell types that are required for more complex functional tissues. In addition to the 3D requirements for tissue structure it must also be considered that tissue growth is a dynamic, non-steady state process where biological parameters, physical rate

J.B. Chaudhuri & M. Al-Rubeai (eds.), Bioreactors for Tissue Engineering, 1 - 18.

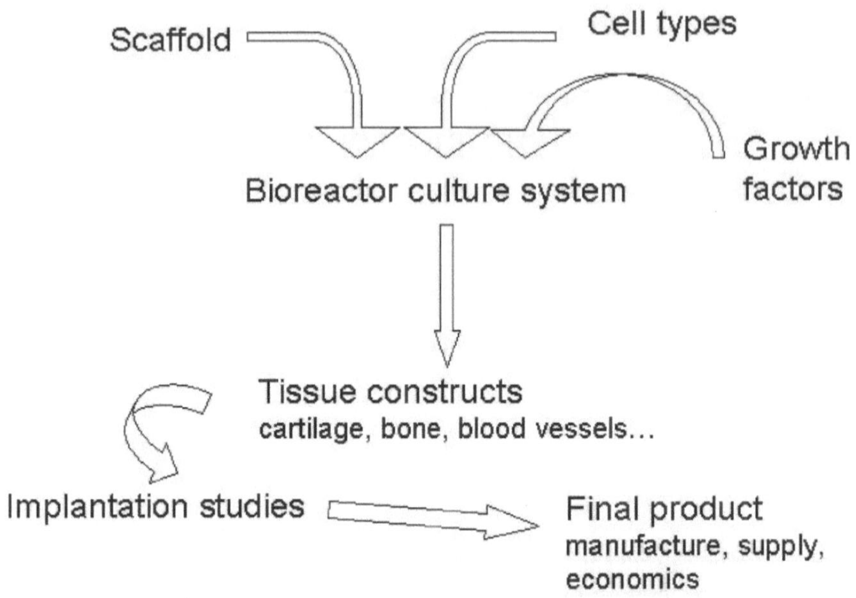

Figure 1: Outline process of tissue engineering

processes and the local mechanical environment are continually changing with time – in essence tissue engineering is a 4D challenge.

Human cells are extremely sensitive to fluctuations in culture environment and the traditional depletion and toxic by-product accumulation associated with batch culture can adversely affect them. Tissue culture traditionally follows the methods of batch or semi-batch culture, with the carbon source (glucose, glutamine) present at the start of the culture. The majority of 2D tissue culture systems are T-flasks and well-plates in which cells are grown statically without any mixing. These static systems have limitations that arise from their non-homogeneous nature. Without mixing, concentration gradients of pH, nutrients, toxins, gases and growth factors will occur in the culture medium (Collins et al. 1998). This will give rise to different cell behaviour in different spatial locations in the culture and will also reduce culture reproducibility.

The alternative to static growth is to use dynamic culture systems where mixing gives rise to homogeneous concentrations of nutrients, toxins and other components thus removing the problems associated with gradients and stabilising the culture environment. Other advantages of mixing in tissue cultures are evident: for example, in the growth of chondrocytes for cartilage tissue, the mixing provided by a bioreactor system improved the yield and uniformity of cell seeding onto scaffolds, and increased cell proliferation rates (Freed and Vunjak-Novakovic 1997).

In this chapter we introduce the role of bioreactors for tissue engineering. The chapter is divided into two main sections. The first section considers general bioreactor design requirements, specific issues such as nutrient transport and the use of mechanical forces to condition growing tissues, and describes some reactor types used for tissue engineering. The second part illustrates some of these concepts with respect to the creation of bone tissue.

2. TISSUE ENGINEERING BIOREACTORS

2.1 The Role of Tissue Engineering Bioreactors

A bioreactor is the general term applied to a closed culture environment, that is usually mixed, that enables control of one or more environmental or operating variables that affect biological processes. In industrial biotechnology the use of a bioreactor (or fermenter) is to grow eukaryotic or prokaryotic cells to high cell densities in order to produce a metabolic product, enzyme, or to overexpress a recombinant gene product. In these applications several types of bioreactor system have been utilised, the most common being the batch stirred tank reactor. Other examples include continuous stirred tank reactors, packed beds, and membrane bioreactors (Asenjo and Merchuk 1995).

In tissue engineering, bioreactors are used to give controlled and reproducible cell and tissue proliferation. The variables that are controlled include temperature; pH, gas concentration, media flowrate, shear stress, hydrodynamics and mechanical forces. For tissue construction the endpoints that we are trying to achieve include construct size, shape/architecture, biochemical composition, biological function and morphology. In tissue engineering the desired product is 3D functional tissue and the bioreactors used here differ from conventional cell culture bioreactors in that they must provide an environment that prompts the cells into proliferating and differentiating as they would *in vivo*. Whilst proliferation of cells (mitosis) is required to form the bulk of the tissue, there is also a separate distinct cell cycle phase (interphase) of differentiation that allows the cells to specialise (Strehl et al. 2002). Furthermore, many tissue culture products involve the co-culture of cell types, as opposed to single cell types grown in fermenters. In addition, increased complexity arises as tissue products are grown on a solid scaffold.

In tissue engineering, bioreactors may be applied in several areas. Firstly, we need bioreactor systems for the expansion of cells for transplantation, for example, haematopoietic cells, embryonic stem cells, mature blood cells (Cabral 2001). Secondly, we use bioreactors to grow 3D tissues prior to implantation in humans, eg skin, cartilage, bone, blood vessels etc. Finally, organ support devices, for example for liver and kidney, may also be considered as bioreactor systems. This chapter focuses on the second of these applications, that is for growing 3D tissues. In summary, the main challenge of tissue growth for tissue engineering is not the mass culture and bulk production of one specific cell or factor as achieved by immobilised cell techniques, but the sufficient growth of a multitude of cell types in an organised, interactive 3D structure.

2.2 Bioreactor Systems Design for Tissue Engineering

2.2.1 General Design Requirements

Although the use of bioreactor systems for tissue engineering is in its infancy we can consider some general features to allow us to rationally design reactors. A bioreactor should perform, at least one of, the following: create an environment that enables cells to proliferate and differentiate as *in vivo*; establish spatially uniform cell distributions on 3D scaffolds; maintain desired concentrations of nutrients; provide efficient mass transfer to tissue; and expose tissues to physical stimuli (Freed and Vunjak-Novakovic 1997). These aspects will be considered in detail below.

For the growth of all types of mammalian cells (cell lines, primary cells, progenitor cells) precise control of the *in vitro* physiological environment during culture is essential. The intention is to maintain key variables at values that mimic in vivo conditions. Primarily, this means controlling the temperature, nutrient and oxygen concentrations and pH values within close limits. The system should also be designed to operate under sterile conditions such that the only cell type/s present are those required and preventing entry of infecting microorganisms. This requirement can be further broken down into the following: pre-sterilization of equipment; preparation of sterile media; maintenance of sterility during culture (Sinclair and Ashley 1995). Because of the extended length of tissue culture a bioreactor system should allow aseptic feeding and sampling to minimise the probability of infection.

In order to control the reactor environment it is necessary to measure key parameters. For example these could include temperature, pH, O_2 and CO_2 concentrations, nutrient composition, media flowrate, metabolite concentrations, and specific tissue markers. It must be remembered that the cultivation and tissue-specific parameters are changing with time. Tissue culture is a non-steady state process with some parameters that continually change. Currently it is not possible to easily measure all of these variables on-line, especially the biological factors. Thus there is a requirement to develop sensors for on-line measurement, or alternatively to remove samples for fast, near-on line analysis.

As we try to regenerate more complex tissues comprising multiple cell types, it will be necessary for the bioreactor to enable the simultaneous culture of two or more cell types. For example, most tissue engineered blood vessels are co-cultures of endothelial and smooth muscle cells. This aspect of tissue culture will require an appropriate scaffold to allow co-culture, but this must be considered in tandem with the bioreactor design. For example, co-culture may require different media and growth factors for each cell type. It may be necessary to maintain the two cell types under different culture conditions to expand cell numbers and then at longer times switch to a common cultivation protocol.

In addition to tissue-related considerations mentioned above, there are also manufacturing considerations that must be resolved. For the tissue constructs to be acceptable for clinical trials and subsequent therapy, cultivation must be carried out to appropriate good manufacturing practice (GMP) and quality assurance (QA) guidelines to ensure consistent cell and tissue quality. It should be noted that these GMP and QA issues will also apply to cell sourcing and expansion processes and scaffold production.

2.2.2 Scaffold Seeding

Initially, the scaffold will need to be seeded with cells. Seeding may be carried out in a separate operation where the cells are allowed to attach on the scaffold and then the seeded scaffold is inserted into the bioreactor, or the cells are seeded directly onto the scaffold within the bioreactor. In either case scaffold seeding is a batch operation to ensure maximum cell attachment. A key challenge for seeding is to be able to uniformly distribute high initial cell numbers onto a 3D scaffold. Static seeding is most widely used, however, there are advantages to be found with the use of dynamic seeding. These include higher attachment efficiencies and more uniform cell distribution on the scaffold leading to higher quality tissue in terms of the structure and composition (Freed and Vunjak-Novakovic 1997; Martin et al. 2004). A variety of dynamic seeding methods have been reported including simple mixing, seeding in spinner flasks, and convective seeding in perfusion reactors (Martin et al. 2004).

2.2.3 Nutrient Transport

The provision of nutrients and gases to nascent tissue is probably the limiting factor for the creation and maintenance of multi-layer constructs. *In vivo,* tissues are on average no further than 100 μm from a capillary providing nutrients (Yang et al. 2001). It is useful therefore to define a microenvironment based on this typical length scale for tissue engineering. We can consider a cube with sides of 100 μm as being a fundamental microenvironment for tissue (Palsson 2000). This microenvironment is characterised by neighbouring cells, chemical environment and geometry. A tissue section of this size represents the size of organ functional subunits, and contains between 500 – 1000 cells, usually a mixed cell population. This number can vary with tissue type, for example, in cartilage there is approximately one cell per 100 μm^3 (Palsson 2000). The *in vitro* culture of tissue a few millimeters thick will therefore result in cells within the constructs being greater than 100 μm from a nutrient source and subject to low nutrient values. *In vivo* this problem is overcome by the vascular network: *in vitro* this will require porous scaffolds to allow high nutrient fluxes to the cells. Advances in scaffold design will contribute to solving this problem. In addition, a porous scaffold must be placed in a dynamic culture environment to maximise fluid and molecular mass transfer in the scaffold. Thus, appropriate design of the scaffold and bioreactor should be undertaken in order to maximise the mass transfer processes. The design of bioreactors must be such that concentrations of nutrients, oxygen and growth factors must be homogeneous at this 100 μm scale.

Mass transfer in tissue engineering involves the transport of nutrients and gases from the place that they are supplied to the site of the cells. Generally, two processes govern mass transfer of nutrients to cells: convective mass transfer or the transport of molecules in the bulk motion of the media flow; and diffusion, the movement of species down a concentration gradient. In tissue engineering, convective mass transfer will get the nutrients to the vicinity of the scaffold and the cells. From this point it will be molecular diffusion that governs transfer through the scaffold structure and cell layers. The ratio between convective transport and diffusive transport is a key consideration (Peclet number). At the cell surface the nutrients will be taken up and here there is a balance between the rate at which the nutrients are taken up and the rate at which they are supplied by diffusion (Damkohler number). When this ratio is large, nutrients are consumed more rapidly than they can diffuse to the surface, and the surface concentration will fall. When it is small, the uptake rate is slow and diffusion maintains the nutrient concentrations as nearly constant.

Another mass transfer consideration is the choice of a suitable media flowrate to ensure there is continuous supply of fresh nutrient and removal of toxins. We must consider the balance between the residence time of the media in the reactor, the resulting shear rate and its effect on the nascent tissue, and nutrient demand requirements. Overall *in vivo* rates of oxygen uptake of 25 – 250 μmol $O_2/cm^3/h$ and perfusion rates of 0.07 $mL/cm^3/min$, based on an average cellularity of 500 million cells per cm^3, may be good starting points (Palsson 2000).

2.2.4 Biomechanical Stimuli

Many tissues have mechanical functions *in vivo* (eg bone, cartilage, ligament, blood vessels). There is evidence that mechanical input (i.e. hydrostatic pressure, fluid flow induced shear) during in vitro culture gives tissues that have characteristics more like in vivo structures. At its most simple, fluid flow affects cell shape and cell function (Freed and Vunjak-Novakovic 1997). Cellular morphology can be affected by the environment and so some reactors have incorporated mechanical shear forces to replicate *in vivo* stresses. Pulsed bioreactors with adjustable stroke volumes and rates have been used to culture cardiac patches made from PGA or collagen (Hoerstrup et al. 2000). Carver and Heath (1999) used a perfusion chamber with a PGA matrix, but also incorporated an intermittent compression cycle using hydrostatic pressure (5 seconds on, 15 s off, for a 20 min period every 4 h). This continued for cultures of 5 weeks and appeared to increase cartilage synthesis. Changes in mechanical load may also regulate ECM gene expression by fibroblasts. Stretched fibroblast matrix upregulated ECM proteins typical to tendons and ligaments (tenascin-C, collagen XII) (Chiquet 1999). Specific reactors are being developed to induce stimuli as cultures progress (Dennis and Kosnik 2000).

2.2.5 A Systems Approach

Figure 2 shows some of the potential interactions that should be appreciated when engineering tissues in a bioreactor system. The figure represents a simple systems approach to tissue engineering; *ie* it is important to consider a complex interacting system and that we cannot only focus on single components working alone. Since many of the interactions can be studied on a biological, physical or chemical level, these interactions demonstrate the complexity of tissue engineering. Some authors have derived data or constructed models for specific parts of this model (Endy and Brent 2001; Lauffenburger and Griffith 2001). The increasing use of bioreactors in tissue engineering means that control and manipulation of the culture environment will become as important as cell and scaffold issues.

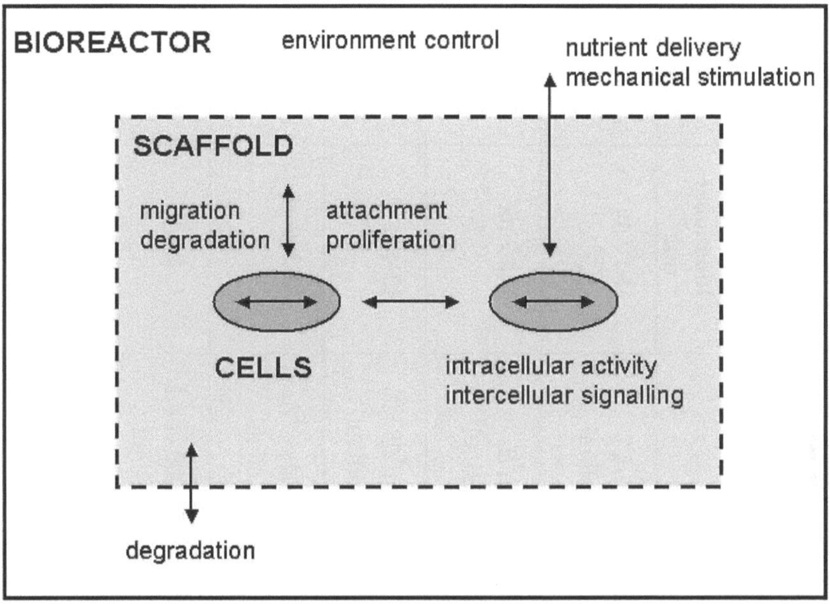

Figure 2: A systems view of tissue engineering processes

2.3 Common Bioreactor Types

Recently, the need to culture specific types of tissue has seen the introduction of more elaborate reactor systems to replicate a particular organ and accommodate the accompanying use of scaffolds to guide and support cell growth. For a given clinical application culture conditions may need to be optimised with respect to cell source, scaffold, media composition and fluid dynamics (Freed and Vunjak-Novakovic 1997). Thus we should consider that it is unlikely that a single bioreactor design will be used for all cell and tissue culture operations, but that bespoke reactor designs will be needed for individual tissues or classes of tissue. In terms of classifying tissue bioreactors, we can divide reactors up by their main mode of operation, for example perfusion reactors, stirred vessels and rotating reactors. We can then superimpose onto the base mechanism the attributes required by the specific tissue, ie geometry, fluid dynamics, mechanical forces etc. Table 1 summarises the characteristics of typical bioreactors used for tissue engineering, estimating the reactor sizes required for production of an functional unit of tissue and a complete organ.

Table 1. Summary of bioreactor characteristics

Reactor type	**Media change**	**Mixing & shear characteristics**	**Tissue structure**	**Culture dimensions[1]**	**Reactor size needed to grow a functional unit[2]**	**Reactor size needed to grow an organ[3]**
Tissue culture flask	Batch	Poorly mixed, no shear, diffusion	2D sheet	290 cm^2/L 1 x 10^5 cell/ml	1-10 ml	10-1000 L
Agitated vessels	Batch or continuous	Well mixed, turbulent, convection	2D or 3D	2,800 cm^2/L 5 x 10^5 cell/ml	0.2-2 ml	2-200 L
Packed beds	Continuous feed (perfusion)	Well mixed, laminar flow, convection	3D	18,000 cm^2/L 2.5 x 10^6 cell/ml	40-400 µl	0.4-40 L
Fluidised bed	Continuous feed (perfusion)	Well mixed, laminar flow, convection	3D	25,000-70,000 cm^2/L 5-6 x 10^6 cell/ml	20-200 µl	0.2-20 L
Membrane bioreactors	Continuous feed (perfusion)	Well mixed, laminar flow, convection and solution diffusion	3D	100,000-200,000 cm^2/L 2 x 10^8 cell/ml	0. 5-5 µl	0.005-0.5 L

(1) (Scragg 1991)
(2) A typical functional subunit contains 10^2-10^3 cells (Palsson 2000)
(3) A typical organ contains a few hundred million subunits or 10^9-10^{11} cells

2.3.1 Perfusion Culture for Tissue Engineering

Bioreactors for tissue engineering have evolved from a combination of traditional systems and novel ideas. The existing bioreactor system that is most relevant to fixed, adherent human tissue is the perfusion bioreactor. The culture of cells that were fed with a continuous flow of medium was first seen as an alternative as early as 1912 (Burrows 1912), with further significant developments made in the following years (Carrel 1923; Lindbergh 1939; Rose 1954).

Perfusion culture is the steady flow of media over or through a cell population/bed of cells. Cells are usually retained within a chamber, rather than removed as in the continuous flow chemostat (Griffiths 1990). The development of *in vitro* perfusion culture was a response to the observation of the physiological supply of nutrients to tissue. Perfusion removes the feed or famine cycles that usually occur in static and batch cell cultures and the continual supply of nutrient allows the culture of potentially higher final cell densities. Homogeneous perfusion cultures have minimal concentration gradients within the system and are typically cell filtration or sedimentation reactors. The reactor in Figure 3 is a heterogeneous perfusion system in that there are spatial variations in cell, nutrient and metabolite concentrations within the chambers.

The development of a bioreactor for culturing human cells is heavily influenced by the fact that the cells are anchorage dependent and must be attached to a matrix. Traditional mammalian cell bioreactors have immobilised cells using a variety of cell attachment systems, such as: fiber beds (Chiou et al. 1991), hollow fibers (Tharakan and Chau 1986) and encapsulation (Lee and Palsson 1990). Tissue engineering bioreactors are different in that the immobilisation system is actually part of the product. Indeed, the initial seeding and attachment of human cells is a crucial prerequisite to their growth, it also helps orientation, increased growth surface area and prevention of cell wash-out from the reactor.

2.3.2 Stirred Vessels

Stirred vessels, mostly in the form of spinner flasks, have been used in tissue engineering (Figure 4). One of the most widely reported uses is for dynamic seeding of scaffolds, where the scaffolds are suspended into the culture medium which is then inoculated with the cells. As a result of convective mixing and collisions with the scaffold, cell attachment on the matrix occurs (Freed and Vunjak-Novakovic 1997). Using cartilage as a model it was found that mixing during culture in a spinner flask gave rise to higher amounts of collagen and GAG synthesis (Gooch et al. 2001). Because of the shear fields generated by such reactors there is an upper limit to the intensity of mixing that can be used. Nevertheless spinner flasks combined with an appropriate scaffold offer a useful tool for tissue engineers.

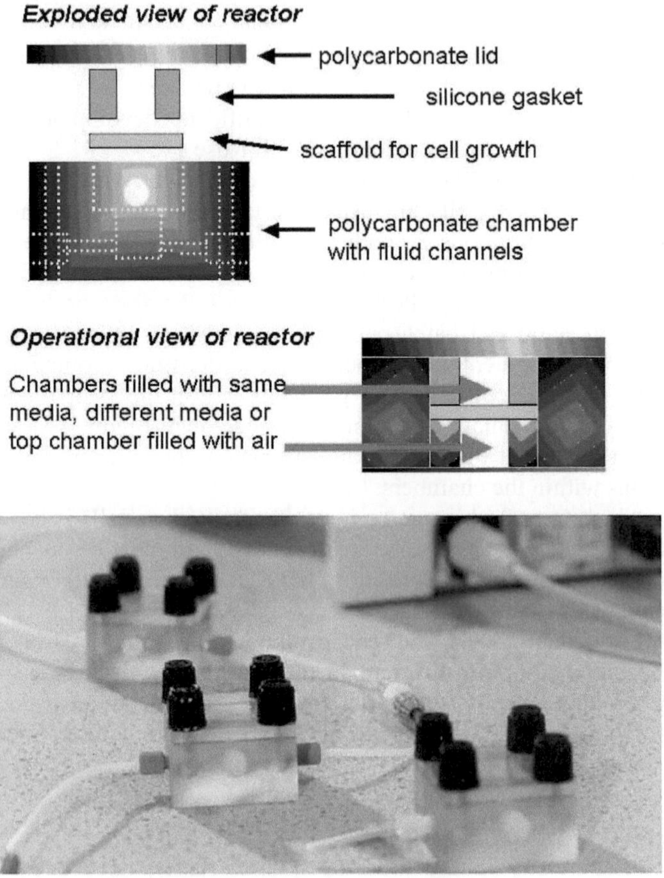

Figure 3: Schematic figure and photograph of a perfusion bioreactor from the authors' laboratory

2.3.3 Low Gravity Bioreactors

The reduction in physical shear and its effects on morphology have most significantly been examined by the use of the NASA slow turning lateral vessel (STLV) or rotating wall perfusion vessel (RWPV) (Unsworth and Lelkes 1998). This reactor recreates microgravity and effectively suspends the cell-matrix construct (typically PGA or PLLA-PGA) and creates Reynolds numbers more conducive to a minimal boundary layer and hence better mass transfer (~Re = 150). Much of the work incorporating the NASA bioreactor has been in the area of cartilage replacement (Begley and Kleis 2000; Freed and Vunjak-Novakovic 1995). However, the effects of microgravity on skin have been examined where increased proliferation and altered epithelial morphology was observed (Doolin et al. 1999).

Figure 4: Typical spinner flask used for seeding tissue constructs

3. CASE STUDY: SCAFFOLDS AND BIOREACTORS FOR CULTURE OF BONE TISSUE

3.1 Introduction

Bone cell culture *in vitro* is well developed in the traditional 2D static culture environment (Beresford and Owen 1998). However, bone tissue is not two dimensional and has a defined 3D architecture. As stated above, bone cells are no further than 100-200 μm from a capillary and therefore a nutrient source. Mimicking or reproducing vascular networks *in vitro* is not easy and so cells further than 200 μm from the surface of the scaffold will not proliferate for long because of nutrient limitations. Bone tissue is complex: there are two forms of bone, namely cortical and cancellous which differ in morphology due to their different primary functions. Adult skeleton consists of 80% cortical bone and 20% cancellous bone and proportions vary at different locations in the skeleton. Cortical bone has 10% porosity and gives mechanical support and protection to internal organs, whereas cancellous bone has 50-90% porosity and provides the environment for metabolic functions such as the haemopoietic system (Buckwalter et al. 1996; Marks and Hermey 1996).

Bone tissue engineering aims to create a cell-scaffold construct suitable to treat trauma injuries, osteoporosis, arthritis and bone cancers (Rose and Oreffo 2002). Considering a cortical bone graft as an example, the scaffold and bioreactor must provide a large surface area to volume ratio to give cell densities representative of native bone. This will be highly dependent on the scaffold architecture. The scaffold material must also provide a suitable surface for attachment, proliferation and function. In addition, a mechanically strong scaffold is required to support the culture of cells *in vitro*, during dynamic culture in a bioreactor, then *in vivo* after implantation where load bearing functions may be required in relatively short time scales as compared to soft tissues. Thus the scaffold design will be as important as controlled cultivation in a bioreactor.

3.2 Scaffold Considerations for Bone Tissue Engineering

The ideal scaffold for bone tissue would mimic the properties of the natural extracellular matrix produced by the osteoblasts in addition to having mechanical properties as described above. A good overview of scaffold materials for tissue engineering is given by Yang (2001). When considering structure, pore size and porosity is of particular importance for the relatively large construct required for bone. The diameter of cells in suspension dictates the minimum pore size, which varies with cell type (Whang et al. 1999).

Optimum pore size must take potential vascularisation into account as well as the change in pore size with degradation and cell invasion (Cohen et al. 1991). Whang et al (1999) reported that the optimum pore size for osteoid ingrowth on bioresorbable scaffolds is 40-100 μm, and is 100-350 μm for regeneration of bone on porous ceramics (Klawitter and Hulbert 1971). It has also been reported that 50-150 μm induces osteoid formation, and 150-500 μm pores lead to mineralized bone (Hulbert et al. 1970). However, Holy *et al* (2000) reported bone formation in vitro on PLGA with pores ranging from 1.5-2.2 mm. The choice of optimum particle size is under the same consideration. The study by Mankini *et al* (2002) showed particles of 0.1-0.25 mm resulted in the greatest bone formation when implanted sub-cutaneously in mice.

The surface topography of the scaffold also effects bone growth. Microporosity due to the molecular structure of the material as well as well grooves and roughness appear to affect bone growth (Boyan et al. 1996; Schwartz et al. 1999; Schwartz et al. 2001). The effect of topography has also been reported, where a study was made of rat osteoblast cultures on Phyester dentine, with or without a grid of grooves cut into it (Gray et al. 1996). The results showed that the topography affects the siting, timing and extent of new bone formation, and also orientation-osteocyte lacunae were often aligned along the axis of the groove or the edge of the dish or slab of tissue.

Suitable chemical and physicochemical properties of the scaffold are vital for good cell attachment, migration, proliferation and function of the cells in culture. Integrin-substrate interactions are vital because of their role in the adhesion. Biomaterials that may be unsuitable only because of poor cell adhesion may be coated with extracellular matrix components such as fibronectin or modified with the general adhesion peptide sequence arginine-glycine-aspartic acid (RGD) or other more cell specific sequences. It should be noted that modification with such molecules does not mean all substrates will show the same cell development. In a recent study three substrates (untreated polystyrene, tissue culture grade polystyrene and type-I collagen coated tissue culture grade polystyrene) were coated with fibronectin (Stephansson et al. 2002). While the results of cell adhesion, morphology, and cell proliferation were similar, cell differentiation was different for each substrate.

There are several scaffold architectures and pore structures that may be suitable for bone culture. Honeycomb structures have straight pores of defined shape that give some mechanical strength depending on the material. For example hydroxyapatite with 110 μm tunnels (Jin et al. 2000) and collagen with pores of around 100 μm (Ishii et al. 2001). Various gel and sponge scaffolds may be used which give a a loose but defined network that gives the scaffold some degree of elastic firmness. In order to obtain adequate mechanical properties, porous blocks and particles of ceramics such as hydroxyapatite may be used (Jin et al. 2000). However, it should be noted that these materials are not biodegradable and thus will not be remodelled *in vivo*. For irregular and non-load bearing defects, the use of injectable polymers that polymerise in situ may be considered (Peter et al. 1998; Salem et al. 2003). Collagen fibers are naturally found in the extracellular matrix, and a fiber structure provides another possible scaffold. Fibers can form a random matrix, or mesh, such as PET felts. Alternatively, hollow fibers with a defined structure and porosity may provide benefits for mimicking vascular networks. Figure 5 shows a PLGA hollow fiber (from the authors' laboratory) which supports the growth of human osteoblast cell lines and primary cells.

3.3 Bioreactor Design for Bone Tissue Engineering

Once the scaffold material and architecture has been decided, the bioreactor configuration can be considered. Ideally, a bioreactor for bone tissue should mimic and control every parameter experienced by bone cells *in vivo*. These include pressure variations, mechanical and hydrodynamic forces, distance from capillaries, nutrient and waste transport, and temperature. It is important to determine the relevant mechanical force to apply to the proliferating cells. For example, whilst bone tissue *in vivo* is clearly influenced by mechanical loading forces, a detailed examination of different loading models on osteoblasts *in vitro* concluded that it is interstitial fluid shear that is the most potent stimulus of bone cell metabolism (Basso and Heersche 2002).

If a single-piece porous scaffold is used, such as a porous block, sponge or PET mesh, then a fixed bed or spinner flask are suitable. Another possibility is a hollow fiber bioreactor. In order to utilise the PLGA fibers shown in Figure 5 we have developed a hollow fiber bioreactor for bone tissue engineering (Figure 6).

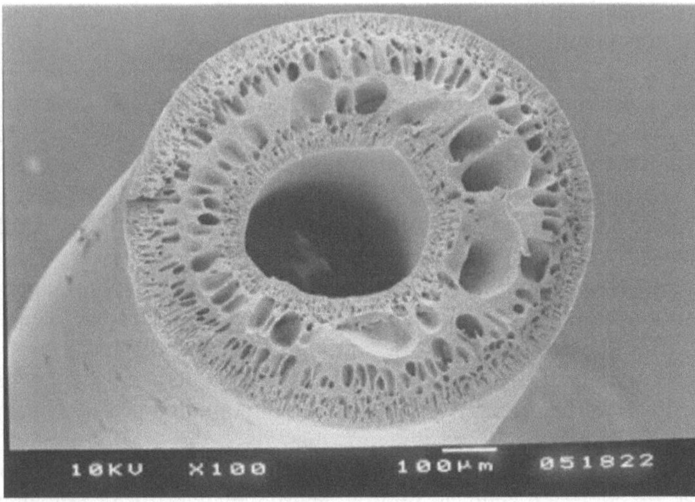

Figure 5: Scanning electron micrograph of a PLGA hollow fiber

The hollow fiber bioreactor in Figure 6A shows a similar structure to cortical bone where the fiber bundles (Figure 6B) represent the Haversian canals. Cells can be seeded and cultured onto the outside of the fibers to mimic osteon formation with nutrient supply being provided through the lumen of the fibers (Figure 6C). The major advantage of the hollow fiber bioreactor is a large surface area:volume ratio as well as reduction of mass transfer limitations as the distance from the nutrient supply for every cell can be kept to a minimum.

If however particles are used as a scaffold, a bioreactor that suspends them individually such as a low gravity bioreactor or fluidised bed can be used to prevent abrasion and cell damage. Fixed and packed bed bioreactors can support both proliferation and differentiation that can be further encouraged by the application of an external physical force. Reactors containing suspended particles are more likely to encourage proliferation as opposed to differentiation due to the low gravity environment.

Each of these bioreactor types has their advantages and disadvantages with respect to bone tissue culture. Whichever configuration is selected, a detailed evaluation of the effect of the operation parameters on the tissue construct must be quantified to determine the limits of the scaffold-bioreactor system. For fixed bed bioreactors, the scaffold should be as uniform as possible to obtain uniform bed characteristics that are important for good nutrient distribution throughout the construct. A single-piece scaffold with an open structure can be operated at low flow rates and will give low pressure drops. But, as with all packings, there will be is a minimum flowrate for effective 'wetting' of the scaffold, ie the flowrate that will deliver nutrients to all parts of the scaffold. Flowrates, operating pressures and pressure drops, temperature control and fluid dynamics including shear stresses need

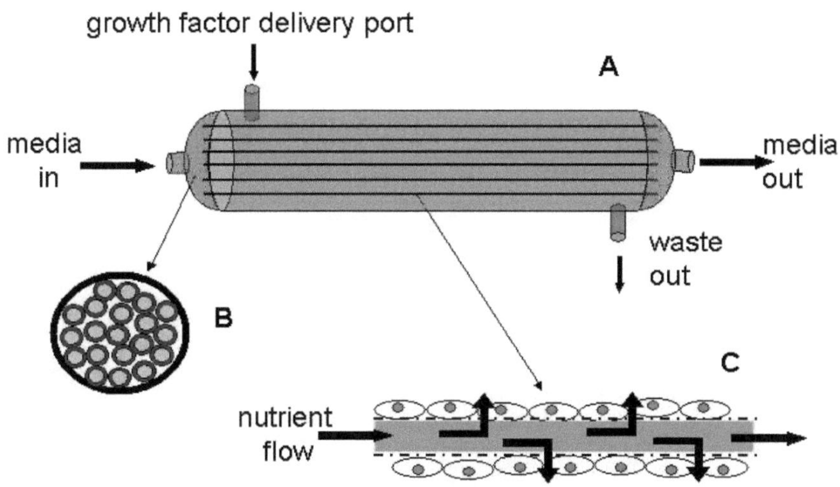

Figure 6: Hollow fiber bioreactor for bone tissue engineering

to be characterised. When using particles, there is a possibility of channelling (non-uniform fluid (media) distribution across the cross-section). Not only will the bulk of the fluid pass through these channels and not reach the extremities of the packing but also the immediate cells will experience increased shear.

The major problem of growing clinical quantities of bone tissue (and in fact all tissues) is nutrient diffusion limitation. As vascularisation is left to occur after implantation, pores to allow flow of media are needed. Once cells start to grow in the pores, the size is restricted (Yang et al. 2001). It is still common for good growth to be achieved on the outer surfaces and in the outer pores only, with little or no growth in the centre of the scaffold. This leads to an incomplete and hence useless bone section (Granet et al. 1998). The bioreactor of choice should house the chosen scaffold and be able to provide the entire culture with the necessary nutrient supply for the duration of the construct development.

4. CONCLUSIONS

The in vitro creation of three-dimensional tissues will require well-controlled culture tools to maximise nutrient mass transfer, allow the culture of multiple cell types, and assert mechanical forces on the cells. The development of bioreactor technologies will help greatly in this respect. Although still in its infancy, there are some basic design rules and general biological and physical considerations that we can integrate to create bioreactor systems that will manage the complex interactions that exist in tissue between individual cells, between cells and the matrix and cells and their environment. Furthermore, it is also becoming clear that a single bioreactor type will not be suitable to grow all tissue types. Bespoke bioreactor systems will be required for specific tissues or classes of tissue. This chapter has not detailed all of the issues related to tissue engineering bioreactors; its aim was to outline the key features that will contribute to the development of dynamic culture systems for the growth of human tissues. The following chapters in this book provide a source of knowledge on different aspects of bioreactor design and operation that we hope will provide a foundation for the future successful development of tissue engineering.

5. REFERENCES

Abbott A. 2003. Biology's new dimension. Nature 424:870-872.

Asenjo JA, Merchuk JC, editors. 1995. Bioreactor System Design. New York: Marcel Dekker.

Basso N, Heersche JNM. 2002. Characteristics of in vitro osteoblastic cell loading models. Bone 30(2):347-351.

Begley C, Kleis S. 2000. The fluid dynamic and shear environment in the NASA/JSC rotating-wall perfused-vessel bioreactor. Biotechnology & Bioengineering 70(1.):32-40.

Beresford JN, Owen ME, editors. 1998. Marrow stromal cell culture. 1 ed. Cambridge: Cambridge University Press. 153 p.

Bock G, Goode J, editors. 2003. Tissue Engineering of Cartilage and Bone. Chichester: John Wiley & Sons Ltd.

Boyan BD, Hummert TW, Dean DD, Schwartz Z. 1996. Role of material surfaces in regulating bone and cartilage cell response. Biomaterials 17(2):137-146.

Buckwalter A, Glimcher MJ, Cooper R, Recker R. 1996. Bone biology. Journal of Bone Joint Surgery 77A:1256-1289.

Burrows MT. 1912. A method of furnishing a continuous supply of new medium to a tissue culture. In Vitro Anat. Rec 6:141.

Cabral JMS. 2001. Ex vivo expansion of hematopoietic stem cells in bioreactors. Biotechnology Letters 23:741-751.

Carrel A. 1923. A method for the physiological study of tissues. In Vitro. J. Exper. Med. 38:407.

Carver S, Heath C. 1999. Semi-continuous perfusion system for delivering intermittent physiological pressure to regeneration cartilage. Tissue Engineering 5:1-.

Chiou T, Murakami S, Wang D, Wu W. 1991. A fibre-bed bioreactor for anchorage dependent animal cell cultures. part 1 - bioreactor design and operations. Biotechnology and Bioengineering 37:755-761.

Chiquet M. 1999. Regulation of extracellular matrix gene expression by mechanical stress. Matrix Biology 18(5):417-426.

Cohen S, Yoshioka T, Lucarelli M, Hwang LH, Langer R. 1991. Controlled Delivery Systems for Proteins Based on Poly(Lactic Glycolic Acid) Microspheres. Pharmaceutical Research 8(6):713-720.

Collins PC, Miller WM, Papoutsakis ET. 1998. Stirred culture of peripheral and cord blood hematopoietic cells offers advantages over traditional static systems for clinically relevant applications. Biotechnology and Bioengineering 59:534-543.

Dennis R, Kosnik P. 2000. Excitability and isometric contractile properties of mammalian skeletal muscle constructs engineered in-vitro. In-vitro Cell. Dev. Biol. Anim 36(5):327-335.

Doolin E, Geldziler B, Strande L, Kain M, Hewitt C. 1999. Effects of microgravity on growing cultured skin constructs. Tissue Engineering 5:573.

Endy D, Brent R. 2001. Modelling cellular behaviour. Nature 409:391-395.

Freed LE, Vunjak-Novakovic G. 1995. Cultivation of cell-polymer tissue construct in simulated microgravity. Biotechnology & Bioengineering 46:306-313.

Freed LE, Vunjak-Novakovic G. 1997. Tissue culture bioreactors: chondrogenesis as a model system. In: Lanza RP, Langer R, Chick WL, editors. Principles of Tissue Engineering: R.G. Landes Company. p 151-165.

Gooch KJ, Kwon JH, Blunk T, Langer R, Freed LE, Vunjak-Novakovic G. 2001. Effects of mixing intensity on tissue-engineered cartilage. Biotechnology and Bioengineering 72(4):402-407.

Granet C, Laroche N, Vico L, Alexandre C, Lafage-Proust MH. 1998. Rotating-wall vessels, promising bioreactors for osteoblastic cell culture: comparison with other 3D conditions. Medical & Biological Engineering & Computing 36(4):513-519.

Gray C, Boyde A, Jones S. 1996. Topographically Induced Bone Formation In Vitro: Implications for Bone Implants and Bone Grafts. Bone 18(2):115-123.

Griffiths B. 1990. Perfusion systems for cell cultivation. . In: Freshney R, editor. Animal Cell Culture - A Practical Approach. Oxford: IRL Press.

Hoerstrup S, Sodian R, Sperling J, Vacanti J, Mayer J. 2000. New pulsatile bioreactor for in vitro formation of tissue engineered heart valves. Tissue Engineering 6:75-.

Holy CE, Shoichet MS, Davies JE. 2000. Engineering three-dimensional bone tissue in vitro using biodegradable scaffolds: Investigating initial cell-seeding density and culture period. Journal of Biomedical Materials Research 51(3):376-382.

Hulbert SF, Young F, Matthews RS, Klawitter JJ, Talbert CD, Stelling FH. 1970. The potential of ceramic materials as permanently implantable skeletal prostheses. Journal of Biomedical Materials Research 4:433-456.

Ishii I, Tomizawa A, Kawachi H, Suzuki T, Kotani A, Koshushi I, Itoh H, Morisaki N, Bujo H, Saito Y and others. 2001. Histological and functional analysis of vascular smooth muscle cells in a novel culture system with honeycomb-like structure. Atherosclerosis 158(2):377-384.

Jin Q-M, Takita H, Kohgo T, Atsumi K, Itoh H, Kuboki Y. 2000. Effects of geometry of hydroxyapatite as a cell substratum in BMP-induced ectopic bone formation. Journal of Biomedical Materials 52:491-499.

Klawitter JJ, Hulbert SF. 1971. Application of porous ceramics for the attachment of load-bearing internal orthopedic applications. Journal of Biomedical Materials Research Symp 2:161.

Lauffenburger D, Griffith LG. 2001. Who's got the pull around here? Cell organization in development and tissue engineering. Proceedings of the National Academy of Science of the USA 98(8):4282-4284.

Lee G, Palsson B. 1990. Immobilisation can improve the stability of hybridoma antibody productivity in serum free media. Biotechnology & Bioengineering 36:1049-1055.

Lindbergh C. 1939. A culture flask for the circulation of a large quantity of fluid medium. J. Exper. Med. 70:231.

Marks SC, Hermey DC. 1996. The structure and development of bone. In: Bilezikian JP, Raisz LG, Rodan GA, editors. Principles of Bone Biology. San Diego: Academic Press.

Martin I, Wendt D, Heberer M. 2004. The role of bioreactors in tissue engineering. Trends in Biotechnology 22(2):80-86.

Palsson B. 2000. Tissue Engineering. In: Enderle JD, Blanchard SM, Bronzino JD, editors. Introduction to Biomedical Engineering. San Diego: Academic Press. p 579-655.

Peter SJ, Miller ST, Zhu G, Yasko AW, Mikos AG. 1998. In vivo degradation of a ploy(propylene fumarate)/beta-tricalcium phosphate injectable composite scaffold. Journal of Biomedical Materials Research 41:1-7.

Rose FRAJ, Oreffo ROC. 2002. Bone tissue engineering: Hope vs hype. Biochemical and Biophysical Communications 292:1-7.

Rose G. 1954. Separable and multipurpose tissue culture chamber. Texas Rep. Biol. And Med. . 12:1074.

Salem AK, Rose F, Oreffo ROC, Yang XB, Davies MC, Mitchell JR, Roberts CJ, Stolnik-Trenkic S, Tendler SJB, Williams PM and others. 2003. Porous polymer and cell composites that self-assemble in situ. Advanced Materials 15(3):210-+.

Schwartz Z, Lohmann CH, Oefinger J, Bonewald LF, Dean DD, Boyan BD. 1999. Implant surface characteristics modulate differentiation behaviour of cells in the osteoblastic lineage. Advances in Dental Research 13:38-48.

Schwartz Z, Lohmann CH, Sisk M, Cochran DL, Sylvia VL, Simpson J, Dean DD, Boyan BD. 2001. Local factor production by MG63 osteoblast-like cells in response to surface roughness and 1,25-(OH)(2)D-3 is mediated via protein kinase C- and protein kinase A-dependent pathways. Biomaterials 22(7):731-741.

Scragg A. 1991. Bioreactors in biotechnology: A practical approach. Chichester: Ellis Horwood Ltd.

Sinclair A, Ashley MHJ. 1995. Sterilization and containment. In: Asenjo JA, Merchuk JC, editors. Bioreactor System Design. New York: Marcel Dekker. p 553-588.

Stephansson SN, Byers BA, Garcia AJ. 2002. Enhanced expression of the osteoblastic phenotype on substrates that modulate fibronectin conformation and integrin receptor binding. Biomaterials 23:2527-2534.

Strehl R, Schumacher K, de Vries U, Minuth W. 2002. Proliferating cells versus differentiated cells in tissue engineering. Tissue Engineering 8(1):37-42.

Tharakan J, Chau P. 1986. A radial flow hollow fibre bioreactor for the large scale culture of mammalian cells. Biotechnology and Bioengineering 28:329-342.

Unsworth BR, Lelkes PI. 1998. Growing tissues in microgravity. Nature Medicine 4(8):901-907.

Whang K, Healy KE, Elenz DR, Nam EK, Tsai DC, Thomas CH, Nuber GW, Glorieux FH, Travers R, Sprague SM. 1999. Engineering bone regeneration with bioabsorbable scaffolds with novel microarchitecture. Tissue Engineering 5(1):35-51.

Yang S, Leong K-F, Du Z, Chua C-K. 2001. The design of scaffolds for use in tissue engineering. Part I. Traditional factors. Tissue Engineering 7(6):679-689.

CHAPTER 2

MICROREACTOR OPTIMISATION FOR FUNCTIONAL TISSUE ENGINEERING

W.W. MINUTH, R. STREHL AND K. SCHUMACHER

Department of Anatomy, University of Regensburg, Regensburg, Germany

1. INTRODUCTION

Many applications in tissue engineering aim at developing functional tissues from cells in combination with a suitable matrix to support and accelerate regenerative healing (Langer and Vacanti 1999; Vacanti and Langer 1999). Stem cells (Bianco and G 2001; Hori et al. 2002; Stanworth and Newland 2001; Xu and Reid 2001) as well as cells from adult tissues and organs (Vacanti et al. 1998) can generally be used for these projects. In a first step cells are isolated and expanded in culture to obtain a sufficient amount of cells. In the following step cells are seeded onto a three-dimensional scaffold to serve as a growth matrix which determines the three-dimensional structure of the construct. That way the four basic tissue groups with their specific functional characteristics can be produced to regenerate loss of epithelium (Jahoda and Reynolds 2001; Nerem and Seliktar 2001), connective tissue injury (Erickson et al. 2002; Fisher et al. 2002; Kaps et al. 2002; Obradovic et al. 2001) as well as degeneration of muscle (Luyten et al. 2001; Shimizu et al. 2002) and neural tissue (Dillon et al. 1998; Woerly 2000).

While the expansion of isolated cells in Petri-dishes poses no great difficulties, developing tissue constructs often show severe morphological, physiological and biochemical changes caused by dedifferentiation (Bhadriraju and Hansen 2000; Murphy and Sambanis 2001; Ziegelaar et al. 2002). It has been demonstrated in numerous studies that the quality of artificial tissue is highly dependant on the scaffold material (Doll et al. 2001; Hutmacher 2000; Kuberka et al. 2002; Ojeh et al. 2001), cellular attachment (Linhart et al. 2001; Ohgushi and Caplan 1999), intercellular communication (Boyan et al. 1996; Folch and Toner 2000; Francis and Palsson 1997) and the culture conditions (Feng et al. 1994; Korke et al. 2002; Yamato et al. 2002). All these factors have to complement one another in order to prevent the development of atypical characteristics by dedifferentiation. Characteristic examples are the expression of atypical collagen in tissue-engineered

J.B. Chaudhuri & M. Al-Rubeai (eds.), Bioreactors for Tissue Engineering, 19 - 45.

cartilage and bone constructs (Bradamante et al. 1991; Merker et al. 1978; Schmidt et al. 1986), the calcification of artificial heart valves (Korossis et al. 2000), the loss of their endothelial coating (Hoerstrup S.P. et al. 2000; Steinhoff et al. 2000) as well as the downregulation of specific cellular functions observed in liver (Kim and Vacanti 1999; Takezawa et al. 2000), pancreatic (Kaufmann et al. 1999; Papas et al. 1999; Tziampazis and Sambanis 1995), and kidney (Humes 2000; Humes et al. 1999) constructs.

Many years of experiments have shown that the possibilities to generate fully functional tissue within the static environment of a culture dish are very limited. This can be attributed to the fact that tissue development in-vivo is governed by a multitude of cellular and extracellular influences which cannot be reproduced in a culture dish (Figure 1). Hence the need for a culture system that allows the simulation of individual environments for specific tissues in order to achieve a high degree of cellular differentiation in-vitro. To overcome the technical limitations of the culture dish different kinds of perfusion culture container have been constructed in the past (Jockenhoevel et al. 2002; Kremer et al. 2001; Minuth et al. 2000; Mizuno et al. 2001).

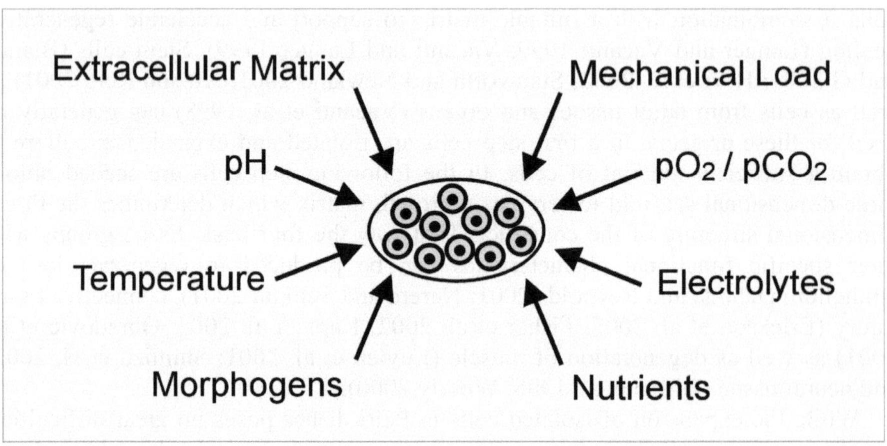

Figure 1: Schematic illustration demonstrating the influence of multiple factors on functional tissue development.

2. MODULAR TISSUE CULTURE SYSTEM

Although remarkable progress in the generation of artificial tissues has been made, further technical development is necessary. With respect to the multitude of factors governing tissue maturation we have been developing a culture system that allows us to adjust the culture environment to the very specific needs of individual tissues. Based on our technical and experimental knowledge we would like to present an improved perfusion culture system for the generation of functional artificial tissues (Figures 2, 3, 4).

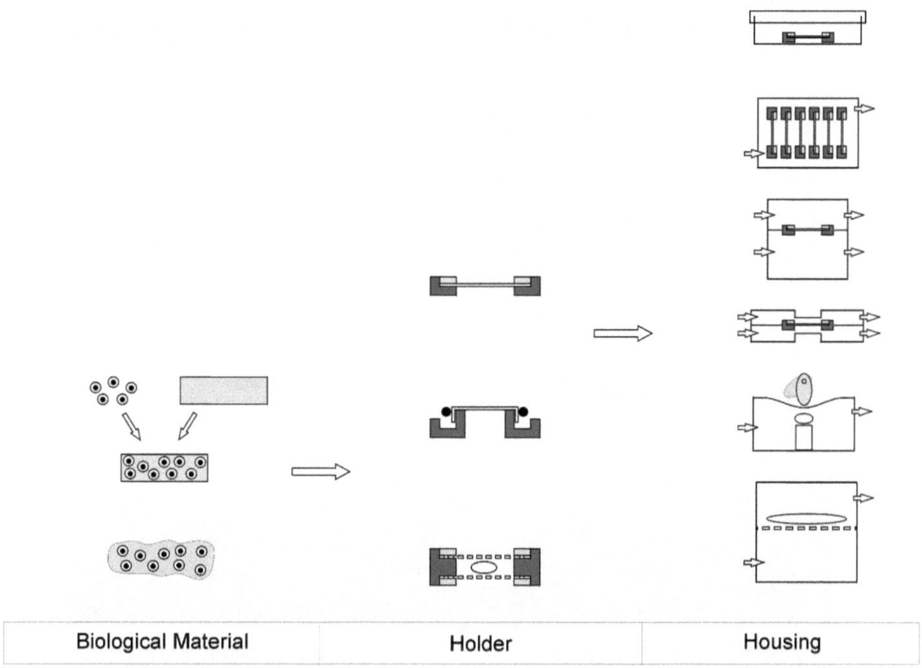

Figure 2: Principal features of perfusion culture in an advanced environment. As biological material, cells on a scaffold or whole pieces of tissues are used, different types of holders are used to avoid damage to the growing tissue. The tissue within the holder is then either placed into a culture dish or in a perfusion culture container. In one type of perfusion container the tissue is bathed in medium, in a different type it is exposed to a luminal/basal medium gradient. In a microscope chamber with a transparent lid and base growing tissue constructs can be observed during culture. A flexible lid and a rotating excenter allow the exposure of tissues to liquid pressure differences during culture. Large tissue constructs are cultured and shaped on a holder in a specialized housing.

Our concept comprises different tissue holders (Figure 5), newly designed culture containers (Figures 3, 6), improved medium transport, gas exchanger and gas expander modules (Figure 7). The modular design of the culture system provides a highly flexible basis for the production of tissue constructs.

Our strategy to generate tissues under in-vitro conditions consists of three principal steps (Figure 2):

1. The starting material for the cultures can either be an intact piece of tissue or single cells growing within a synthetic or natural scaffold.
2. To protect the developing tissue from mechanical damage during handling and culture the biological material is placed into a suitable holder.

3. To ensure optimal nutrition during culture the tissue construct within the holder is continuously supplied with fresh culture medium and oxygen in a specially designed perfusion container.

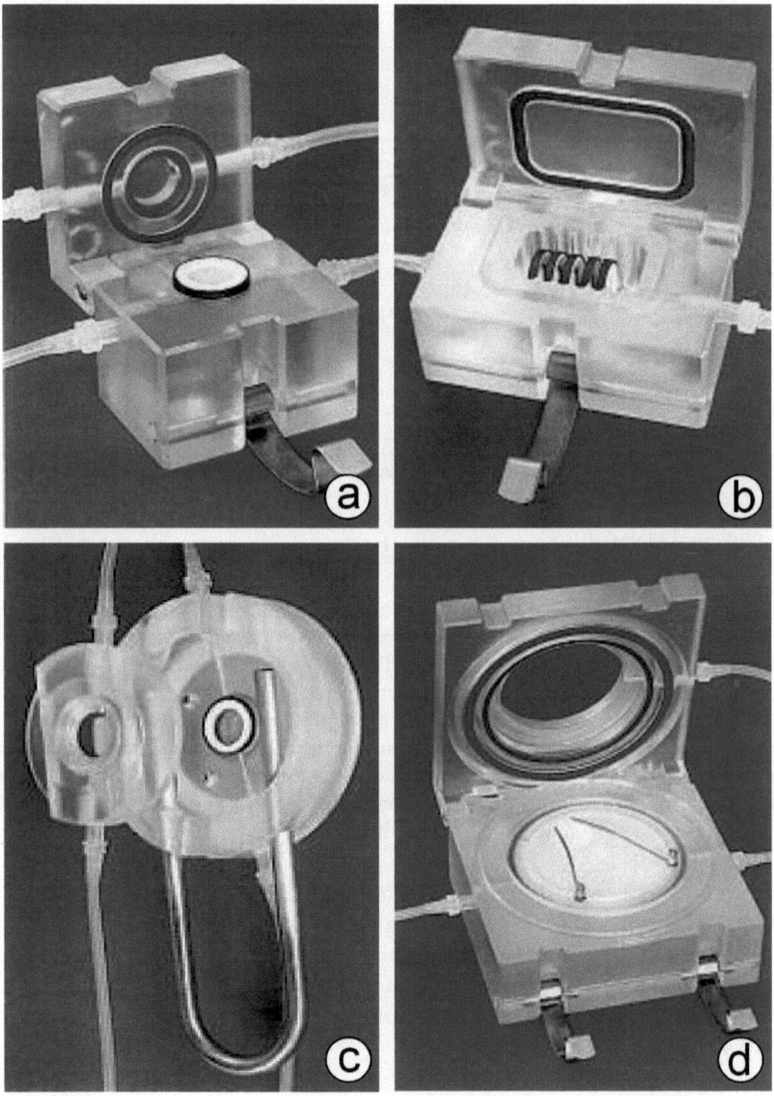

Figure 3: Photographic illustration of different types of perfusion containers. a) perfusion container with 6 holders (13 mm diameter) b) Gradient container with 1 tissue holder (13 mm diameter). c) Gradient container with 1 tissue holder (13 mm diameter) for microscopic observation. d) Tissue engineering container with a 47 mm tissue holder and specific clamps to hold the construct in position.

Fresh culture medium is supplied to the microreactors in a continuous flow or in pulses, so that during long term culture the growing tissue is always exposed to fresh medium. The waste medium is collected in separate bottles and not recirculated. This technique allows the simple bathing of tissue under continuous medium perfusion (Figures 3a, 3d, 6a) or the exposure of epithelia to a gradient with different fluids at the luminal and basal sides (Figures 3b, 3c, 6b). Another culture container is very flat and made of transparent material to allow the microscopical observation of the developing tissue (Figure 3c). A special container model features a flexible silicon lid. By applying force to this lid an a rotating extender mimics a mechanical load for developing cartilage or bone tissue. Finally, a broad range of specifically shaped tissue constructs can be generated in a housing, where holder characteristics and medium supply are adapted to the requirements of an auricle, cartilage or bone construct (Figure 3d) (Haisch et al. 2002; Risbud and Sittinger 2002).

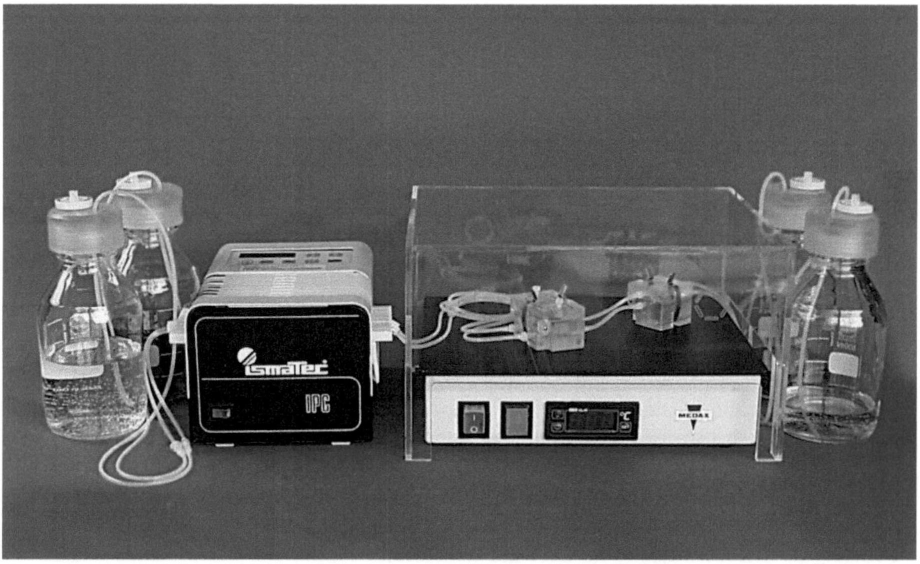

Figure 4: Perfusion culture setups are self-contained and can be used on a laboratory table. A peristaltic pump transports the media (1 ml/h) to a gas expander module and then to a gradient container. The waste medium is collected in the bottles on the right side. A heating plate and a Plexiglas lid maintain the desired temperature.

Despite earlier progress in tissue engineering we believe that it is still important to learn from nature. As a consequence the microenvironment inside a microreactor has to be adapted to the specific needs of individual tissues. In-vivo tissue development depends on optimal intercellular interaction, close communication between cells, the extracellular matrix, specific morphogens, suitable nutrients, mechanical load or rheological stress, pH, temperature, pO_2/pCO_2 and a favorable electrolyte environment (Figure 1).

2.1 Individual Holder for the Developing Tissue

When a piece of tissue is cultured in a Petri-dish, the side of the tissue resting on the bottom of the dish receives a seriously reduced supply of nutrients and oxygen. To prevent damage caused by this unilateral supply biological materials are placed in suitable holders machined out of polycarbonate or Procan® (Figure 5). The use of a holder further prevents mechanical damage and facilitates the handling during transportation. This method also allows an exact geometrical placement of the developing tissue within any type of perfusion culture container.

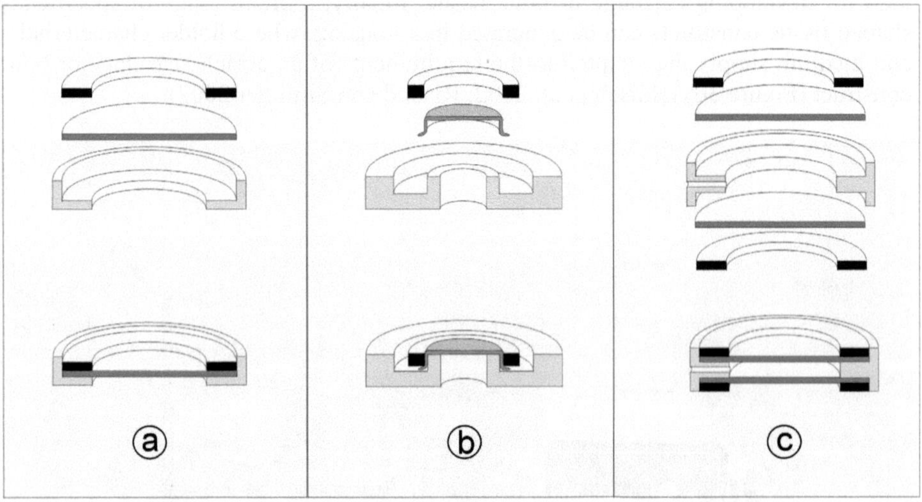

Figure 5: Schematic illustration of different types of tissue holders. a) Inflexible scaffolds or filters with 13 respectively 47 mm diameter are placed in a basal ring and fixed by a tension ring. b) Flexible matrices are held like the skin on a drum. c) Two nets form a space within the holder into which a piece of tissue is placed.

A variety of different tissue holders are possible:

1. One type of tissue holder can hold any type of flat or three-dimensional scaffold, filter or net with a diameter of 13 or 47mm. The selected scaffold material is placed into the holder and is held in place by a tension ring (Figure 5a).
2. The next type of carrier holds pieces of natural extracellular matrices such as the fibrous capsule of neonatal rabbit kidney or other organs. The matrix with adherent tissue is held in position on the holder like the skin of a drum (Figure 5b).
3. Another type of carrier holds intact pieces of tissue between two nets to allow permanent exchange of medium (Figure 5c).

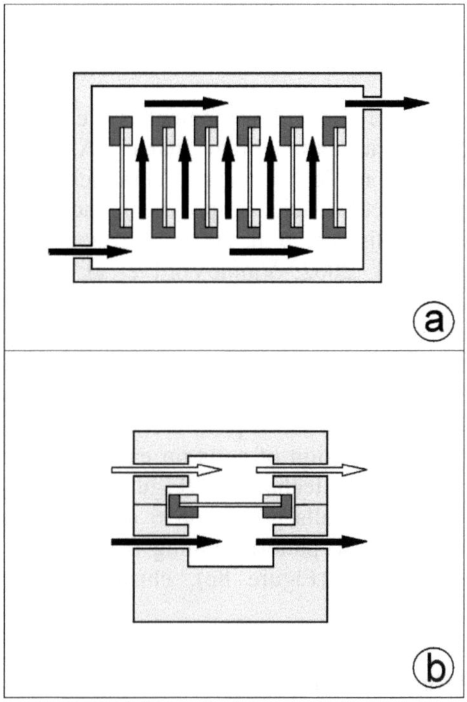

Figure 6: Schematic illustration of the flow path in perfusion culture containers. a) Medium enters at the basal side of a container, rises between the tissue holders and leaves at the top. b) In a gradient container the tissue holder separates the container into an upper and basal part. Tissue can be superfused with different media at the luminal and basal side.

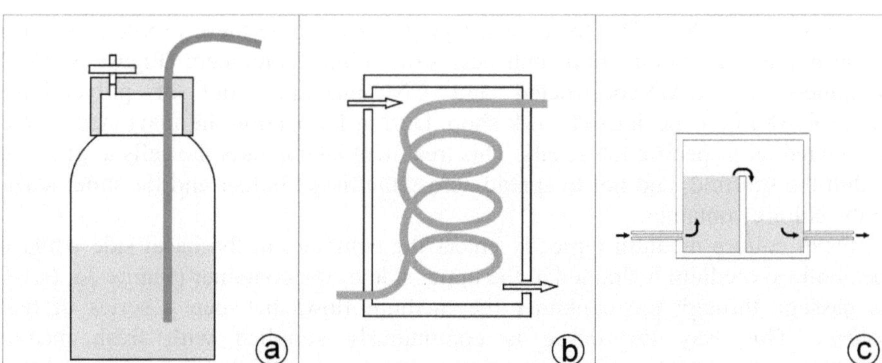

Figure 7: Schematic illustration of innovative tools for perfusion culture. a) Medium is transported without contact with screw cap material. b) Respiratory gases are exchanged by diffusion through a silicone tube coil. c) Gas bubbles are eliminated in a gas expander module.

2.2 Finding the Best Scaffold

One prerequisite for optimal tissue development is the positive interaction of cells with an artificial or natural extracellular matrix. A huge variety of supports, biomatrices and scaffolds made of metals, ceramics or various biodegradable polymers such as polyglycolic acid or polylactic acid are available for tissue engineering applications (Elisseeff et al. 2001; Murray and Spector 2001; Passaretti et al. 2001; Tessmar et al. 2002; VandeVord et al. 2002). Consequently, it is important to find the most suitable extracellular matrix. This can be done by testing different kinds of supports for an individual application (Figure 5a, 8a). Cells are seeded onto the biomaterial and cultured in a 24-well plate in a CO_2–incubator. Medium containing serum and growth factors or morphogens can be used for a limited period of time to stimulate cell proliferation or differentiation of cells. The distribution and degree of adhesion of cells on each matrix can be analysed at the end of culture by fluorescence techniques (Figures 8b-g). The illustration shows fluorescent nuclei of MDCK cells grown on 6 different scaffolds. Each scaffold shows a very individual growth pattern ranging from perfect confluent attachment (Figure 8b), dome formation (Figure 8c), clustering (Figures 8d-f) to blister formation (Figure 8g).

2.3. Construction of Optimised Microreactors

Under in-vivo conditions most tissues are constantly supplied with nutrients by capillaries. This physiological system cannot sufficiently be mimicked in the stagnant environment of conventional culture dishes. Due to this fact many of our early experiments did not perform as expected and showed a loss of function (Bhadriraju and Hansen 2000; Murphy and Sambanis 2001; Schumacher et al. 2002; Ziegelaar et al. 2002). We have been trying to overcome these problems by using tissue holders in combination with perfusion culture containers (Figure 3, 6). All containers are CAD-constructed and CNC-machined out of polycarbonate (Makrolon®) in a specialized work shop. During fabrication the machined surfaces are treated with specific lubricants. This treatment later causes the cells to grow only within the scaffold, and not to spread across the tissue holder and the inner surface of the culture container.

Fresh culture medium typically enters the container at the basal side while the metabolized medium is drained at the upper side of the container (Figure 3a, 6a). On its passage through the container the medium flows between a series of tissue holders. This way the tissue is continuously supplied with fresh medium, guaranteeing constant nutrition and preventing an unphysiological accumulation of metabolic products and an overshoot of paracrine factors. Metabolized medium is not reperfused in our experiments. In a gradient container the tissue carrier is placed between the base and the lid of the container (Figures 3b, 6b) so that the luminal and basal side of the tissue can be superfused with individual media, mimicking a natural and thus ideal environment for epithelia. For example, the typical environment of

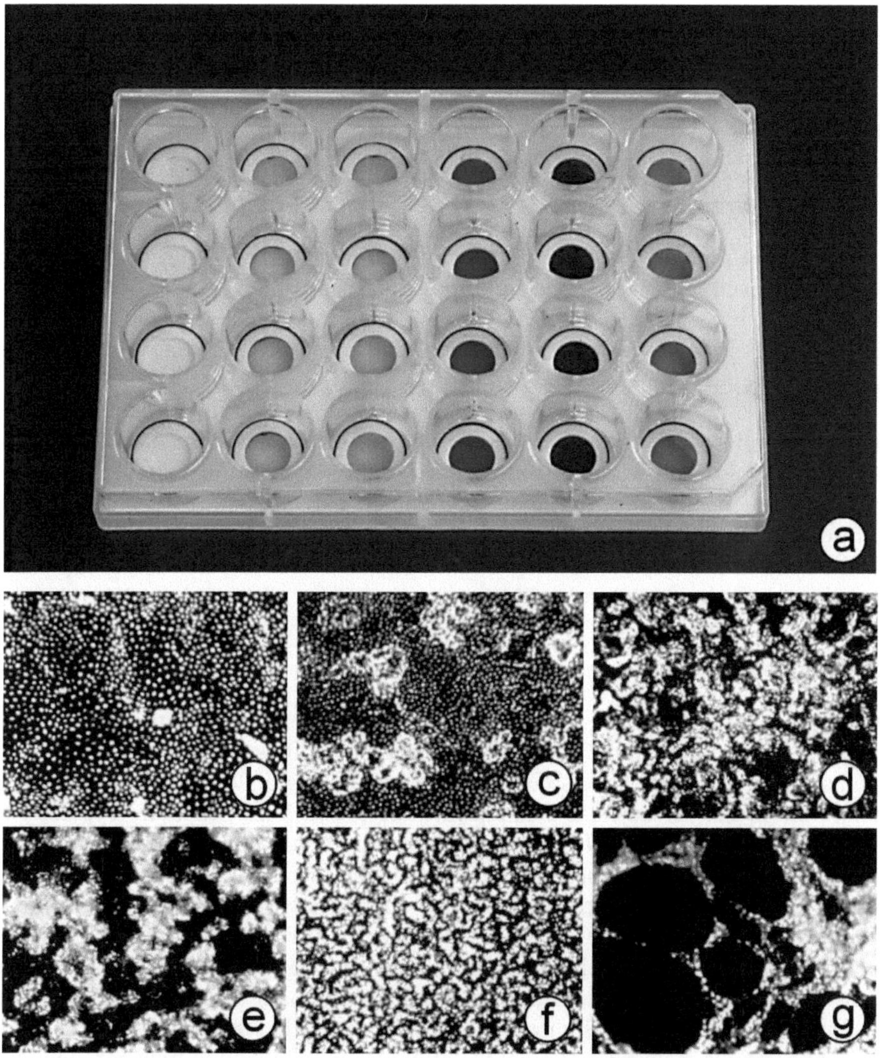

Figure 8: Tissue holders with 6 different 13 mm diameter scaffolds are shown within a 24-well culture dish. The growth pattern of MDCK cells is demonstrated after a 4 day culture period by labeling the nuclei with propidiumiodide. Each scaffold shows a very individual growth pattern ranging from perfect confluent attachment (b), dome formation (c), clustering (d-f) to blister formation (g).

the stomach, small intestine, gall bladder or kidney can be mimicked by using a hyper- or hypotonic medium on one side, while an interstitial-like fluid is used on the other side. A cartilage environment can be simulated by using culture medium resembling synovial fluid at one side and conventional medium at the other side. A gradient container typically can hold 1 (Figure 3a) or 2 x 3 tissues carriers (not shown) so that three samples can be used as controls, while the other three are utilized in an experimental series. Perfusion culture experiments under visual control can be performed using a microscope chamber (Figure 3c) in which the tissue holder with a diameter of 13 or 47 mm is mounted between a transparent lid and base part of the container, so that microscopical observation is possible.

2.4 Temperature Control

Tissue culture must be performed at a constant temperature of usually 37°C. This can be achieved by placing the culture system into an incubator. However, the disadvantage is the limited accessibility inside the incubator. A better solution is to place all modules of the system under atmospheric air onto a heating plate and cover them with a removable lid (Figure 4).

Another solution is to incorporate an electronically controlled heating element into the modules. Such an element could be placed into the module's wall or glued onto the module. Low voltage heating units and circuit breakers should be used in order to avoid electric shock. The heating elements should not be in contact with the culture medium and the cultured tissue. The heating system can be computer monitored continuously concerning power usage and heat output to register potential irregularities during long term culture.

2.5 Transport of Medium

Developing tissue has to be supplied with a continuous or pulsating flow of fresh, oxygenated culture medium in order to avoid unstirred layers of fluid and to achieve optimal differentiation. This is best accomplished using a slowly rotating peristaltic pump able to deliver adjustable pump rates of 0.1 to 5 ml per hour. Typically the medium is drawn up from the bottom of a storage bottle through tubing and passes through the bottle cap before reaching the continuing tubing. In this scenario the pump's suction pressure has to overcome the difference in elevation within the storage bottle as well as the capillary force resulting from thin tubing. This usually does not cause problems at high pump rates. At low pump rates, however, medium is insufficiently aspirated from the bottle and the amount of air bubbles increases. The culture medium is usually transported through bottle screw caps and tubing that consist of various materials. Typical materials along the medium transport path are a glass tube and large diameter silicon tubing within the storage bottle, polysulfonate bottle caps, Luer® connectors made from polypropylene and small diameter silicon tubing through the pump and towards the culture container. Due to the large diameter tubing the dead volume in this application is unnecessarily large. Therefore it can take hours for medium to reach the culture container at low transport rates of a few milliliters per hour.

The culture medium is saturated with oxygen during its passage through silicon tubing to guarantee optimal supply for the developing tissue (Ma et al. 2001). However, the problem in perfusion cultures is that gas bubbles preferentially form at material transitions. These microscopic gas bubbles are transported along with the culture medium, increase in size and eventually form an embolus that massively impedes medium flow. Gas bubbles can also accumulate in the culture container where they lead to a regional shortage of medium supply.

Air bubbles cause additional problems in both storage and waste bottle where growing air accumulations cause erratic breaks in the fluid continuum. In small diameter tubing this causes massive pressure changes (Schumacher et al. 2002). In a gradient culture setup where two media have to be transported at exactly the same speed and pressure such effects can lead to pressure differences which in turn can destroy the growing tissue.

Consequently, to address the problem of gas bubble formation the medium transport path has been redesigned using only one material and tubing of a small and constant diameter. New bottle caps maintain sterility within the storage bottles (Figure 7a). One opening in the cap is designed for a continuous piece of tubing reaching from the bottom of the storage bottle to the inlet of the culture container thus avoiding material transitions along the fluid path. This considerably reduces bubble formation. Another opening holds a sterile filter allowing sterile air to enter the storage bottle as medium is drawn from it. This opening can also be used as an inlet for medium or gases.

3. GAS SUPPLY

Tissues have very individual oxygen requirements (Ma et al. 2001; Murphy and Sambanis 2001). For that reason it is important to regulate the respiratory gas content. A popular method for medium oxygenation is to blow a pressurized gas mixture into the storage bottle. The disadvantage of this method is the formation of gas bubbles in the transported medium. These bubbles accumulate along the medium transport path and cause pressure differences within the system. This method can also introduce infections caused by contaminated gases. Thus, the technical dilemma is to obtain maximum oxygen saturation while avoiding gas bubble formation.

Our solution is a newly developed gas exchanger module housing a long thin-walled silicon tube for medium to pass through (Figure 7b). The tubing is wound up into a spiral inside the exchanger module which features a gas inlet and outlet. The tubing is highly gas-permeable and guarantees optimal diffusion of gases between culture medium and surrounding atmosphere. The desired gas atmosphere within the exchanger is maintained by a constant flow of a specific gas-mixture through the module. This way the content of oxygen, carbogen or any other gas can be increased or decreased in the medium by diffusion making it possible to adjust the gas partial pressures within the medium under absolutely sterile conditions. By maintaining a defined carbon dioxide concentration in the medium this method can also be employed to control medium pH via the bicarbonate buffer.

3.1 Elimination of Gas Bubbles

In perfusion culture applications culture medium with high oxygen content has to be transported while the formation of gas bubbles has to be avoided. Using advanced oxygenation techniques as described before the occurrence of harmful bubbles can be greatly reduced but nevertheless some bubbles will remain in the medium. Consequently, we developed an additional gas expansion module that removes remaining gas bubbles from the culture medium (Figure 7c). Medium is pumped into a specially designed container, where it rises within a small reservoir. Here the medium can expand and equilibrate before it drops down a barrier. During this process gas bubbles are separated from the medium and collected at the top of the container so that the medium which leaves the container is gas-saturated but bubble-free.

The gas expansion module itself can be ventilated through a port at the top. This ventilation port can be coupled with the port on a parallel gas expansion module. This way the individual channels of a gradient culture setup can be bridged to obtain identical pressures at the luminal and basal line (Figures 3b, 3c; 4).

3.2 Constant pH

Cultures in a CO_2-incubator are usually buffered by a system consisting of a relatively high amount of $NaHCO_3$ and a 5% CO_2 atmosphere to maintain a constant pH of 7.4. If such a specially formulated medium is used for perfusion culture outside a CO_2-incubator, the pH shifts from the physiological range to alkaline values due to the low content of CO_2 (0,3%) in atmospheric air. For that reason any medium used outside a CO_2-incubator has to be stabilized by reducing the $NaHCO_3$-concentration or by adding biological buffers such as HEPES or BUFFER ALL (Sigma-Aldrich-Chemie, Deisenhofen, Germany). The necessary amount of buffer can be determined experimentally by adding increasing amounts of buffer to the medium and letting it equilibrate on a thermo plate under laboratory atmosphere. Addition of 50 mmol/l HEPES or an equivalent of BUFFER ALL (ca. 1%) to IMDM for example will maintain a pH of 7.4 throughout long-term perfusion culture.

3.3 Metabolic Parameters in the Medium

In most of the experiments described herein Iscove`s modified Dulbecco`s medium (IMDM; order # 21980-032; Gibco BRL-Life Technologies, Eggenstein, Germany) without serum is used as the standard medium for embryonic renal collecting duct epithelia. Aldosterone (1 x 10^{-7}M; Sigma-Aldrich-Chemie) as survival factor and 1% antibiotic-antimycotic solution (Gibco BRL-Life Technologies, Karlsruhe, Germany) is added to all culture media. Since the generation of artificial tissues requires prolonged culture periods, the early registration of alterations in the environment that may damage the cells is important. Consequently, metabolic activity is monitored by analyzing the superfused culture medium. A T-connector is mounted in the outlet tube of the container and medium samples are collected using a sterile syringe. Media parameters such as pH, pCO_2, pO_2, lactate, osmolarity and electrolyte concentrations of Na^+, K^+, Cl^- and Ca^{2+} are determined in undiluted 200 µl samples of the culture medium (Table 1). The samples are analyzed immediately with a Stat Profile 9 Plus analyzer (Nova Biomedical, Rödermark, Germany) according to the manufacturer's instructions. Solutions with defined electrolyte concentrations serve as controls.

Table 1. Medium parameters in perfusion culture measured by a Stat Profile analyzer

pH	7.4
pO_2	191.8 mmHg
pCO_2	11.9 mmHg
Na^+	117.7 mmol/l
K^+	3.96 mmol/l
Cl^-	79.8 mmol/l
Ca^{++}	1.15 mmol/l
Glucose	446 mg/dl
Lactate	0 mmol/l
Osmolarity	253 mOsm

3.4 Lactate Liberation

Metabolically active cells but also biodegradable scaffolds liberate lactate, which is secreted into the medium. Unphysiologically high concentrations of lactate can lead to a shift in pH that harms the cultured cells. As long as only a small amount of tissue is present in a large volume of medium lactate liberation can usually be neglected. When the amount of tissue increases and the volume of medium decreases, however, concentrations of lactate may reach unphysiological levels. Addition of serum to the culture medium can buffer lactate secretion to a certain degree. In perfusion cultures the constant exchange of medium eliminates the need for serum as an additional buffer.

3.5 Modulation of Oxygen Content

Optimal oxygen supply is the basis for the development of artificial tissues. In a CO_2-incubator (95% air/5% CO_2) average partial pressures of 33 mm Hg CO_2 and 150 mm Hg O_2 are maintained in fully equilibrated IMDM medium. To obtain optimal equilibration of the pH, pO_2 and pCO_2 in a perfusion setup media are pumped through long, small-diameter, gas-permeable silicone tubes, which allow a continuous and optimal exchange of gases. The gas analysis of IMDM (3024 mg/l $NaHCO_3$, 75 mmol/l HEPES) equilibrated against atmospheric air in such a setup shows partial pressures of 11.9 mm Hg CO_2 and 191 mm Hg O_2 (Table 1). Compared to stagnant cultures in a CO_2-incubator the oxygen partial pressure in the perfusion culture medium is considerably higher. Obviously the gas-permeable silicone tubes provide a large surface for gas exchange by diffusion due to the small inner diameter of the tubes (1 mm) and its extended length (1 m).

To study the influence of different respiratory gas concentrations on functional tissue development we performed a series of experiments in which embryonic collecting duct cells from neonatal rabbit kidney were cultured on an organ-specific collagen support in a perfusion container while being exposed to lowered oxygen concentrations (Figure 9). During its passage through gas permeable silicon tubing the medium was exposed to atmospheric air (controls) or to a gas mixture with decreased oxygen content. Equilibration of the culture medium with atmospheric air resulted in a partial pressure of 155 mmHg O_2, while absence of oxygen in the gas mix resulted in a partial pressure of 43 mmHg O_2 in the medium.

Immunohistochemical analysis of the epithelia after a culture period of 14 days showed that independent of oxygen content perfectly polarized epithelium with typical cytokeratin 19 (CK-19) expression had developed (Figure 9a, b). Reduction of oxygen, however, increased cyclooxygenase (COX-2) expression as a stress response (Figure 9c) and decreased Na/K-ATPase expression (Figure 9e). In controls COX-2 expression was low (Figure 9d) while Na/K- ATPase expression was high (Figure 9f). The result clearly indicates that oxygen in the demonstrated concentration range has no influence on the development of a polarized CD epithelium. However, as seen by the upregulation of COX-2 and the downregulation of Na/K-ATPase, low concentration of oxygen evokes a stress response within the epithelial cells, while normal oxygen concentration does not.

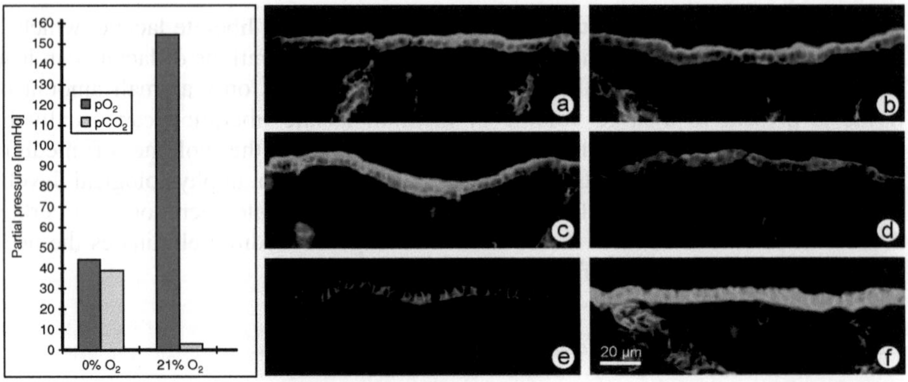

Figure 9: Development of embryonic renal CD epithelium features cultured for 14 days in high and low oxygen atmosphere. In both cases a perfect epithelium is developed. At normal oxygen concentration immunohistochemistry shows typical CK-19, low COX-2 and high Na/K-ATPase expression. In contrast, at low oyxgen concentrations the CD epithelia develop typical CK-19 but high COX-2 and low Na/K-ATPase expression.

4. INDIVIDUAL CULTURE MEDIA

For decades cells traditionally have been cultured in a relatively small selection of commercially available culture media. The application of different culture media can teach us in how far tissue differentiation can be triggered by electrolytes and nutritional factors. Valuable results were obtained by culturing embryonic renal collecting duct (CD) epithelia in 6 different culture media under serum-free conditions in perfusion culture for 14 days (Figure 10) (Schumacher et al. 1999). In each series a perfect epithelium had developed. No morphological differences could

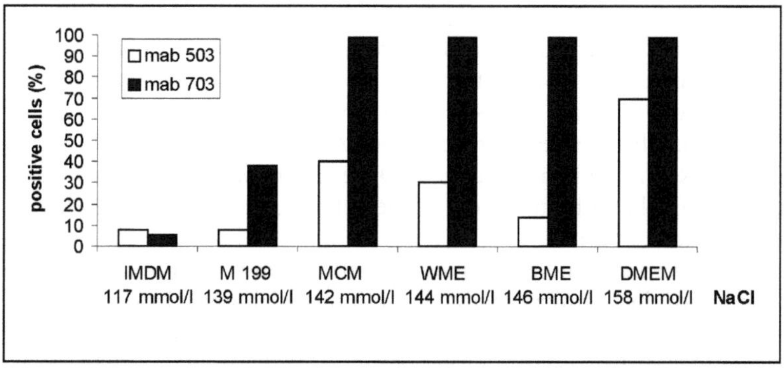

Figure 10: Culture of embryonic CD epithelia in 6 different culture media. In each series a perfect epithelium is established. IMDM with the lowest Na^+ content (117 mmol/l) produces only 5 to 10% mab 703 binding cells. M 199 medium with intermediate Na^+ content results in 30 to 40% mab 703 binding cells. In contrast, all the other media such as MCM, WME, BME and DMEM with a high Na^+ content ranging between 142 to 158 mmol/l produce 95 to 100% mab 703 binding cells. The development of mab 503 binding as an IC cell feature does not correlate with the increasing Na concentrations of the different media.

be observed between the different experimental series. In contrast, monitoring of tissue development by monoclonal antibodies revealed striking differences. Mab 703 recognizes principal (P) cells and mab 503 labels intercalated (IC) cells within the renal collecting duct (CD). It was found that P cell development varied depending on the culture media used. The degree of differentiation further suggested a possible correlation with the Na^+ content in the different culture media. IMDM with the lowest Na^+ content (117 mmol/l) produced only 5 to 10% mab 703 binding cells. M 199 medium with intermediate Na^+ content resulted in 30 to 40% mab 703 binding cells. In contrast, all the other media such as MCM, WME, BME and DMEM with a high Na^+ content ranging between 142 to 158 mmol/l produced 95 to 100% mab 703 binding cells. In contrast, the development of mab 503 binding as an IC cell feature did not correlate with the increasing Na concentrations in the different media. It is concluded that in the absence of growth factors in perfusion culture under serum-free conditions each of the media triggers development of a specific type of epithelium.

4.1 Effects of NaCl Load in Culture Media

One may argue that various culture media are different in composition, which in turn may influence the profile of differentiation. Consequently, in the following experimental series IMDM was the only culture medium used. However, NaCl and

Na-gluconate were added to investigate the influence of electrolytes on differentiation (Minuth et al. 1999; Steiner et al. 1997).

Within a culture dish epithelial cells are exposed to the same medium on the luminal and basal side. This is an unphysiological situation for the epithelia because it represents a permanent biological short circuit, which frequently results in sub optimal development of polarization and a loss of specific features (Minuth et al. 1999). In-vivo epithelia are exposed to an environment consisting of the same fluid at the luminal and basal side during embryonic development only. During functional maturation this environment changes. Once epithelia have formed a functional barrier, the luminal and basal aspect of the cells are confronted with different fluids and a gradient is built up.

To mimic an embryonic environment in-vitro epithelia within a tissue holder can be perfused with the identical media at the luminal and basal side within a gradient container (Figure 3a, 11a). An adaption to the environment of the adult organism can be simulated by culturing embryonic CD epithelia with increased Na^+-concentrations on the luminal side, while standard medium is used on the basal side of a gradient container (Figure 3b, 11b).

The electrolyte parameters of the culture medium are altered by simply adding NaCl and Na-gluconate (Steiner et al. 1997). Comparing IMDM with the serum of neonatal rabbits as a model fluid for the interstitial space unexpectedly large differences in the concentration of electrolytes are observed. Serum, for example, contains 137 mmol/l Na^+ while IMDM only contains 117 mmol/l Na^+ (Table 1). Similarly large differences are seen when comparing K^+, Cl^-, and Ca^{2+}-concentrations in serum to IMDM or other media. Differences up to 25% are found.

Simulation of an embryonic environment for renal CD epithelia in a gradient container for 14 days with standard IMDM on both the luminal and basal side revealed only 5 to 10% of cells positive for the renal collecting duct specific monoclonal antibodies mab 703 and mab 503 from day 1 until day 19 (Figure 11a). In contrast, mimicking an adult environment in a gradient container using IMDM containing additional 12 mmol/l NaCl/17 mmol/l Na-gluconate on the luminal side dramatic changes occurred in the CD epithelium (Figure 11b). Mab 703 binding was found on less than 5% of the cells within the cultured epithelium on day 1. Until day 3 to 6 only 10% of the cells were positive. Then on day 9 more than 70% and on day 19 more than 90% of the cells became positive for mab 703. Mab 503 was found on less than 5% of the cells on day 1, then on day 3 and on day 6 only 30% immunopositive cells were present. A further increase of immunopositive cells was registered at day 9 and day 14 so that up to 70% respectively 80% of the cells within the epithelium were positive for the antibody after day 19. For control, the expression of Na/K ATPase was not altered in epithelial cells confronted with identical (Figure 11a) or different electrolyte concentrations (Figure 11b). The development of PNA-binding on the majority of cells was dependent on aldosterone application and was independent of a NaCl load in this series of experiments.

The time course of development showed that mab 703 and mab 503 label on cells of the embryonic CD epithelia was developed to a minor degree until day 5 (Figure 11b). Then after day 6 the development increased and reached a maximum after day 10. It is obvious that the development of CD cell features started after an unexpectedly long latent period. This latent period is paralleled by a phase of downregulation of mitotic activity in the epithelium as observed by the immunohistochemical markers Ki67 or MIB1 in the maturing kidney (Minuth et al. 1999). This result is a clear indication that the development of features cannot be expected within a few days but takes at least 10 to 14 days.

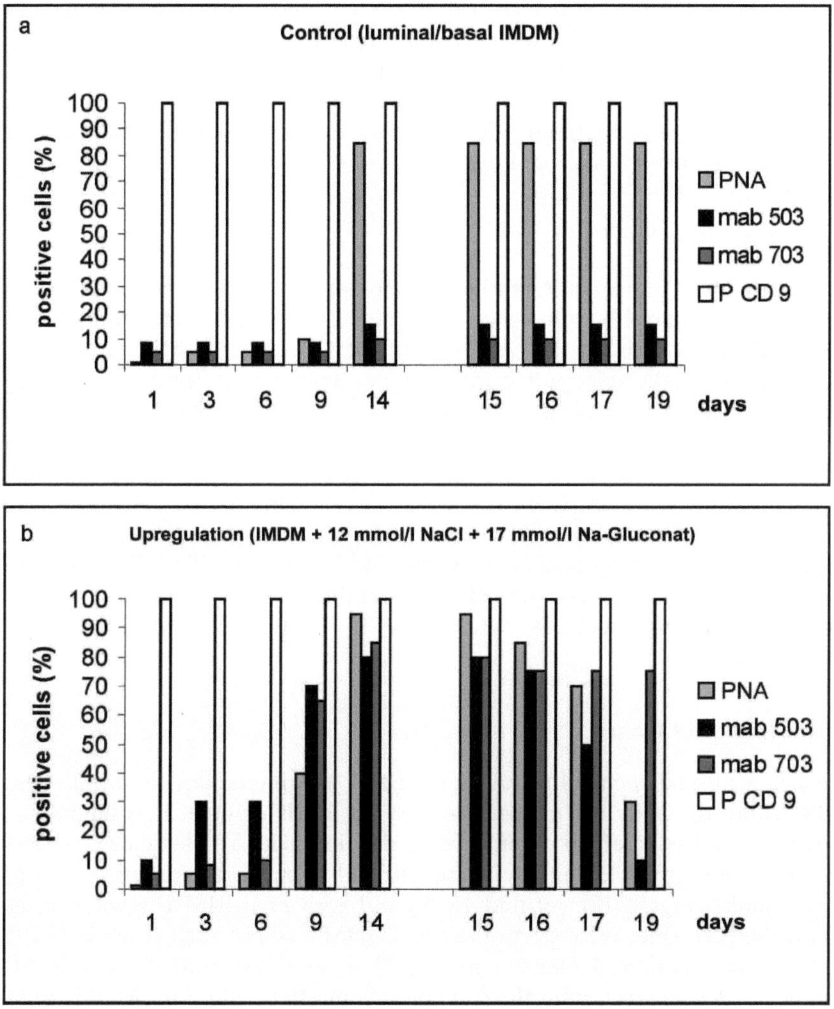

Figure 11: Development of individual features of embryonic CD epithelia in a gradient container. a) IMDM is superfused at the luminal and basal side. b) IMDM is used at the basal side, while IMDM with additional 12 mmol/l NaCl + 17 mmol/l Nagluconate is used at the luminal side for 14 days. Only culture of the epithelium in a gradient evokes the development of specific features.

When at day 14 the culture medium containing additional NaCl was replaced by standard IMDM for another 5 days (Figure 11b), a considerable decrease of mab 503 binding cells was found, while the amount of mab 703 binding cells within the epithelia remained constant. Thus, not only the upregulation but also the preservation of mab 503 antigen expression is controlled by the extracellular electrolyte environment. This finding shows that without the use of growth factors the development of epithelia can be triggered by the electrolyte environment (Minuth et al. 1999; Steiner et al. 1997). We conclude that on the one hand adapting the electrolyte composition of the culture medium to the specific requirements of each individual tissue should considerably improve the quality of constructs generated in tissue engineering. On the other hand uncontrolled liberation of electrolytes from biomaterials and scaffolds by biodegradation may influence the proper development of cultured tissues.

4.2 Sensitivity to NaCl

In the previous gradient culture experiments rather high concentrations of 12 mmol/l NaCl and 17 mmol/l Na-gluconate were added to IMDM at the luminal side, while pure IMDM was used at the basal side of the CD epithelium (Figure 11b). To test the effect of smaller electrolyte gradients on development embryonic CD epithelia were exposed to increasing changes of NaCl ranging from 3 to 24 mmol/l in a gradient container (Figure 12) (Schumacher et al. 2001). The experiments showed that CD epithelia can be stimulated by increasing concentrations of NaCl in the luminal culture medium. Even low doses of 3-6 mmol/l NaCl lead to a qualitative change in differentiation. The cell biological changes in facultative protein expression were dose-dependent in a range of 3 to 15 mmol/l NaCl as indicated by a constant increase of mab 703 and 503 binding cells. In contrast, constitutively expressed proteins such as cytokeratin 19 (CK-19) or Na/K-ATPase remain unchanged.

4.3 Influence of Urea

Urea is a normal metabolite found in serum and interstitial fluid. All tissues and cells within the body are exposed to varying levels of urea. It is unknown though whether urea has an effect on embryonic development. To elucidate possible effects of urea, embryonic CD epithelia were cultured in perfusion reactors in the presence of 9 mmol/l (Figure 13b, e) und 18 mmol/l urea in IMDM (Figure 13c, f), while control experiments were performed in IMDM without urea (Figure 13a, d). The histochemical pattern demonstrated that PNA- and SBA-binding was developed in nearly all of the cells within the CD epithelium after culture in IMDM (Figure 13a, d). However, presence of 9 mmol/l urea in the culture medium lead to a decrease of PNA label (Figure 13b) while application of 18 mmol/l urea resulted in a complete loss of PNA binding (Figure 13c), while SBA-binding remained unaffected (Figure 13f). Thus the presence of 18 mmol/l urea in the culture medium is able to change the profile of differentiation in embryonic CD epithelia.

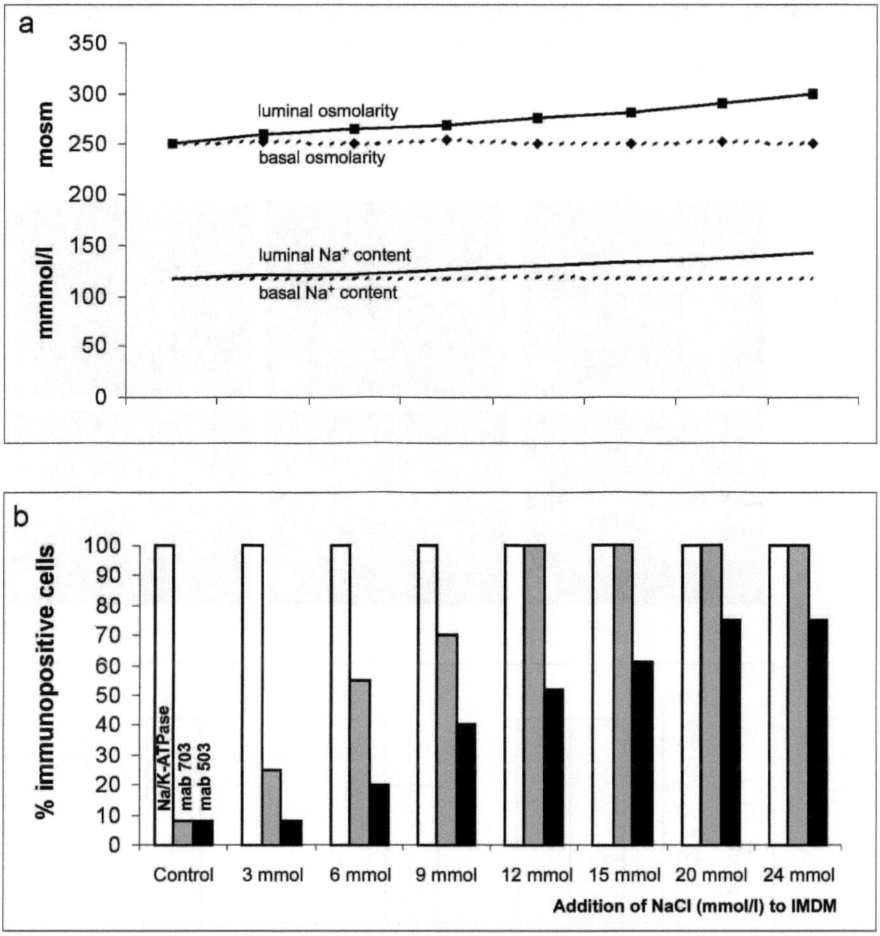

Figure 12: Development of CD epithelia cultured in a fluid gradient for 14 days. Medium at the basal side is IMDM. Medium at the luminal side is IMDM, but additionally contains between 3 and 24 mmol/l NaCl. Immunohistochemistry shows that minimal alteration of 6 mmol/l NaCl in the luminal medium is sufficient to influence cellular differentiation.

5. FUTURE CONSTRUCTIONAL ASPECTS

To date all perfusion culture containers show a significant dead volume around the tissue construct that is taken up by culture medium (Figure 14a, b). This fluid filled space can cause substantial hydraulic effects and passes pressure differences directly on to the cultured tissue. In addition this dead volume is a preferred site for gas bubbles to accumulate. The aggregation of gas bubbles poses a great problem because the supply of culture medium is shortened regionally and even more so because the bubbles' surface tension can damage the developing tissue. We aim to reduce the dead volume by decreasing the height of the container (Figure 14c), but in order to practically eliminate the dead volume the culture container has to be filled with an artificial interstitium (Figure 14d).

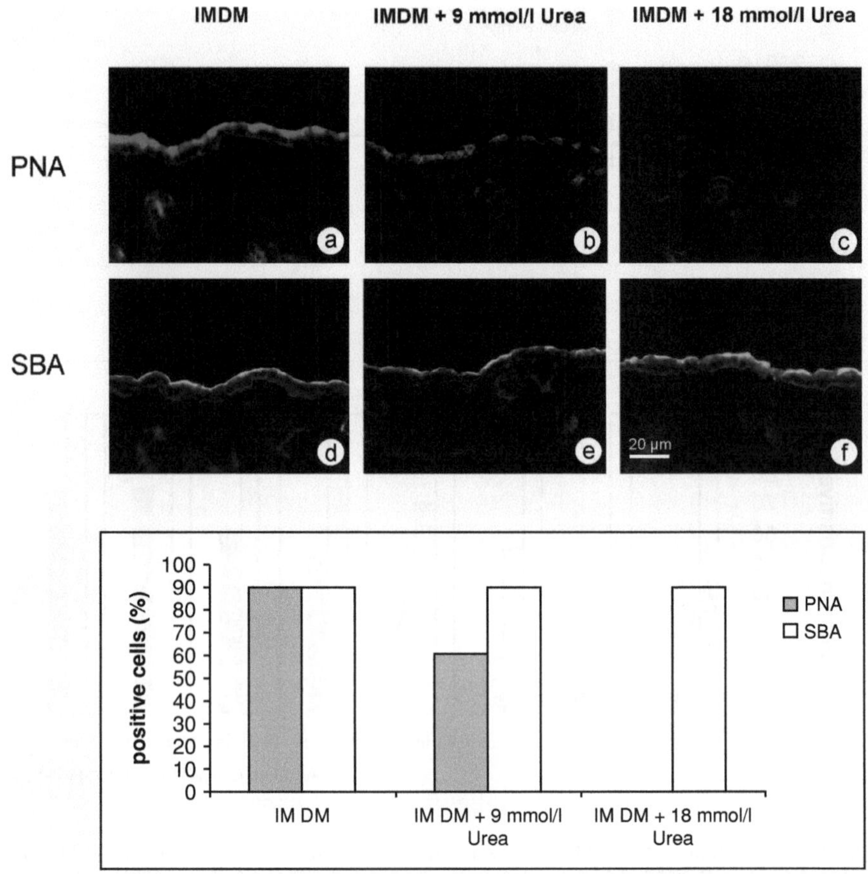

Figure 13: Culture of renal CD epithelia for 14 days in IMDM containing addional urea. Control experiments with IMDM show that 90% PNA- (a) and SBA- (d) positive cells are found within the epithelium. Addition of 9 mmol/l urea leads to complete maintenance of SBA- (e) positive cells, but to a slight reduction of PNA- (b) binding cells. In contrast, application of 18 mmol/l urea results in a complete loss of PNA- (c) binding cells, but no alteration of SBA- (f) positive cells is observed.

The technical solution is to fill the culture container with a highly porous biocompatible material that decreases fluid pressure and distributes pressure more evenly across the construct by capillary effects. Such a material provides mechanical protection to the tissue and minimizes the dead volume within the culture container.

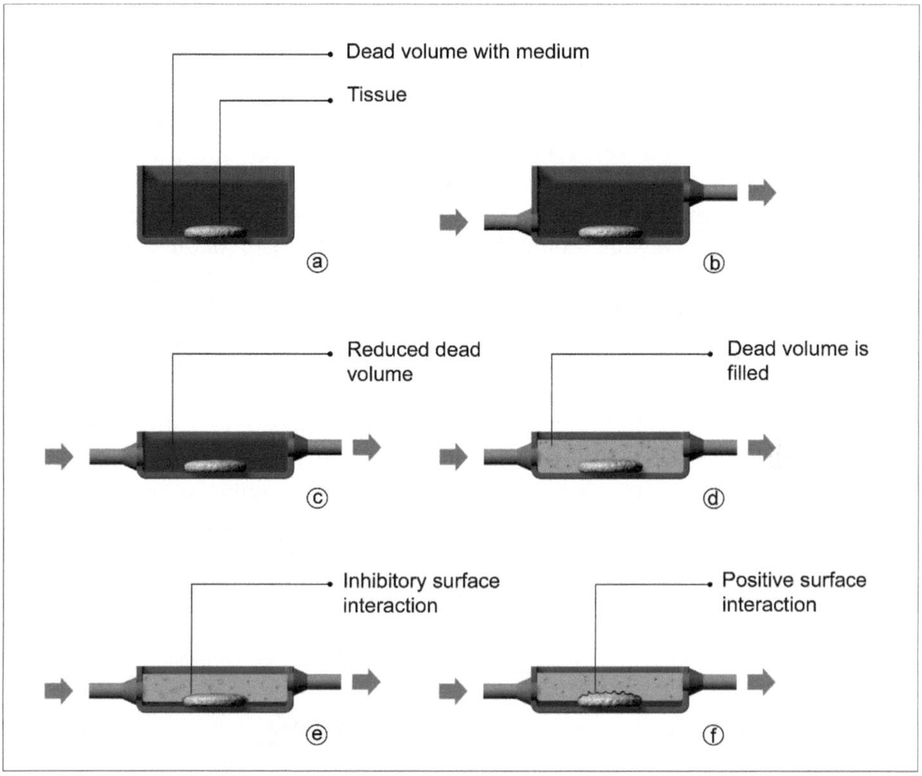

Figure 14: Future development of perfusion culture containers. To date all culture dishes (a) and perfusion culture container (b) show a significant dead space around the tissue construct that is taken up by culture medium. In future constructions the dead volume will be reduced by decreasing the height of the container (c) and filling it with an artificial interstitium (d). Possible filling materials are fleeces from cellulose or fiberglass as well as synthetic sponges from biopolymers or natural cell-free biomatrices. These materials could be used in direct contact with the construct and could be engineered to specifically modulate growth and differentiation. A negative surface interaction could be growth inhibitory (e), while a positive surface interaction could lead to directional growth of the construct (f). It is possible to link morphogens, growth factors or hormones to the artificial capillary network.

Possible filling materials are fleeces from cellulose or fiberglass as well as synthetic sponges from biopolymers or natural cell-free biomatrices. These materials could be used in direct contact with the construct and the material's surface could be engineered to specifically modulate growth and differentiation. A negative surface interaction could be growth inhibitory (Figure 14e), while a positive surface interaction could lead to directional growth of the construct (Figure 14f). It is possible to link morphogens, growth factors or hormones to the artificial interstitium

(Bhadriraju and Hansen 2000; Burdick et al. 2001; Mann et al. 2001; Mann and West 2002; Rowley and Mooney 2002; van Susante et al. 2001). During matrix degradation these substances could be released in immediate vicinity to the developing tissue and contribute to its functional maturation. By releasing bound growth factors and providing extracellular guide structures the artificial interstitium could promote outgrowth of cells from the tissue construct into the surrounding matrix thus leading to an increase in size. It is imaginable that growth factors do not have to be added to the culture medium in the future but can be linked exclusively to the artificial interstitium. This will increase bioavailability and will save cost. The artificial interstitial network could be coated locally with defined extracellular matrix proteins to induce guided growth and development of three-dimensional tissue structures. A combination of individually coated materials could form three-dimensional superstructures within the culture container that provide complex information for cell attachment, cell migration, mitosis and differentiation.

One important prerequisite for the generation of optimal constructs is the ability to recreate a physiological and tissue-typical environment in culture. Therefore suitable culture containers are required in order to optimise culture conditions for differentiated tissues. An artificial interstitial material could be molded to any desired shape in order to perfectly surround the tissue construct and tissue carriers could be constructed in various shapes or sizes. The basic container design consists

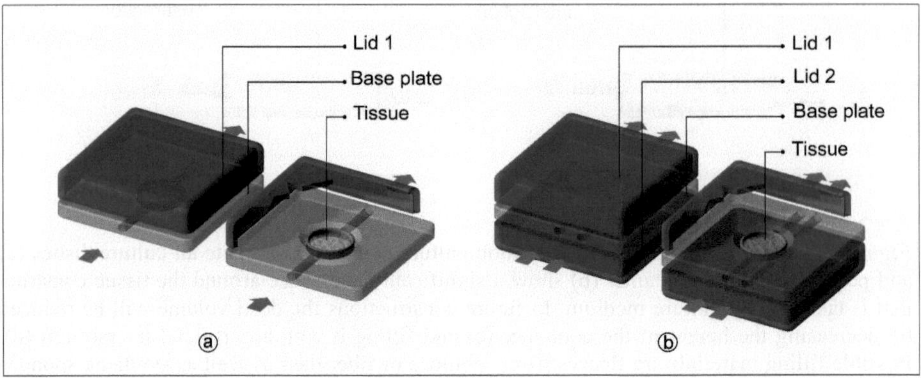

Figure 15: The basic container design consists of a base plate and a lid. (a) In a perfusion chamber the cell carrier or tissue construct is mounted onto the base plate. Then a lid filled with artificial interstitium is secured on top. Lid and/or base plate feature a medium inlet and a medium outlet on opposing ends. The container is sealed by clamping the lid onto a silicon gasket on the base plate. (b) A gradient culture container can be assembled from two identical lids and a modified base plate. In the gradient culture container the cell holder with tissue construct is secured in an opening of the base plate where it separates luminal from basal compartment.

of a base plate and a lid (Figure 15a). In a perfusion container the cell holder or tissue construct is mounted onto the base plate. Then a lid filled with artificial interstitium is secured on top. The lid and/or base plate feature a medium inlet and a medium outlet on opposing ends. The container is sealed by clamping the lid onto a silicon gasket on the base plate. Within the closed container the artificial interstitium is in direct contact with the tissue construct. The advantage of these new containers is a decreased overall height that entails a reduction in dead volume and optimizes medium exchange.

A gradient culture container can be assembled from two identical lids and a modified base plate. In the gradient culture chamber the cell carrier or tissue construct is secured in an opening of the base plate where it separates luminal from basal compartment (Figure 15b). All tissue carriers and filter inserts used today can be integrated (Figure 16). Lids are mounted onto top and bottom of the base plate with the help of a clamp. Upper and lower compartments have a set of medium in- and outlets each so that individual culture media can be applied. That way tissues can be experimentally exposed to fluid gradients.

The complete system consists of several modules that are assembled into a working line for tissue culture (Figure 17). The peristaltic pump maintains a flow of fresh culture medium from the storage bottle through the gas exchanger. The medium continues through the gas expansion module where remaining gas bubbles are eliminated and on to the perfusion culture container that holds the tissue construct. Used medium that leaves the culture container is not recycled but is collected in separate waste bottles. Gas exchanger, gas expansion module and culture container are placed on the surface of a heating unit that maintains a constant ambient temperature. All modules and heating plate are covered with a protective plexiglass lid. This setup is self-contained to maintain sterility and allows culture of tissue constructs under reproducible conditions for extended periods of time. Several culture lines can be run in parallel using a multi-channel peristaltic pump.

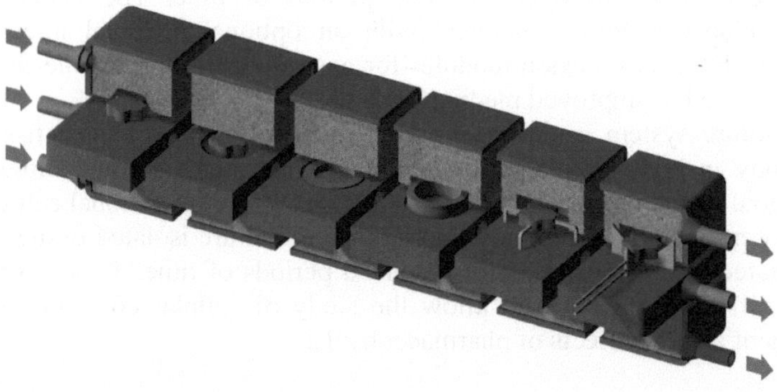

Figure 16: All types of tissue holders or filter inserts used today can be integrated. Lids are mounted onto top and bottom of the base plate with the help of a clamp. Upper and lower compartments have a set of medium in- and outlets each so that individual culture media can be applied.

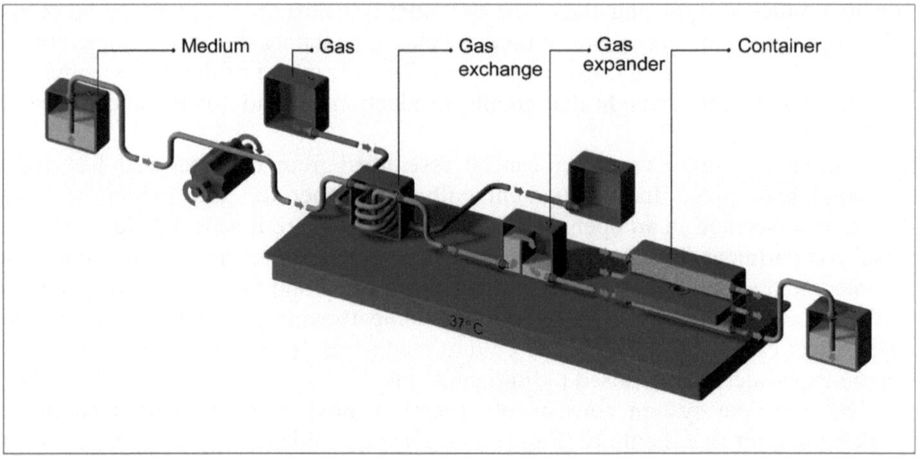

Figure 17: Tissue factor. The complete setup for the generation of artificial tissue consists of a storage bottle for culture medium, a peristaltic pump, a gas exchanger, a gas expander to trap gas bubbles, a culture container and a bottle to collect the waste medium. The temperature of the whole system is maintained by a heating plate. Using a multichannel pump several culture lines can be run in parallel.

5.1. Prospects

We present a modular system designed for the culture of various tissue under physiological conditions over extended periods of time. The system comprises newly developed culture containers with an optional artificial interstitium, gas exchangers and gas expansion modules for gas saturation and bubble-free supply to the tissue as well as improved medium transport.

The culture system is designed for the generation of artificial tissues for the human body in regenerative medicine and for long-term pharmacological testing. Another goal is to close the experimental gap between conventional cell cultures and animal experiments by providing a possibility to culture isolated tissue, organ parts and generated tissue constructs for extended periods of time. Tissue constructs can combine different cell types to allow the study of cellular communication, tissue development and the effects of pharmaceuticals.

Note: The development of the concepts presented here was supported by Minucells and Minutissue (www.minucells.de; D-93077 Bad Abbach, Germany).

6. REFERENCES

Bhadriraju K, Hansen LK. 2000. Hepatocyte adhesion, growth and differentiated function on RGD-containing proteins. Biomaterials 21(3):267-272.

Bianco P, G RP. 2001. Stem cells in tissue engineering. Nature 414(6859):118-121.

Boyan BD, Hummert TW, Dean DD, Schwartz Z. 1996. Role of material surfaces in regulating bone and cartilage cell response. Biomaterials 17(2):137-146.

Bradamante Z, Kostovic-Knezevic L, Levak-Svajger B, Svajger A. 1991. Differentiation of the secondary elastic cartilage in the external ear of the rat. Int J Dev Biol 35(3):311-320.

Burdick JA, Mason MN, Anseth KS. 2001. In situ forming lactic acid based orthopaedic biomaterials: influence of oligomer chemistry on osteoblast attachment and function. J Biomater Sci Polym Ed 12(11):1253-1265.

Dillon GP, Yu X, Sridharan A, Ranieri JP, Bellamkonda RV. 1998. The influence of physical structure and charge on neurite extension in a 3D hydrogel scaffold. J Biomater Sci Polym Ed 9(10):1049-1069.

Doll B, Sfeir C, Winn S, Huard J, Hollinger J. 2001. Critical aspects of tissue-engineered therapy for bone regeneration. Crit Rev Eukaryot Gene Expr 11(1-3):173-198.

Elisseeff J, McIntosh W, Fu K, Blunk BT, Langer R. 2001. Controlled-release of IGF-I and TGF-beta1 in a photopolymerizing hydrogel for cartilage tissue engineering. J Orthop Res 19(6):1098-1104.

Erickson GR, Gimble JM, Franklin DM, Rice HE, Awad H, Guilak F. 2002. Chondrogenic potential of adipose tissue-derived stromal cells in vitro and in vivo. Biochem Biophys Res Commun 18: 290(2):763-769.

Feng M, Peng J, Song C, Wang Y. 1994. Mammalian cell cultivation in space. Microgravity Sci Technol 7(2):207-210.

Fisher JP, Vehof JW, Dean D, van der Waerden JP, Holland TA, Mikos AG, Jansen JA. 2002. Soft and hard tissue response to photocrosslinked poly(propylene fumarate) scaffolds in a rabbit model. J Biomed Mater Res 59(3):547-556.

Folch A, Toner M. 2000. Microengineering of cellular interactions. Annu Rev Biomed Eng 2:227-256.

Francis K, Palsson BO. 1997. Effective intercellular communication distances are determined by the relative time constants for cyto/chemokine secretion and diffusion. Proc Natl Acad Sci U S A 94(23):12258-12262.

Haisch A, Klaring S, Groger A, Gebert C, Sittinger M. 2002. A tissue-engineering model for the manufacture of auricular-shaped cartilage implants. Eur Arch Otorhinolaryngol 259(6):316-321.

Hoerstrup S.P., Sodian R., S D, Wang J, Bacha EA, Martin DP, Moran AM, Guleserian KJ, Sperling JS, Kaushal S and others. 2000. Functional living trileaflet heart valves grown in vitro. Circulation 102(19,3):III44-9.

Hori Y, Nakamura T, Kimura D, Kaino K, Kurokawa Y, Satomi S, Y S. 2002. Experimental study on tissue engineering of the small intestine by mesenchymal stem cell seeding. J Surg Res 102(2):156-160.

Humes HD. 2000. Bioartificial kidney for full renal replacement therapy. Semin Nephrol 20(1):71-82.

Humes HD, MacKay SM, Funke AJ, Buffington DA. 1999. Tissue engineering of a bioartificial renal tubule assist device: in vitro transport and metabolic characteristics. Kidney Int 55(6):2502-2514.

Hutmacher DW. 2000. Scaffolds in tissue engineering bone and cartilage. Biomaterials 21(24):2529-2543.

Jahoda CA, Reynolds AJ. 2001. Hair follicle dermal sheath cells: unsung participants in wound healing. Lancet 358(9291):1445-1448.

Jockenhoevel S, Zund G, Hoerstrup SP, Schnell A, Turina M. 2002. Cardiovascular tissue engineering: a new laminar flow chamber for in vitro improvement of mechanical tissue properties. ASAIO J 48(1):8-11.

Kaps C, Bramlage C, Smolian H, Haisch A, Ungethum U, Burmester GR, Sittinger M, Gross G, Haupl T. 2002. Bone morphogenetic proteins promote cartilage differentiation and protect engineered artificial cartilage from fibroblast invasion and destruction. Arthritis Rheum 46(1):149-162.

Kaufmann PM, Fiegel HC, Kneser U, Pollok JM, Kluth D, Rogiers X. 1999. Influence of pancreatic islets on growth and differentiation of hepatocytes in co-culture. Tissue Eng 5(6):583-596.

Kim SS, Vacanti JP. 1999. The current status of tissue engineering as potential therapy. Semin Pediatr Surg 8(3):119-123.

Korke R, Rink A, Seow TK, Chung MC, Beattie CW, Hu WS. 2002. Genomic and proteomic perspectives in cell culture engineering. J Biotechnol 94(1):73-92.

Korossis SA, Fisher J, Ingham E. 2000. Cardiac valve replacement: a bioengineering approach. Biomed Mater Eng 10(2):83-124.

Kremer M, Lang E, Berger A. 2001. Organotypical engineering of differentiated composite-skin equivalents of human keratinocytes in a collagen-GAG matrix (INTEGRA Artificial Skin) in a perfusion culture system. Langenbecks Arch Surg 386(5):357-363.

Kuberka M, D vH, Schoof H, Heschel I, Rau G. 2002. Magnification of the pore size in biodegradable collagen sponges. Int J Artif Organs 25(1):67-73.

Langer RS, Vacanti JP. 1999. Tissue engineering: the challenges ahead. Sci Am 280(4):86-89.

Linhart W, Peters F, Lehmann W, Schwarz K, Schilling AF, Amling M, Rueger JM, Epple M. 2001. Biologically and chemically optimized composites of carbonated apatite and polyglycolide as bone substitution materials. J Biomed Mater Res 54(2):162-171.

Luyten FP, Dell'Accio F, De Bari C. 2001. Skeletal tissue engineering: opportunities and challenges. Best Pract Res Clin Rheumatol 15(5):759-769.

Ma T, Yang ST, Kniss DA. 2001. Oxygen tension influences proliferation and differentiation in a tissue-engineered model of placental trophoblast-like cells. Tissue Eng 7(5):495-506.

Mann BK, Gobin AS, Tsai AT, Schmedlen RH, West JL. 2001. Smooth muscle cell growth in photopolymerized hydrogels with cell adhesive and proteolytically degradable domains: synthetic ECM analogs for tissue engineering. Biomaterials 22(22):3045-3051.

Mann BK, West JL. 2002. Cell adhesion peptides alter smooth muscle cell adhesion, proliferation, migration, and matrix protein synthesis on modified surfaces and in polymer scaffolds. J Biomed Mater Res 60(1):110-117.

Merker HJ, Lilja S, Barrach HJ, Gunter T. 1978. Formation of an atypical collagen and cartilage pattern in limb bud cultures by highly sulfated GAG. Virchows Arch A Pathol Anat Histol 380(1):11-30.

Minuth WW, Schumacher K, Strehl R, Kloth S. 2000. Physiological and cell biological aspects of perfusion culture technique employed to generate differentiated tissues for long term biomaterial testing and tissue engineering. J Biomater Sci Polym Ed 11(5):495-522.

Minuth WW, Steiner P, Strehl R, Schumacher K, de Vries U, Kloth S. 1999. Modulation of cell differentiation in perfusion culture. Exp Nephrol 7:394-406.

Mizuno S, Allemann F, Glowacki J. 2001. Effects of medium perfusion on matrix production by bovine chondrocytes in three-dimensional collagen sponges. J Biomed Mater Res 56(3):368-375.

Murphy CL, Sambanis A. 2001. Effect of oxygen tension and alginate encapsulation on restoration of the differentiated phenotype of passaged chondrocytes. Tissue Eng 7(6):791-803.

Murray MM, Spector M. 2001. The migration of cells from the ruptured human anterior cruciate ligament into collagen-glycosaminoglycan regeneration templates in vitro. Biomaterials 22(17):2393-2402.

Nerem RM, Seliktar D. 2001. Vascular tissue engineering. Annu Rev Biomed Eng 3:225-243.

Obradovic B, Martin I, Padera RF, Treppo S, Freed LE, Vunjak-Novakovic G. 2001. Integration of engineered cartilage. J Orthop Res 19(6):1089-1097.

Ohgushi H, Caplan AI. 1999. Stem cell technology and bioceramics: from cell to gene engineering. J Biomed Mater Res 48(6):913-927.

Ojeh NO, Frame JD, Navsaria HA. 2001. In vitro characterization of an artifical dermal scaffold. Tissue Eng 7(4):457-472.

Papas KK, Long RCJ, Sambanis A, Constantinidis I. 1999. Development of a bioartificial pancreas: I. long-term propagation and basal and induced secretion from entrapped betaTC3 cell cultures. Biotechnol Bioeng 66(4):219-230.

Passaretti D, Silverman RP, Huang W, Kirchhoff CH, Ashiku S, Randolph MA, Yaremchuk MJ. 2001. Cultured chondrocytes produce injectable tissue-engineered cartilage in hydrogel polymer. Tissue Eng 7(6):805-815.

Risbud MV, Sittinger M. 2002. Tissue engineering: advances in cartilage generation. Trends Biotechnol 20(8):351-356.

Rowley JA, Mooney DJ. 2002. Alginate type and RGD density control myoblast phenotype. J Biomed Mater Res 60(2):217-223.

Schmidt J, Livne E, Erfle V, Gossner W, Silbermann M. 1986. Morphology and in vivo growth characteristics of an atypical murine proliferative osseous lesion induced in vitro. Cancer Res 46(6):3090-3098.
Schumacher K, Klotz-Vangerow S, Tauc M, Minuth WW. 2001. Embryonic renal collecting duct cell differentiation is influenced in a concentration-dependent manner by the electrolyt environment. Am J Nephrol 21:165-175.
Schumacher K, Strehl R, de Vries U, Minuth WW. 2002. Advanced technique for long term culture of epithelia in a continuous luminal-basal medium gradient. Biomaterials 23(3):805-815.
Schumacher K, Strehl R, Kloth S, Tauc M, Minuth WW. 1999. The influence of culture media on embryonic renal collecting duct cell differentiation. In Vitro Cell Dev Biol Animal 35:465-471.
Shimizu T, Yamato M, Isoi Y, Akutsu T, Setomaru T, Abe K, Kikuchi A, Umezu M, Okano T. 2002. Fabrication of pulsatile cardiac tissue grafts using a novel 3-dimensional cell sheet manipulation technique and temperature-responsive cell culture surfaces. Circ Res 90(3):e40.
Stanworth SJ, Newland AC. 2001. Stem cells: progress in research and edging towards the clinical setting. Clin Med 1(5):378-382.
Steiner P, Strehl R, Kloth S, Tauc M, Minuth WW. 1997. In vitro development and preservation of specific features of collecting duct epithelial cells from embryonic rabbit kidney are regulated by the electrolyte environment. Differentiation 62:193-202.
Steinhoff G, Stock U, Karim N, Mertsching H, Timke A, Meliss RR, Pethig K, Haverich A, A. B. 2000. Tissue engineering of pulmonary heart valves on allogenic acellular matrix conduits: in vivo restoration of valve tissue. Circulation 102(19,3):III50-5.
Takezawa T, Inoue M, Aoki S, Sekiguchi M, Wada K, Anazawa H, Hanai N. 2000. Concept for organ engineering: a reconstruction method of rat liver for in vitro culture. Tissue Eng 6(6):641-650.
Tessmar JK, Mikos AG, Gopferich A. 2002. Amine-reactive biodegradable diblock copolymers. Biomacromolecules 3(1):194-200.
Tziampazis E, Sambanis A. 1995. Tissue engineering of a bioartificial pancreas: modeling the cell environment and device function. Biotechnol Prog 11(2):115-126.
Vacanti JP, Langer R. 1999. Tissue engineering: the design and fabrication of living replacement devices for surgical reconstruction and transplantation. Lancet 354(1):SI32-134.
Vacanti JP, Langer R, Upton J, Marler JJ. 1998. Transplantation of cells in matrices for tissue regeneration. Adv Drug Deliv Rev 33(1-2):165-182.
van Susante JLC, Pieper J, Buma P, van Kuppevelt TH, van Beuningen H, van Der Kraan PM, Veerkamp JH, van den Berg WB, Veth RPH. 2001. Linkage of chondroitin-sulfate to type I collagen scaffolds stimulates the bioactivity of seeded chondrocytes in vitro. Biomaterials 22(17):2359-2369.
VandeVord PJ, Matthew HW, DeSilva SP, Mayton L, Wu B, Wooley PH. 2002. Evaluation of the biocompatibility of a chitosan scaffold in mice. J Biomed Mater Res 59(3):585-590.
Woerly S. 2000. Restorative surgery of the central nervous system by means of tissue engineering using NeuroGel implants. Neurosurg Rev 23(2):59-77.
Xu AS, Reid LM. 2001. Soft, porous poly (D, L-lactide-co-glycotide) microcarriers designed for ex vivo studies and for transplantation of adherent cell types including progenitors. Ann N Y Acad Sci 944:144-159.
Yamato M, Konno C, Utsumi M, Kikuchi A, Okano T. 2002. Thermally responsive polymer-grafted surfaces facilitate patterned cell seeding and co-culture. Biomaterials 23(2):561-567.
Ziegelaar BW, Aigner J, Staudenmaier R, Lempart K, Mack B, Happ T, Sittinger M, Endres M, Naumann A, Kastenbauer E and others. 2002. The characterisation of human respiratory epithelial cells cultured on resorbable scaffolds: first steps towards a tissue engineered tracheal replacement. Biomaterials 23(6):1425-1438.

CHAPTER 3

TAYLOR-VORTEX BIOREACTORS FOR ENHANCED MASS TRANSPORT

S.J. CURRAN AND R.A. BLACK

UK Centre for Tissue Engineering, University of Liverpool, Liverpool, UK

1. INTRODUCTION

The route by which engineered tissue products are manufactured *in vitro* commonly utilises dynamic tissue culture bioreactors in which cell culture media and proliferating cells interact. Very often, the cells are supported on biodegradable polymer constructs that assist in the development and formation of the 3-D architecture of the tissue. It has been observed experimentally that a core of necrotic tissue often forms within these constructs, and this has been attributed to poor mass transport of nutrients and oxygen from the culture media (Freed *et al.* 1996; Vunjak-Novakovic *et al.*, 1998, Obradovic *et al.*, 1999).

Mixing within the liquid media provides a more uniform concentration of nutrients and oxygen and facilitates their delivery to cells through enhanced convective transport. The degree of mixing can be problematic, however, as cells have a thin permeable membrane and are sensitive to shear forces. As the mixing intensity increases, the size of circulating eddies decreases, which leads to an increase in the shearing forces that are generated within the fluid, and on the cells and constructs that it contains (Papoutsakis, 1991). Exposure to critically high shear may result in irreparable damage to the cell membrane, leading to cell death, as can prolonged exposure to sub-critical shear stresses. Nevertheless, shear stress has an important role to play in the stimulation of mechanotransduction pathways, which can, in turn, influence the physiology of the cell (Enfors *et al.*, 2001). Ideal mixing in the bioreactor is difficult to achieve in practice, therefore a compromise between shear and mass transport must be made (Vunjak-Novakovic *et al.*, 1999). For a functional tissue sample to be engineered *in vitro,* a uniform physico-chemical environment must be maintained over time for normal morphogenesis of the cells (Spier, 1995). Individual cells might be considered to grow within their own microenvironment, where they experience dynamic physical and chemical changes,

J.B. Chaudhuri & M. Al-Rubeai (eds.), Bioreactors for Tissue Engineering, 47 - 85.

which must be controlled within acceptable limits. It follows that the bioreactor can be considered as a unit that is comprised of these microenvironments (Tramper, 1995). In practical terms, it is not possible to maintain this degree of uniformity, even in a very small bioreactor, but this remains a long-term goal. Traditional spinner flask reactors suffer considerably from the fact that they cannot generate a uniform flow and mass transport environment (Vunjak-Novakovic *et al.*, 1996). Well-mixed high shear regions exist around the rotating impeller whereas laminar regions, or even dead zones exist far from the impeller, exacerbated at larger scales of reactor (Leib *et al.,* 2001). Techniques to overcome this can be employed but have a tendency to make the bioreactor design more complex and, importantly, more difficult to scale correctly such that the transport phenomena are replicated. The long-term term goal of tissue engineering is to commercialise the *in-vitro* production of functional tissues and for this to be realised scaling of the production process is of paramount importance (Langer and Vacanti, 1993; Lewis, 1995; Naughton, 1998).

The Taylor-vortex bioreactor is a promising alternative to achieve many of these goals and comprises two coaxial cylinders, either of which can rotate. Media, cells and constructs occupy the annular space between the cylinders. The system has a simple geometry making it relatively easy to scale. The flow regimes that can be achieved within this type of bioreactor are diverse from laminar through to highly turbulent flows, with many unique regimes in between as a consequence of the geometry. Importantly, these regimes are uniform throughout the bioreactor with no dead zones and there is a high degree of control over the maintenance and transition between regimes.

This review chapter addresses the design aspects of a Taylor-vortex bioreactor and describes approaches used to control and quantify the flow and mass transport environment with a view to developing mass transport correlations for scaling purposes.

2. HYDRODYNAMICS OF TAYLOR-VORTEX BIOREACTORS

2.1. Early Studies of Coaxial Cylinder Flows

The symmetry of flow of fluid contained between two coaxial cylinders appears to have been first recorded in Isaac Newton's famous 1687 work Principia where he hypothesised that the flow was composed of concentric streamlines. In 1848 Stokes suggested the use of dust particles would provide a method to visualise the phenomenon and confirm Newton's hypothesis. Stokes also suggested that the rotation of the inner cylinder at a speed greater than the outer cylinder would produce circulating eddies in the flow. In 1881, Margules recognised that this flow type could be used to measure viscosity, and, in 1888, Mallock constructed the first concentric cylinder viscometer. Mallock observed that the flow was always unstable when the speed of the inner cylinder exceeded that of the outer cylinder; when only the outer cylinder rotated, the flow remained stable up to a critical speed, beyond which the flow became turbulent. In 1890, Couette developed a concentric cylinder viscometer where only the outer cylinder could rotate. It is in recognition of this work that the laminar, solid body rotation of fluid in the gap between the cylinders is referred to as Couette flow.

2.2. Centrifugal Instability for Inviscid Fluids: Rayleigh Criterion

Rayleigh first addressed the problem of hydrodynamic instability in Couette flow between 1916 and 1920. Mallock and Couette had suggested that for the same shear rate in the annulus the flow would be laminar below a critical speed of outer cylinder rotation when the inner cylinder was stationary. However, for the same shear rate using inner cylinder rotation and a stationary outer cylinder, the flow could be unstable. Rayleigh developed a criterion to explain this based on an energy analysis of the flow for an inviscid fluid. The Rayleigh criterion, which states that there is never a stable flow if the inner wall rotates and the outer wall is stationary, no matter how slow the inner wall rotates, has been rigorously proved for inviscid flow. In the case of real fluids, the effects of fluid viscosity mean that the Rayleigh criterion is only observed as an outer limit, and consequently the flow remains stable when the inner wall rotates at low angular velocities. For high rotation rates of the outer cylinder, the viscous forces are small relative to the dynamic forces and the Rayleigh criterion is approached asymptotically, effects that have been observed experimentally.

2.3. Centrifugal Instability for Viscous Fluids: Taylor Number

Whilst the Rayleigh criterion is a useful guideline for inviscid instability in coaxial cylinder systems, the pioneering work of G. I. Taylor (1923) established the fundamental theory for the onset and development of the instability of viscous fluids. Qualitative studies on the flow transitions were made by injecting a dye into the rotating fluid in the annulus. Taylor theoretically addressed the centrifugal instability of a viscous fluid by solving the equations of motion and continuity for the stable laminar Couette flow and then superimposed a small disturbance flow. His assumptions included axisymmetry, an axially imposed periodic disturbance flow and negligible gravitational effects. His analysis reduced to

$$\mathrm{Ta} = \frac{4\Omega_i r_i^{\,4}}{\nu^2}\frac{\eta^2-\mu}{1-\eta^2}\left(\frac{1-\eta}{\eta}\right)^4 = \varepsilon\,\mathrm{Re}^2 + \mathrm{O}(\varepsilon)^2 \qquad (1)$$

where Ω_i is the angular speed of the inner cylinder (rad/s), r_i is the inner cylinder radius (m), ν is the fluid kinematic viscosity (m^2/s), μ is the ratio of angular speeds of the cylinders, η is the ratio of radii of the cylinders and ε is the ratio of the gap width to the outer cylinder radius. The dimensionless Taylor number, Ta, is a measure of centrifugal to viscous forces and is used to characterise the flow regime in coaxial cylinder flows employing real fluids. The Reynolds number, Re, is a measure of inertial to viscous forces and the two are related through equation (1), namely that $\mathrm{Ta} \propto \mathrm{Re}^2$. The basis for the characteristic length scale in Taylor's analysis was the annulus gap width. When alternative length scales are chosen as the basis, the definition of the Taylor number is subject to change. A number of alternative definitions for Taylor number (and indeed Reynolds number) can be found in the literature, the most common of which are given in Table 1. For inner cylinder rotation only, the Taylor number according to Rudman *et al.* (1994) is defined as

$$\mathrm{Ta} = \frac{\Omega_i r_i d}{\nu}\left(\frac{d}{r_i}\right)^{0.5} \tag{2}$$

where d is the annular gap width.

2.4. Visualisation of Flow Transitions

The dye tracer injection technique employed by Taylor (1923) is simple and has been used extensively by many researchers in their observations of coaxial and other flows. The technique is limited, though, since the dispersion of the dye can happen quickly in more agitated flows. Furthermore, this technique, while useful in studying dramatic transitions in the flow, is not so well suited to more subtle and complex transitions. Here, the use of neutrally buoyant metallic flakes seeded in the fluid is to be preferred. The technique was first described in detail by Schultz-Grunow and Hein (1956) who studied a wide range of Taylor-vortex flows and this visualisation technique has been universally adopted since.

Particles that are suitable for flow visualisation studies include metallic coated polymeric flakes (eg. Kalliroscope^{TM}), titanium coated particles (eg. Timiron Supersilk^{TM}), aluminium powder and fish scales. Particle density relative to fluid density may require consideration and matching these to ensure that sedimentation effects of the particles is not excessive and that the particle motion is due entirely to the fluid motions and not gravitational or other effects is desirable.

The first attempt at studying all of the possible transitions attainable with independent cylinder rotation was made by Coles (1965). He suggested that transition could take two forms. The first, transition takes place by spectral evolution, when the inner cylinder rate exceeds the outer cylinder rate. The most general case of this, which has been widely investigated, is exemplified when the outer cylinder is at rest and the inner cylinder rotates, examples of which are shown in Figure 1 (Curran, 2002). As the speed of the inner cylinder increases from rest, the flow is laminar and circumferentially stable up to a Ta_i of approximately 41.2, and the suspended particles experience a solid body rotation in the annulus (a). $\mathrm{Ta}_i = 41.2$ is the critical Taylor number, Ta_{crit} where the primary hydrodynamic transition to laminar Taylor-vortex flow occurs (b). Here, secondary flow phenomena are superimposed on the Couette flow and a series of counter-rotating toroidal vortices that are periodic in the axial direction form in the annulus. Each vortex has equal size and equal rotational speed. According to Savas (1985), the dark areas that are observed are the sinks and indicate radial inward motion of the particles. The thin dark lines are the sources and indicate radial outward motion of the particles. The light areas indicate motion in the azimuthal (circumferential) direction. Laminar Taylor-vortex flow is represented schematically in Figure 2 where U_i is the rotational speed of the inner cylinder. With further speed increments, travelling waves that are periodic in the circumferential direction develop and become superimposed on the laminar Taylor-vortex flow, thereby producing a doubly periodic or wavy-vortex flow phenomena (c).

Table 1. Alternative Definitions of the Taylor Number

Author	*Characteristic*	*Taylor Number*
Lee (1995)	r_i length; Ω_i only	$\mathrm{Ta} = \frac{r_i \Omega_i^3 d^3}{\nu^2}$ (a)
Takhar (1994)	a general length; Ω_o only	$\mathrm{Ta} = \frac{\Omega_o a^2}{\nu}$ (b)
Sugata (1991)	ζ fluid depth; Ω average annular rotation rate	$\mathrm{Ta} = \frac{4\Omega^2 (r_o - r_i)^5}{\nu^2 \zeta}$ (c)
White (1991)	D gap width; Ω_i only	$\mathrm{Ta} = \frac{r_i D^3 (\Omega_i^2 - \Omega_o^2)}{\nu^2}$ (d)
Mobbs & Ozogan (1984)	Eccentric cylinders; Ω_i; r_i	$\mathrm{Ta} = \frac{2\Omega_i^2 r_i d^3}{\nu^2 (r_i - r_o)}$ (e)
Sherman (1991)	r_i length; Ω_i only	$\mathrm{Ta} = \frac{2\Omega_i^2 r_i^2 d^3}{\nu^2}$ (f)
Schlichting (1987)	r_i length; Ω_i only	$\mathrm{Ta} = \frac{\Omega_i d}{\nu} (r_i d)^{0.5}$ (g)
Roberts (1965)	η ; Ω_i only	$\mathrm{Ta} = \frac{2\eta^2}{1-\eta^2} \frac{\Omega^2 d^4}{\nu^2}$ (h)

With still further speed increments, a modulated vortex flow ensues as a result of higher modes being generated by harmonics of the two fundamental frequencies in the doubly periodic wavy-vortex flow. A discrete spectrum to higher frequencies is achieved with continued speed increments and more complex stable flow structures form, however, on reaching this point these structures are now no longer stable (d). At sufficiently high speeds the sharp spectral lines become broadened and a continuous spectrum gradually reforms. This second continuous spectrum indicates featureless turbulent flow.

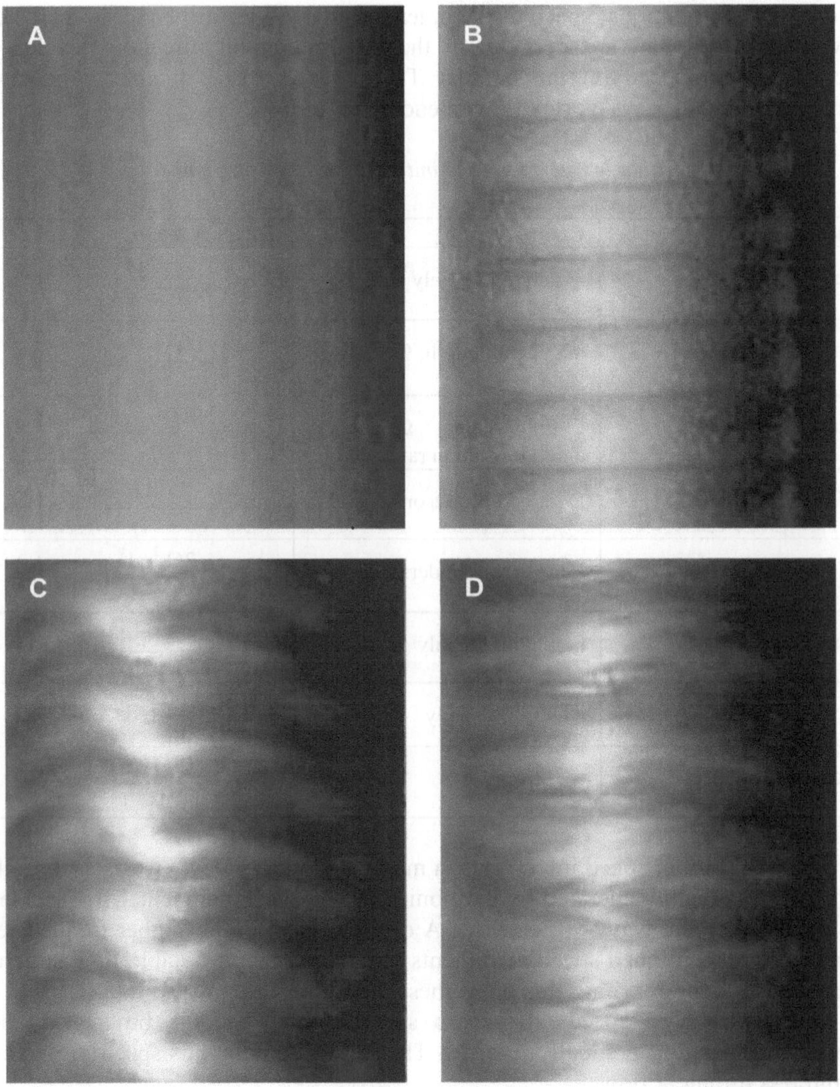

Figure 1: Flow visualisation of coaxial cylinder flow using 0.2wt% 8μm Timiron SupersilkTM particles seeded in water: (a) laminar Couette flow at Tai < 41.2; (b) laminar Taylor-vortex flow at Tai = 41.3; (c) wavy-vortex flow at Tai = 118.7; (d) turbulent vortex flow at Tai = 300.4.

The second type of transition proposed by Coles is catastrophic transition and occurs when the outer cylinder speed exceeds the inner cylinder speed. In this case, the transition from laminar Couette to turbulent flow is nearly immediate with no formation of any vortex flows in between. A state of intermittency will be observed between the limiting laminar and turbulent regimes when finite regions of laminar and turbulent flow coexist and distinct, well-defined interfaces between the two flow conditions are clearly visible. The interfaces themselves may be at points of transition from laminar to turbulent flow or vice-versa. The most clear cut case of intermittency is spiral turbulence. Here, a spiralling turbulent ribbon rotates through the predominantly laminar flow at approximately the mean velocity of the two cylinders. At certain speeds of rotation, intermittency is observed as turbulent spots or bursts in the predominantly laminar flow which appear and disappear at random.

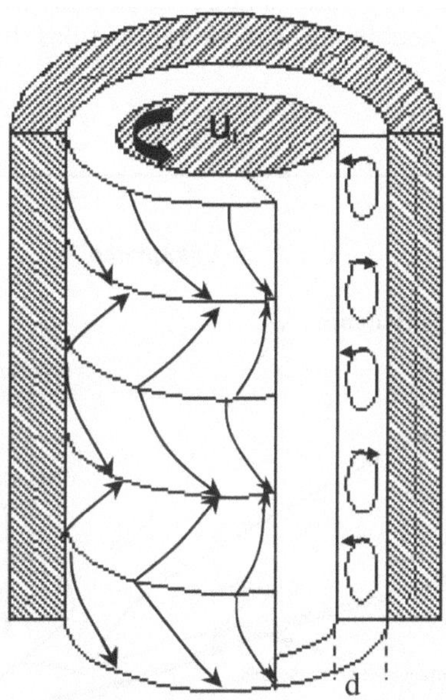

Figure 2: Schematic representation of laminar Taylor-vortex flow (redrawn from Schlichting, 1987).

A more comprehensive study of the transitions in coaxial cylinder flows with independent cylinder rotation was made by Andereck *et al.* (1986), who constructed a map showing the distinct regimes of flow that are attainable and their transition points over a range of inner and outer cylinder Reynolds numbers, R_i and R_o, respectively. This map is reproduced in Figure 3. Note that laminar Taylor-vortex flow can be maintained over a broader range of Reynolds number when both cylinders rotate independently, whereas for inner cylinder rotation ($R_o = 0$) the flow quickly changes from laminar Taylor-vortex to wavy-vortex flow for a relatively small increase in R_i. For a fixed geometry, this would correspond to a small change in inner cylinder speed, which could prove to be difficult to control.

2.5. Factors Influencing the Flow Transition Point

The transition from one flow regime to another is determined theoretically by the Taylor number (or, alternatively, the Reynolds number), however, there are a number of experimental factors that can give rise to considerable variation in the point of transition. Not surprisingly, perhaps, these are related to either the geometry of the system or the speed of rotation. In general, the cell culture medium is an incompressible Newtonian fluid, and so does not significantly influence the Taylor number. In bioreactor conditions where the cell concentration becomes high or the number of constructs per unit volume is large, deviations from Newtonian behaviour may have to be taken into account. For most applications, however, these effects are not likely to be significant.

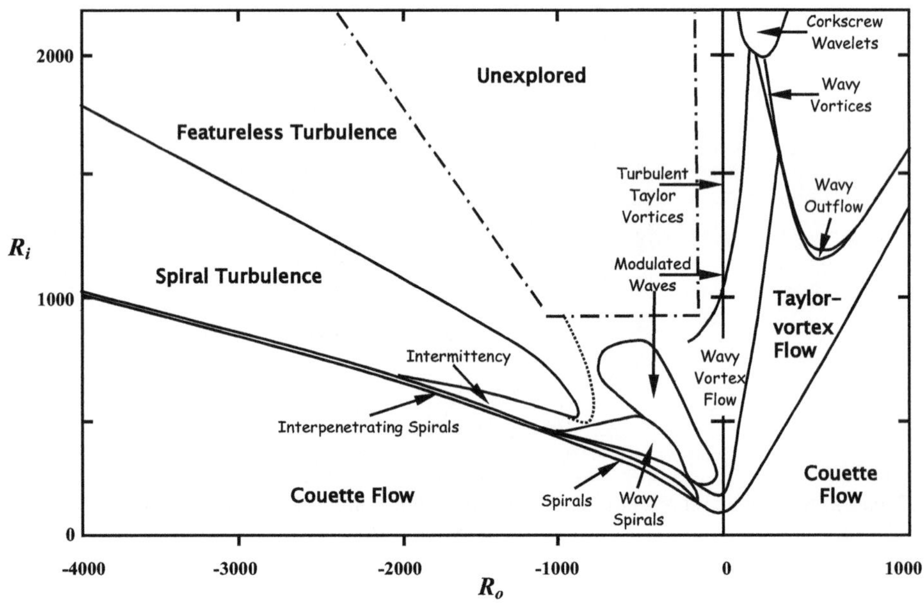

Figure 3: Flow regimes and transition points for independent coaxial cylinder rotation at varying inner (R_i) and outer (R_o) Reynolds numbers (negative sign denotes rotation in the opposite sense to the inner cylinder). Redrawn from Andereck (1986).

2.5.1. Aspect Ratio

The aspect ratio in a concentric cylinder system, Γ, is the ratio of the height of the fluid contained in the annular space between the walls of the inner and outer cylinders, h to the width of the annular gap, d:

$$\Gamma = \frac{h}{d} \quad (3)$$

The aspect ratio appears to only be significantly important once the second transition is reached, namely to wavy-vortex flow. In laminar Couette flow, and the transition to Taylor vortex flow, it appears that the dependence on the radius ratio η is only significant from the geometric perspective. The aspect ratio appears to play no part in the outcome of the hydrodynamic structure or stability. This conclusion is on the basis of the experiments of Cole (1976) who examined the onset of wavy vortex flow for $1 < \Gamma$,107 and $0.894 < \eta < 0.954$ using torque measurements and flow visualisation techniques. In both sets of experiments, Cole established that Ta_{crit} for transition to wavy vortex flow increased rapidly for $\Gamma < 40$. This was in contrast to the value for Ta_{crit} for the first transition to laminar Taylor-vortex flow, which was the same for columns where $\Gamma > 8$. It was suggested that for $\Gamma < 8$ the discrepancy could be due to the formation of Eckman layers (see §2.5.2 below). The number of azimuthal waves developed at the onset of wavy vortex flow was also found to be dependent on the aspect ratio, with fewer waves for large Γ and vice-versa. The radius ratio appeared to have little effect on the structure of the wave formation however, for columns where $\eta \leq 0.7$ the value of Ta at the onset of wavy flow was very much higher than the primary $Ta_{crit.}$. Similar general observations have been made by Lorenzen (1983) for columns where $\eta = 0.507$ and $10 \leq \Gamma \leq 17$ and from numerical studies by Edwards (1991) where $\eta = 0.87$ and $8 \leq \Gamma \leq 34$.

2.5.2. End Effects

Much of the theory pertaining to concentric cylinder flows stems from a superimposed disturbance on the initial laminar velocity profile for Couette flow. The equation for this profile is true for an infinite fluid. For a concentric cylinder flow, this literally means that $\Gamma = \infty$. When the aspect ratio is large, the fluid contained in the annular space is approximated to an infinite fluid i.e. $\Gamma \rightarrow \infty$. In real systems, of course, Γ has a finite value due to the horizontal boundaries at the top and bottom of the column and the question arises as to how large Γ can become before the assumption of an infinite fluid becomes valid. The influence of the horizontal boundaries on the hydrodynamics is collectively referred to as 'end effects', which have a significant influence on the flow even at subcritical Taylor numbers. The fluid layers adjacent to these boundaries are called the Eckman layers where vortices form at $Ta \ll Ta_{crit}$. This effect was evident in the flow visualisation studies of Andereck *et al,* (1986), but has been observed in earlier studies too (e.g. Coles, 1965). As Ta increases, further vortices develop on top of the Eckman layers until vortices developed from the top and bottom of the column meet in the middle of the column where upon Taylor-vortex flow ensues. The point at which the vortices meet in the middle is Ta_{crit}. When Γ is large, the formation of vortices in the middle section of the annulus is observed to occur near instantaneously as Ta_{crit} is reached and this supports the linear stability theory. When the length of the fluid column, h, is not an even integer multiple of the gap width, d, there may be an initial mismatch of the converging vortices from the top and bottom of the column. This presents itself as a transient delay to laminar Taylor-vortex flow even when Ta_{crit} is reached. This was observed experimentally by Park and Donnelly (1981) for large Γ columns, and by Mullin (1982) for small Γ columns. In the latter case, a stagnant fluid region could be observed in the region of mismatch, and vortices could be made to appear or disappear depending on the rate of increment of Ta. For a well-proportioned column, and if Ta is increased slowly, the number of vortices remains constant, as does the average wavelength, which is given by

$$\alpha_{av} = \frac{2h}{Nd} \tag{4}$$

where N is the number of vortices. There is systematic uncertainty in the numbers of vortices that will form due to the quantization condition that an integer number of vortices will fill the annulus. This percentage systematic uncertainty in wavelength, $\partial\alpha$ is given by

$$\frac{\partial\alpha}{\alpha_{av}} = \pm 2\left(\frac{100}{N}\right) \tag{5}$$

Therefore, even for columns where h is not an even integer of d, eventually, and in all cases, a regular spacing of identical vortices is generated except in the Eckman layers that may differ slightly from the annular vortices, again due to the geometry. One would expect that the influence of the end effects would become more pronounced as Γ becomes smaller and this has been observed to be the case. Blennerhassett and Hall (1979) showed, theoretically, that Ta_{crit} increases substantially as Γ decreases to the order of 1 up to the primary instability of laminar Taylor-vortex flow. Furthermore, Hall (1980) demonstrated anomalous modes of flow when Γ becomes very small.

2.5.3. Rate of Acceleration

With knowledge of the Taylor number, it has been suggested that the hydrodynamic state can be predicted. A further question that arises, however, is the uniqueness of the flow structure: can different structures be generated given the same initial and final Taylor number? Coles (1965) graphically illustrated that the hydrodynamic structure was not dependent on just the three parameters influencing the Taylor number, namely, geometry, rheology and speed of rotation, but was also dependent on the path taken between the initial and final values. Coles observed that up to 26 distinct stable flow states could be observed for the same final Taylor number when the rate of increase of the Taylor number from Ta = 0 was varied. The rate of increase of Ta is entirely due to the acceleration rate of the inner cylinder from rest in this case. Each of the states generated in these experiments was reproduced at least 3 times with replication of the Taylor number to 2-3%, and speed ranges of up to 14 times Ta_{crit}. The proof of non-uniqueness was provided by Snyder (1969), who produced 26, 24, 23 or 22 vortices at a fixed supercritical Taylor number in the same column depending on the initial conditions used and the path to the final condition. Koschmeider (1979) made similar experiments in a tall column ($\Gamma = 123$ and $\eta = 0.896$) using very slow acceleration rates of $7x10^{-4}$ rad/s^2 and sudden starts to 100 Ta_{crit}; similar observations were made that the final state differed depending on the acceleration. The acceleration rate appears to affect the wavelength established after time dependent starts, which is of course a function of Ta at which the start is made and the final Ta. It has been observed that the shortest wavelengths are generated for sudden start experiments. With progressively decreasing acceleration the wavelength can increase up to the critical wavelength $\alpha_{crit} = 3.12$ (Burkhalter and Koschmeider, 1974).

2.5.4. Gap Size

The foregoing discussion on Taylor numbers for concentric cylinder flows has been restricted to the case where a narrow gap exists. A narrow gap is defined as being very much smaller than the mean gap radius,

$$d << \frac{1}{2}(r_o - r_i)$$

For the case of a wide gap, which is the condition for $\eta \geq 0.5$, the Taylor number is redefined as

$$Ta = \frac{64}{9}\left(\frac{\Omega_i r_i^2}{\nu}\right)^2 k$$

where

$$k = (1-\mu)(1-4\mu)$$

These definitions apply for η=0.5 and the Rayleigh criterion for this case is $\mu >$ ¼ From a numerical solution to the flow problem for the wide gap, Chandrasekhar (1961) has shown that

$$Ta_{crit} \to 1.533x10^4 \; and \; \alpha_{crit} \to 6.2$$
$$as \; \mu \to 1/4 \; and \; k \to 0$$

where Ta_{crit} is calculated using equation (1). In terms of angular velocity of the rotating walls, the Rayleigh criterion for the wide gap case may be expressed as

$$\varpi_2 = \frac{1}{4}\varpi_1$$

$$\varpi_1 = \frac{\Omega_i r_i^2}{\nu} = \frac{0.375}{1-\mu}\sqrt{\frac{\mathrm{Ta}}{\mathrm{k}}}$$

$$\varpi_2 = \frac{\Omega_o r_i^2}{\nu} = \mu\varpi_1$$

For the critical case and as $\mu \rightarrow 1/4$,

$$\varpi_1 \rightarrow \frac{53.61}{\sqrt{(1-4\mu)}}$$

Roberts (1965) calculated Ta_{crit} and α_{crit} for progressively increasing gap width (η) and for the particular case of inner cylinder rotation only (μ=0). The Taylor number was defined as

$$\mathrm{Ta} = \frac{2\eta^2}{1-\eta^2}\frac{\Omega_i^2 d^4}{\nu^2}$$

in the range $0.500 < \eta < 0.975$. A summary of the results found in this study is given in Table 2.

The study of Coles (1965) suggested that the progression of flow states described in section 2.4 are evident only if the radius ratio of the cylinders $\eta > 0.714$. For $\eta < 0.714$ the transition from laminar Taylor-vortex flow to turbulent vortex flow is immediate with none of the other intervening states being clearly observed from flow visualisation studies.

Table 2. Critical Taylor number and Critical Wavelength as a Function of Gap Width (after Roberts, 1965)

η	Ta_{crit} *(from equation (1))*	α_{crit}
0.975	1723.89	3.1268
0.950	1754.76	3.1276
0.925	1787.93	3.1282
0.900	1823.37	3.1288
0.875	1861.48	3.1295
0.850	1902.40	3.1302
0.750	2102.17	3.1355
0.650	2383.96	3.1425
0.500	3099.57	3.1631

3. SHEAR FORCES IN TAYLOR-VORTEX FLOWS

3.1. Critical Shear Stress on Cells

Animal cells are sensitive to shear forces (Thomas, 1990), which is due to their relatively large size of 10 to 30 μm and the fact the cell contents are protected only by the cell membrane (Hua *et al.*, 1993). Whilst some cells cultures can be grown in free suspension, most require attachment to a biocompatible surface for controlled proliferation and differentiation. This is particularly so in the context of tissue engineering. In such circumstances, if the attachment surface is fixed in place, the cells appear to be more susceptible to shear stress as they cannot freely rotate or translate to accommodate the imposed forces (Croughan *et al.*, 1987), which may mean that laminar Couette flows must be imposed to minimise these forces (Williams *et al.*, 2002). Although excessive shear forces will be detrimental to the cell, some fluid shear may be desirable as it can lead to increased cell permeability (Fry, 1968) and increased secretion of extracellular proteins (Abu-Reesh and Kargi, 1989). It can also influence the shape of the cell (Levesque and Nerem, 1985). These effects all have a bearing on the cell physiology (Frangos *et al.*, 1988), which is of paramount importance in tissue engineering. An alternative to a fixed scaffold support onto which cells can proliferate is the use of microcarrier beads (Butler, 1988). These are generally microporous spheres of an appropriate biomaterial onto and into which cells proliferate. They are typically 100–450 μm in diameter and provide a very large surface area per unit culture volume for cell attachment. The microcarriers should be neutrally buoyant in the media so that they behave, from a hydrodynamic viewpoint, like freely suspended cells in the bioreactor. Mechanisms of shear-induced damage to cells on microcarriers have been reviewed by Cherry and Papoutsakis (1988) and Lakhotia and Papoutsakis (1992).

In principle, it is prudent to establish first a critical shear stress level for a given culture of cells and a given bioreactor configuration. A number of experimental investigations have been made to establish a 'critical shear stress' and it is evident that this is dependent on the type of cell, the method of culture (free suspension/attached) and the time of exposure to the shear stress. A summary of some of the studies made to assess critical shear stress is presented in Table 3. In this table, a range of cell types are given for varying degrees of shear stress exposure and exposure times. The various types of test apparatus that are used to stimulate cells has been reviewed by Brown (2000) and these include viscometers (concentric cylinder Couette and cone and plate types), laminar flow (LF) chambers and spinner flasks. Individual designs are common, as are modifications to the basic designs described by Brown. These geometric changes will clearly have a bearing on the outcome of the shear experiments. It is evident from the wide range of data available that there is no unique critical shear stress that can be applied in the design of a cellular bioreactor. At best, guideline values can be used for a given set of experimental conditions. Furthermore, the culture time becomes critically important. At the present time, the majority of data available for the viability of cells (which is generally associated with critical shear stress) is for short-term experiments (< 24h). It seems probable that for longer-term culture operation at shear stress levels far below critical is advantageous. It is also apparent that cells in free suspension can withstand higher stress levels than adherent cells. The reasons for this were addressed, from a theoretical viewpoint, by Cherry and Papoutsakis (1986) and Papoutsakis (1991), and experimentally by McQueen *et al.* (1987) and Croughan *et al.* (1989). The size of eddies in the flow relative to the size of the cells/microcarriers appears to be important. Once the flow becomes turbulent it is possible to estimate the size of the smallest (Kolmogorov) eddies. If these are larger than the cells, the cells maybe simply transported around inside the eddy, however, when the eddy is smaller than the size of the cell high shearing on the cell will occur. As the turbulence intensity increases, the smallest eddies become smaller and so the shearing damage will increase. Cells, which are fixed onto a stationary support, do not have this ability to move with the fluid and thereby experience significantly higher stresses.

An extensive review on the shearing effects on cells in bioreactors from a theoretical, experimental and commercial viewpoint has been compiled by Schugerl & Kretzmer (2000). As a guideline, on the basis of the numerous experimental studies done on shear effects on cells, stresses in the range 0 to 10 dyn/cm^2 appear to be of most interest for the long term culture of freely suspended cells in bioreactors.

Table 3. Experimental Studies on the Application of Shear Stress to Cells

Cell Type	*Test Apparatus*	*Exposure Level*	*Exposure Time*	*Reference*
BHK-21, CHO-K1, Vero	Spinner Flask	70, 110, 200 rpm	0-250 h	Wu (1999)
BHK	LF Chamber	0.26-5.4 N/m^2	2-24 h	Stathopoulos (1985)
BHK	LF Chamber	0-2.5 N/m^2	0-24 h	Ludwig (1992)
FS-4 Fibroblasts	Spinner Flask	0-30 dyn/cm^2	0-180 h	Croughan (1989)
FS-4 Fibroblasts	Spinner Flask	1-12 s^{-1} (shear rate)	0-150 h	Croughan (1987)
HDP-1 Mouse Hybridoma	Viscometer	50-1000 dyn/cm^2	0.5-3 h	Abu-Reesh (1989)
CRL-8018 Hybridoma	Viscometer	0-50 dyn/cm^2	0-10 min	Petersen (1988)
TB/C3 Mouse Hybridoma	Viscometer	0.019-600 N/m^2	10 min to 24 h	Born (1992)
6 Cell Types	Viscometer	1, 10, 100 N/m^2	5-20 min	Mardikhar (2000)
S hassjoo	71 STR	2-40 N/m^2	0-14 d	Chen (2000)
VERO	Spinner Flask	3-50 N/m^2	0-140 h	Wu (2000)
hFOB Human Fetal Osteoblasts	LF Chamber	0-2 N/m^2	0-5 min	Jacobs (1998)
Leukocyte	LF Chamber	0.5-20 dyn/cm^2	0-2 h	Dong (2000)
CHO	21 STR	0-50N/m^2	0-100 h	Michaels (1992)
Platelets	Viscometer	500-4000s^{-1} (shear rate)	100 s	Rhodes (1997)
HUVEC	Cone Plate	4-400 N/m^2	0-250 min	Schnittler (1993)
HUVEC	LF Chamber	0-1 N/m^2	0-8 h	Yoshikawa (1997)
SMC	LF Chamber	0.5-2.5 N/m^2	0-24 h	Papadaki (1996)

Table 3 (cont.)

Table 3 (cont.)

Cell Type	*Test Apparatus*	*Exposure Level*	*Exposure Time*	*Reference*
EC	Static Culture	0-2 N/m^2	0-24 h	Sato (2000)
EC	Cone-Plate	1-5 dyn/cm^2	up to 8 d	Dewey (1981)
EC	LF Chamber	0.2-3 N/m^2	0-24 h	Thoumine (1995)
EC	LF Chamber	0.2-2.5 N/m^2	0-30 h	Nollert (1991)
EC	LF Chamber	1-8.5 N/m^2	0-24 h	Levesque (1985)

3.2. Measured Shears in Taylor Flows

3.2.1. Laminar Couette Flows

The fluid shear rate in a fluid rotating in a concentric annulus can be predicted theoretically for the specific case of *laminar Couette flow* using equation (6) below:

$$\frac{dV}{dr} = \frac{r_o^2\Omega_o - r_i^2\Omega_i}{r_o^2 - r_i^2} + \frac{1}{r^2}\frac{r_o^2 r_i^2(\Omega_o - \Omega_i)}{r_o^2 - r_i^2} \tag{6}$$

When only the inner cylinder rotates, $\Omega_o = 0$ and (6) is more simply written as

$$\frac{dV}{dr} = \frac{r_i^2\Omega_i}{r_o^2 - r_i^2} + \frac{1}{r^2}\frac{r_o^2 r_i^2\Omega_i}{r_o^2 - r_i^2} \tag{7}$$

Three cases are normally of interest with respect to the shear rate: the inner wall shear rate ($r = r_i$)

$$\left|\frac{dV}{dr}\right|_{r=r_i} = \frac{\Omega_i\left(r_i^2 + r_o^2\right)}{r_o^2 - r_i^2}$$

and the outer wall shear rate ($r = r_o$)

$$\left|\frac{dV}{dr}\right|_{r=r_o} = \frac{2r_i^2\Omega_i}{r_o^2 - r_i^2}.$$

The magnitude of the shear rate at the mid-point of the gap ($r = (r_o+r_i)/2$) is given by

$$\left|\frac{dV}{dr}\right|_{r=(r_o+r_i)/2} = \frac{r_i^2\Omega_i}{r_o^2 - r_i^2} + \frac{4}{(r_o - r_i)^2}\left[\frac{r_o^2 r_i^2\Omega_i}{r_o^2 - r_i^2}\right].$$

In coaxial cylinder bioreactors employing freely suspended cells, or cells supported on microcarrier beads in laminar Couette flow, the particles will have a circular trajectory at the azimuthal velocity of the fluid. The particle paths will cover a range of different radii within the annulus and the shear rates on them can be determined by choosing the appropriate value of r in either equations (6) or (7) depending if only one or both cylinders rotate.

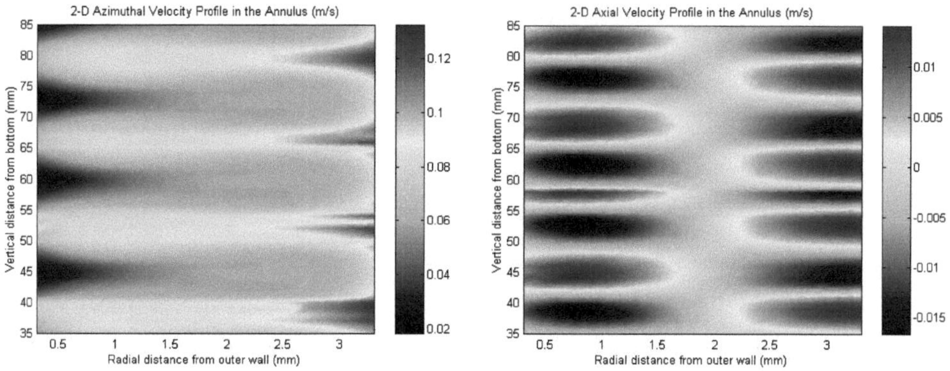

Figure 4: Laser Doppler anemometry data for wavy-vortex flow at $Ta_i = 148$.

3.2.2. *Vortex Flows*

When the critical Taylor number is reached and laminar Taylor vortex flow ensues in the annulus there are axial and radial velocity vectors in addition to the azimuthal component. This introduces additional shearing forces on the particles in the flow in addition to those in the radial plane from the azimuthal velocity alone. In order to estimate these and calculate the total shear experienced by particles it is necessary to quantify the velocity field in the annulus.

In our laboratory, Laser Doppler Anemometry (LDA) has been used to quantify the 2-D velocity field in coaxial cylinder systems at two scales; the smaller system designated as CC-25 has a 25mm rotating inner cylinder, a stationary outer cylinder, a 1mm annular gap width and an aspect ratio of 37.5 whilst the larger system designated as CC-75 has a 75mm rotating inner cylinder, 3mm gap width and aspect ratio 37.6 (Curran and Black, 2000). Azimuthal and axial velocity components are measured to generate velocity flow maps for each component. Representative data are shown for wavy-vortex flow at Tai = 148 in CC-25 in Figure 4.

The spatial resolution of the velocity data points is dependent on the capabilities of the LDA measuring probe. Our data has a resolution of 100μm axially and 50μm radially. This allows for accurate computation the shearing forces on particles anywhere in the annular flow domain. The rate of strain tensor (equation (8)) and a pseudo shear rate (equation (9)) allow for calculation of the total shear rate. The velocity components in equation (8) are azimuthal (u_j) and axial (u_z). The total shear stress is then calculated with knowledge of the fluid rheology obtained from rheometry measurements.

$$e_{ij} = \begin{bmatrix} 0 & \frac{1}{2}\left(\frac{\partial u_\phi}{\partial r}\right) & \frac{1}{2}\left(\frac{\partial u_z}{\partial r}\right) \\ \frac{1}{2}\left(\frac{\partial u_\phi}{\partial r}\right) & 0 & \frac{1}{2}\left(\frac{\partial u_\phi}{\partial z}\right) \\ \frac{1}{2}\left(\frac{\partial u_z}{\partial r}\right) & \frac{1}{2}\left(\frac{\partial u_\phi}{\partial z}\right) & \frac{\partial u_z}{\partial z} \end{bmatrix} \tag{8}$$

$$\gamma = \left(2\sum_{i=1}^{3}\sum_{j=1}^{3} e_{ij}e_{ij}\right)^{\frac{1}{2}} \tag{9}$$

3.3. Effect of Scaling

Geometric scaling of coaxial cylinder systems is best achieved by selecting the desired gap width of the annulus and then maintaining the aspect ratio in the scaling process. The same flow regime is replicated in the scaled system by maintaining Ta as in the original system. In general, as the system is scaled up, the speed of rotation required to maintain the same Ta decreases. Since the gap width will increase on scale-up, the overall shear on similarly sized cells and constructs will also decrease. Maximum shear stresses were evaluated in systems CC-25 and CC-75 based on LDA measurements for the near wall regions and in the majority of the annular core. The results are shown in Figure 5 for a broad range of flow regimes. Shear stresses experienced by particles in the larger system CC-75 are significantly reduced compared with CC-25 as Ta increases. In both systems, for Ta < 100, the shear stresses are in an acceptable range for longer-term cell culture (< 1 Pa) and in the Taylor-vortex regime (Ta = 41.2) the shear stresses are less than 0.3Pa. The highest stresses occur on the edge of the boundary layer extending into the annular core region from either wall (Curran and Black, 2000). Based on these data, the relative change in shear rate for similarly sized constructs with change in geometric scale of vessel has been found to follow

$$\left(\frac{\varphi_1}{\varphi_2}\right)^{\alpha} = \left(\frac{\gamma_2}{\gamma_1}\right) \tag{10}$$

where φ is geometric scale factor, γ is shear rate and the subscripts 1 and 2 are for the smaller scale and larger scale respectively.

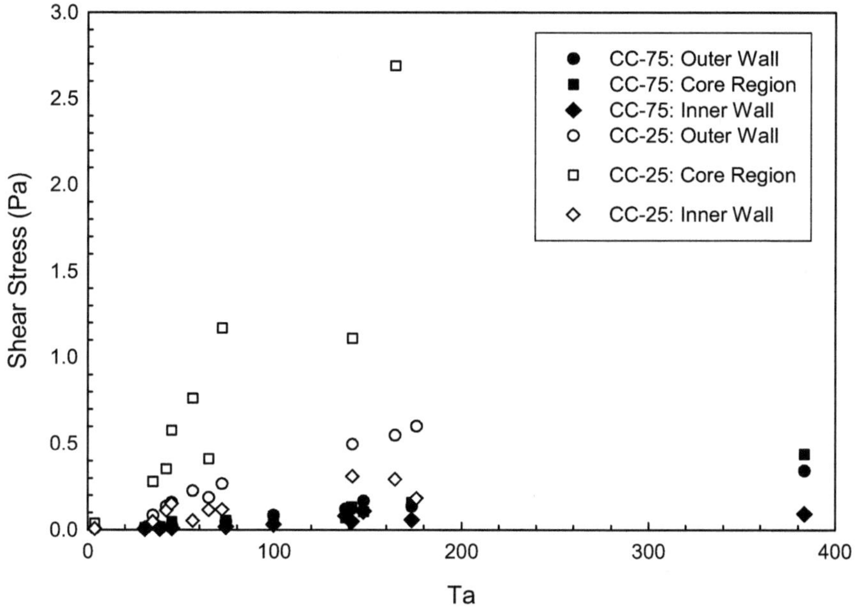

Figure 5: Maximum shear stresses calculated near the inner and outer walls and the majority annular core region in CC-25 and CC-75 from LDA measurements.

4. MIXING AND MASS TRANSPORT IN TAYLOR-VORTEX FLOWS

4.1. Transport Mechanism in Taylor-Vortex Flows

With the rotational speed of the inner cylinder slowly increasing, the distinguishable flow states observed are progressively, laminar Couette flow, laminar Taylor-vortex flow, wavy vortex flow and turbulent vortex flow in a coaxial cylinder system provided h > 0.714. Taylor vortex reactors have been employed for a number of varied chemical and biochemical applications where they can be operated as batch systems but are typically operated in a continuous mode. In this mode, continuous operation might be achieved by feeding into one end of the annulus and removing at the other end thereby generating an axial flow superimposed on the Taylor vortex flow. Reactors of this type are referred to as Taylor-Poiseuille flow reactors. These have been studied extensively (Inamura *et al.,* 1993; Kataoka *et al.* 1975, 1977, 1981; Ogihara *et al.* 1995). Alternatively, the reactants can be fed into the top and bottom of the annulus at equal rates so that there is no axially imposed flow. Reactors operating in this way have been studied by Campero and Vigil (1997), Tam and Swinney (1987), and Vastano *et al.* (1990). This type of Taylor-vortex reactor approximates to a one-dimensional reaction-diffusion device.

Mass transport in concentric cylinder systems occurs by inter-vortex exchange and intra-vortex mixing. In laminar Couette flows the fluid experiences a solid body rotation and no transport is assumed to occur in the radial or axial directions since motion is confined to the azimuthal direction only. In the case of laminar Taylor-vortex flow, the unitary vortex cell consists of two counter-rotating vortices represented schematically in Figure 6. This unitary cell comprises two inflow boundaries that direct fluid radially into the rotating wall. These are located at the centre of the cell and both move together in parallel. At either end of the vortex cell, an outflow boundary exists. This directs fluid away from the rotating inner wall and out towards the stationary outer wall.

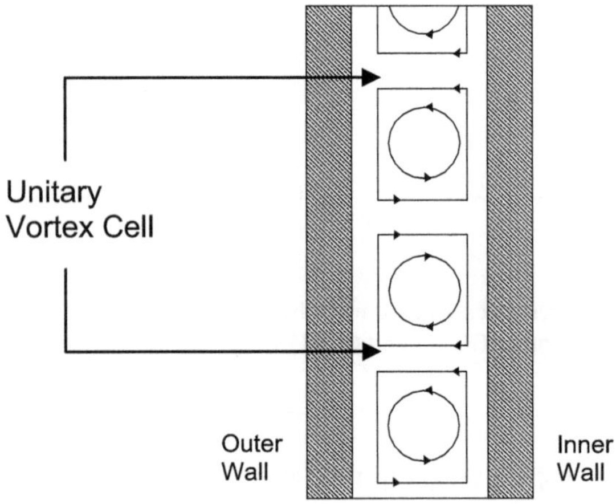

Figure 6: Schematic of the unitary vortex cell in a section of the annulus

Ohmura *et al.* (1997) showed that the inflow boundaries act as a barrier to inter-vortex mixing and that the outflow boundaries promote inter-vortex mixing. The average vortex height, h_v, was found to be important and was non-dimensionalised as h_v/d. When $h_v/d = 1$, the vortex motion is stable with no waves on either inflow or outflow boundaries. When $h_v/d < 1$, small amplitude wave motion was evident at both inflow and outflow boundaries, and, for $h_v/d > 1$, the wave motion was evident only at the inflow boundaries. The acceleration to the equilibrium speed for the rotating inner cylinder is very influential in the value of h_v/d. Intra-vortex mixing consists of a tangential (circumferential) dispersion accompanied by axial and radial mixing. In laminar Taylor vortex flow, the tangential dispersion is weak (Kataoka and Takigawa, 1981; Legrand and Coeuret, 1986), and numerical models of Howes and Rudman (1990) indicate weak radial and axial mixing in the vortex core for Taylor flow. In contrast, layers of fluid near to vortex boundaries are readily mixed. Inflowing boundaries produce excellent micromixing in a thin layer as the inflowing boundaries from each counter rotating part of the vortex come into contact (Liu and

Lee, 1999). Effective mixing is also achieved at the outflowing boundaries, but, at these boundaries, unitary vortex cells meet, and the mixing here is essentially an exchange of material from one unitary vortex cell to the next. It is at these out-flow boundaries that inter-vortex exchange takes place. Carefully controlled dye tracer experiments performed by Desmet *et al.* (1996) showed that dye injected into the centre of a vortex is slowly transported to the outflowing boundaries but concentrated at the inflowing boundaries. Injection near to the outflowing boundaries shows very efficient transfer from one vortex to the next.

As the speed of the inner cylinder is increased, wavy vortex flow is initiated for $h > 0.714$, and the wavy motion forces fluid from the vortex boundaries, directing it into the core. The waviness produces regions of upward and downward vortex deformation corresponding to regions of upward and downward axial flow. Axial (and intra-vortex) mixing is therefore increased. Of course, superimposed on this is the azimuthal motion. Radial mixing is therefore also promoted. Ryrie (1992) showed that with even very small deviations from laminar Taylor vortex flow to wavy vortex flow, fluid trajectories became chaotic even in the absence of molecular diffusivity resulting in an enhanced effective diffusivity or axial dispersion. From numerical simulations of the flow field (which agree with experimental observations), Rudolph *et al.* (1998) showed that up to 50% of the volume of a vortex is transported into and out of a vortex in one azimuthal wave period. They verified Ryrie's findings in terms of chaotic particle paths and suggest that axial scalar transport is substantially increased over molecular diffusion. They proposed that optimum fluid transport between vortices takes place at Ta = 253 and that mixing and transfer decrease at both higher and lower Ta than this for wavy vortex flow. Whilst the axial dispersion coefficient can be up to three times higher in turbulent vortex flow, this will be intrinsic to the turbulence rather than to axial transport between vortices.

4.2. Taylor Vortices as Ideal Mixing Units

The vortex cell (vortex pair) in laminar Taylor-vortex flow behaves as an ideal batch reactor. Evidence of this behaviour was demonstrated by Kataoka *et al.* (1975), who used a salt tracer technique. In this study, a 6.5×10^{-3}M aqueous sodium chloride solution was added as tracer to a working viscous fluid of 40wt.% glycerine, and conductivity measurements were made; a small amount of green dye allowed for visualisation of the transport. A more refined series of experiments was performed by Ohmura *et al.* (1997) to investigate inter-vortex mixing in wavy vortex flow and turbulent vortex flow using a similar technique. They concluded that inter-vortex mixing is dependent on the vortex height, h_v, and the wave motion in wavy vortex flow and turbulence character in turbulent vortex flow and that intra-vortex mixing increases monotomically with increasing Taylor number. Inamura *et al.* (1993) produced near monodispersed sized polystyrene latex particles in a continuous Couette-Taylor-vortex flow. Sodium Lauryl sulphate, potassium persulphate and chemical grade styrene were mixed in the system producing non-coagulated monodisperse beads. A similar series of experiments performed by Kataoka (1975) produced a comparable result. Ogihara *et al.* (1995) produced monodisperse silica particles in this flow regime and accounted this to the ideal batch reactor behaviour of the vortex cells.

4.3. Quantifying the Transport Process

Mass transport correlations are used to relate the transport of the species of interest to the fluid flow regime and the physical properties of the transporting medium. In cases where a liquid or gas is transporting through a liquid medium, the mass transport correlation typically takes the form of equation (11):

$$\mathrm{Sh} = \varphi\,\mathrm{Re}^{\alpha}\,\mathrm{Sc}^{\beta} \tag{11}$$

where α, β and φ are constants, Sh is the dimensionless Sherwood number, Re is the Reynolds number and Sc is the Schmidt number. The Sherwood number is interpreted as the ratio of the actual mass transfer coefficient to the purely diffusive value. The Schmidt number is simply the ratio of the kinematic viscosity to the molecular diffusivity and therefore gives a measure of the relative efficiency of the fluid as a conductor of momentum and solute. When the dimensionless numbers are equal to 1.0, the velocity and concentration profiles are geometrically similar and the degree of similarity decreases as departure is made from 1.0. Mass transport coefficients are often correlated by equations employing these dimensionless numbers. Correlation by this method better serves to scale-up (or down) the mass transport process. The constants of the correlation are determined empirically. For the Taylor-vortex system, the dimensionless numbers are defined as

$$\mathrm{Sh} = \frac{(r_o - r_i)k_l}{D} \tag{12}$$

$$\mathrm{Sc} = \frac{\nu}{D} \tag{13}$$

$$\mathrm{Re} = \frac{(\Omega_i - \Omega_o)dr_i}{\nu} \tag{14}$$

where D is diffusivity of the transporting species (m^2/s) and k_l is the mass transfer coefficient (m/s). The unique flow structure of Taylor vortices lends itself to the quantitative analysis of mass transport within them. The axial length of a vortex pair, Δ, is approximately twice the annular gap size (Kataoka *et al.,* 1975) in the range $41.2 < \mathrm{Ta} < 300$, and has been shown to have little variation beyond this point. Koschmeider (1979) showed that

$$\Delta = 2.4(r_o - r_i) \qquad \text{for } 320 < \mathrm{Ta} < 8200$$

The toroidal vortex cell is considered as a tubular batch reactor of length L,

$$L = \pi(r_o + r_i)$$

with complete recirculation. The flow can be modelled as that in a tube of infinite length in which longitudinal dispersion occurs. The dispersion is characterised by the dimensionless Bodenstein number, Bo,

$$Bo = \frac{u_t L}{D_d}$$

where u_t is the mean azimuthal velocity of the vortex cell and D_d is the dispersion coefficient (m^2/s). This is related to the number of perfect mixing elements, or vortex cells, N, of equal size

$$Bo = 2N \tag{15}$$

The circumferential dispersion coefficient is then calculated from

$$D_d = \frac{u_t L}{2N} \tag{16}$$

Legrand and Coeuret (1986) found the dispersion coefficient related to the Taylor number according to

$$\frac{D_d}{\nu}\left[\frac{r_i d^3}{L^4}\right]^{\frac{1}{2}} = 0.000239Ta \tag{17}$$

From experimental measurements, it is possible to plot the term on the left hand side of equation (17) against Taylor number to evaluate the constant for the given experimental configuration provided the number of vortices and the mean velocity of the vortex is known. For the intermixing vortex region, the dispersion coefficient calculated by this method can be replaced in the forced convection correlations of the form Sh = *f*(Re, Sc) for the mass transfer coefficient.

4.4. Oxygen Transport Studies

Controlled oxygen transport measurements were made in our laboratories in a Taylor-vortex bioreactor with the dimensions of CC-25 described in section 3.2.2. The inner cylinder was allowed to rotate and the outer cylinder was kept stationary.

The annulus was filled with cell culture media and freely suspended C20/A4 human chondrosarcoma cells at a concentration of 2.5×10^5 cells/ml. to a working volume of 15.9 ml. Oxygen entered the annulus naturally from the atmosphere via a 150μm ring around the rotating shaft of the inner cylinder and oxygen concentration measurements were made at 10s intervals over 24h with a fibre-optic oxygen probe with a 50μm tip (PreSens GmbH). This series of experiments deliberately restricted the oxygen delivery to more easily discern the effects of the hydrodynamic regime on axial transport. A schematic of the bioreactor configuration is shown in Figure 7. The fluid contents were maintained at 37 ± 0.2°C.

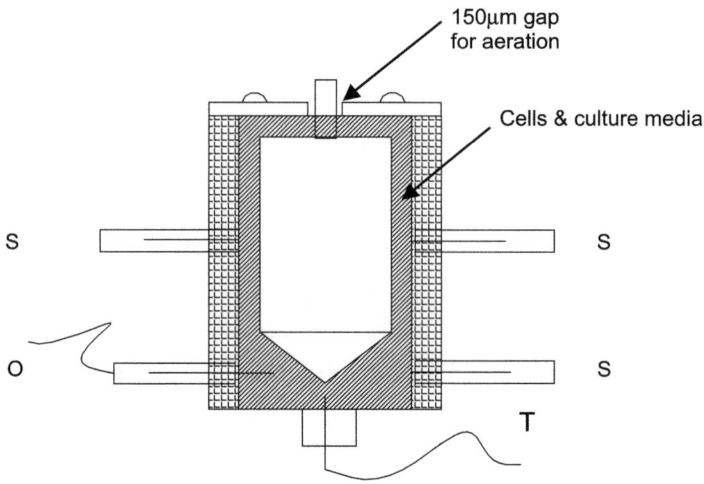

Figure 7: Schematic of CC-25 bioreactor for oxygen transport studies: sampling port (S); oxygen probe (O); thermocouple (T). The fluid volume is 15.9ml.

The range of rotational speeds of the inner cylinder and the corresponding dimensionless numbers for the oxygen transport studies are given in Table 4. This range covers laminar Couette flow, laminar Taylor-vortex flow and a series of progressively more disturbed wavy vortex flow regimes. In each case, a 100% absolute oxygen concentration was reached at the initiation of oxygen concentration measurements and the depletion in concentration over time as a consequence of the respiration of the cells was monitored. The oxygen concentration profile for the different flow regimes is shown in Figure 8. The regime of laminar Couette flow was used as a basis for comparison of the axial transport of oxygen. In this regime of flow no axial transport is assumed to occur due to forced convection as no axial mixing occurs, so depletion at the probe is due only to cell consumption. The depletion follows an exponential decay, and the mass transfer coefficient (k_l) can be estimated from the approach of Zhang *et al.* (1992):

$$k_{\ell}a' = \frac{1}{t_{0.63}} \tag{18}$$

where $t_{0.63}$ is the time taken for the concentration of oxygen to fall to 63% of the initial value and a' is the interfacial area for oxygen to transport into the liquid media per unit reactor volume (m^2/m^3). This occurs at the 150μm gap and a' is calculated as $0.212m^2$ resulting in a mass transfer coefficient of 2.81×10^{-4}m/s from equation (18). The phase equilibrium for the transfer of oxygen across the phase boundary at the gap is governed by the partial pressure of oxygen in the atmosphere and the Henry constant, H. The media in CC-25 varied between 36.8 and 37.2°C and a mean value for H was taken as 5.06×10^9 Pa molfrac^{-1}. The partial pressure of oxygen in air is 0.2095 atm or 2.12×10^4 Pa. The density of cell culture media is 985 kgm^{-3} which results in a mean dissolved oxygen concentration of 0.13 mg per litre across the phase boundary.

Table 4. *Hydrodynamic Operating Parameters for Oxygen Transport Studies*

Speed (rpm)	Ta_i	Re_i
77.81	28.80	101.83
100.96	37.37	132.13
116.81	43.25	152.92
155.64	57.65	203.84
311.30	115.25	407.45
389.13	144.08	509.38
466.96	172.88	611.23
778.28	288.11	1018.63

From the data of Figure 8, it appears that as the Taylor number increases, the oxygen concentration at the base of the reactor more closely approaches that at the free surface of the reactor (100% absolute). This implies that the enhanced hydrodynamic environment of Taylor-vortex and wavy-vortex flow better transports oxygen from the free surface at the top of the reactor down into the bulk fluid.

4.4.1 Laminar Couette Flow

At Ta_i = 28.80, laminar Couette flow prevails. In this regime there is no axial component of velocity, only radial and azimuthal components. Oxygen transport in the radial and azimuthal directions is very good as there is significant forced convection transfer and it is assumed that near to the free surface the absolute relative % O_2 will be close to 100% for the length of the experiment. Axial transport in this flow regime is significantly limited since transport in the absence of an axial velocity component in the flow will be due to molecular diffusion and free convection only. Initially, the cells in free-suspension are assumed to be randomly distributed throughout the fluid volume and are moving in a solid body rotation in an

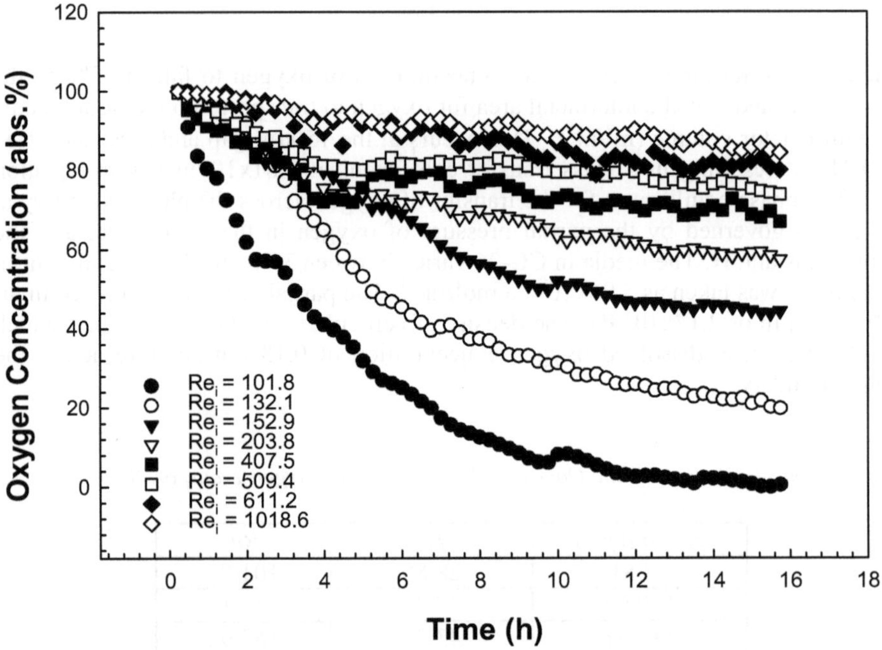

Figure 8: Oxygen concentration profiles in Taylor-vortex bioreactor CC-25 for varying regimes of flow using 2.50x10^6 cells/ml C20/A4 Chondrosarcoma cells in a 15.9ml volume culture medium volume.

azimuthal direction. If viable, these cells are aerobically respiring and consuming oxygen. In the probe location at the base of the vessel, the oxygen will be consumed by the cells and as a consequence of poor transport the concentration will deplete rapidly. Notwithstanding this, the cells have an increased tendency to sediment in laminar Couette flows. Whilst the precise settling velocity of chondrosarcoma cells is not known, it is known that the cells will sediment since this occurs in the tissue culture flasks where monolayers of cells form on the base of the flask and no cells remain in free suspension in the media. In CC-25 over a 16h time frame, the numbers of cells found in the lower part of the reactor in laminar Couette flow increases and this has been demonstrated from flow cytometry studies. Increased cell numbers in this region will further reduce the oxygen concentration.

In laminar Couette flow where a constant layer of fluid is exposed to the free surface and there is no axial disturbance, the penetration theory for mass transport given by equation (19) applies (Coulson *et al.*, 1990).

$$(J_A)_{z=0} = -D\left(\frac{\partial c_A}{\partial y}\right)_{z=0} = (c_{Ai} - c_{Ao})\sqrt{\frac{D}{\pi t}} \qquad (19)$$

The mass transfer coefficient is given by $\sqrt{\frac{D}{\pi t}}$. The diffusivity of oxygen in cell culture media at 37°C is taken to be $1.70 \times 10^{-9} m^2 s^{-1}$ (Kallos and Behie, 1999). Based on a mean dissolved oxygen concentration of 0.13 gm^{-3} for C_{Ai}, the mass flux of oxygen at the free surface can be determined as a function of time.

4.4.2. Laminar Taylor-vortex flow

From Taylor's theory, Ta_{crit} occurs at 41.2 and there is sudden transition from laminar Couette flow to Taylor-vortex flow. From our LDA data for the system CC-25 the results indicate that this is true, although the transition is not as drastic; there is some gradual disturbance on approach to Ta_{crit}. Although this disturbance is very small it might well be significant in terms of gaseous transport effects. In oxygen transport studies hydrodynamic conditions just sub critical (Ta_i = 37.37) and just supercritical (Ta_i = 43.25) have been made. In the former case, there is clearly a marked improvement in the transport of oxygen over the condition for Ta_i = 28.80. This is likely due to a combination of two things. A slightly reduced effect on cellular sedimentation is likely as it observed from the LDA data for CC-25 and Ta_i = 35.07 that there is a very small upward axial velocity to counteract sedimentary effects. Secondly, this slightly increased axial motion will have an improved effect on the transfer of oxygen from the free surface into the bulk of the fluid.

Transport is further enhanced in laminar Taylor-vortex flow at Ta_i = 43.25. The steady laminar Taylor vortices will significantly reduce sedimentation of cells to the base of the vessel and this will reduce consumption of oxygen in this region. The vortices do not intermix in this flow regime so although there will be transferral of oxygen from the free surface into the bulk fluid the overall rate of transference to the base of the vessel will be still quite slow and in general, there will a depletion of the oxygen concentration over time at the measurement point.

In pure laminar Taylor vortex flow, only the top vortex is exposed to the free surface. It is known that there is poor intra-vortex mixing in this flow regime so transport only occurs by inter-vortex exchange and this is known to occur on the vortex flow boundaries. Under this flow regime, the transport can be considered as a tanks-in-series model (Levenspiel, 1972). A number of parameters require calculation to evaluate the transport by this theory. Using Ta_i = 43.25 as the criteria for laminar Taylor-vortex flow, the critical wave number (α_{crit} = 3.12) may be calculated from equation (4). The fluctuation around this is approximately 2% from equation (5) allowing for maxima and minima of 3.18 and 3.06 respectively. The geometry of CC-25 means that the annulus for formation of Taylor vortices is 25mm high and the aspect ratio is 37.5. Rearrangement of equation (4) using maxima and minima values for α_{av} suggests the number of vortices that will form in laminar Taylor vortex flow is between 15.71 and 16.35. From the quantization condition that only a whole number can form it is expected that this will normally be 16 but might reach 17. This has been verified by flow visualisation imaging. Working on the basis of 16 vortices forming, the Bodenstein number, Bo is 32 from equation (15). Because transport occurs on the vortex boundaries and all the vortices are identical, the transport from the free surface to the probe location in the vessel can be conceived a step-wise function illustrated in Figure 9.

At the free surface, elements of fluid in the upper Taylor-vortex are readily exposed to the surface due to the rotational motion of the vortex. Free surface

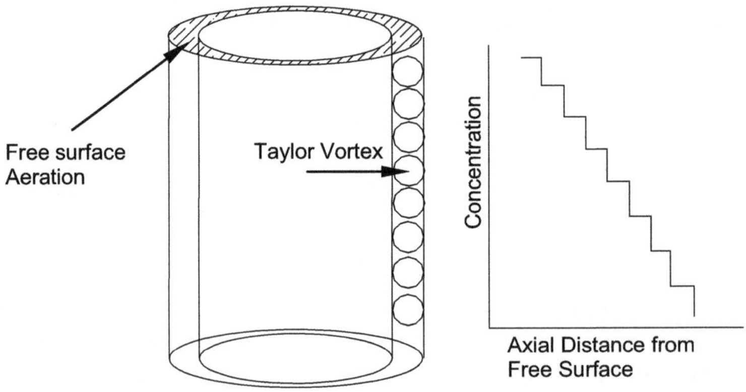

Figure 9: Conceptualisation of the step-wise axial mass transport occurring in Taylor-vortex flow. Instantaneous transport is assumed to occur within each vortex by virtue of their ideal mixing behaviour.

transport here is best described by the surface renewal theory for mass transport (Coulson *et al.,* 1990). Using the step-wise concept of Figure 9, and knowing the oxygen concentration at the free surface and at the probe (base of the vessel), the concentration at the intermediate step can be approximated at any time interval for measurement using 16 steps (Taylor-vortices) as the basis of calculation. Data are shown in Figure 10 for time periods of 30 min, 1h, 5h, 10h and 15h for $Ta_i = 43.25$. The circumferential dispersion coefficient is calculated from equation (16) using a vortex velocity, u_t. The value of u_t is best approximated by the azimuthal component of velocity as the magnitude of this far exceeds any of the other components. From our LDA data, the mean azimuthal velocity is taken as $0.1 ms^{-1}$ for CC-25 at $Ta_i = 45.058$. This best approximates the laminar Taylor vortex flow for oxygen measurements in CC-25R. The dispersion coefficient, D_d is then calculated to be $2.55x10^{-4}\ m^2s^{-1}$. It is observed that this is approximately three orders of magnitude higher than the diffusivity of oxygen in the media. In order to evaluate the mass transfer flux using the surface renewal theory, the time of exposure, t_e of fluid elements in the top vortex requires calculation. This is probably best done as a guideline using

$$t_e \approx \frac{d_{ed}}{u_t}$$

where u_t is the vortex velocity and d_{ed} is its diameter.

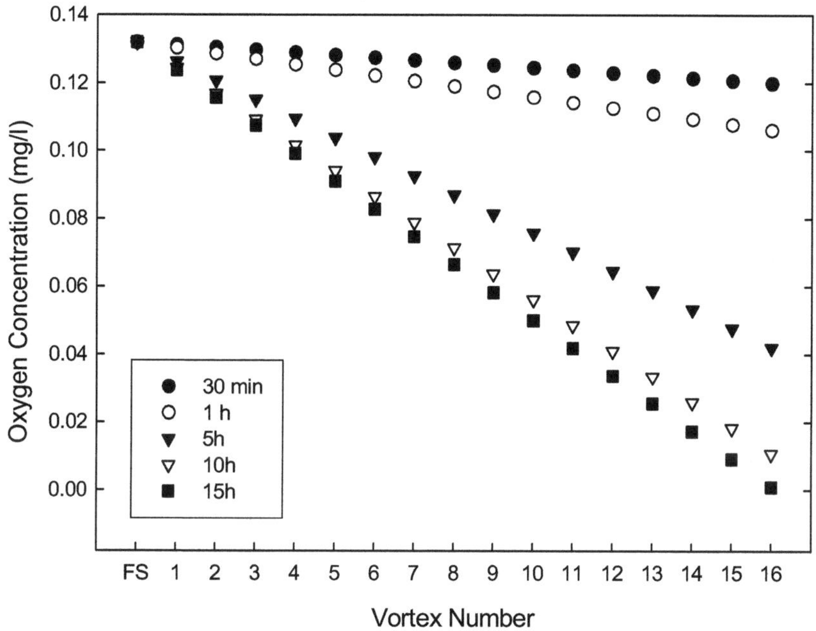

Figure 10: Oxygen concentration profiles at each of the vortex boundaries in Taylor-vortex flow at Tai = 43.25 for five different experimental times.

From our LDA and flow visualisation data in CC-25 at Ta_i = 45.06, the vortex diameter, d_{ed} is approximately 3mm with a mean velocity, u_{ed} of $0.1ms^{-1}$. An average renewal time, t_e of 0.03s is calculated. The average mass flux predicted by equation (18) only accounts for the diffusivity of oxygen in the media, and takes no account of further dispersion by the fluid motion. This motion will further enhance the flux, so modification of equation (19) should be made to account for this and is given by equation (20) here.

$$(J_A)_{av} = 2(c_{Ai} - c_{Ao})\sqrt{\frac{D + D_d}{\pi t_e}} \qquad (20)$$

Our results show that the magnitude of the transfer rate is significantly increased by virtue of the hydrodynamics in laminar Taylor-vortex flow to laminar Couette flow.

When taking account of diffusivity alone, the Taylor-vortex flow provides a transfer rate that is three orders of magnitude higher and when dispersion effects are additionally considered this grows to around six orders of magnitude higher.

4.4.3. Wavy-Vortex Flow

Wavy-vortex flow is developed by $Ta_i = 57.65$ and beyond this, the amplitude and frequency of the waviness will increase. Beyond laminar Taylor-vortex flow, the problem of sedimentation of cells contributing to increased depletion of oxygen at the measuring probe is removed and oxygen consumption is entirely due to the transport of oxygen from the free surface as a function of the hydrodynamics. It is observed in general that as the Taylor number continues to increase the absolute % O_2 continues to increase over the 16h experimental time period, however it is apparent that degree of change becomes less and that there is approach to asymptotic behaviour. It is known from the LDA data that beyond about $Ta_i = 150$ for CC-25, the shear stress exerted on cells in the fluid becomes supercritical. This means that beyond this point the viability of the cells will rapidly decrease. Once this occurs, the consumption of oxygen by cells will decrease and this will influence the oxygen concentration data; a converse situation to the cell sedimentation problem is set up. In the sedimentation problem excessive consumption of oxygen occurs due to an uneven distribution of cells in the reactor and this places bias on the transport profile. For non-viable cells there is reduced consumption. It can be concluded from this that the higher oxygen concentrations shown beyond $Ta_i = 144.08$ will be due to reduced cell viability and not improvement in the transfer process. Conceptually, the optimum transport should occur at around $Ta_i = 141$ from the LDA data, when regular wavy vortex flow prevails. At this point there is both intra and inter vortex mixing to facilitate the transport process. This is the point where the asymptote for transfer is expected. The asymptote is a consequence of the limitation of transfer due to the physico-chemical arrangement imposed on the reactor at the free surface.

Whilst the random surface renewal theory for mass transport (Coulson *et al.*, 1990) might better describe the free surface transport in wavy-vortex flow conceptually, there is little difference in the mathematical theory to provide anything further quantitatively than the regular surface renewal theory. Beyond laminar Taylor-vortex flow, and particularly beyond $Ta_i = 145$, there is interpenetration of the vortices and determination of the dispersion coefficient then becomes difficult due to the irregular flow structure. Up to $Ta_i = 145$, the vortices remain unique with a definite waveform and the dispersion coefficient can still be evaluated using the Bodenstein number (equation 14). In fact the only effect on the dispersion coefficient is that the vortex velocity u_t increases with increasing Ta_i. Similar calculations to those for Ta_i have shown that the transfer rate for oxygen increases but at a progressively slower rate for supercritical Ta_i. This is not really in keeping with the observations of Figure 8, where there is still a significant difference in the levels of oxygen maintenance between laminar Taylor-vortex ($Ta_i = 43.25$) and wavy vortex flow ($Ta_i = 144.08$). It is known that the viability of the cells begins to decrease as the Taylor number increases from flow cytometry studies due to increased shear stresses and this may lead to higher than expected oxygen levels. It is also possible that there is more uniform distribution of the cells from the lower regions of the vessel up towards the free surface that has an influence on the oxygen levels. It is clear however that the hydrodynamic environment does greatly promote transport in the bulk fluid and that asymptotic behaviour is approached for optimal transport. What is clear in this study is that over time, the oxygen concentration is falling, suggesting that consumption by cells in general exceeds supply. However, the system has been designed to reduce the supply and maximise the consumption as a baseline to see how this can be influenced by the hydrodynamics, so this is not surprising. The dispersion coefficient, D_d for varying Ta_i is shown in Figure 11. This is based on equation (15) and our LDA data for u_t with varying Ta_i.

4.5. Mass Transport Correlations

In establishing mean liquid phase mass transfer coefficients for given regimes of flow, a number of assumptions must be firstly made:

- Laminar Couette flow provides a datum level with respect to axial mass transport as a consequence of the hydrodynamic environment. In this regime it is assumed there is no forced convective transport axially.
- It is assumed that consumption of oxygen by cells throughout the bulk fluid is constant.
- At time 15.75h in laminar Couette flow the absolute oxygen concentration has reached zero (Figure 8). For other more agitated regimes of flow at this time point, the oxygen concentration is higher. This is assumed to be due to enhanced mass transport by forced convection alone.
- The relative increase in the concentration of oxygen at time 15.75h relative to Couette flow is assumed to be directly proportional to the relative increase in the liquid phase mass transfer coefficient, k_l.

Table 5. Relative Changes in the Concentration of Oxygen Measured at the Probe in CC-25 for Varying Flow Regimes. The relative increase in concentration is due to the more agitated flow which increases the mass transfer coefficient, k_l.

Re	101.7	132.2	152.6	203.6	407.3	509.2	611.1	1018.6
Ta	28.8	37.4	43.2	57.6	115.2	144.0	172.8	288.1
$\%O_2$ (15.75h)	31.9	53.2	73.1	74.2	75.6	80.2	91.4	93.2
$\frac{\%O_{2(Re)}}{\%O_{2(Re=101.68)}}$	1	1.67	2.29	2.33	2.37	2.51	2.87	2.92
k_l (×E-04)	2.81	4.69	6.45	6.55	6.67	7.07	8.06	8.22

Based on these assumptions, a mean mass transport coefficient in the liquid phase, k_l, can be assigned to any of the flow regimes of Figure 8 using the mean transfer coefficient for the laminar Couette flow condition, 2.81×10^{-4} m/s. Data for four different flow regimes are given in Table 5. For comparison, large and small

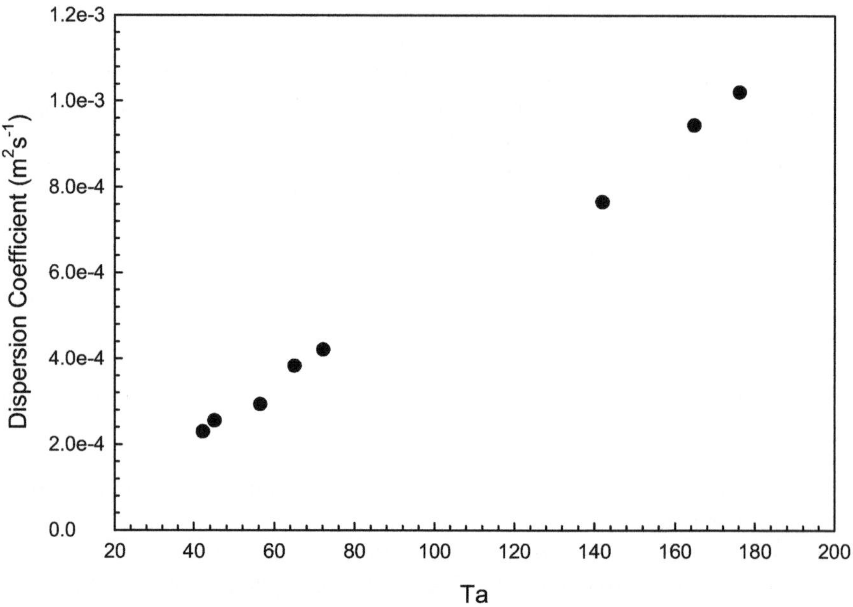

Figure 11: Dispersion coefficient in CC-25 for varying Taylor number.

bubbles of air have mass transfer coefficients of $4.0\text{x}10^{-4}$m/s and $1.0\text{x}10^{-4}$ m/s, respectively (Bailey and Ollis, 1986).

The mass transport correlation for oxygen transport in CC-25 is based on equation (11) with definitions from equations (12) through (15). The dimensionless numbers for our study are given in Table 6. Applying a multiple non-linear least squares regression to the data of Table 6, the general global mass transport correlation is:

$$\text{Sh} = 9.21\,\text{Re}^{0.28}\,\text{Sc}^{0.33} \tag{21}$$

Comparison of the experimentally determined values of the Sherwood number and the Sherwood number calculated from the correlation of equation (21) reveals that this equation does not adequately describe the oxygen transport process for all regimes of flow investigated. The primary transition from laminar Couette flow to laminar Taylor-vortex flow occurs at approximately $Re_i = 150$. It is apparent from Figure 12 that separate correlations might better represent the transport process, namely (1) laminar Couette flow; and (2) beyond the onset of vortex flow.

Table 6. Calculated Values for the Dimensionless Numbers Based on Experimental Oxygen Transport Data in CC-25 for Varying Flow Regimes.

Sh	*Sc*	Re_i
165.47	588.24	101.68
276.01	588.24	132.24
379.58	588.24	152.62
385.35	588.24	203.56
392.40	588.24	407.32
416.40	588.24	509.19
474.39	588.24	611.07
483.63	588.24	1018.59

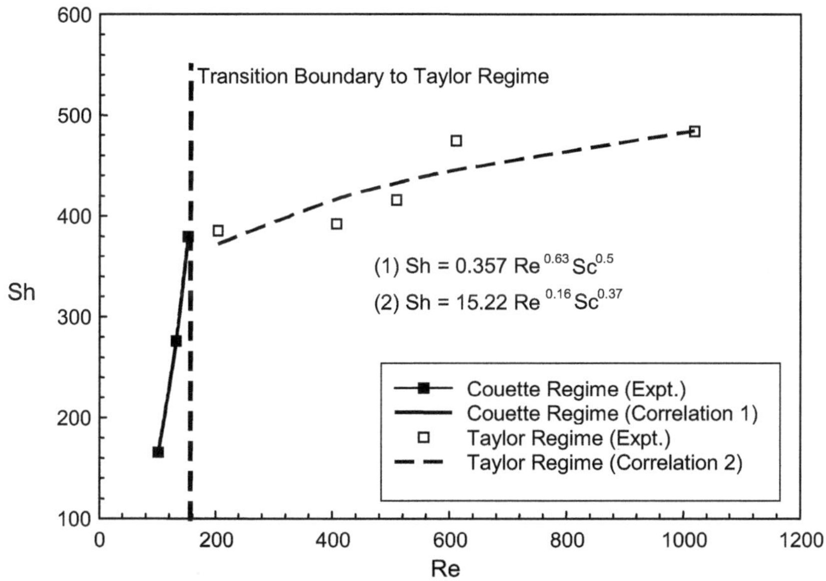

Figure 12: Comparison of Sherwood number calculated from experimental measurements of k_l and from the mass transport correlation obtained from least squares regression in CC-25 for (1) laminar and (2) Taylor vortex flow (R^2=0.8). After Bailey and Ollis, (1986).

The correlations, which are shown in Figure 12, are given by

$$Sh = 0.357\ Re^{0.63}\ Sc^{0.5}\ \textit{(laminar Couette flow)} \tag{22}$$

$$Sh = 15.22\ Re^{0.16}\ Sc^{0.37}\ \textit{(vortex flow)} \tag{23}$$

The data shown in Figure 12 indicate that there is a significant increase in the Sherwood number with only a small change in Taylor number up to the point of Taylor-vortex transition and this suggests significantly improved axial mass transport in the annulus for the Taylor flow regime. There is some smaller degree of increased transport in the further flow transitions that occur to wavy and turbulent vortex regimes but this will be at the expense of increased shear.

4.6. Viability of Cells in Taylor Flows

The viability of cells in the CC-25 bioreactor for varying flow regime has been assessed using flow cytometry. Two samples of 200μl volume were withdrawn from the bioreactor at hourly intervals at two locations, midway down the annulus and at the base of the annulus (see Figure. 7) and stained with ethidium homodimer and Calcein AM (L-3224 Viability Assay, Molecular Probes USA). The hydrodynamic regimes investigated are shown in Table 7, with oxygen transport corresponding to that flow regime described above.

Table 7. *Hydrodynamic Operating Parameters for Cell Viability Studies in CC-25.*

Mean Shear $(s^{-1})^{+}$	Ta_i	Re_i	*Regime*
100	28.80	101.83	Laminar Couette
150	43.25	152.92	Laminar Taylor-Vortex
400	115.25	407.95	Wavy-Vortex
1000	288.11	1018.63	Turbulent Wavy-Vortex

The viability of cells sampled at various time points from CC-25 are shown in Figure 13 as a percentage of the initial cell concentration.

5. CONCLUSIONS

The coaxial cylinder bioreactor is a promising tool for the cultivation of cells on biodegradeable constructs for tissue engineering applications. Such systems are commercially available from Synthecon in the USA and Cellon in Europe where they are typically operated in a laminar Couette regime of flow with outer cylinder rotation only. This is primarily to maintain low shears in the annulus, and in lateral rotation maintains free suspension of the constructs in the culture media providing microgravity cultivation (Begley, 2000). Whilst there has been some success in regenerating small tissue samples in these systems, the tissue is imperfect and often suffers from deep core necrosis and calcification on the construct edges (Unsworth, 1998; Hammond, 2001).

[+] Mean shear from linear theory used as the basis for control with a Rheologica Stresstech GmbH rheometer

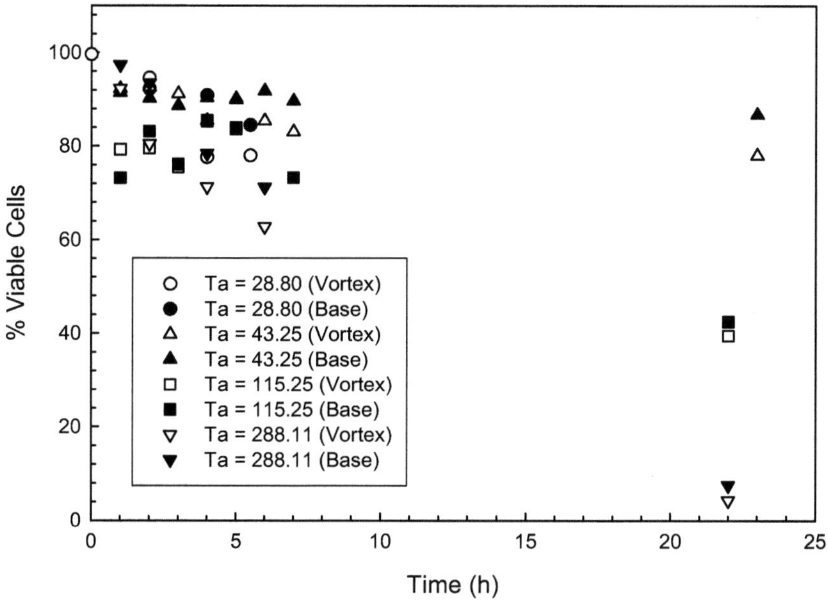

Figure 13: Viability data for C20/A4 chondrosarcoma cells cultured in CC-25 bioreactor sampled from vortex region in the midpoint of the annulus and from the base region below the annulus.

Our laboratory data suggest that operation of coaxial cylinder systems in a Taylor-vortex regime of flow will provide near ideal mixing and significantly improve axial mass transport whilst still maintaining very low shears. A recent study operating a Synthecon bioreactor in the Taylor regime of flow has verified our findings that the shears remain low whilst significantly enhancing the mass transport (Haut *et al.*, 2003). The viability of cells in Taylor flow regimes can be maintained even when oxygenation to the system is limited. The shear forces are significantly reduced on the cells and constructs as the geometric scale of the system is increased and this suggests that operation of larger vessels in regimes of flow that enhance mass transport will provide very low shears on the cells that might contribute to improved longer-term growth deep within the constructs.

6. ACKNOWLEDGEMENTS

The authors are grateful for the financial support of University of Liverpool through the award of a University postgraduate studentship to SJC. The UK Centre for Tissue Engineering, an interdisciplinary research collaboration between the Universities of Liverpool and Manchester, was established by the joint UK Research Councils (BBSRC, EPSRC and MRC) in 2001.

7. REFERENCES

Abu-Reesh I, Kargi F. 1989. Biological responses of hybridoma cells to defined hydrodynamic shear stress. J Biotechnol 9:167-178.

Andereck CD, Liu SS, Swinney HL. 1986. Flow regimes in a circular couette system with independently rotating cylinders. J Fluid Mech 164:155-183.

Bailey JE, Ollis DF. 1986. Biochemical Engineering Fundamentals. 2nd Edition. McGraw-Hill Inc. New York.

Begley CM, Kleis SJ. 2000. The fluid dynamic and shear environment in the NASA/JSC rotating-wall-perfused vessel bioreactor. Biotech Bioeng 70:32-40.

Blennerhassett PJ, Hall P. 1979. Centrifugal instabilities of circumferential flows in finite cylinders: linear theory. Proc Roy Soc London A 365:191-207.

Born C, Zhang Z, Al-Rubeai M, Thomas CR. 1992. Estimation of disruption of animal cells by laminar shear stress. . Biotech Bioeng 40:1004-1010.

Brown TD. 2000. Techniques for mechanical stimulation of cells in vitro: A Review. J Biomech 33:3-14

Burkhalter JE, Koschmeider EL. 1974. Steady supercritical taylor vorticesafter sudden starts. Phys Fluids 17:1929-1935.

Butler M. 1988. A comparative review of microcarriers available for the growth of anchorage-dependent animal cells. Animal Cell Biotechnology 3:283-303.

Campero RJ, Vigil RD. 1997. Axial dispersion during low-reynolds number taylor-couette flow: intra-vortex mixing effects. Chem Eng Sci 52:3303-3310.

Chandrasekhar S. 1961. Hydrodynamic and Hydromagnetic Instability. Clarendon Press, Oxford.

Chen SY, Huang SY. 2000. Shear stress effects on cell growth and l-dopa production by suspension culture of Stizolobium hassjoo cells in an agitated bioreactor. Bioproc Eng 22:5-12.

Cherry RS, Papoutsakis ET. 1986. Hydrodynamic effects on cells in agitated tissue culture reactors. Bioproc Eng 1:29-41.

Cherry RS,, Papoutsakis ET. 1988. Physical mechanisms of cell damage in microcarrier cell culture bioreactors. Biotech Bioeng 32:1001-1014.

Cole JA. 1976. Taylor-vortex instability and annulus length effects. J Fluid Mech 75:1-15.

Coles D. 1965. Transition in circular couette flow. J Fluid Mech 21:385-425

Coulson JM, Richardson JF, Backhurst JR, Harker JH. 1991. Chemical Engineering Vol 1. 4th Edition, Pergamon Press, Oxford.

Croughan MS, Hamel JF, Wang DIC. 1987. Hydrodynamic effects on animal cells grown in microcarrier cultures. Biotech Bioeng 29:130-141.

Croughan MS, Sayre ES, Wang DIC. 1989. Viscous reduction of turbulent damage in animal cell culture. Biotech. Bioeng. 33:862-872.

Curran SJ, Black RA. 2000.Mass transport and hydrodynamic modelling for an annular flow bioreactor. ASME BED Advances in Bioengineering 51.

Curran SJ. 2002. Hydrodynamics and Mass Transport in an Annular Flow Bioreactor. PhD Thesis University of Liverpool, UK.

Debler W, Fuhner E, Schaaf B. 1969. Torque and flow patterns in supercritical taylor-vortex flow. In 12th Int. Congr. Appl. Mech. Springer, Berlin, 158-178.

Desmet G, Verelst H, Baron GV. 1996. Local and global dispersion effects in Couette-Taylor flow [I]: Decription and modelling of the dispersion effects. Chem. Eng. Sci. 51:1287-1298.

Desmet G, Verelst H, Baron GV. 1996. Local and global dispersion effects in Couette-Taylor flow [II]: Quantitative measurements and discussion of the reactor performance. Chem Eng Sci 51:1299-1309.

Dewey CF, Bussolari SR, Gimbrone MA, Davies PF. 1981. The dynamic response of vascular endothelial cells to fluid shear stress. J Biomech Eng 103:177-185.

Dong C, Lei XX. 2000. Biomechanics of cell rolling: shear flow, cell surface adhesion and cell deformability. J Biomech 33:35-43.

Drazin PG, Reid WH. 1981. Hydrodynamic Stability. Cambridge University Press.

Edwards WS, Beane SR, Varma S. 1991. Onset of wavy-vortices in the finite length Taylor-Couette problem. Phys Fluids A 3:1510-1518.

Enfors SO, Jahic M, Rozkov A, Xu B, Hecker M, Jurgen B, Kruger E, Schweder T, Hamer G, O'Beirne D, Noisommit-Rizzi N, Reuss M, Boone L, Hewitt C, McFarlane C, Nienow A, Kovacs T, Tragardh C, Fuchs L, Revstedt J, Friberg PC, Hjertager B, Blomsten G, Skogman H, Hjort S, Hoeks F, Lin

HY, Neubauer P, van der Lans R, Luyben K, Vrabel P, Manelius A. 2001. Physiological responses to mixing in large scale bioreactors. J Biotechnol 85:175-185
Frangos JA, McIntire LV, Eskin SG. 1988. Shear stress induced stimulation of mammalian cell metabolism. Biotech Bioeng 32:1053-1061.
Freed LE, Hollander AP, Martin I, Barry JR, Langer R, Vunjak-Novakovic G. 1998. Chondrogenesis in a cell-polymer-bioreactor system. Exp Cell Res 240:58-65.
Fry DL. 1968. Acute vascular endothelial changes associated with increased blood velocity. Circ Res 22:165-197.
Goodwin T, Jessup J, Wolf D. 1992. Morphologic differentiation of colon carcinoma cell lines HT-29 and HT-29KM in rotating wall vessels. In Vitro Cell Dev Biol 28A: 47-60.
Hall P. 1980. Centrifugal instabilities of circumferential flows in finite cylinders: non-linear theory. Proc Roy Soc London A 372:317-356.
Hammond TG, Hammond JM. 2001. Optimised suspension culture: the rotating wall vessel. Am J Physiol Renal Physiol 281:F12-F25.
Haut B, Ben Amor H, Coulon L, Jacquet A, Halloin V. 2003. Hydrodynamics and mass transfer in a Couette-Taylor bioreactor for the culture of animal cells. Chem Eng Sci 58:777-784.
Howes T, Rudman M. 1998. Flow and axial dispersion simulation for traveling axisymmetric Taylor vortices. AIChE J 44:255-262.
Hu WS, Peshwa MV. 1991. Animal cell bioreactors: recent advances and challenges to scale-up. Can J Chem Eng 69:409-420.
Hua J, Erickson LE, Yiin TY, Glasgow LA. 1993. A review of the effects of shear and interfacial phenomena on cell viability. Critical Rev Biotech 13:305-328.
Inamura T, Saito K, Ishikura S. 1993. A new approach to continuous emulsion polymerization. Polym Int 30:203-206.
Kallos MS, Behie LA. 1999. Inoculation and growth conditions for high density expansion of mammalian neural stem cells in suspension bioreactors. Biotechnol Bioeng 63:473-483.
Kataoka K, Doi H, Hongo, T, Futagawa M. 1975. Ideal plug flow properties of Taylor vortex flow. J Chem Eng Jap 8:472-476.
Kataoka K, Doi H, Komai T. 1977. Heat and mass transfer in Taylor vortex flow with constant axial flowrates. Int J Heat Mass Transfer 20:57-63.
Kataoka K, Takigawa T. 1981. Intermixing over cell boundary between Taylor vortices. AIChE J 27:504-508.
Koschmeider EL. 1979. Turbulent Taylor-vortex flow. J Fluid Mech 93:515-527.
Lakhotia S, Papoutsakis ET. 1992. Agitation induced cell injury in microcarrier cultures. Protective effect of viscosity is agitation intensity dependent: experiments and modelling. Biotech Bioeng 39:95-107.
Langer R, Vacanti J. 1993. Tissue engineering. Science 260:920-926.
Lee SHK, Sengupta S, Wei T. 1995. Effect of polymer additives on Gortler vortices in Taylor-Couette Flow. J Fluid Mech 282:115-129.
Legrand J, Coeuret F. 1986. Circumferential mixing in one-phase and two-phase Taylor vortex flows. Chem Eng Sci 41:47-53.
Leib TM, Pereira CJ, Villadsen J. 2001. Bioreactors: a chemical engineering perspective. Chem Eng Sci 56:5485-5497.
Levenspiel O. 1972. Chemical Reaction Engineering. 2nd Edition, Wiley Publications, New York, pp 253-314.
Levesque MJ, Nerem RM. 1985. The elongation and orientation of cultured endothelial cells in response to shear stress. J Biomech Eng 107:341-347.
Lewis JW. 1928. Observed structures in rotating cylinder flows. Proc Roy Soc A London 117:388-406.
Lewis R. 1995. Tissue engineering now coming into its own as a scientific field. The Scientist 9:12-15.
Liu CI, Lee DJ. 1999. Micromixing effects in a Couette flow reactor. Chem Eng Sci 54:2883-2888.
Lorenzen A, Pfister G, Mullin T. 1983. End-effects on the transition to time-dependent motion in the Taylor experiment. Phys Fluids 26:10-13.
Ludwig A, Kretzmer G, Schugerl K. 1992. Determination of a "critical shear stress level" applied to adherent mammalian cells. Enz Microb Technol 14:209-213.
Mardikhar SH, Niranjan K. 2000. Observations on the shear damage to different animal cells in a concentric cylinder viscometer. Biotechnol. Bioeng. 68:697-704 .
McQueen A, Meilhoc E, Bailey J. 1987. Flow effects on the viability and lysis of suspended animal cells. Biotechnol Lett 9:831-839.

Michaels JD, Kunas KT, Papoutsakis ET. 1992. Fluid mechanical damage of freely suspended animal cells in agitated bioreactors: effects of dextran, derivatized celluloses and polyvinyl alcohol. Chem Eng Commun 118:341-360.
Mobbs FR, Ozogan MS. 1984. Study of sub-critical Taylor-vortex flow between eccentric rotating cylinders by torque measurements and visual observations. Int J Heat & Fluid Flow 5:251-253.
Mullin T. 1982. Mutations of steady cellular flows in the Taylor experiment. J Fluid Mech 121:207-218.
Naughton G. 1998. Tissue engineering-new challenges. ASAIO J 115-116.
Nollert MU, Diamond SL, McIntire LV. 1991. Hydrodynamic shear stress and mass transport modulation of endothelial cell metabolism. Biotech Bioeng 38: 588-602.
Obradovic B, Carrier RL, Vunjak-Novakovic G, Freed LE. 1999. Gas exchange is essential for bioreactor cultivation of tissue engineered cartilage. Biotech Bioeng 63:197-205.
Ogihara T, Matsuda G, Yanagawa T, Ogata N, Fujita K, Nomura M. 1995. Continuous synthesis of monodispersed silica particles using Couette-Taylor vortex flow. J Soc Ceram Jpn Int Ed 103:151-154.
Ohmura N, Kataoka K, Shibata Y, Makino T. 1997. Effective mass diffusion over cell boundaries in a Taylor-Couette flow system. Chem Eng Sci 52:1757-1765.
Papadaki M, McIntire LV, Eskin SG. 1996. Effects of shear stress on the growth of aortic smooth muscle cells in vitro. Biotech Bioeng 50:555-561.
Papoutsakis ET. 1991. Fluid-mechanical damage of animal cells in bioreactors. Trends in Biotech 9:427-437.
Park K, Donnelly RJ. 1981. Study of the transition to Taylor-vortex flow. Phys Rev A 24:2277-2279.
Petersen JF, McIntire LV, Papoutsakis ET. 1988. Shear sensitivity of cultured hybridoma cells CRL-8018 depends on mode of growth, culture age and metabolite concentration. J Biotechnol 7:229-246.
Rhodes NP, Shortland AP, Rattray A, Black RA, Williams DF. 1997. Activation status of platelet aggregates and platelet microparticles shed in sheared blood. J Mat Sci Mat Med 8:747-751.
Roberts PH. 1965. The solution of the characteristic value problems. Proc Roy Soc London A 283:550-556.
Rudman M, Thompson MC, Hourigan K. 1994. Particle shear rate history in a Taylor-Couette column. ASME Fluids Eng Div FED 189:23-30.
Rudolph M, Shinbrot T, Lueptow RM. 1998. A model of mixing and transport in wavy Taylor-Couette flow. Physica D 121:163-174.
Ryrie S. 1992. Mixing by chaotic advection in a class of spatially periodic flows. J Fluid Mech 236:1-19.
Sato M, Nagayama K, Kataoka N, Sasaki M, Hane K. 2000. Local mechanical properties measured by atomic force microscopy for cultured bovine endothelial cells exposed to shear stress. J Biomech 33:127-135.
Savas O. 1985. On flow visualization using reflective flakes. J Fluid Mech 152:235-248.
Schugerl K, Kretzmer G (editors). 2000. Influence Of Stress On Cell Growth And Product Formation. Adv Biochem Eng Biotechnol vol 67. Springer.
Schlichting H. 1987. Boundary Layer Theory. New York: McGraw-Hill Inc. 7th Edition.
Schnittler HJ, Franke RP, Akbay U, Mrowietz C, Drenckhahn D. 1993. Improved in vitro rheological system for studying the effect of fluid shear stress on cultured cells. Am J Physiol 265:C289-C298.
Schultz-Grunow F, Hein H. 1956. Beitrag zur Couettestromung. Z Flugwiss 4:28-30.
Sherman HS. 1991. Viscous Flow. New York: McGraw-Hill Inc.
Snyder HA. 1969. Wave-number selection at finite amplitude in rotating Couette flow. J Fluid Mech 26:545-562.
Spier RE. 1995. Gradients in animal and plant cell technology systems. Enz Microb Technol 17:91-92.
Stathopoulos NA, Hellums JD. 1985. Shear stress effects on human embryonic kidney cells in vitro. Biotech Bioeng 28:1021-1026.
Sugata S, Yoden S. 1991. Effects of centrifugal force on stability of axisymmetric viscous flow in a rotating annulus. J Fluid Mech 229:471-482.
Takhar HS, Ali MA, Soundalgekar VM. 1992. The effect of radial temperature gradient and axial magnetic field on the stability of Couette flow: the narrow gap problem. Int J Energy Res 16:597-621.
Tam WY, Swinney HL. 1987. Mass transport in turbulent Couette-Taylor flow. Phys Rev A 36:1374-1381.
Taylor GI. 1923. Stability of a viscous liquid contained between two rotating cylinders. Phil Trans Roy Soc London A 157:565-578.

Temenoff JS, Mikos AG. 2000. Tissue engineering for regeneration of articular cartilage. Biomaterials 21:431-440.

Thomas CR. 1990. Problems of shear in biotechnology. In: Chemical Engineering Problems in Biotechnology. London: Elsevier.

Thoumine O, Ziegler T, Girard PR, Nerem RM. 1995. Elongation of confluent endothelial cells in culture: the importance of fields of force in the associated alterations of their cytoskeletal structures. Exp Cell Res 29: 427-441.

Tramper J. 1995. Oxygen gradients in animal cell bioreactors. Cytotechnology 18:27-34.

Unsworth BR, Lelkes PI. 1998. Growing tissues in microgravity. Nature Medicine 4:901-907.

Vastano JA, Russo T, Swinney HL. 1990. Bifurcation to spatially induced chaos in a reaction-diffusion system. Physica D 46:23-42.

Vunjak-Novakovic G, Freed LE, Biron RJ, Langer R. 1996. Effects of mixing on the composition and morphology of tissue engineered cartilage. AIChE J 42:850-860.

Vunjak-Novakovic G, Obradovic B, Martin I, Bursac PM, Langer R, Freed LE. 1998. Dynamic cell seeding of polymer scaffolds for cartilage tissue engineering. Biotechnol Prog 14:193-202.

Vunjak-Novakovic G, Martin I, Obradovic B, Treppo S, Grodzinsky AJ, Langer R, Freed LE. 1999. Bioreactor cultivation conditions modulate the composition and mechanical properties of tissue engineered cartilage. J Orthop Res 17:130-138.

White, FM. 1991. Viscous Fluid Flow. New York: McGraw Hill Inc.

Williams KA, Saini S, Wick TM. 2002. Computational dynamics modelling of steady state momentum and mass transport in a bioreactor for cartilage tissue engineering. Biotechnol Prog 18:951-963.

Wu SC. 1999. Influence of hydrodynamic shear stress on microcarrier attached cell growth: cell line dependency and surfactant protection. Bioproc Eng 21:201-206.

Wu SC, Huang GYL. 2000. Hydrodynamic shear forces increase Japanese encephalitis virus production from microcarrier grown VERO cells. Bioproc. Eng. 23:229-233.

Yoshikawa N, Ariyoshi H, Ikeda M, Sakon M, Kawasaki T, Monden M. 1997. Shear stress causes polarized change in cytoplasmic calcium concentration in human umbilical vein endothelial cells. Cell Calcium 22:189-194.

Zhang S, Hand-Corrigan A, Spier RE. 1992. Oxygen transfer properties of bubbles in animal cell culture media. Biotech. Bioeng. 40:252-259.

CHAPTER 4

PACKED BED BIOREACTORS

J. N. WARNOCK, K. BRATCH AND M. AL-RUBEAI

Department of Chemical Engineering, University of Birmingham, Birmingham, U.K.

1. INTRODUCTION

Intensive cell culture systems, such as packed bed bioreactors, have attracted considerable interest for the commercial production of biopharmaceutical products from mammalian cells (Thelwell and Brindle 1999) and their potential for long term cultivation of high cell densities makes them an attractive option for tissue engineering applications. The large cell numbers obtained can subsequently be used for transplant, or the bioreactor can serve as a surrogate organ such as a bioartificial liver (BAL) (Jauregui *et al.* 1996). However, to fully utilise these systems a comprehensive understanding of the influence of reactor design on cell growth and physiology is required (Banik and Heath 1995; Knight 1989).

The principle behind a packed bed bioreactor (PBR) is that cells are immobilised within a suitable stationary matrix that forms the bed. Several types of matrices have been used, including macroporous microcarriers (Fassnacht *et al.* 1999; McTaggart and Al Rubeai 2000), porous ceramic beads (Park and Stephanopoulos 1993), porous glass beads (Racher and Griffiths 1993), glass fibres (Rodrigues *et al.* 1999), and polyester discs (Hu *et al.* 2000; Kadouri and Zipori 1989; Kaufman *et al.* 2000; Wang *et al.* 1992; Warnock 2003). These provide a large surface area to allow cell attachment, resulting in cell concentrations as high as 5 x 10^8 cells ml^{-1} of matrix (Park and Stephanopoulos 1993). The porous structures also offer a protective environment for cells from local shear forces (Cong *et al.* 2001). Culture medium is perfused through the bed to supply cells with nutrients and to remove toxic metabolites. This allows the reactor to be run in batch, fed-batch or a continuous mode of operation, enabling long term cultivation of cells, and production runs in excess of 30 days are not uncommon (Cong *et al.* 2001; Ducommun *et al.* 2001; Fassnacht *et al.* 1999; McTaggart 2000).

The main problem associated with PBR is the heterogeneity of the reactor caused by concentration gradients of nutrients and waste products (Chresand *et al.* 1988;

J.B. Chaudhuri & M. Al-Rubeai (eds.), Bioreactors for Tissue Engineering, 87 - 113.

Drury *et al.* 1988; Piret and Cooney 1990; Piret and Cooney 1991). These will have an adverse effect on the function of a bioartificial tissue, such as a BAL (bioartificial liver), where the metabolic activity of the cells is intrinsically linked to their function (Thelwell and Brindle 1999). Transport limitations are the main limiting factor in the linear scale up of PBRs. Additionally, direct sampling of cells is often not possible (Ducommun *et al.* 2001; Kaufman *et al.* 2000) which causes problems in the calculation of culture parameters such as specific growth rate. Instead, cell numbers are estimated indirectly from either the glucose or oxygen uptake rates or the lactate production rate (Mancuso *et al.* 1990; Pörtner *et al.* 1994; Rodrigues *et al.* 1999). Unfortunately, these parameters can vary greatly with changes in growth conditions (Lin *et al.* 1993; Ljunggren and Häggström 1994; Neermann and Wagner 1996). PBRs are often used to culture anchorage dependent cells. Therefore, cells for inoculation are prepared from roller bottles. Hence, the development of large-scale systems is difficult due to the number of cells required for seeding (Cong *et al.* 2001).

Since the development of the packed bed bioreactor by Spier and Whiteside (1976), many advances have been made. This review will present the general characteristics of packed beds with respect to mass transport, fluid flow and changes in pressure over the bed and how these affect cellular activity. It also describes the various types of reactors that have been developed and their application to the engineering of living tissues.

2. MASS TRANSPORT

Adequate mass transfer is necessary in packed bed bioreactors in order to achieve maximal cell growth. It is also desirable that no detrimental effects, such as damage from hydrodynamic forces, result from techniques used to achieve efficient transport of nutrients. Proficient mass transfer is easily attainable at laboratory scales but the nutrient and metabolite gradients that develop often limit the scale-up of immobilised cell bioreactors (Piret and Cooney 1991). The environment surrounding a cell will determine the level of cellular activity. Hence, mass transfer often has a decisive role in the selection of a suitable bioreactor (Bratch 2000).

There are two criteria for mass transport in a bioreactor. Firstly, the system must be capable of transferring nutrients to cells. Cells require glucose as their primary carbon source for energy, as well as various other nutrients contained in culture medium. In addition, oxygen concentration will affect cellular viability and function. Secondly, glucose is metabolised into pyruvate, which in turn is converted into lactate. Similarly, L-glutamine is metabolised into ammonia. At high concentrations these may become toxic and it is therefore necessary that they be transported away from cells. Transport to and from cells is especially important in the design of a BAL assist device as it needs to ensure the successful detoxification of blood borne endogenous toxins, the unhindered return of liver specific factors back into patient's blood and the long term maintenance of cell viability and function (Bratch 2000).

Oxygen and glucose limitations have both been correlated with necrotic regions in tumours (Tannock 1972; Vaupel 1977), as well as spheroids of packed cells

(Mueller-Klieser and Sutherland 1982). Glucose, and other easily soluble nutrients, should be maintained above limiting levels. Oxygen, however, is poorly soluble (0.22 mM at 37°C under atmospheric pressure) and is the main limiting factor in packed bed bioreactor design (Kalogerakis and Behie 1997; Park and Stephanopoulos 1993; Piret and Cooney 1991). Oxygen supply to cells and oxygen concentration at the cell site may limit the survival and functions of liver cells in a BAL device. Hypoxic conditions have both short-term and long-term detrimental effects on liver cells. During cell attachment to solid substrata low oxygen concentrations hinder attachment and adaptation to the substratum. Long-term exposure to low oxygen concentrations affects cell viability and enzymatic activities (Catapano 1996). It has also been demonstrated that dissolved oxygen (DO_2) concentrations above 70% air saturation become toxic to cells. Excess levels enhance the formation of superoxide radicals, peroxides and hydroxyl radicals, which can damage DNA and cell membranes, reduce cell viability, induce cell death (apoptosis) and create the condition of "oxidant stress" (Ellis 1991; Williams *et al.* 1997).

2.1 Principles of Oxygen Transfer

It has been estimated that mammalian cells at a density of 10^6 cells ml^{-1} require between 2 and 10 kg O_2 m^{-3} h^{-1} (Othmer 1991). The actual oxygen demand will vary with culture conditions, but is at its highest when the cells are most biologically active (Bailey and Ollis 1986). In addition to the oxygen demand, a relationship exists between the oxygen uptake rate and the level of dissolved oxygen (Atkinson and Mavituna 1983). In essence, for a given bioreaction a critical oxygen concentration, C_{crit}, can be determined, above which the rate of reaction is independent of the oxygen concentration i.e. zero order with respect to oxygen. This is illustrated in Figure 1. Therefore, DO_2 should be maintained above this value to prevent hypoxia occurring. This becomes more complex in packed bed bioreactors, where cells are immobilised in particles. Biological activity is determined by the amount of DO_2 at the biological site within the solid particle, not by the level of DO_2 in the bulk liquid medium. Thus, efficient oxygenation requires that the liquid medium be suitably aerated and also that the medium can be transported to the cells.

The basic principles of oxygenation of the medium in a packed bed bioreactor are essentially the same as those for any bioreactor where gas-liquid mixing is required. The main difference is that medium is conditioned in a separate vessel from the cells, reducing the impact of mechanical damage to the cells caused by either stirring or bubble burst. Aeration is usually achieved in one of three ways. Surface aeration is the simplest but requires a large surface area and is therefore not suitable for large scales. The commonest method is to directly sparge the medium with air/oxygen. The third method is bubble-free aeration where gases flow through silicone tubing immersed in the culture medium. This is often used as an alternative to sparging when cells are in direct contact with gases as they are susceptible to damage caused by bubble burst. The rate of oxygen transfer is dependent on four primary parameters; the area of contact between the gas and the liquid, the driving force (i.e. the difference in oxygen concentration between the two phases), the two-phase fluid dynamics and the composition of the culture medium (Othmer 1991).

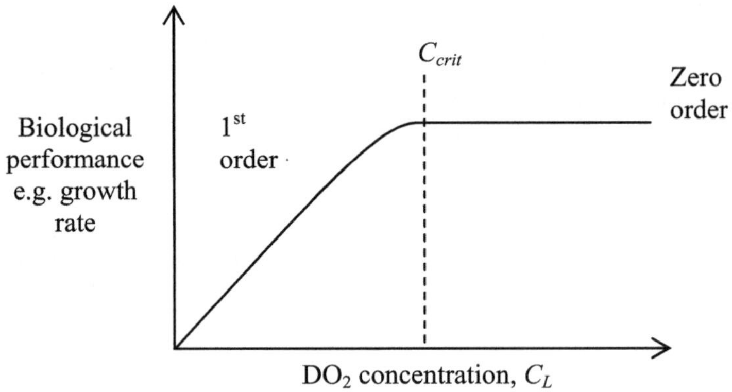

Figure 1. The relationship between biological performance and the dissolved oxygen concentration, where C_{crit} is the critical oxygen concentration.

The rate of mass transfer, J, is assumed to be proportional to the concentration difference that exists between each phase, the surface area between each phase, A, and a coefficient (the liquid film mass transfer coefficient, k_L), which relates the three. Thus

$$J = k_L A\left(C_g^* - C_L\right) \tag{1}$$

Where C_g^* is the solubility of oxygen in the medium that is in equilibrium with the gas phase partial pressure of oxygen (i.e. the highest possible oxygen concentration), and C_L is the actual concentration of oxygen in the medium. Thus, the aeration rate per unit volume of the conditioning vessel, N, is given by

$$N = \frac{J}{V} = k_L\left(\frac{A}{V}\right)\left(C_g^* - C_L\right) = k_L a\left(C_g^* - C_L\right) \tag{2}$$

Where a is the interfacial area per unit volume. At steady state, N must be the same as the OUR, which should be that demanded by the cell population in the packed bed and in the range $C_L > C_{crit}$. The oxygen uptake rate (OUR) of hepatocytes is therefore an important parameter for the design of BAL devices.

2.2 *Cellular Oxygen Uptake Rate*

The oxygen consumption rate of hepatocytes during attachment is important since, during this period, there is little or no bulk flow of the medium and therefore oxygen transport will be diffusion controlled. This will have an impact on bioreactor design especially in high density/low liquid volume designs, which are likely to constitute BAL devices. The oxygen consumption rates for rat hepatocytes during and post-attachment to microcarriers have been measured. Oxygen consumption was higher during attachment compared to post-attachment (Smith *et al.* 1996).

In studies conducted by Bratch (2000), it was seen that primary rat hepatocytes in suspension have a much higher OUR compared to the HepZ cell line in suspension. The OUR of cells in suspension is higher than cells attached to microcarriers for both cell types as illustrated in Figure 2.

Typical demands for oxygen by animal cells during batch culture are in the range 2×10^{-16} to 2×10^{-15} mol O_2 $cell^{-1}$ min^{-1} (Emery *et al.* 1995). The mean OUR for primary rat hepatocytes in suspension was found to be 2.18×10^{-14} mol O_2 $cell^{-1}$ min^{-1} which is much higher than the value quoted for typical animal cells. The OUR of primary rat hepatocytes attached to microcarriers is much lower with a mean value of 2.12×10^{-15} mol O_2 $cell^{-1}$ min^{-1}. This is comparable to published values of 3.54×10^{-15} (Smith *et al.* 1996) and 1.50×10^{-15} (Foy *et al.* 1994). Both these values are for attached cells.

OUR values obtained for the HepZ cell line attached to microcarriers are very similar to those obtained for attached primary cells. However, the OUR of the HepZ cell line in suspension, although higher compared to attached cells, are not as high as primary cells in suspension and are more comparable to the demands of typical animal cells.

Cells in suspension may have a higher OUR due to the oxygen demanding metabolic processes involved in cell attachment. As these cells are unable to survive without a substratum it is possible that cells in suspension are likely to try and attain attached status in order to survive. Other factors that may have contributed to hepatocytes in suspension having elevated OUR compared to attached cells include cells undergoing increased shear stress and dying as stressed cells are known to consume more oxygen than cells in normal conditions.

The OUR of cells has implications on the number of cells a bioreactor can support. It is evident from this study that there are differences in the OUR of attached cells compared to cells in suspension. Therefore, when considering the number of cells a reactor can maintain, it is important to consider the appropriate OUR. In a packed bed reactor where cells are immobilised on microcarriers, attached cell OUR would be more appropriate in determining the number of cells the reactor can support. The higher oxygen demand of cells in suspension must be considered for the initial attachment period when the reactor is inoculated with cells. Failure to provide adequate oxygenation during this period may result in the hindrance of the attachment process and lead to cell death. It is essential, though, that the oxygen uptake rate (OUR) of the attached cells does not exceed the oxygen transfer rate of the packed bed system (McTaggart 2000).

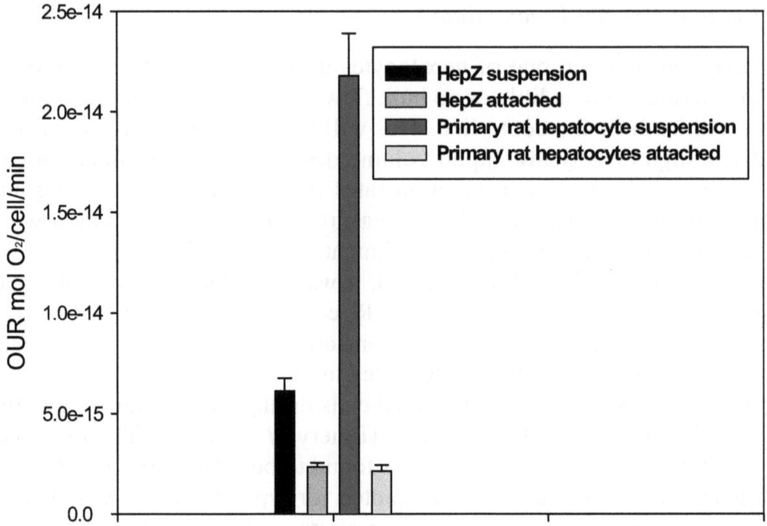

Figure 2. The oxygen uptake rate of attached and suspended cells. Primary rat hepatocytes were suspended in Williams'E medium following isolation. For the HepZ cell line OUR was measured following trypsinisation of cells from 60% confluent cultures and cells were resuspended in DMEM + 5% FCS. Medium containing no cells was used as a control enabling a background reading to be taken. Cells were seeded onto microcarriers to determine the OUR of attached cells. The number of attached cells was estimated by counting the number of cells remaining in suspension and by MTT assay of the discs. Samples of 15 discs were suspended in 4ml of medium and placed into the chamber of the Rank Brothers' device for OUR measurement. Data represent mean ± S.D. of 6 samples from 2 independent studies (Bratch 2000).

2.3 *Nutrient and Metabolite Gradients*

Oxygen depletion has been identified as the critical, axial scale-limiting factor in packed bed bioreactors (Piret *et al.* 1991;Pörtner *et al.* 1999). If it is assumed that the fluid flow through the packed bed shows no variation in velocity across the cross-section then the characteristics can be equated to plug flow. The mass balance for the reactor can then be formulated using the *differential-section* approach. As suggested in Figure 3, the packed bed can be divided into numerous thin slices, perpendicular to axial flow, each at steady state.

A mass balance can be taken on the thin section in order to determine the rate of formation of component *c*, r_{fc}, in terms of amount per unit volume per unit time. This can be expressed in its final form as (for full derivation see Bailey and Ollis 1986)

$$\frac{d}{dz}(uc) = r_{fc} \tag{3}$$

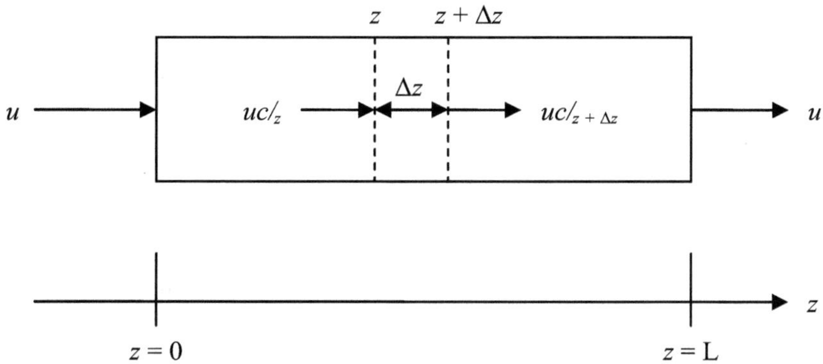

Figure 3. Plug-flow reactor, where u is the fluid velocity, z is the axial distance and c is an arbitrary component.

As cells are immobilised within the matrix of the packed bed there will be no change in fluid density and the axial velocity will remain constant. Hence, the equation will become:

$$u\frac{dc}{dz} = r_{fc} \tag{4}$$

The quantity *z/u* is equal to the time taken for a small slice of fluid to move from the reactor entrance to the axial position *z*. If the transit time, *t*, is used as a new independent variable, the mass balance equation can be rewritten as

$$\frac{dc}{dt} = r_{fc} \tag{5}$$

This is *exactly* the same as the mass balance for a batch reactor. Therefore, under ideal flow conditions the packed bed reactor can be divided into thin slices where there is no interaction between neighbouring slices, and each segment behaves as a batch reactor. Consequently, providing the residence time, *L/u*, in the packed bed is the same as a batch reaction time, the productivity of the PBR will be identical to a batch reactor. However, as PBRs employ medium recirculation, they are analogous to multiple batch reactions.

The problem, though, is that for each batch reaction, the cells consume nutrients such as glucose and oxygen, and produce toxic metabolites such as lactate and ammonia. It is reasonable to assume that mammalian cells will follow Monod growth kinetics and the rate of substrate consumption will be

$$\frac{ds}{dt} = -\left(\frac{1}{Y}\right)\left(\frac{\mu_{\max} xs}{s+K_s}\right) + mx \tag{6}$$

where Y is the yield factor, μ_{max} is the maximum growth rate, K_s is the substrate concentration when μ is equal to ½ μ_{max} and m is the specific maintenance rate. The effluent substrate concentration will be the s value that corresponds to $t=L/u$. Similarly, the substrate concentration at point z in the packed bed will be the s value when $t=z/u$. Therefore, as the distance between z_0 and z increases, s will decrease and a nutrient gradient appears along the axial length of the reactor. From this it can be seen that both reactor length and fluid velocity are key parameters in PBR design.

As mentioned earlier, the dissolved oxygen concentration must be maintained above C_{crit} in order to avoid a loss of biological function and the formation of hypoxic regions within the bed. The level of DO_2 should also be kept below 70% saturation to avoid the detrimental effects associated with oxidant stress. As a result, culture medium is aerated in the conditioning vessel and the level of dissolved oxygen will theoretically be the same as that of the medium at the reactor inlet. The length of the reactor should be such that the DO_2 in the effluent stream remains greater than C_{crit}. The problem is further compounded by the fact that oxygen is respired by cells to produce CO_2, which will cause a drop in pH of the culture medium. Therefore, even if cellular activity is not inhibited by a low level of oxygen concentration it may be affected by acidic pH. This imposes a severe limitation on axial length if a uniform distribution of cells is to be attained, especially as cells proliferate and may reach tissue like densities. It has been reported that the maximum length for axial flow through a PBR is approximately 15cm before oxygen limitation occurs (Pörtner *et al.* 1999). As a consequence, numerous researchers have focused on reactor designs that are able to overcome this limitation (see section 4).

2.4 *Intraparticle Mass Transfer*

The plug-flow reactor model is an adequate representation of substrate consumption and product formation in a PBR provided that there is a negligible mass transfer limitation between the bulk liquid and the immobilised cells. This assumption is not necessarily correct and the mass transport and biochemical reactions within the matrix should be considered before the overall activity can be determined. Figure 4 shows a cross-sectional schematic diagram of a cell immobilised within a particle in contact with culture medium flowing over the particle. Far from the immobilised cells, the liquid medium has the same substrate concentration and other variables as the bulk liquid medium. These are the parameters that are usually measured in the conditioning vessel.

Cells are immobilised within the porous structure, and this is where substrate is consumed and metabolites are produced, leading to the generation of a concentration gradient between the bulk medium and intraparticle medium. If the medium is stagnant, e.g. during attachment of cells to the matrix, transport can only take place

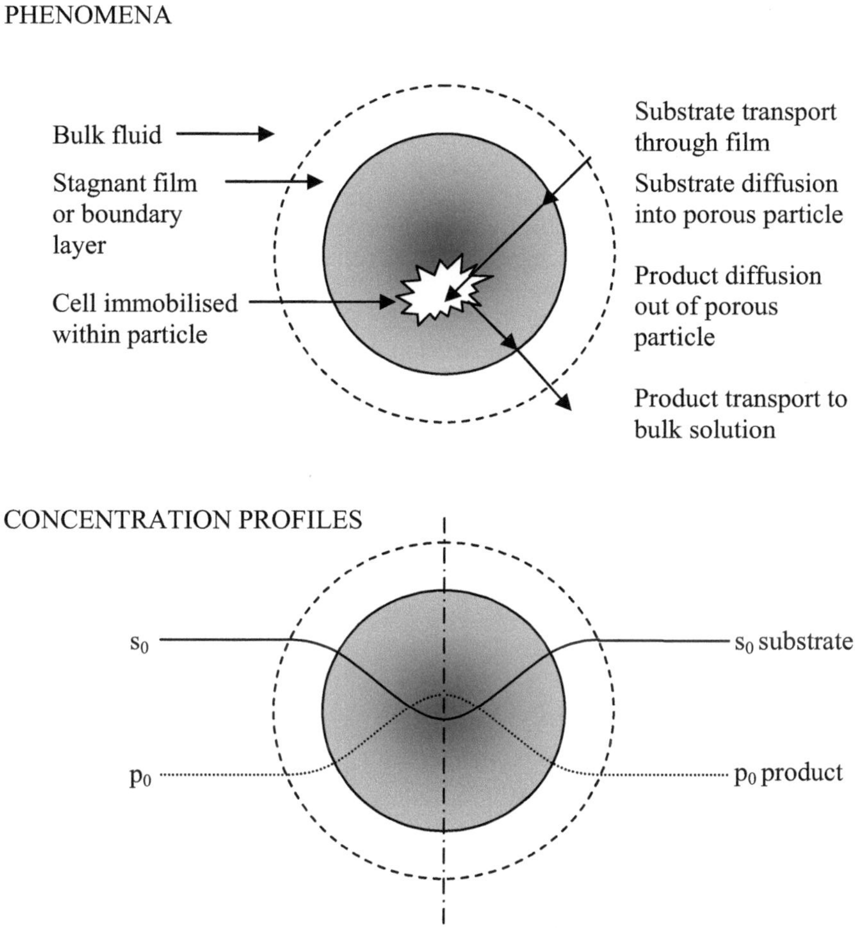

Figure 4. Illustration of mass transport in a symmetrical porous particle containing immobilised cells. The phenomena summarised in the top diagram results in the substrate and product profiles sketched in the bottom diagram.

by molecular diffusion. During operation, the fluid flow aids intraparticle convection, allowing more efficient mass transfer.

Cellular activity is determined by the chemical concentrations in the particle, not in the bulk medium. Due to the concentration gradients, cellular viability and functionality can vary greatly, depending on their internal position. The OUR and glucose consumption rates for a system are the sum of all the individual uptake rates for each particle.

Following the growth of cells to tissue-like densities, the glucose consumption rate has been seen to decrease (Kennard and Piret 1994). Cells initially penetrate relatively deeply into the porous matrix. However, over time an outer shell of highly dense cells forms, limiting transfer of oxygen and nutrients to the centre of the particle (Al Rubeai *et al.* 1990; Yamaji and Fukuda 1992). Additionally, toxic metabolites are unable to diffuse out of the particle and CO_2 production causes a drop in the local pH. Consequently, this microenvironment induces cell death and a necrotic core develops, reducing the overall cell density and viability. This can be clearly seen in Figures 5 and 6.

Transport by diffusion alone limits the size of particles to a few hundred microns in diameter before hypoxic regions develop in the centre of the microcarriers (Park and Stephanopoulos 1993). In a study by Preissmann *et al.* (1997) an assessment of limiting parameters for cells growing on macroporous microcarriers was carried out. As no significant concentration gradients could be detected, oxygen transfer to and into the carriers was examined. They found a significant transfer resistance within the laminar boundary layer at the surface of the carrier and that 40% air saturation in the bulk liquid could not provide efficient oxygenation during the exponential growth phase.

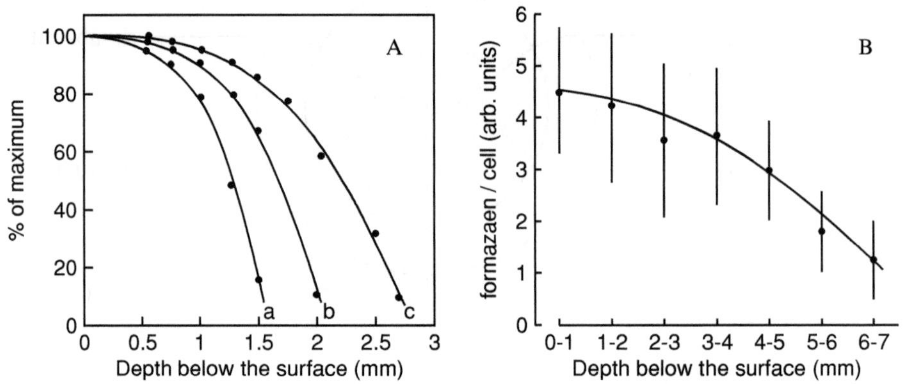

Figure 5. A. Metabolic activity profile within the gel expressed as a percentage of the values at the surface of the gel. Cells were immobilised at 9 x 10^6 (a), 4.5 x 10^6 (b) and 2.25 x 10^6 (c) ml^{-1} media and incubated for 46 hours. B. Quantification of MTT reduction by microdensitometry. Cells were immobilised at 9 x 10^6 ml^{-1} media and measurements were taken after 4 hours incubation (Used with permission, Al Rubeai and Spier 1989).

Figure 6. Time course of distribution of viable cell population within 3 mm3 Biomass Support Particles (Kanebo Kasei, Japan). Viable cells were stained with MTT (1mg ml^{-1}) for 4 hours at 37°C. The BSPs were washed with PBS and cut with a razor. Observations of cross sections were made using a stereomicroscope (Olympus, Tokyo, Japan). (Used with permission, Yamaji and Fukuda 1992).

3. FLUID FLOW

3.1 Pressure Drop-Flow Relationship

The Carmen-Kozeny equation is often used to relate the pressure drop in a packed bed of particles to the liquid flow rate in the laminar flow range (Re<10). The packed bed is considered to be equivalent to many tubes of identical diameter following tortuous paths of identical length, carrying fluid at a velocity *U*, which is the superficial velocity divided by the void fraction of the packed bed (Rhodes 1998). The equation for pressure drop is therefore:

$$\frac{-\Delta p}{H} = \frac{K(1-\varepsilon)^2}{\varepsilon^3}\mu U S^2 \tag{7}$$

where p is pressure, H is the height of the bed, K is a constant that depends on particle shape and surface properties, ε is the void fraction, μ is the viscosity of the fluid, U is the fluid velocity and S is the surface area per unit volume of particles. However, this is not an applicable model in the case of packed bed reactors that utilise porous particles as intraparticle convection is expected to influence the hydraulic behaviour of the system. A simple model was developed by Park and Stephanopoulos (1993) to describe fluid flow through a porous packed bed, based on the Blake-Kozeny equation. This was used to describe the extent of intraparticle convection at varying operating conditions and bioreactor design parameters.

Pressure-induced convection in porous packed bed reactors was first reported by Stephanopolous and Tsiveriotis (1989). The distribution of DO2 was determined following the derivation of the intraparticle flow field, by solving the combined diffusion, convection and oxygen uptake problem. They showed that a relationship exists between the intraparticle Peclet number, which is proportional to medium flow rate, and the critical Thiele modulus2 defined at the threshold of the formation of an anoxic region at the centre of the particle. The Thiele modulus2 is also proportional to cell number. Therefore, under conditions of intraparticle convection, the viable cell number increases proportionately with medium flow.

In a porous packed bed, the majority of the fluid will flow through the interstitial spaces, but in response to the pressure drop some medium will also flow through the pores of the carriers, as depicted in Figure 7. Medium exiting the particles will be depleted of limiting nutrients which are quickly replenished by mixing with the bulk medium. If particle diameter is increased, greater interstitial spaces form and the amount of fluid that flows through them increases. To maintain intraparticle flow the total fluid flow rate has to be increased, accordingly. Naturally, there is a limit to the extent that intraparticle flow can be sustained by increasing the total flow rate. Increasing flow rate will also increase the shear stress on attached cells, which will affect viability. Intraparticle flow can also be enhanced by increasing the pore size.

However, this reduces the available surface area for cell growth. Particle diameter and pore size are therefore important considerations in the selection of a suitable matrix for packed bed reactors.

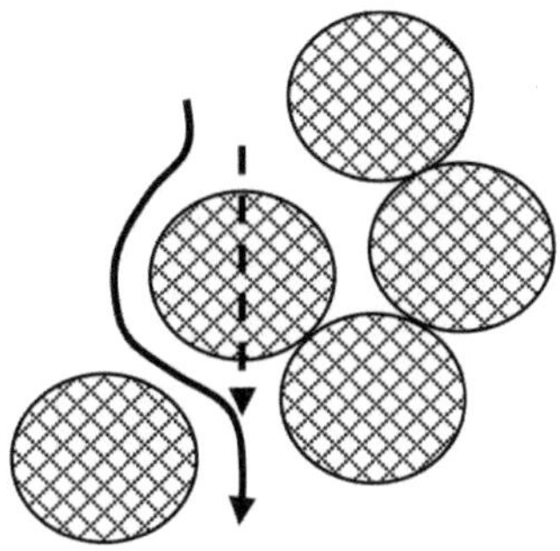

Figure 7. Presumed medium flow through a packed bed of porous beads. The solid line shows the main interstitial flow, while the intraparticle flow is represented by the dotted line.

The dominant mechanism of oxygen transport to immobilised cells is intraparticle convection. The extent of this is determined by bead size, which determines the amount of flow through the particle pores, and the total medium flow rate (Park and Stephanopoulos 1993). Consequently, a linear relationship exists between total flow rate and the total number of viable cells in the reactor.

3.2 Non-Ideal Fluid Flow

As mentioned in the previous section, packed bed reactors can be described as following plug flow patterns, provided that the axial fluid velocity remains constant through the cross-section. In reality, reactors never fully follow these flow patterns. Ideal flow is illustrated in Figure 8a. Deviations from ideal plug flow are caused by fluid channelling through the packed bed and by dead zones where fluid is essentially stagnant (Figure 8b).

In reactors following ideal flow patterns all the species in the reactor are assumed to have the same residence time although in reality there is a spread of times and the mean value is taken as the residence time for ideal flow conditions. In all other reactor types, the various species in the feed spend different times inside the reactor. There is a distribution of residence times of material within the reactor. If we consider an element of fluid entering a continuous flow reactor, because of the mixing in the vessel, this fluid will be dispersed into smaller elements. These smaller elements will separate and disperse throughout the vessel. Therefore some parts of the fluid will rapidly find their way to the outlet while other portions will travel about the vessel for varying times before exiting. The determination of the distribution of these residence times is a valuable indicator of the mixing and flow patterns within the vessel.

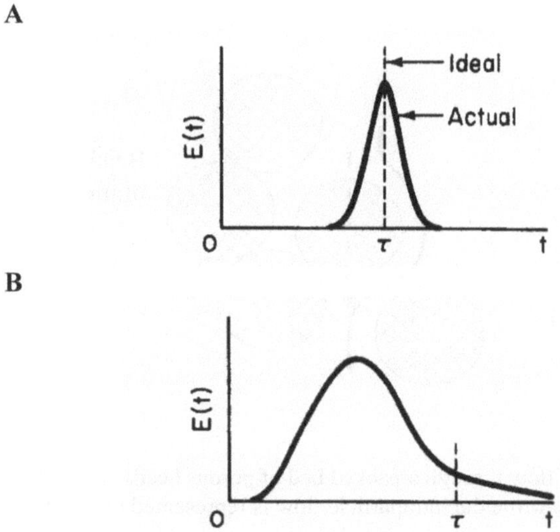

Figure 8. E(t) (residence time distribution) against t for an ideal plug flow reactor (A) and a tubular reactor with dead zone channelling (B) (Fogler 1986).

In bioreactors containing microcarriers the flow pathways will be influenced by the distribution of microcarriers within the vessel. This distribution varies greatly and is influenced by particle geometry, the packing density (i.e. volume of particles contained in the bed) and cell density. It is conceivable that at high cell densities, bridging may occur between closely packed particles. Bridging is the phenomenon where cells attached to the outer surface of carriers form a bridge with other carriers, resulting in an aggregate. This is more commonly observed with solid microcarriers. The appearance of bridging in a packed bed would have the same effect as increasing the particle size, and thus reducing the intraparticle convective flow. This results in channelling and the formation of stagnant dead zones.

Wall effects can also occur in packed bed reactors and generally give rise to increased flow rates along the walls of the vessel compared to a longer pathway through the microcarriers packing the reactor, which offer greater resistance. The presence of wall effects and the emergence of dead zones can be seen in Figure 9. Here, the fluid flow through a packed bed reactor filled with polyester discs was examined. The impact of these observations on cells within the reactor can be predicted. Cells on polyester discs in stagnant regions will be deprived of nutrients if diffusion alone does not provide sufficient essential molecules. Wall effects will result in medium entering and leaving the reactor without coming into contact with microcarriers, particularly in the upper horizontal region of the reactor, but the proportion of medium not undergoing wall effects may be sufficient to provide nutrients to these regions. It is unrealistic to expect ideal flow patterns in real reactors. Modification of the reactor design may serve to reduce the effects of channelling along reactor walls and existence of dead zones but will not totally eliminate the problems.

4. ALTERNATIVE PACKED BED SYSTEMS

The high demand for animal cell derived products stimulated the development of bioreactors during the 1980's (Hu and Peshwa 1991). The potential of packed bed systems has also seen the development of various designs, each attempting to overcome the characteristic problems of mass transport, nutrient gradients and fluid flow. The key types of packed bed bioreactors are depicted in Figure 10.

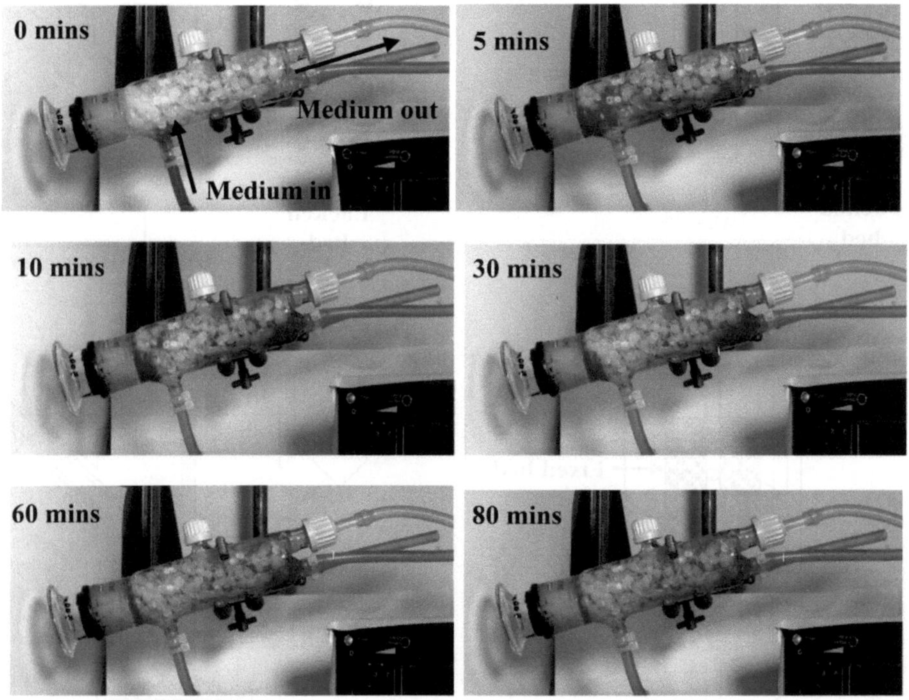

Figure 9. Photographs visually representing the distribution of the tracer at specific time points. A stimulus response technique was employed to look at flow in the reactor. The stimulus was a tracer put into the system and the response was a time record of the tracer leaving the system. The reactor was packed with 6 g of Fibra-Cel discs and the reservoir contained 200ml of water. The reactor was allowed to fill and recirculation ceased once the reactor was full. Phenol red was used as the tracer as it is a constituent of culture medium, has a similar density and its deep red coloration makes it detectable even at low concentrations. The experiments were conducted using a flow rate of 8ml per minute. The more intense the red coloration the greater the concentration of dye in any particular area. Areas within the reactor are seen to exist where dye accumulates over the 80 minute period observed.

4.1 Conventional Packed Bed Reactors

The traditional packed bed reactor, as shown in Figure 10A, consists of two vessels; the packed bed unit containing macroporous particles for cell immobilisation, and a vessel for conditioning the culture medium. Oxygen enriched medium is pumped from the conditioning vessel through the fixed bed and back. Medium exiting the packed bed is depleted of nutrients, such as glucose and oxygen. However, as the medium re-enters the conditioning vessel it mixes with the medium reservoir where

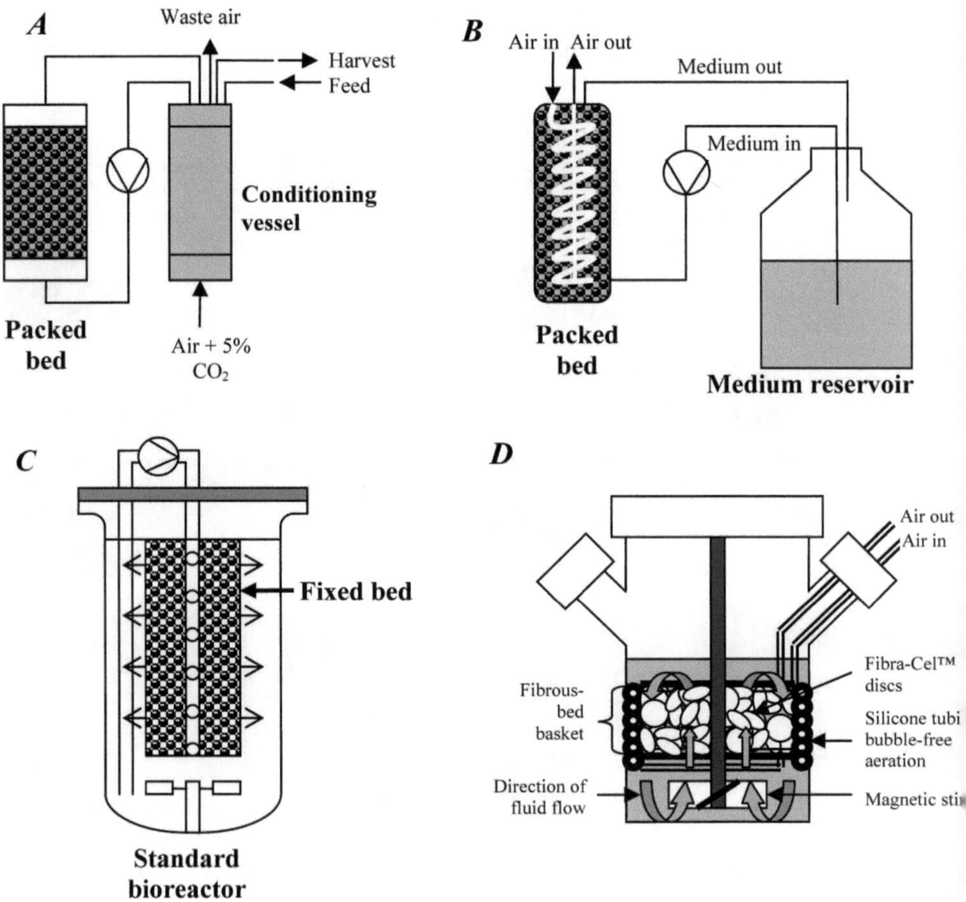

Figure 10. Schematic diagrams of different packed bed reactor systems: A, conventional packed bed reactor with medium conditioning vessel; B, packed bed with internal silicone tubing for direct aeration of the packed bed; C, radial flow packed bed suspended in a standard stirred tank reactor; D, basket type packed bed reactor.

the concentration of nutrients are higher. Toxic metabolites are also diluted, oxygen is replenished and pH can be adjusted to optimal levels. Hence, the medium being pumped into the packed bed is rich in nutrients. Samples can be taken from the conditioning vessel at regular intervals to monitor substrate concentrations. For long term cultivation, continuous cultures can be run by pumping fresh medium into the conditioning vessel and removing spent medium, thus preventing substrate limitation.

4.2 Packed Bed With Internal Silicone Tubing

Fluid flow through conventional packed bed systems is axial, which greatly limits their scale-up potential. Nutrient gradients develop along the length of the reactor, resulting in substrate limitation (primarily oxygen) at the downstream end of the vessel. In order to overcome this, a novel bioreactor system has been developed that employs internal aeration of the packed bed. Silicone tubing is coiled through the centre of the vessel to provide a uniform oxygen concentration along the axial length of the vessel. The system has been tested using several cell types and has been shown to sustain high densities of primary hepatocytes, HepZ cells (Bratch 2000), and the human cell lines FLYRD (McTaggart and Al Rubeai 2000) and TEFLYRD (Warnock 2003). The mass transfer characteristics of this reactor have been assessed and a summary of the results are presented in Table 1.

Table 1. Summary of data obtained in the characterisation of a packed bed bioreactor with internal silicone tubing for aeration. Total volume (V_T) was 500ml in all experiments and the reactor contained 6 g of discs for all studies with the exception of experiments where the reactor contained no discs. The values used for 'a' were 0.2998 and 0.3996cm^2 ml^{-1} for silicone tubing wall thickness of 1.7mm and 1.2mm, respectively.

Gas Flow (ml/min)	Fluid Flow (ml/min)	Silicone tubing wall thickness (mm)	Slope (sec^{-1})	Mass transfer coefficient K_L (cm/sec^{-1})
16	8	1.2	0.00083	0.0021
16	15	1.2	0.00085	0.0021
16	23	1.2	0.00100	0.0025
8	15	1.2	0.00048	0.0012
16	15	1.2	0.00072	0.0018
35	15	1.2	0.00111	0.0028
16	15	1.7 (no discs)	0.00062	0.0034
16	15	1.2 (no discs)	0.00135	0.0034

Higher gas flow rates resulted in elevated K_L values that are likely to be attributed to the increase in oxygen supply accompanying increased gas flow. Silicone tubing wall thickness had no effect on the mass transfer coefficient; this is essentially an alteration in K_d (tubing characteristics). When considering the effect of fluid flow rate on K_L the degree of agitation in the reactor (K_W) is considered. It was found that K_L only altered at higher flow rates (23ml/min), there is little difference in the values obtained for 8ml min^{-1} and 15ml min^{-1}. Increased agitation or flow rate is likely to result in better mixing and therefore better mass transfer.

The presence of microcarriers in the reactor resulted in a significant decrease in K_L. The value of K_L in the absence of discs was 0.0034, whereas in the same conditions in the presence of discs K_L values were 0.0021 or 0.0018. Mass transfer is expected to be lower in the presence of discs as the discs themselves will effect agitation in the vessel and also form an additional barrier for the transfer of oxygen.

Based on the results presented in Figure 2, the maximum number of viable cells, X_V, for the most oxygen demanding cells considered, i.e. primary rat hepatocytes in suspension, was 7.355 x 10^{18} cells. This is the maximum number of cells that can be supported by the reactor in the present mass transfer conditions (medium flow rate 8ml min^{-1} and gas flow 16ml min^{-1}). This value is dependant on the accuracy of the mass transfer assessment. The X_V total value is important for the determination of the number of cells that can be used to inoculate the bioreactor. It also has implications on whether it will be necessary to scale up the vessel if it were to be employed in clinical conditions. The need for scaling up will be dependant on the type of cells used and the number of cells needed to provide artificial support.

4.3 Radial Flow Reactors

In order to attain scale-up in a packed bed reactor radial flow geometry must be applied, as oxygen will not be limited along the radius (Bohmann *et al.* 1995; Kurosawa *et al.* 1991). In this configuration, the fixed bed is contained in an inner chamber within the conditioning vessel. The microcarriers are retained in the bed, and separated from the bulk medium, by a dialysis membrane. Medium is pumped out of the conditioning vessel into the packed bed, as with conventional packed bed systems, but it is able to diffuse out of the matrix into the bulk medium due to the concentration gradients that are present between the circulating medium, medium in the packed bed and the bulk medium. As the radial distance is much less than the axial distance, substrate limitation does not occur.

Until recently only little has been known about the scale-up of radial flow bioreactors. In order to determine the scale-up criteria a mathematical model was developed to establish the oxygen profile within the packed bed. The model corresponded well to the experimental data. It was also used to study the influence of several parameters on the performance of the fixed-bed system, such as the carrier size, the dissolved oxygen concentration, or the superficial flow velocity. By adapting the model it was shown that reaction-diffusion limitation is generally not a problem for other substrates such as glucose or glutamine (Fassnacht and Pörtner 1999). Based on these findings the authors were able to design a 5 litre radial flow packed bed bioreactor, which was integrated into a 10 litre glass bioreactor. The cell densities reached in this system were analogous to those of a 50-100 litre suspension culture, thus demonstrating the high cell densities achievable in packed bed systems (Pörtner *et al.* 1999).

4.4 *Basket Type Packed Bed Reactors*

The main disadvantages of conventional packed bed bioreactors are the nutrient gradient that appears along the length of the vessel and the existence of non-ideal flow patterns. These problems are eliminated in basket-type packed bed bioreactors by the fact that the bed is contained within the conditioning vessel. The simplest form of these reactors is shown in Figure 10D. Here the basket is contained in a spinner flask. Oxygenation of the bulk medium is achieved through silicone tubing, and flow through the bed is achieved with an axial flow impeller. This system is limited in scale to 500ml. In work done by Warnock (2003), sub-optimal cell densities were attributed to poor mass transfer within the bed. The level of fluid flow achieved by mixing was unable to induce the required convection within the bed, leading to mass transport limitations. This was reflected in the cellular metabolism, demonstrated by an elevated glucose consumption rate.

Good scale-up is achieved in the Celligen® bioreactor (New Brunswick Scientific, New Jersey, USA). Mixing in larger bioreactors is accomplished with either a cell-lift impeller, which allows bubble-free medium to flow through the packed bed (Wang *et al.* 1992) or a double screen concentric cylindrical cage impeller. Shi *et al.* (1992) have shown that the latter impeller is able to increase convective mass transfer, resulting in higher cell and product concentrations. The significant improvements were attributed to the increased surface area allowing for convective oxygen transfer and protection of cells from hydrodynamic stresses. Both methods create a homogeneous environment for cells to grow and product can be directly removed from the culture vessel, which allows for easier downstream processing and product purification.

5. APPLICATIONS IN TISSUE ENGINEERING

Packed bed bioreactors have been used almost exclusively in tissue engineering as bioartificial liver support systems. Although they have the potential to generate vast quantities of cells, complications arise in harvesting from the macroporous matrices (Nielsen 1999). While this is problematic for subsequent transplantation, it is of benefit for artificial organs, where cell retention is desirable. The ability to support viable and functional hepatocytes makes it an ideal component for a bioartificial liver.

5.1 *Bioartificial Liver Support Systems*

Aside from liver transplantation there is no established effective treatment for acute hepatic failure (Bratch and Al Rubeai 2001). Liver transplantation is severely limited, though, by the shortage of cadaveric organ donors and the short time period available to find a suitable liver. As a result, mortality in acute Fulminant Hepatic Failure (FHF) has been reported to be as high as 65-70% (Chamuleau 2002).

An artificial hepatic support device that could serve as a bridge to transplantation or ideally spontaneous recovery is very appealing. Mechanical devices such as haemodialysis and haemoperfusion do not correct metabolic abnormalities that exist in end stage liver disease and biologically active devices have been limited by the availability and viability of cultured liver cells (Sussman *et al.* 1994).

The development of an artificial liver support system is an area currently being pursued by many investigators (Allen and Bhatia 2002; Patzer 2001). Artificial materials cannot solely replace metabolic functions. Therefore, hybrid devices consisting of both biological and artificial materials are required. The system should be able to provide sufficient hepatic mass to restore the body's homeostasis and have the ability to detoxify substances causing hepatic coma and supply plasma proteins.

Hybrid artificial organs are designed to duplicate the original functions of the damaged organ by the actions of both types of materials. Isolated hepatocytes are used as the biological component because they are the most important functioning cells comprising the liver (Bratch and Al Rubeai 2001), and these cells are easy to handle and maintain. Artificial materials are used to maximise the function of biological materials, increase the density of cultured cells, promote the effective exchange of substances and reduce immunological hazards.

The development of a bioartificial liver (BAL) is a major challenge. Unlike the heart, kidney or lungs which have one primary function the liver has several complex functions essential to sustain life. Consequently, simple mechanical approaches to replacing liver function have proved unsuccessful.

Several problems need to be solved in order for a BAL device to be of any benefit to patients suffering from FHF. Firstly the system needs to express and maintain liver specific functions at high levels *in vitro*. Also, the module needs to be designed on a practical scale, which will be assisted by achieving a high cell density culture of hepatocytes. The actual capabilities of an artificial liver are very difficult to define, as it is not clear what is essential to keep liver failure patients alive.

Primary hepatocytes of either human or animal origin are the ideal biological component, as they express highly differentiated functions (Morsiani *et al.* 2002). Porcine hepatocytes are among the most popular biological constituents of potential BAL devices as cells can be easily obtained and one pig liver provides a large number of cells, which can be utilised for several patient treatments. Furthermore, pigs have a similar physiology to man, possibly rendering porcine hepatocytes more compatible for use in a clinical device. The potential of primary rat hepatocytes has also been examined (Bratch and Al Rubeai 2001; Takagi *et al.* 2002). A disadvantage of using such cells is that animal hepatocytes synthesise animal proteins, creating the problem of inter-species differences in enzymatic activity of complex networks like coagulation or the complement system (Stange *et al.* 1993). There is also the potential transmission of animal retrovirus and prions which may potentially cause major pandemics. Accordingly, the use of porcine hepatocytes, or other animal cells, in BAL devices is only recommended for acute liver failure patients as a bridge treatment to liver transplantation (Morsiani *et al.* 2002).

The majority of BAL support systems currently employ hollow-fibre reactors as the artificial component (Morsiani *et al.* 2000). However, in recent years promising results have been obtained from packed bed bioreactors. Shiba *et al.* (2003) have presented results for a radial flow bioreactor, used to culture primary porcine hepatocytes. They examined the effects of ascorbic acid 2-phosphate and dissolved oxygen levels on cellular maintenance and function. The ammonium metabolic rate was used as a measure of functionality, and they demonstrated that cells were able to maintain activity at 7.7μmol hr^{-1} after 30 days, using 1.5mmol l^{-1} Asc2-P and 10mg l^{-1} dissolved oxygen. This demonstrates the possibility of using radial flow bioreactors for the long-term culture of hepatocytes. Radial-flow packed bed bioreactors have also been shown to outperform hollow-fibre bioreactors as a suitable BAL (Morsiani *et al.* 2000). The results of this study are presented in Figure 11. At the conclusion of these experiments the cell viability in the PES-RFB was 85%, compared to 40% and 57% in the CNA-HFB and the PES-HFB, respectively. The authors concluded that the RES-RFB was superior to the HFBs in maintaining hepatocytes differentiated functions when perfused with human plasma. Additionally, only the PES-RFB allowed cell respiratory function for the entire duration of the study, signifying the feasibility of this configuration for clinical BAL support.

Bratch (2000) evaluated a packed bed bioreactor with internal silicone tubing as a BAL, using primary rat hepatocytes. The results obtained from this study are summarised in Table 2. Primary rat hepatocyte viability decreased over the course of the experiments at all flow rates considered, but the rate of decrease was greatest at higher flow rates. The results suggested that fluid flow had a detrimental effect on cells. At higher flow rates the effects of shear forces on cells are increased resulting in increased cell death. Primary rat hepatocytes appear to be more susceptible to these effects compared to the HepZ cell line. This is probably due to primary cells being larger in size and more fragile compared to the HepZ cell line.

A packed bed reactor containing 9.0 x 10^7 primary rat hepatocytes demonstrated little fluctuation in medium glucose and pH levels with time. Evidence of cellular functionality was demonstrated by albumin production and lidocaine metabolism by cytochrome p450 resulting in MegX production after 67 hours, but no MegX production was observed at the end of the run (148 hours). Cellular albumin production can be derived by estimating the number of viable cells within the reactor at a particular time point using MTT absorbance values. In the first 24 hours albumin production was approximately 0.208pg $cell^{-1}h^{-1}$ if it is assumed that the cell number is equal to that at zero time but, this may not be a reasonable assumption to make in light of the sensitivity of primary rat hepatocytes. Albumin production decreased to 0.134pg $cell^{-1}h^{-1}$ at 67 hours where total cell number had been estimated from MTT absorbance. The values obtained for Albumin production by primary rat hepatocytes in the packed bed reactor are similar to previously published values (Hu *et al.* 1997).

Due to indications that over the 148 hour period cell condition, and hence viability deteriorated, a second study looked at cellular ultrastructure and functionality over a culture period of 48 hours. Medium glucose and pH levels remained stable over the time period studied, whereas urea, albumin and MegX levels were seen to increase indicating the presence of cellular functionality. Urea production amounted to 0.0073mmol $l^{-1}h^{-1}$ for 10^8 cells, assuming that the initial cell density was maintained. Albumin production after 18 hours in culture equalled 0.25pg $cell^{-1}h^{-1}$. This was comparable to previously published data which reported values of 0.6pg $cell^{-1}h^{-1}$ (Shatford *et al.* 1992) and 0.13pg $cell^{-1}h^{-1}$ (Hu *et al.* 1997). MegX production was 0.47ng $ml^{-1}min^{-1}$ after 48 hours for 10^8 cells.

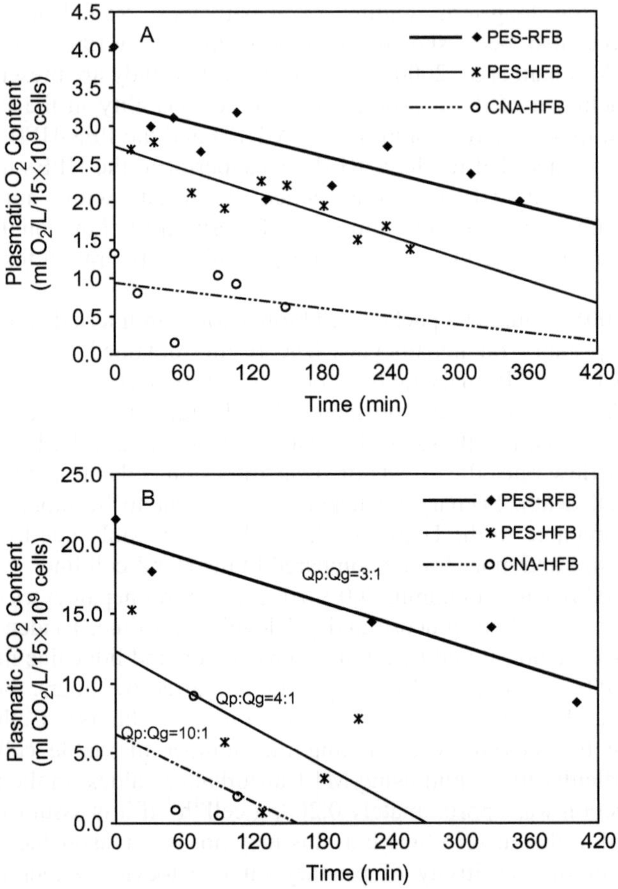

Figure 11. Linear trend of plasma O_2 and CO_2 contents during 7 hours of porcine hepatocyte perfusion with recirculating 90 to 95% O_2/5 to 10% CO_2-enriched human plasma within three different bioreactors: (1) cellulose nitrate/cellulose acetate hollow-fibre bioreactor (CNA-HFB); (2) polyethersulphone hollow-fibre bioreactor (PES-HFB); and (3) polyethersulphone radial-flow bioreactor (PES-RFB). (A) plasmatic O_2 contents (mL O_2/L/15 × 10^9 cells) expressed as the difference between bioreactor's incoming and outcoming plasma O_2 concentrations; (B) plasma CO_2 contents (mL CO_2/L/15 × 10^9 cells) expressed as the difference between bioreactor's incoming and outgoing plasma CO_2 concentrations. Plasma/gas flow ratio (Qp:Qg) was 3:1 in PES RFB, 4:1 in PES-HFB, and 10:1 in CNA-HFB. (Used with permission, Morsiani et al. 2000)

Table 2. Summary of data obtained from packed bed reactors. (+/- indicates the presence or absence respectively of the parameters measured). The reactor was filled with 6g Fibra-Cel™ discs and inoculated with primary rat hepatocytes, suspended in DMEM + 10% Newborn Calf Serum (Bratch 2000).

Number of cells (Duration of study)	Medium flow (ml/min)	Glucose uptake	Ureagenesis	Albumin secretion	MegX production (Time point at which test conducted)
6.0×10^7 (96 hours)	4	-	-	-	+ (72 hours)
6.0×10^7 (96 hours)	8	-	-	-	+ (72 hours)
9.0×10^7 (148 hours)	8	-	-	+	+ (67 hours)
1.0×10^8 (48 hours)	8	-	+	+	+ (48 hours)
2.0×10^8 (163 hours)	8	+	+	-	-
4.0×10^8 (168 hours)	12	+	-	-	-
7.0×10^8 (96 hours)	8	-	-	-	-

5.2 Haematopoietic Cell Expansion

The *ex vivo* expansion of haematopoietic cells is an enabling technology with many potential applications in bone-marrow transplantation, immunotherapy, gene therapy and the production of blood products (Nielsen 1999). Haematopoietic tissue is located in the bone-marrow of adults and is the site of haematopoiesis, the production of mature red and white blood cells. As haematopoiesis becomes better understood, researchers have been able to develop systems that mimic the process *ex vivo*, leading to the generation of haematopoietic tissue (Audet *et al.* 1998; Collins *et al.* 1996; Nielsen 1999).

Traditional chemo- and radiation therapies are often limited by the haematological toxicity associated with high doses (Maraninchi 1993). In order to overcome this problem, autologous bone-marrow transplantation is performed, following high-dose therapy, to restore haematopoiesis. Prior to treatment, bone-marrow is collected under spinal or general anaesthetic. The pelvic bones are punctured and aspirated several times to collect between 0.1 and 1.0 litres of fluid containing bone-marrow and blood cells. This material is then stored in liquid nitrogen until required (McAdams *et al.* 1996). *Ex vivo* expansion of haematopoietic cells has the potential to generate sufficient amounts of tissue from significantly smaller starting samples, though, and has several advantages over bone-marrow transplantation (see review by Cabral 2001).

Packed bed bioreactors have been limited in there use for the expansion of haematopoietic cells. However, Highfill *et al.* (1996) reported the large-scale cultivation of murine bone-marrow cells in a packed bed bioreactor that was able to mimic the bone-marrow environment. A stromal cell population was initially seeded into the reactor to generate the required microenvironment. Once this was established, bone-marrow cells were inoculated into the reactor. Standard culture medium was inadequate for the initiation of haematopoiesis. However, stomal cell conditioned medium was able to instigate cell production. During the 11 week culture period, the 500 ml perfusion culture produced 3.6×10^8 cells. Unfortunately, packed bed reactors have limited use in *ex vivo* expansion of cells due to difficulties in harvesting (Nielsen 1999).

6. SUMMARY

Packed bed bioreactors are a promising tool for tissue engineering applications that have not been fully utilised to date. They are capable of supporting various cell lines for long culture periods under low shear conditions, due to the immobilisation of cells within macroporous matrices. Various configurations have been designed that deal with the problems traditionally associated with immobilised cell cultures, such as non-ideal fluid flow and mass transport limitations. These have enabled good scale-up and superior productivity to alternative systems such as hollow-fibre reactors. Packed bed reactors are currently being employed in studies to develop bioartificial liver support systems (Morsiani *et al.* 2000; Shiba *et al.* 2003) but have also been used for the *ex vivo* expansion of bone marrow cells (Highfill *et al.* 1996).

7. ACKNOWLEDGEMENTS

The authors would like to thank Dr Hideki Yamaji for the contribution of Figure 5.

8. REFERENCES

Al Rubeai, M., Rookes, S. and Emery, A. N. 1990. Studies of Cell Proliferation And Monoclonal Antibody Synthesis And Secretion In Alginate-Entrapped Hybridoma Cells. In: de Bont, J. A. M., Visser, J., Mattiasson, B., and Tramper, J. editors. Physiology of Immobilized Cells. Amsterdam: Elsevier Science Publishers. p181-188.

Al Rubeai, M. and Spier, R.E. 1989. Quantitative cytochemical analysis of immobilised hybridoma cells. Appl Microbiol Biotechnol 31:430-433.

Allen, J.W. and Bhatia, S.N. 2002. Improving the next generation of bioartificial liver devices. Seminars in Cell & Developmental Biology 13: 447-454.

Atkinson, B. and Mavituna, F. 1983. Biochemical Engineering and Biotechnology Handbook. London: Macmillan.

Audet, J., Zandstra, P.W., Eaves, C.J. and Piret, J.M. 1998. Advances in hematopoietic stem cell culture. Current Opinion in Biotechnology 9:146-151.
Bailey, J. E. and Ollis, D. F. 1986. Biochemical Engineering Fundamentals. New York: McGraw Hill Book Co.
Banik, G.G. and Heath, C.A. 1995. Hybridoma growth and antibody production as a function of cell density and specific growth rate in perfusion culture. Biotechnol Bioeng. 48: 289-300.
Bohmann, A., Portner, R. and Markl, H. 1995. Performance of a membrane-dialysis bioreactor with a radial-flow fixed-bed for the cultivation of a hybridoma cell-line. Applied Microbiology and Biotechnology 43: 772-780.
Bratch, K. 2000. The Development of a Bioartificial Liver Support System. The University of Birmingham.
Bratch, K. and Al Rubeai, M. 2001. Culture of primary rat hepatocytes within a flat hollow fibre cassette for potential use as a component of a bioartificial liver support system. Biotechnology Letters 23: 137-141.
Cabral, J.M.S. 2001. Ex vivo expansion of hematopoietic stem cells in bioreactors. Biotechnology Letters 23: 741-751.
Catapano, G. 1996. Mass transfer limitations to the performance of membrane bioartificial liver support devices. International Journal of Artificial Organs 19: 18-35.
Chamuleau, R.A. 2002. Bioartificial liver support anno 2001. Metab Brain Dis 17: 485-491.
Chresand, T.J., Gillies, R.J. and Dale, B.E. 1988. Optimum fiber spacing in a hollow fiber bioreactor. Biotechnol Bioeng. 32: 983-992.
Collins, P.C., Papoutsakis, E.T. and Miller,W.M. 1996. Ex vivo culture systems for hematopoietic cells. Current Opinion in Biotechnology 7: 223-230.
Cong C., Chang Y., Deng J., Xiao C. and Su Z. 2001. A novel scale-up method for mammalian cell culture in packed-bed bioreactor. Biotechnology Letters 23: 881-885.
Drury, D.D., Dale, B.E. and Gillies, R.J. 1988. Oxygen transfer properties of a bioreactor for use with nuclear magnetic resonance spectrometer. Biotechnol Bioeng. 32: 966-974.
Ducommun, P., Ruffieux, P.-A. and Kadouri, A. 2001. Process Development In A Packed Bed Bioreactor. In: Linder-Olsson, E. editor. Animal Cell Technology: From Target to Market. Netherlands: Kluwer Academic Publishers p410-411.
Ellis, L.C. 1991. Free radicals in tissue culture. Part IV. Effects on cells in culture. Art Sci 10: 1-5.
Emery, A.N., Jan, D.C. and Al Rubeai, M. 1995. Oxygenation of intensive cell-culture system. Appl. Microbiol. Biotechnol. 43: 1028-1033.
Fassnacht D., Rössing S, Singh R.P., Al Rubeai, M. and Pörtner R. 1999. Influence of bcl-2 on antibody productivity in high cell density perfusion cultures of hybridoma. Cytotechnology 30: 95-105.
Fassnacht, D. and Portner, R. 1999. Experimental and theoretical considerations on oxygen supply for animal cell growth in fixed-bed reactors. J. Biotechnol. 72: 169-184.
Fogler, H. S. 1986. Elements of Chemical Reaction Engineering. New Jersey: Prentice-Hall. p. 632-668.
Foy, B.D., Rotem, A., Toner, M., Tompkins, R.G. and Yarmush, M.L. 1994. A device to measure the oxygen uptake rate of attached cells: importance in bioartificial organ design. Cell Transplant. 3: 515-527.
Highfill, J.G., Haley, S.D. and Kompala, D.S. 1996. Large-scale production of murine bone marrow cells in an airlift packed bed bioreactor. Biotechnology and Bioengineering 50: 514-520.
Hu, W.S., Friend, J.R., Wu, F.J., Sielaff, T., Peshwa, M.V., Lazar, A., Nyberg, S.L., Remmel, R.P. and Cerra, F.B. 1997. Development of a bioartificial liver employing xenogeneic hepatocytes. Cytotechnology 23: 29-38.
Hu, W.S. and Peshwa, M.V. 1991. Animal-cell bioreactors - recent advances and challenges to scale-up. Canadian Journal of Chemical Engineering 69: 409-420.
Hu,Y.C., Kaufman,J., Cho,M.W., Golding,H. and Shiloach,J. 2000. Production of HIV-1 gp120 in packed-bed bioreactor using the vaccinia virus/T7 expression system. Biotechnol. Prog. 16, 744-750.
Jauregui, H.O., Chowdhury, N.R. and Chowdhury, J.R. 1996. Use of mammalian liver cells for artificial liver support. Cell Transplant. 5: 353-367.
Kadouri, A. and Zipori D. 1989. Production of anti-leukemic factor from Stroma cells in a stationary bed reactor on a new cell support. In: Spier, R.E., Griffiths, J.B., Stephenne, J. and Rooy, P.J editors. Advances In Animal Cell Biology And Technology For Bioprocesses. Tiptree, Essex: Courier International Ltd. p. 327-330.

Kalogerakis, N. and Behie, L.A. 1997. Oxygenation capabilities of basket-type bioreactors for microcarrier cultures of anchorage-dependent cells. Bioprocess Engineering 17: 151-156.
Kaufman, J., Wang, G., Zhang W, Valle M.A. and Shiloach, J. 2000. Continuous production and recovery of recombinant Ca^{2+} binding receptor from HEK 293 cells using perfusion through a packed bed bioreactor. Cytotechnology 33: 3-11.
Kennard, M.L. and Piret, J.M. 1994. Glycolipid membrane-anchored recombinant protein-production from cho cells cultured on porous microcarriers. Biotechnology and Bioengineering 44: 45-54.
Knight P. 1989. Hollow fiber bioreactors for mammalian cell culture. Bio/Technology 7: 459-461.
Kurosawa, H., Markl, H., Niebuhrredder, C. and Matsumura, M. 1991. Dialysis bioreactor with radial-flow fixed-bed for animal-cell culture. Journal of Fermentation and Bioengineering 72: 41-45.
Lin, A.A., Kimura, R. and Miller, W.M. 1993. Production of tPA in recombinant CHO cells under oxygen-limited conditions. Biotechnol Bioeng. 42: 339-350.
Ljunggren, J. and Häggström, L. 1994. Catabolic control of hybridoma cells by glucose and glutamine limited fed batch cultures. Biotechnol Bioeng. 44: 808-818.
Mancuso, A., Fernandes, E.J., Blanch, H.W. and Clarke, D.S. 1990. A nuclaer magnetic resonance technique for determining hybridoma cell concentration in hollow fibre bioreactors. Bio/Technology 8: 1282-1285.
Maraninčhi, D. 1993. The clinical consequences of hematological and nonhematological toxicity following bone-marrow transplantation and the possible impact of hematopoietic growth-factors. Bone Marrow Transplantation 11: 12-22.
McAdams, T.A., Miller, W.M. and Papoutsakis, E.T. 1996. Hematopoietic cell culture therapies .2. Cell culture considerations. Trends in Biotechnology 14: 341-349.
McTaggart, S. 2000. Retroviral Vector Production for Gene Therapy Applications. University of Birmingham.
McTaggart, S. and Al Rubeai, M. 2000. Effects of culture parameters on the production of retroviral vectors by a human packaging cell line. Biotechnol. Prog. 16: 859-865.
Morsiani, E., Brogli, M., Galavotti, D., Pazzi, P., Puviani, A.C. and Azzena, G.F. 2002. Biologic liver support: Optimal cell source and mass. International Journal of Artificial Organs 25: 985-993.
Morsiani, E., Galavotti, D., Puviani, A.C., Valieri, L., Brogli, M., Tosatti, S., Pazzi, P. and Azzena, G. 2000. Radial flow bioreactor outperforms hollow-fiber modules as a perfusing culture system for primary porcine hepatocytes. Transplantation Proceedings 32: 2715-2718.
Mueller-Klieser, W.F. and Sutherland, R.M. 1982. Oxygen tensions in multicell spheroids of two cell lines. Br. J. Cancer 45: 256-264.
Neermann, J. and Wagner, R. 1996. Comparative analysis of glucose and glutamine metabolism in transformed mammalian cell lines, insect and primary liver cells. J. Cell Phys 166: 152-169.
Nielsen, L.K. 1999. Bioreactors for hematopoietic cell culture. Annu. Rev. Biomed. Eng 1: 129-152.
Othmer, K. 1991. Encyclopedia of Chemical Technology. Wiley.
Park S. and Stephanopoulos G. 1993. Packed bed bioreactor with porous ceramic beads for animal cell culture. Biotechnol. Bioeng. 41: 25-34.
Patzer, J.F. 2001. Advances in bioartificial liver assist devices. Ann. N. Y. Acad. Sci. 944: 320-333.
Piret, J.M. and Cooney, C.L. 1990. Mammalian cell and protein distributions in ulltrafiltration hollow-fiber bioreactors. Biotechnol. Bioeng. 36: 902-910.
Piret, J.M. and Cooney, C.L. 1991. Model of oxygen transport limitations in hollow fiber bioreactors. Biotechnol Bioeng. 37: 80-92.
Piret, J.M., Devens, D.A. and Cooney, C.L. 1991. Nutrient and metabolite gradients in mammalian-cell hollow fiber bioreactors. Canadian Journal of Chemical Engineering 69: 421-428.
Pörtner, R., Bohmann, A., Ludemann, I. and Markl, H. 1994. Estimation of specific glucose uptake rates in cultures of hybridoma cells. J. Biotechnol 34: 237-246.
Pörtner, R., Fassnacht, D. and Märkl, H. 1999. Immobilization of mammalian cells in fixed bed reactors. BIOforum International 4: 140-141.
Preissmann, A., Wiesmann, R., Buchholz, R., Werner, R.G. and Noe, W. 1997. Investigations on oxygen limitations of adherent cells growing on macroporous microcarriers. Cytotechnology 24: 121-134.
Racher, A.J. and Griffiths, J.B. 1993. Investigation of parameters affecting a fixed bed bioreactor process for recombinant cell lines. Cytotechnology 13: 125-131.
Rhodes, M. 1998. Fluid Flow through a Packed Bed of Particles. In Introduction to Particle Technology pp. 81-85. Wiley.

Rodrigues M.T.A., Vilaça P.R., Garbuio A., Takagi M., Barbosa Jr S., Léo P., Laignier N.S., Silva A.A.P. and Moro A.M. 1999. Glucose uptake rate as a tool to estimate hybridoma growth in a packed bed bioreactor. Bioprocess Engineering 21: 543-546.

Shatford, R.A., Nyberg, S.L., Meier, S.J., White, J.G., Payne, W.D., Hu, W.S. and Cerra, F.B. 1992. Hepatocyte function in a hollow fiber bioreactor - a potential bioartificial liver. Journal of Surgical Research 53: 549-557.

Shi Y., Ryu, D.D.Y. and Park S. 1992. Performance of mammalian cell culture bioreactor with a new impeller design. Biotechnol. Bioeng. 40: 260-270.

Shiba, Y., Koyama, T., Wang, C., Zhang, Q., Okamura, A., Aoki, S., Mukaiyama, T., Yang, D. and Kodama, M. 2003. Culture of porcine hepatocytes using radial flow bioreactor system. In: Yagasaki, K. Animal Cell Technology for Innovative Life Sciences. Netherlands: Kluwer Academic Publishers.

Smith, M.D., Smirthwaite, A.D., Cairns, D.E., Cousins, R.B. and Gaylor, J.D. 1996. Techniques for measurement of oxygen consumption rates of hepatocytes during attachment and post-attachment. Int. J. Artif. Organs 19: 36-44.

Spier, R.E. and Whiteside, J.P. 1976. The production of foot-and-mouth disease virus from BHK 21 C 13 cells grown on the surface of glass spheres. Biotechnol Bioeng. 18: 649-657.

Stange, J., Ramalov, W. and Mitzner, S. 1993. A new procedure for the removal of protein bound drugs and toxins. Transactions of the American Society for Artificial Internal Organs 39: 621-625.

Stephanopoulos, G. and Tsiveriotis, K. 1989. The effect of intraparticle convection on nutrient transport in porous biological pellets. Chemical Engineering Science 44: 2031-2039.

Sussman, N.L., Gislason, G.T. and Kelly, J.H. 1994. Extracorporeal liver support. Application to fulminant hepatic failure. J. Clin. Gastroenterol. 18: 320-324.

Takagi, M., Kondo, H. and Yoshida, T. 2002. In vitro proliferation of primary rat hepatocytes expressing ureogenesis activity by coculture with STO cells. Journal of Bioscience and Bioengineering 94: 212-217.

Tannock, I.F. 1972. Oxygen diffusion and the distribution of cellular radiosensitivity in tumours. Br. J. Radiol. 45: 515-524.

Thelwell, P.E. and Brindle K.M. 1999. Analysis of CHO-K1 cell growth in a fixed bed bioreactor using magnetic resonance spectroscopy and imaging. Cytotechnology 30, 121-132.

Vaupel, P. 1977. Hypoxia in neoplastic tissue. Microvasc. Res. 13: 399-408.

Wang, G., Zhang, W., Jacklin, C., Freedman, D., Eppstein, L. and Kadouri, A. 1992. Modified CelliGen-packed bed bioreactors for hybridoma cell cultures. Cytotechnology 9: 41-49.

Warnock, J. N. 2003. Optimisation of retrovirus production systems for gene therapy applications. The University of Birmingham.

Williams, S.N.O., Callies, R.M. and Brindle, K.M. 1997. Mapping of oxygen tension and cell distribution in a hollow-fiber bioreactor using magnetic resonance imaging. Biotechnology and Bioengineering 56: 56-61.

Yamaji, H. and Fukuda, H. 1992. Growth and death behavior of anchorage-independent animal-cells immobilized within porous support matrices. Applied Microbiology and Biotechnology 37: 244-251.

CHAPTER 5

DESIGN AND OPERATION OF A RADIAL FLOW BIOREACTOR FOR RECONSTRUCTION OF CULTURED TISSUES

M. KINO-OKA AND M. TAYA

Division of Chemical Engineering, Graduate School of Engineering Science, Osaka University, Toyonaka, Osaka, Japan

1. INTRODUCTION

Generally, the losses of tissue and organ are repaired by the use of a mechanical type substitute or by transplantation of an organ from a donor, through surgical treatments. Organ donation is frequently limited. The alternatives for surgical reconstruction, even in use of the patient's own tissues, are not yet fully developed at many situations and are also associated with the problems to be solved. Mechanical devices, such as kidney dialysis machines, have provided appreciable therapeutic achievements. These mechanical approaches can offer interim relief for many patients, but ultimately, the ideal treatments involve regenerating or repairing the organ or tissue by using the techniques based on tissue engineering (du Moulin and Morohashi, 2000; Lanza et al., 2000, Lysaght and Reyes, 2001; Lalan *et al.*, 2001; Sun and Lal, 2002; McIntire, 2003).

Tissue engineered systems will play a key role in moving away from conventional surgeries by providing a new solution to tissue loss. Unlike traditional approaches for treatment of lost tissue or damaged organ function, tissue engineering enables to replace damaged tissue with regenerated tissue that is designed and constructed to meet the needs of each individual patients. The accelerated development of tissue engineered systems is based on various supporting technologies. For example, the use of various growth factors as therapeutics has proved to be effective both in the clinics and the marketplaces (Mayhew *et al.*, 1998; Naughton ,2002; Lodie *et al.*, 2002). The relationships of families of growth factors and the cellular functions impacted by their presence over specific time periods of cell growth cycle yields tissue engineered products altering tissue growth both *ex*

J.B. Chaudhuri & M. Al-Rubeai (eds.), Bioreactors for Tissue Engineering, 115 - 133.

vivo and *in vivo*. The expanding facts of cell regulation functions are providing new opportunities for intervention and for creating tissues that provide missing functions. Concurrently, advancements in polymer technologies and controlled release technologies allow cells to be grown and supported by an appropriate scaffold that can be tailored to the requirements of a particular cell type.

These advances will realize the appreciate growth and differentiation of target cells during tissue cultures under designed environment. One of the fields of extensive improvements is optimization of conditions for *in vitro* cell seeding and tissue cultivation for cell attachment, cell division, and differentiation in bioreactor systems. This review will outline the feature of manufacturing process for cultured tissues, and the recent development in bioreactor systems accompanied with circulating medium flow.

2. CHARACTERISTIC OF MANUFACTURING PROCESS FOR CULTURED TISSUE

The manufacturing processes for cultured tissue have to be constructed on the basis of a novel strategy when compared with conventional process for petrochemicals and pharmaceuticals. Figure 1 shows the historical flow in development of manufactures. In the field of petrochemical or heavy industries, the industrial principle is always toward unity and vastness with the terms of "heavy, big and large". This principle might be originated from the production process based on continuous operation on a large scale (outdoor manufacturing). The industrial field of pharmaceuticals quests for the principle based on multiplicity and smallness with a representative word of "fine", which is characterized as an indoor batch process. The principles described above, however, are not justified for the manufacturing process of cultured tissues. The raw materials and products for the manufacturing are cells themselves obtained from the patients (or the donors) and cultured tissues, respectively. The raw materials have heterogeneity depending on the state of patients and location of cell harvest, and the products are varied in size for individual patients. These features request the unique strategy in manufacturing process with minimum of raw materials and maximum of products. This particularity required for both raw materials and products adds new key words of "maximum, minimum and individual", which forces tailor-made production belonging to the batch process with less reproducibility as well as to "box-scale manufacturing". Figure 2 shows the general scheme for tissue production. The operations including isolation of cells, primary and secondary cultures, and the preparation of cultured tissues are manually performed with tedious and repetitious tasks. Therefore, the reliable and robust manufacturing process is required for certification of high-product quality. The engineering problems have not been tackled so far although the studies on kinetic analysis, growth-model development, culture operation and bioreactor design should be required to reduce labor and time for *in vitro* production of cultured tissues with high quality.

Tissue Engineering including a broad spectrum of technology platforms is defined as an integrated field based on the principles and methodologies of engineering and life sciences toward fundamental understanding of structure-function relationships in normal and pathological mammalian tissues, and the development of biological substitutes to restore, maintain, or improve tissue function. This is an emerging interdisciplinary area of research and technology development that has the potential to revolutionize our methods of health care systems. Some products are already in use clinically and their number will assuredly increase rapidly in the future with dramatic improvement in the quality of life for millions of people throughout the world. Chemical Engineers have potentials concerning the designs of culture operation (manufacturing process) and vessels (bioreactor), and their evaluation for quality control of products. Next section describes the functions of bioreactor and design of bioreactor accompanied with circulating medium flow.

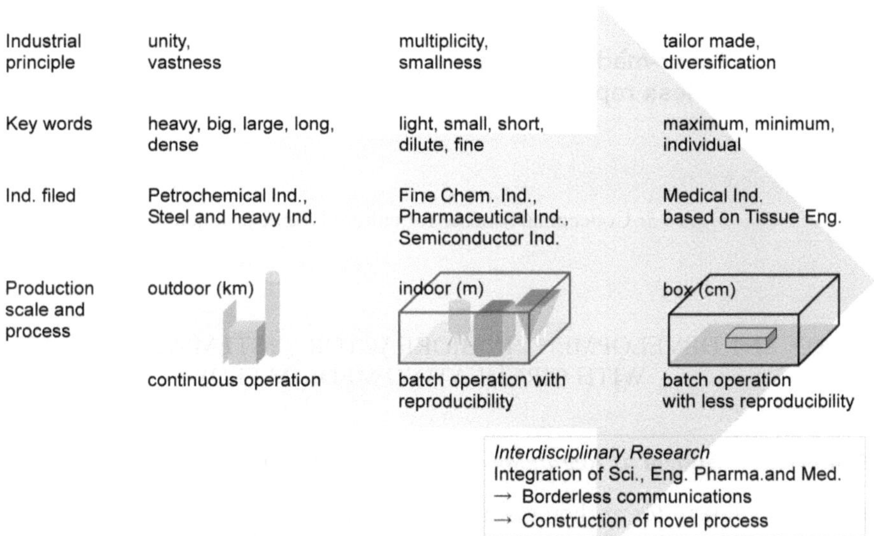

Figure 1: Conceptual features of some manufacturing processes

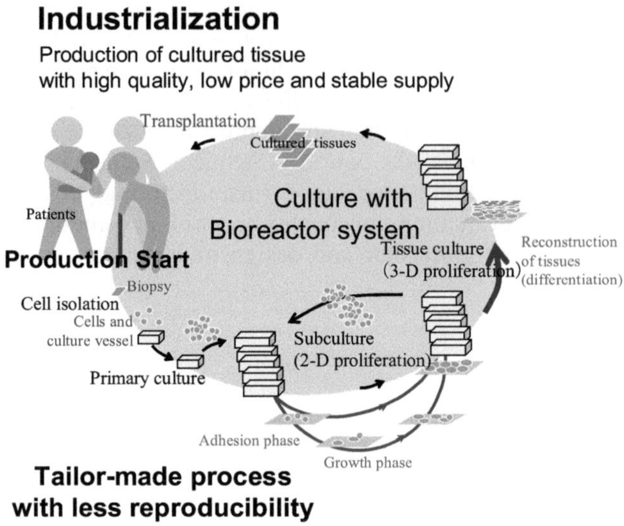

Figure 2: Conceptual scheme for cultured tissue production

3. RECENT DEVELOPMENT IN BIOREACTOR SYSTEM ACCOMPANIED WITH CIRCULATING MEDIUM FLOW

3.1. Bioreactor System as Regulator of Culture Conditions

The culture strategy for anchorage-dependent cells is classified into four categories in terms of medium flow and scaffold as shown in Figure 3. The bioreactor design is mainly required for culture of cells attached on scaffold under the medium flow conditions. The environment surroundings of cells consist of three phases; gas (O_2, CO_2), liquid (medium) and solid (cells, tissues, scaffolds), as illustrated in Figure 4.

Static conditions in Petri dishes or flasks can be used for cell proliferation and for seeding them on thin (<1 mm) polymer scaffolds. Thicker scaffolds require mixed (stirring) conditions to achieve seeding uniformity. Spinner flasks and rotating vessels can meet those conditions. The medium circulating bioreactor has been successfully used in production of cartilage and blood vessel. They can provide mechanical regulatory signals, such as directly applied compression or pulsatile flow. This type of bioreactors has been recently introduced into tissue engineering laboratories (Ratcliffe and Niklason, 2002). The bioreactors bring great advantages compared to standard static seeding conditions. Flow and mixing within the bioreactor can be controlled to enhance mass transfer of nutrients, gases, metabolites, and regulatory molecules, to regulate the size and structure of the reconstructive tissue. A closed bioreactor system that continuously supplies nutrients and automatically controls tissue culture parameters would be ideal for the cultivation of tissues for actual clinical use because the moderate conditions realize the desired proliferation and differentiation. Mass transfer plays a crucial role in the

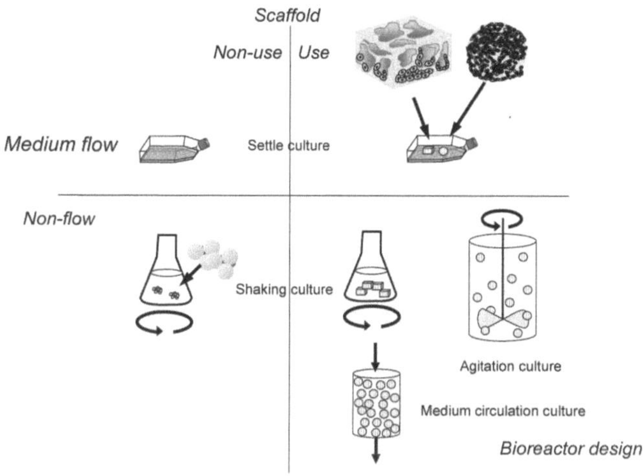

Figure 3: Classification of culture system for tissue production

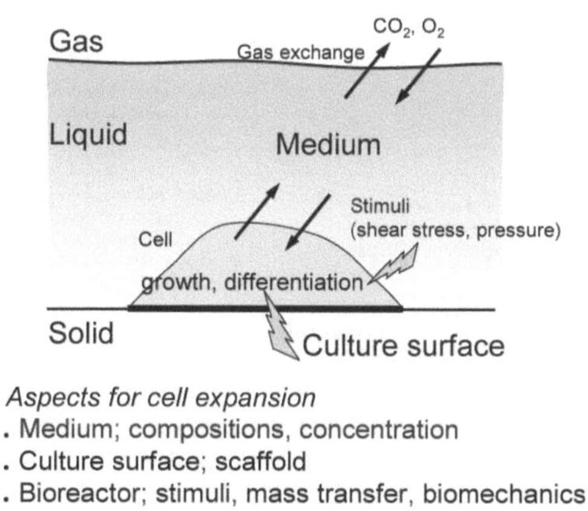

Figure 4: Phases and events appearing in tissue culture system

survival and function of tissues *in vivo* that are composed of cells having high metabolic activities and that are normally well vascularized. The application of vascularization to tissue-engineered constructs from these cells have been conducted, which requires adequate supply of nutrients and oxygen, removal of toxic wastes and end products, and release of intracellular metabolites. In addition, many of the tissues are required to have a mechanical function. However, the mechanical properties of many of these tissues have not been precisely defined. In addition, it is unclear which of the properties is important to use as design parameters for the engineered tissue replacements. Increasing evidence in biological studies indicates cells mediate the response of mechanical signals. Externally applied mechanical signals are regulators in the development and expression of functions of many tissues.

3.2. Bioreactor System as Integrator of Culture Tools

Bioreactor systems can be employed for many applications at various scales of operation in cell therapies and tissue engineering, including cell and tissue production, transduction on both patient-specific clinical-scale (*e.g.*, cartilage), and large-scale (*e.g.*, embryonic stem cells, skin grafts). Many different types of bioreactors, including homogeneous and heterogeneous systems, have been developed for these and other applications. In the manufacture of cultured tissues, as shown in Figure 5, bioreactor system leads to the stabilization of culture process and improved quality of products by prehension and assessment of cellular phenomena. Biochemical engineering can contribute to the system construction with integration of the techniques including manipulation (bioreactor), measurement (culture monitoring tool), and prediction (growth kinetic simulation). All integrated systems

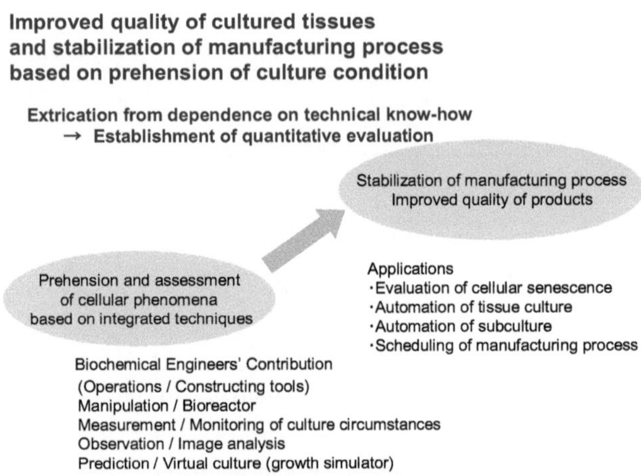

Figure 5: Contribution of chemical engineers to improved quality of cultured tissues

should be designed to meet a number of requirements.

3.2.1. Bioreactor

The manipulation of tissue culture has following requirements; (i) to control the physicochemical environment (e.g., dissolved oxygen (DO) conc., pH, pCO_2, shear rate); (ii) to ensure aseptic feeding and sampling; (iii) to make effective supply of expensive growth factors and medium components; (iv) to facilitate monitoring of cell and/or tissue "quality" (e.g., cell function, tissue structure);. (v) to ensure consistent cell and/or tissue quality in the face of donor cell variability. (vi) to use materials which are compatible with the cells and processing steps; (vii) to use automated processing steps to increase reproducibility of bioreactor operation. Suspension culture systems such as homogeneously agitated reactor are well documented for addressing many of the design requirements. Stirred reactors are ideally suited to maintaining a uniform physicochemical environment, and to obtaining representative cell samples. In addition, Freed *et al.* (Freed *et al.*,1993,1997,1998; Vunjak-Novakovic *et al.*, 2002) proposed agitated culture system which is employed to prepare tissue constructs such as bioartificial cartilage. These systems exhibit added complexity because cells in the interior of the construct are exposed to DO concentration lower than those at the construct surface. There will be gradients in DO conc., metabolites, and growth factors across the tissue constructs in these systems. Therefore, the potential effects of DO conc. on proliferation and maturation are important to design cultured tissues, and many tissue engineering applications require heterogeneous bioreactor systems with one or a few tissue constructs in each reactor chamber.

Depending on the medium circulation pattern and flow rate, there may also be substantial gradients across the reactor. These external gradients can be minimized by optimizing the flow distribution within the bioreactor chamber and by increasing the medium recirculation rate. The medium recirculation rate should be adjusted such that the conditions of pH, metabolite and growth factor concentrations, etc. at the exit of the chamber provide a similar growth and differentiation environment as those in vivo. However, the chamber design must be such that the required recirculation rate does not result in an adverse environment of high shear rate. Oxygen is normally limiting nutrient due to its low solubility in culture medium. Thus, one way to decrease the required recirculation rate is to provide high oxygen transport along the flow in reactor chamber.

3.2.2. Monitoring system

Measurement of the components in medium exiting a culture vessel can provide valuable information regarding a tissue construct. Metabolic activity may be used to estimate the cell concentration. However, metabolic patterns will vary with DO conc. across a tissue construct, and may also vary with cell differentiation. Tissue-specific metabolic activity may also be evaluated using medium samples. For example, bioartificial liver constructs can be evaluated in terms of albumin synthesis, ammonia clearance, urea synthesis, galactose elimination, and cytochrome P450 related hepatic oxygenase activity.

Evaluation of DO conc., pH gradients and flow rate in the local tissue structure encounters difficult problems. The promising methods using of fluorescence reagents and NMR are recently developed to obtain high-resolution measurements of interstitial pH, DO conc. and flow rate between adjacent blood vessels in tissue, using fluorescence ratio imaging and phosphorescence quenching microscopy (Dellian *et al.*, 1996; Helmlinger *et al.* 1997; Neves *et al.*, 2003).

3.2.3. Image analysis for cell observation

Animal cell morphology has been intensively researched in biological and medical fields since cultured cells were observed simply through a microscope. In recent years much attention has been paid to the relationship between the morphological changes and the intracellular phenomena of cells such as orientation of cytoskeleton and nucleus. The cellular morphology is considered to include fundamental information of cellular potential. Computer can be a tool for extracting images and viewing the motion of objects in a visual way, and accomplishes the conversion of optical images to digitized ones, which can be converted to the numeral variables or parameters expressing the time-elapsed changes of objects. In recent years, the computer-aided systems have applied to the analysis of cellular motions in mammalian cell cultures, although they are rather sophisticated and still less developed. Some researchers analyzed the phenotypic features of human cells as a sequence of intracellular reactions, and proposed parameters for characterizing the morphological and physiological conditions of cells (Kino-oka *et al.*, 2000; Hirai *et al.*, 2002; Umegaki *et al.*, 2002).

3.2 4. Growth kinetics and simulation for prediction and evaluation

Model predictions were presented and analyzed to study the complex dynamics of large cell populations (Miura and Shiota, 2000) and to determine how the observed proliferation rates are affected by the initial spatial cell distribution (Kino-oka *et al.*, 2002), the seeding density (Kino-oka *et al.*, 2000), and the characteristics of growth surface (Yashiki *et al.*, 2001). The development of cell populations of anchorage-dependent cells on surfaces poses interesting problems in modeling the sequences of operations including cell division and contact inhibition. The detailed understanding of the dynamics of whole cell population and the prediction of their behavior is very important since we often study intracellular functions and mechanisms. This would be especially advantageous for tissue constructs seeded with cells from patients or donors, because these constructs may exhibit substantial patient-to-patient variability. In such case of tailor-made process accompanied with inhomogeneity, instability and fluctuation in conditions of the raw materials, a feasible method to evaluate and predict culture states is needed to support the operators who have to judge the cessation of tissue cultures from information calculated by a kinetic model.

4. BIOREACTOR WITH CIRCULATING MEDIUM FLOW

4.1. Configuration of Bioreactor with Circulating Medium Flow

Tissue engineering represents a new and promising concept to create functional and viable tissue from autologous cells for surgical application. Culture of mammalian cells in monolayer on tissue plates or in suspension is a common technique in the field of cell biology and the investigation of pathogenesis of human disease. The bioreactor system allows control over *in vitro* culture conditions. A promising cell culture system is the circulating medium flow bioreactor in which cells are maintained. A large number of engineering design aspects must be considered to obtain fully functional tissues.

4.2. Bioreactor for Mechanical Environment

Tissue engineering approaches have used implanted cells, scaffolds, DNA, protein and/or protein fragments to replace or repair injured or diseased tissues and organs. Despite early successes, tissue engineers have faced challenges in repairing or replacing tissues that serve predominantly as biomechanical function in a mechanical environment, such as cartilage, bone, blood vessels, and heart valves. An evolving discipline called "functional tissue engineering" seeks to address these challenges.

One approach to tissue engineering is to develop *in vitro* conditions to ultimately fabricate functional cardiovascular structures prior to final implantation. Sodian *et al.* (2001, 2002) developed a new pulsatile flow system that provides biochemical and biomechanical signals to regulate autologous patch-tissue development *in vitro*. The bioreactor is connected to an air-driven respirator pump, and the cell culture medium continuously circulates through a closed-loop system. The mechanical stimulation by periodically stretching the tissue-engineered patch constructs were conducted by adjusting the stroke volume, the stroke rate, and the inspiration/expiration time of the ventilator. The newly developed bioreactor provides optimal biomechanical and biodynamical stimuli for controlled tissue development and *in vitro* conditioning of an autologous tissue-engineered patch.

The culture of articular cartilage includes the general principles of functional tissue engineering that may serve tissues for the replacement or repair of load-bearing structures of the body. Various physical and/or physicochemical factors influence the development, growth, and metabolic function of cartilage (Mow *et al.*, 1992). Dynamic compressive loading of cartilage disks introduces deformations and other changes within the tissue, such as hydrostatic pressure, interstitial fluid flow, and streaming potential (Murata *et al.*, 1998; Mizuno *et al.*, 2002).

The growth of chondrocytes on scaffolds has been augmented by the use of medium flow type bioreactors to increase transport of nutrients to and waste away from cells. Various designs have been aimed at increasing mass transfer rates, including shaker flasks and stirring bioreactors with magnetic impeller and rotating bioreactors. The limitations of such designs include unregulated and unknown fluid

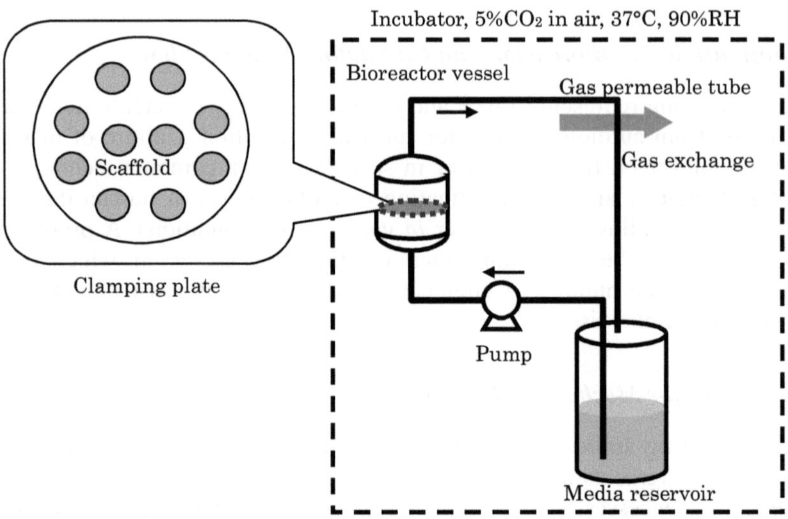

Figure 6: Schematic drawing of bioreactor system with medium circulating flow (Pazzano et al., 2000)

flows through the construct in culture. A fluid flow of 1 μm/s has been shown to accelerate cartilage extracellular matrix assembly. This phenomenon has not been exploited in the process of bioreactor culture. As shown in Figure 6, Pazzano *et al.* (2000) constructed the pressure/perfusion three dimensional culture system which is not regarded as a model for evaluation of physiology and pathophysiology for mature cartilage tissue, but it has distinct advantages for evaluating the regulation of anabolic and catabolic functions of cells in a deformation-free pressure environment. Externally applied hydrostatic pressure, oxygen tension, and interstitially generated osmotic and swelling pressures influence the metabolic function of cartilage disks (Davisson *et al.* 2002). Mizuno *et al.* reported a method of producing collagen sponges that supports the viability and activity of cells *in vitro*, including maintenance of chondrocytic phenotype (Mizuno *et al.*, 1998).

5. RADIAL FLOW BIOREACTOR SYSTEM

5.1. Oxygen Supply in Bioreactor

In general, the oxygen requirement of animal cells is not as high as that of microorganisms, and then in the suspension culture of animal cells, it is considered that the problem of oxygen supply in bioreactor is relatively easy to be solved because the positional homogeneity of oxygen supply can be feasible in bioreactor.

In the case of tissue culture, however, the spatial heterogeneity of animal cells in three dimensional structure is an obstacles to oxygen supply to the cells. Consequently, it is a reasonable choice that the cells are fixed on the supports of scaffolds and medium is forced to flow into or around the scaffolds. In this way of tissue culture, owing to the change in the cell density in bioreactor, the level of DO in scaffolds is varied.

Oxygen gradient and the distance at which oxygen can diffuse before decreasing to lethal levels depends upon the oxygen consumption rate per unit volume of tissue and the packing density of tissue. Thus, the problem is most serious under the conditions of high packing density of tissue. Theoretical analyses have been carried out concerning oxygen consumption and diffusion in rectilinear, cylindrical, and spherical geometries to determine how much tissue can be supported under defined conditions.

One promising system for high cell density culture is hollow fiber bioreactor. Gloeckner and Lemke (2001) have developed a miniaturized hollow-fiber bioreactor system for mammalian cell culture with a volume of 1 mL. Compartments holding cells and medium are separated by a semipermeable membrane installed in the bioreactor, and oxygenation of the cell compartment is accomplished using an oxygen-permeable membrane. By means of the geometry of the transparent housing, cells can be observed microscopically during culture. The leukemic cell lines were cultivated up to densities of 3.5×10^7cells/mL without medium change or manipulation of the cells. This new miniaturized hollow fiber bioreactor offers advantages in tissue engineering because it enables continuous nutrient supply for cells in high density, retention of added or autocrine-produced factors, and undisturbed long-term culture in a closed system.

5.2. Application of Radial Flow Bioreactor System

There are several systems to achieve a high cell density *in vitro*. A radial flow bioreactor for cultivating mammalian cells was developed to produce antibody by Kirin Brewery Co. in Japan (Yoshida *et al.*, 1993). Figure 7 shows the system of radial flow bioreactor (Able Co. in Japan) which is commercialized for cell cultures. Using recombinant CHO and BHK cells, the performance of the bioreactor was presented. Since medium flows radially across the bed, a supply of nutrients and oxygen with low shear stress can be achieved. The advantages of fixed bed bioreactor lie on the point that the system can be made compact, simple, and easy to operate. The medium feed rate was controlled so that the glucose was not depleted, and the circulation rate was controlled to keep DO at the exit from the reactor at above 1.0 ppm. The circulation rate reached 530 L/day, most of the cells being remained in the reactor. Then, the cell concentration in beds was 8×10^7cells/mL-beads. In interleukin-6 production using BHK cells, the radial flow reactor was scaled up to 5 L in volume (Yoshida *et al.*, 1998).

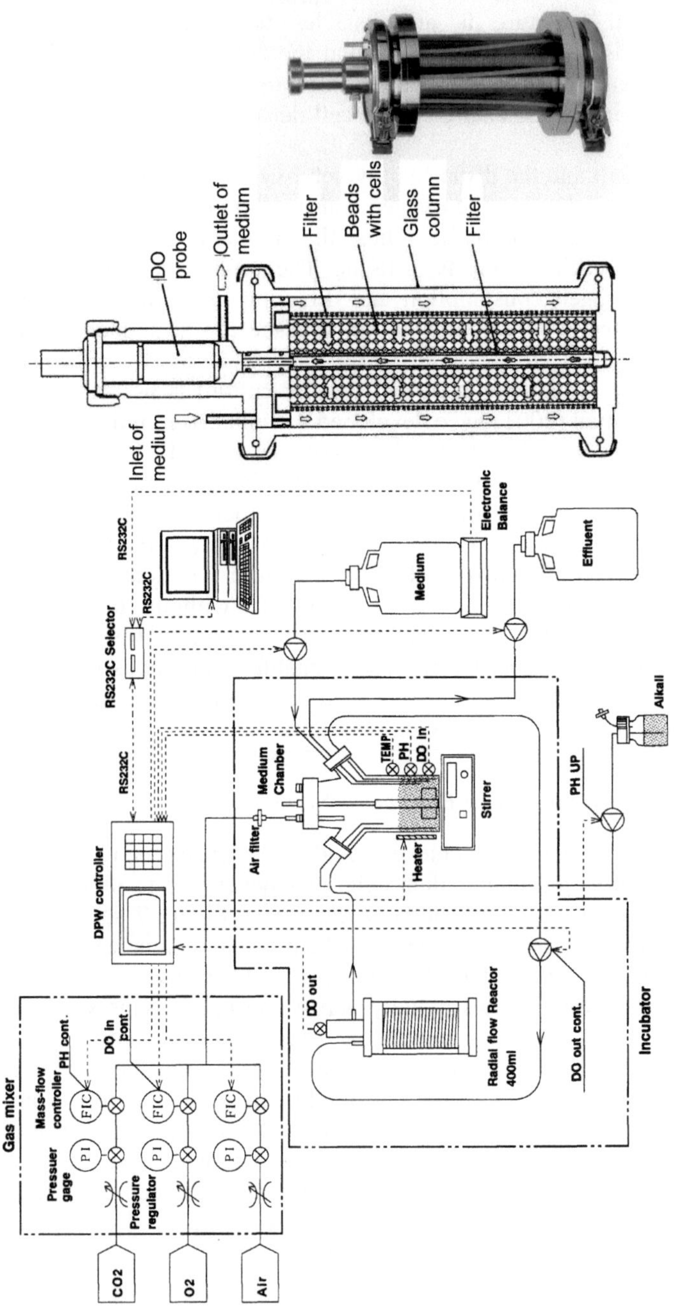

Figure 7: Schematic drawings of flow radial reactor system for antibody production

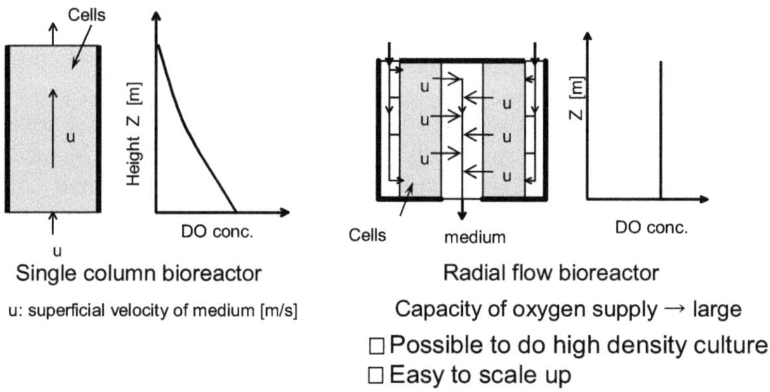

Figure 8: Conceptual illustration showing features of radial flow bioreactor

As shown in Figure 8, by considering the necessity of the high oxygen supply with low shear stress for bioreactor suitable to tissue culture, the construction strategy should be described as follows; (1) increase in specific cross-sectional area (A_C=A/V_g) which is the ratio of cross-sectional area to volume in growth unit, (2) increase in effective volume for growth (V_e= V_g/V) which is ratio of volume in growth unit to working volume. Supplying the required amount of oxygen while avoiding cell damage is a critical point in successful bioreactor design. Since the radial flow bioreactor has a large cross-sectional area and short flow path in the same bed volume as that of an axial packed-bed bioreactor, it has advantages in terms of a small pressure drop and low concentration gradient. The critical value of shear stress for cell damage of BHK cells is 1.0-5.0 N/m^2. The shear stress experienced by the cells in a packed-bed bioreactor can be determined as a shear stress per unit surface area of matrix, F_D/S, given by the following equation.

$$\frac{F_D}{S} = \frac{\Delta P}{L} \frac{d}{6(1-\varepsilon)} \tag{1}$$

where ΔP, L, d, and ε are pressure drop, length of flow path, diameter of beads, and void fraction, respectively.

In the case of laminar flow in a packed-bed bioreactor, the following Kozeny-Carman equation is derived.

$$q = \frac{\varepsilon^3}{kS^2(1-\varepsilon)^2} \frac{\Delta P}{\mu L} \tag{2}$$

where q and μ are superficial velocity of medium and viscosity of medium, respectively.

From Eqs. (1) and (2), the following equation is given.

$$q = \frac{\varepsilon^3 d}{6k(1-\varepsilon)\mu} \frac{F_D}{S} \tag{3}$$

When the inlet and outlet dissolved oxygen concentrations of the packed bed bioreactor are 20 and 2 ppm, respectively, and oxygen uptake rate (Q_{O2}) is 5 x10^{-10} mmol-O_2/(cell•h), the value of *L* is calculated to be 6.6 cm at cell density of 3 x10^8 cells/mL. Therefore, a radial flow bioreactor with a bed radius of 6.6 cm can supply oxygen to cells without cell damage caused by shear stress less than F_D/S = 1.0 N/m^2. The scale up in the direction of the axis of rotation has no limitation. Moreover, Bohmann *et al.* (1992, 1995) developed the radial flow bioreactor with membrane dialysis in which hybridoma cells were cultivated continuously over a period of 6 weeks. The dialysis membrane enabled the supply of low-molecular-weight nutrients and removal of toxic metabolites, while high-molecular-weight nutrients and products (e.g., monoclonal antibodies) are retained and accumulated. According to their report, antibodies was accumulated at an average of 100 mg L^{-1}, being approximately 10 times more than in fixed bed cultures without dialysis membrane.

The radial flow bioreactor system was applied to cultures of hepatocytes for construction of artificial liver (Kawada *et al.*, 1998; Morsiani *et al.*, 2001). Since the shear stress of this bioreactor is lower than the other type, high density culture without cell damage is possible. Thus, high density culture on a large-scale of human functional hepatoma was successful (Nagamori *et al.*, 2000), and the cells produced large amounts of human albumin and other liver specific proteins. In addition, Ledezma *et al.* (1999) presented analyses of the oxygen tension, shear stress, and pressure drop in a bioreactor for an artificial liver composed of plasma-perfused chambers containing porcine hepatocytes. Morsiani *et al.*(2000, 2002) indicated that the hepatocyte cells were metabolically active in the radial-flow bioreactor, leading to a potential use of this system for temporary extracorporeal liver support in acute hepatic failure patients.

Hongo (2003) in Able Co. (Tokyo, Japan) proposed the miniaturized radial flow bioreactor (Figure 9) which was fabricated to conduct the high density culture of hepatocyte cells. This bioreactor compromised of a cylindrical vessel with radial flow for cell growth and a medium-conditioning tank for oxygen enrichment in the medium. The medium subjected to oxygen enrichment in the tank was circulated between the chamber and tank using a tubing pump. The total volume of medium in the system was 30 mL (working volume). DO concentration in bulk liquid was measured with a DO probe installed in the tank. In the growth chamber, the cells were cultivated in a compartment of the growth chamber partitioned by two cylindrical meshes with a concentric arrangement (volume of growth chamber: 5 mL). The cells were inoculated uniformly by anchoring on porous beads (hydroxyapatite, particle size; 0.6-1.0 mm, porosity; 0.8) set in the growth chamber. During the culture the medium was introduced into the outer cylinder of the vessel in a radial flow manner and discharged from a port situated at the center of the top of the vessel. The superficial velocity in the chamber could be varied by the tube pump (ca. 7 mL/min).

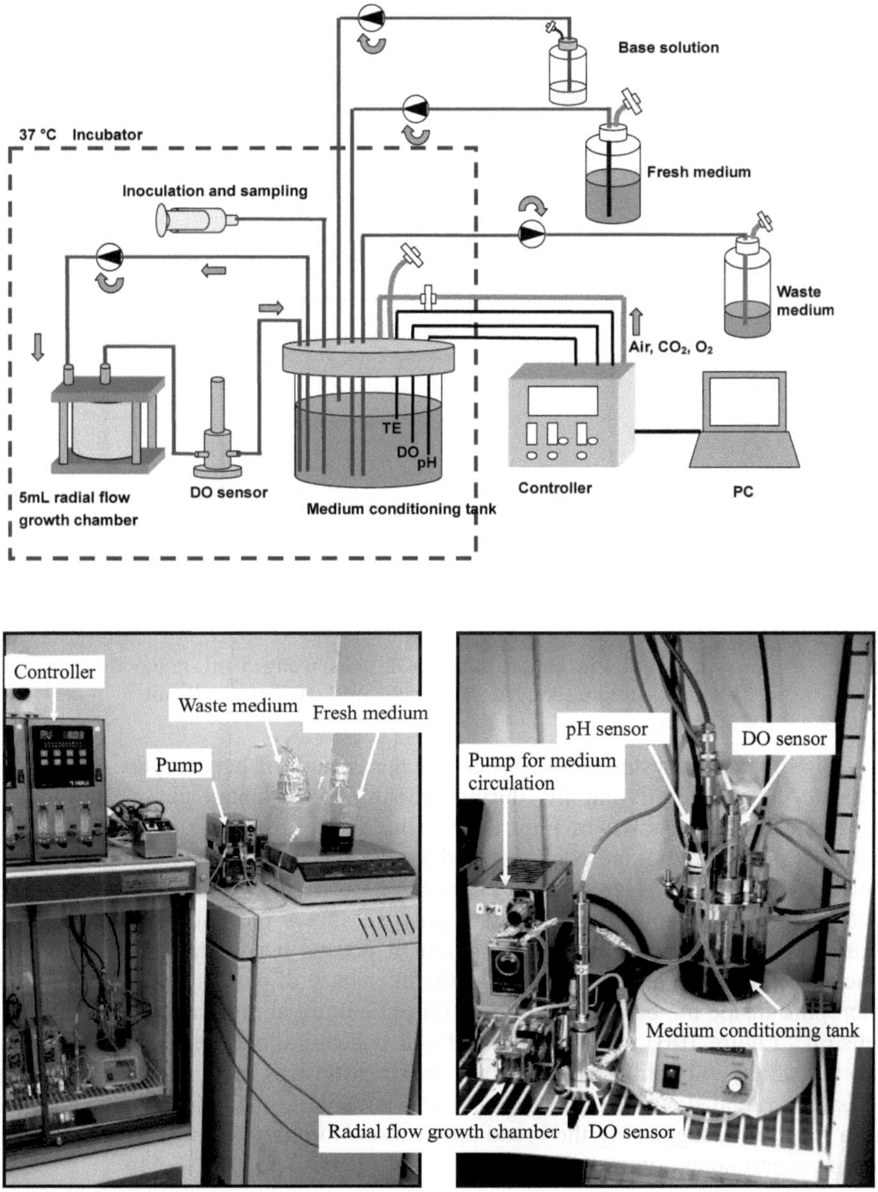

Figure 9: Schematic drawings of miniaturized radial flow reactor system for tissue cultures

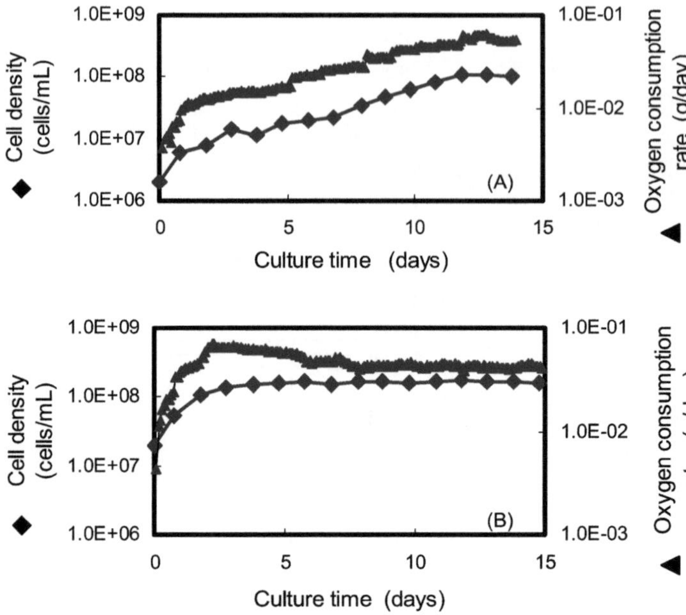

Figure 10: Growth profiles in culture of HepG2 cells using 5 mL-radial flow bioreactor system. Inoculum size; (A) $2x10^6$ cells/mL, (B) $2x10^7$ cells/mL

This bioreactor system was applied for the culture of hepatocytes (Hep-G2) to evaluate the proliferative ability during growth and stationary phases by changing inoculum size. As shown in Figure 10A, the active growth in the case of $2x10^6$ cells/mL of inoculum size was observed giving the final density of cells over 10^8 cells/mL for 12 days. Here, the cell concentration during culture was estimated by measuring oxygen consumption rate. Figure 11 shows the photographs of growth chamber at 12 days and the cross-section of porous beads stained with toluidine blue. The cells were observed to bridge the interspace of porous beads as well as the micropore inside the beads. In addition, during culture at high inoculum size ($2x10^7$ cells/mL, Figure 10B), the saturation of cell concentration occurred within 3 days and the stable cell activity without any deactivation was recognized until 15 days. Thus, this radial flow bioreactor system was suitable to maintain the artificial liver for a long term, suggesting the broad applications to pharmacology as alternative of animal experimentation.

6. SUMMARY

Cell culture is the most important operation to produce tissues endowed with desired functions and constructions. The approaches to induce the functions of tissues have been done by many researchers using various types of stimuli such as physical and chemical factors existing in solid, aqueous and gaseous phases. The bioreactors can prepare accommodative culture conditions, realizing that the well-ordered assemble of the cells allows the success of tissue reconstruction *in vitro*. We exemplified the circulating medium flow bioreactor as the typical regulating system of culture under enforced oxygen supply and mechanical stress. In addition, the radial flow bioreactor was superior to oxygen supply in high cell density culture under low shear stress, resulting from high level of cross-sectional area of medium flow. The precise design of bioreactor based on assessments of shear stress and oxygen supply facilitate the production of well designed tissues *in vitro*.

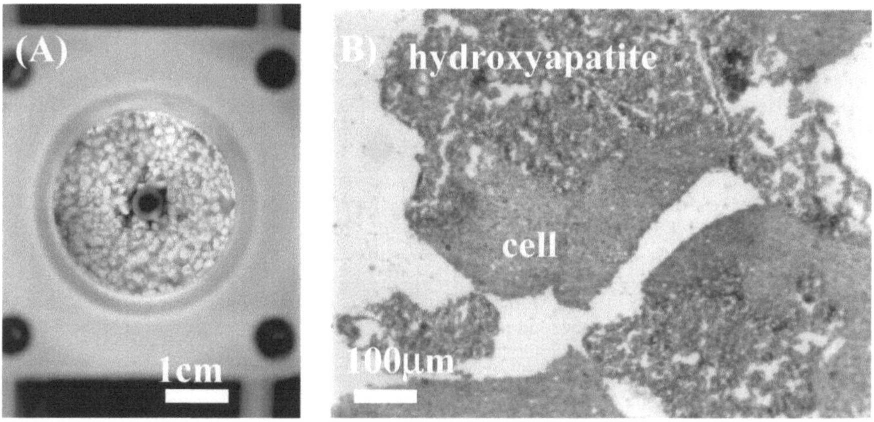

Figure 11: Photographs of (A) growth chamber and (B) the cross-section of porous beads stained with toluidine blue at 12 days in culture of HepG2 cells

7. ACKNOWLEDGEMENTS

The authors would like to thank Mr. Hongo for providing the data and photos used in this article. This work was conducted as one of the Multidisciplinary Research Laboratory System organized in the Graduate School of Engineering Science, Osaka University.

8. REFERENCES

Bohmann A, Pötner R, Mäkl H. 1995. Performance of a membrane-dialysis bioreactor with a radial-flow fixed bed for the cultivation of a hybridoma cell line. Appl Microbiol Biotechnol 43:772-780.

Bohmann A, Pötner R, Schmieding J, Kasche V, Mäkl H. 1992. The membrane dialysis bioreactor with integrated radial-flow fixed bed--a new approach for continuous cultivation of animal cells. Cytotechnol 9:51-57.

Davisson T, Sah RL, Ratcliffe A. 2002. Perfusion increases cell content and matrix synthesis in chondrocyte three dimensional cultures. Tissue Eng 8:807-816.

Dellian M, Helmlinger G, Yuan F, Jain RK. 1996. Fluorescence ratio imaging of interstitial pH in solid tumours: effect of glucose on spatial and temporal gradients British J Cancer 74: 1206-1215.
Freed LE, Vunjak-Novakovic G, Langer R. 1993. Cultivation of cell-polymer cartilage implants in bioreactors. J Cell Biochem 51:257-264.
Freed LE, Langer R, Martin I, Pellis NR, Vunjak-Novakovic G. 1997. Tissue engineering of cartilage in space. Proc Nat Acad Sci 94:13885-13890.
Freed LE, Hollander AP, Martin I, Barry JR, Langer R, Vunjak-Novakovic G. 1998. Chondrogenesis in a cell-polymer-bioreactor system. Exp Cell Res 240:58–65.
Gloeckner H, Lemke H-D. 2001. New miniaturized hollow–fiber bioreactor for in vivo like cell culture, cell expansion, and production of cell-derived products. Biotechnol Prog 17:828-831.
Helmlinger G, Yuan F, Dellian M, Jain RK. 1997. Interstitial pH and pO_2 gradients in solid tumors in vivo: high-resolution measurements reveal a lack of correlation. Nature Medicine 3:177-182.
Hirai H, Umegaki R, Kino-oka M, Taya M. 2002. Characterization of cellular motions through direct observation of individual cells at early stage in anchorage-dependent culture. J Biosci Bioeng 94:351-356.
Hongo T. 2003. Three-dimensional, high density cell culture in radial-flow bioreactor. The Cell 35:472-473 (in Japanese).
Kawada M, Nagamori S, Aizaki H, Fukuya K, Niiya M, Matsuura T, Sujino H, Hasumura AS, Yashida H, Mizutani S, Ikenaga H. 1998. Massive culture of human liver cancer cells in a newly developed radial flow bioreactor system: ultrafine structure of functionally enhanced hepatocarcinoma cell lines. InVitro Cell Dev Biol-Animal 34:109-115.
Kino-oka M, Prenosil JE. 2000. Development of on-line monitoring system of human keratinocyte growth by image analysis and its application to bioreactor culture. Biotechnol Bioeng 67:234-239.
Kino-oka M, Umegaki R, Taya M, Tone S, Prenosil JE. 2000. Valuation of growth parameters in monolayer keratinocyte culture based on a two-dimensional cell placement model. J Biosci Bioeng, 89:285-287.
Kino-oka M, Hara Y, Yashiki S, Taya M, Yamamoto T. 2002. Model development for spatial growth in culture of chondrocyte cells embedded in collagen gels. Tissue Engineering 5th International Meeting of the Tissue Engineering Society international 8:1172.
Lalan S, Pomerantseva I, Vacanti JP. 2001. Tissue engineering and its potential impact on surgery. World J Surg 25:1458-1466.
Lanza RP, Langer R, Vacanti J. 2000. Principles of Tissue Engineering. San Diego: Academic Press.
Ledezma GA, Folch A, Bhatia SN, Balis UJ, Yarmush ML, Toner M. 1999. Numerical model of fluid flow and oxygen transport in a radial-flow microchannel containing hepatocytes. J Biomechanical Eng 121:58-64.
Lodie TA, Blickarz CE, Devarakonda TJ, He C, Dash AB, Clarke J, Geleneck K, Shihabuddin L, Tubo R. 2002. Systematic analysis of reportedly distinct populations of multipotent bone marrow-derived stem cells reveals a lack of distinction. Tissue Eng 8:739-751.
Lysaght MJ, Reyes J. 2001. The growth of tissue engineering. Tissue Eng 7:485-491.
McIntire LV. 2003. WTEC panel report on tissue engineering. Tissue Eng 9:3-7.
Mayhew TA, Williams GR, Sebuca MA, Kuniholm G, du Moulin GC. 1998. Validation of a quality assurance program for autologous cultured chondrocyte implantation. Tissue Eng 4:325-334.
Miura T, Shiota K. 2000. Extracellular matrix environment influences chndrogenic pattern formation in limb bud micromass culture: experimental verificationof theoretical models. Anat Rec 258:100-107.
Mizuno S, Ushida T, Tateishi T, Glowacki J. 1998. Effect of physical stimulation on chondrogenesis in vitro. Mater Sci Eng C 6:301-306.
Mizuno S, Tateishi T, Ushida T, Glowacki J. 2002. Hydrostatic fluid pressure enhances matrix synthesis and accumulation by bovin chondrocytes in three- dimensional culture. J Cell Physiol 193:319-327.
Morsiani E, Galavotti D, Puviani AC, Valieri L, Brogli M, Tosatti S, Pazzi P, Azzena G. 2000. Radial flow bioreactor outperforms hollow-fiber modules as a perfusing culture system for primary porcine hepatocytes. Transplantation Proc 32:2715–2718.
Morsiani E, Brogli M, Galavotti D, Bellini T, Ricci D, Pazzi P, Puviani AC. 2001. Long-term expression of highly differentiated functions by isolated porcine hepatocytes perfused in a radial-flow bioreactor. Artificial Organs 25:740-748.
Morsiani E, Pazzi P, Puviani AC, Brogli M, Valieri L, Gorini P; Scoletta P, Marangoni E, Ragazzi R, Azzena G. 2002. Early experiences with a porcine hepatocyte-based bioartificial liver in acute hepatic failure patients. Inter J Artificial Organs 25:192-202.

du Moulin GC, Morohashi M. 2000. Development of a regulatory strategy for the cellular therapies: an American perspective. Mater Sci Eng C 13:15-17.
Murata T, Ushida T, Mizuno S, Tateishi T. 1998. Proteoglycan synthesis by chondrocytes cultured under hydrostatic pressure and perfusion. Mater Sci Eng C 6:297-300.
Nagamori S, Hasumura S, Matsuura T, Aizaki H, Kawada M. 2000. Developments in bioartificial liver research: concept, performance, and applications. J Gasteroentoerol 35:493-503.
Naughton GK. 2002. Critical issues in tissue engineering. Ann NY Acad Sci 961:372-385.
Neves A A, Medcalf N, Brindle K. 2003. Functional assessment of tissue engineered meniscal cartilage by magnetic resonance imaging and spectroscopy. Tissue Eng 9:51-62.
Pazzano D, Mercier KA, Moran JM, Fong SS, DiBiasio DD, Rulfs JX, Kohles SS, Bonassar LJ. 2000. Comparison of chondrogenesis in static and perfused bioreactor culture. Biotechnol Prog 16:893-896.
Ratcliffe A, Niklason LE. 2002. Bioreactors and bioprocessing for tissue engineering. Ann NY Acad Sci 961:210-215.
Sodian R, Lemke T, Fritsche C, Hoerstrup SP, Fu P, Potapov EV, Hausmann H, Hetzer R. 2002. Tissue-engineering bioreactors: a new combined cell-seeding and perfusion system for vascular tissue engineering. Tissue Eng 8:863-870.
Sodian R, Lemke T, Loebe M, Hoerstrup SP, Potapov EV, Hausmann H, Meyer R, Hetzer R. 2001 New pulsatile bioreactor for fabrication of tissue-engineered patches. J Biomed Mater Res Appli Biomater 58:402-405.
Sun W, Lal P. 2002. Recent development on computer aided tissue engineering – a review. Computer Methods and Programs in Biomedicine 67:85-103.
Umegaki R, Murai K, Kino-oka M, Taya M. 2002. Correlation of cellular life span with growth parameters observed in successive cultures of human keratinocytes. J Biosci Bioeng 94:231-236.
Vunjak-Novakovic G, Obradovic B, Martin I, Freed LE. 2002. Bioreactor studies of native and tissue engineered cartilage. Biorheology 39:259-268.
Yashiki S, Umegaki R, Kino-oka M, Taya M. 2001. Evaluation of attachment and growth of anchorage-dependent cells on culture surfaces with type I collagen coating. J Biosci Bioeng 92:385-388.
Yoshida H, Mizutani S Ikenaga H. 1993. Production of monoclonal antibodies with a radial-flow bioreactor. In: Kaminogawa S, Ametani A; Hachimura, S (Editors). Animal Cell Technology: Basic & Applied Aspects, volume 5. Dordrecht: Kluwer Academic Publishers. pp 347-353.
Yoshida H, Mizutani S, Ikenaga H. 1997. Scale-up of interleukin-6 production by BHK cells using a radial-flow reactor packed with porous glass beads. J Ferment Bioeng 84:279-281.

CHAPTER 6

CARTILAGE GROWTH IN MAGNETIC RESONANCE MICROSCOPY-COMPATIBLE HOLLOW FIBER BIOREACTORS

J.B. GRECO AND R.G. SPENCER

National Institutes of Health, National Institute on Aging, Baltimore, USA

1. INTRODUCTION

Repair of articular cartilage damaged by traumatic injury or degenerative joint disease represents an important therapeutic challenge. In spite of significant progress in understanding the pathogenesis of these highly prevalent processes, there are still no well-accepted disease-modifying interventions available (Hunziker, 2002).

Articular cartilage is an avascular tissue consisting of an extracellular matrix with low cellular density. These cartilage-specific cells, called chondrocytes, are highly specialized and metabolically active, and are responsible for the formation and maintenance of the extracellular matrix. Collagen, mostly in the form of type II collagen, is the major structural component of cartilage, and provides the tissue with tensile and shear strength. The second major macromolecular component of the extracellular matrix are proteoglycans, which consist of a protein core with glycosaminoglycan (GAG) side chains. The GAG molecules are negatively charged, and this large fixed charge density (FCD) provides compressive strength to the tissue.

The ability to reliably determine biochemical and biomechanical properties of articular hyaline cartilage through non-invasive methods will potentially have a tremendous impact on progress towards the treatment of osteoarthritis (OA). First, it will permit early diagnosis of the disease prior to the occurrence of irreversible end-stage changes and will increase the efficacy of established treatments incorporating lifestyle modifications. Such interventions, including weight loss and exercise, are effective if initiated before the development of end-stage disease (Bischoff and Roos, 2003). Second, it will permit the monitoring of emerging tissue engineering-

J.B. Chaudhuri & M. Al-Rubeai (eds.), Bioreactors for Tissue Engineering, 135 - 163.

based repair protocols, such as the implantation of autologous chondrocytes (Brittberg *et al.*, 1994), as well as the efficacy of pharmacological therapies (Kuritzky and Weaver, 2003).

An important step towards developing biological repair protocols is establishing an appropriate model system for engineered cartilage. The basic rationale behind the use of bioreactors for tissue engineering has been discussed at length elsewhere (Risbud and Sittinger, 2002; Darling and Athanasiou, 2003). The phenotypic instability of chondrocytes in monolayer culture is a significant concern in cartilage tissue engineering. In monolayer culture, chondrocytes typically dedifferentiate to a fibroblastic phenotype both morphologically and in terms of synthesis products, including initiation of type I collagen production and concomitant loss of the expression of type II collagen and aggrecan core protein (Schnabel *et al.*, 2002). In contrast, the bioreactor permits 3-D tissue growth, which supports the true hyaline cartilage phenotype. Furthermore, the macroscopic growth of cartilage in bioreactor systems permits cell-matrix interactions and incorporates the effects of the matrix barrier on substrate delivery and metabolic product efflux more realistically than do monolayer systems. Additionally, bioreactor systems provide great flexibility in growth conditions and geometric configurations. This flexibility extends to the composition of the tissue culture medium (TCM), including incorporation of bioactive agents such as pharmaceuticals and growth factors, the gas mixture provided, growth temperature, and, in some instances, humidity and pressure. Our studies of cartilage tissue engineering have utilized a homemade MRI-compatible hollow-fiber bioreactor (HFBR).

2. MRI MEASUREMENTS OF CARTILAGE

2.1. MRI Measurements Applicable to Cartilage in the HFBR

There are several reasons to incorporate magnetic resonance imaging (MRI) and nuclear magnetic resonance spectroscopy (NMR) studies into cartilage tissue engineering research. MRI is non-invasive and non-destructive, so that it can probe the interior of 3-dimensional tissue without disturbing tissue integrity and without deleterious effect on the sample. Thus, multiple MRI studies can be performed on a given sample, and other types of studies can also be performed on the same sample. In addition, serial studies at multiple time points on a sample are possible. Sample preparation for MRI studies is typically very straightforward, minimizing related artifacts. MRI images can be acquired with resolution on the order of tens of microns using high magnetic field systems. In addition, rather than being a single modality, MRI is in fact a collection of techniques which probe different aspects of tissue biophysics and biochemistry. Thus, depending on the details of the particular MRI experiment performed, information can be obtained on, among other things, hydration, stiffness, permeability, fluid flow, matrix composition including specific matrix molecule components, and cellular properties and viability. For these reasons, MRI studies of cartilage have achieved great importance in orthopedic medicine (Burstein *et al.*, 2000; Gray *et al.*, 2001; Burstein and Gray, 2003).

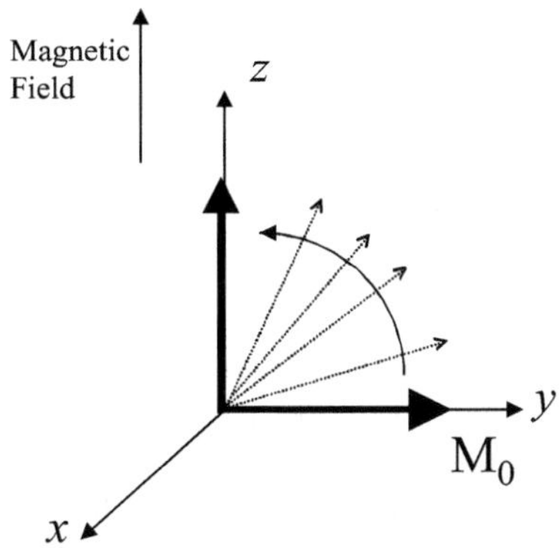

Figure 1: T_1 relaxation. The magnetization vector of length M_0, which has been tipped perpendicular to the magnetic field, relaxes back to its original position along the z axis according to the equation $M_z(t) = M_0(1 - \exp[-t/T_1])$. The curved arrow represents the motion of the magnetization vector during the relaxation time. The case $T_1=T_2$ is illustrated.

The use of MRI-compatible bioreactor systems for cellular studies is fairly well established (Gillies *et al.*, 1993; Evelhoch *et al.*, 2000; Pilatus *et al.*, 2000). However, comparable studies of engineered tissue have been rather limited. Besides our cartilage work, the bioartificial liver system has perhaps been the most extensively evaluated (Macdonald *et al.*, 1998; MacDonald *et al.*, 2001; Wolfe *et al.*, 2002), and recent work describes the use of MR to monitor a meniscal cartilage system (Neves *et al.*, 2003). The limited application of MRI to tissue engineering studies to date may be the result of certain inherent challenges. Among the most problematic of these are the size and construction constraints which result from the requirement of placing the sample within the MRI magnet. Thus, an essential element of our bioreactor system is its compatibility with MRI and NMR studies.

2.2. T_1 Measurements

To understand the type of information available through MRI studies of tissue, we will first briefly describe the basis for some of the imaging parameters used: T_1, T_2, diffusion, magnetization transfer, and gadolinium-enhanced imaging. The vast majority of MR imaging studies are sensitive to physical properties as they relate to tissue water.

T_1 can be described as follows. In the presence of a strong magnetic field, such as in an MRI magnet, tissue water becomes slightly magnetized. This initial magnetization can be visualized by imagining an arrow, or vector, representing the hydrogen nuclei of the water in the tissue, pointing in the direction of the magnetic field. See Figure 1. Part of the MRI procedure involves using a radio frequency electromagnetic pulse to reorient this tissue magnetization vector so that it is perpendicular to the magnetic field. After this re-orientation, the vector gradually moves back into alignment with the magnetic field. For this to happen, it has to recover its length in the direction of the magnetic field. The rate of this process follows an exponential law, with a time constant that has traditionally been called the longitudinal relaxation time, and denoted T_1 from the earliest days of magnetic resonance. The utility of this measurement lies in the fact that different tissues and different parts of a tissue can have distinct T_1 values. By using particular MRI methods, local T_1 values can be mapped and measured. The physics of this process is of less interest for our purposes than what has been learned about the biological significance of T_1 values. In particular, T_1 depends largely upon hydration—the more hydrated tissue is, the longer is its T_1 value; thus, CSF and urine, for example, have T_1 values in the range of 3 seconds or more. In contrast, fatty tissue tends to have a short T_1 value, on the order of 250 milliseconds. The T_1 for articular cartilage is about 1-2 seconds. While T_1 is related to hydration, it is also affected by other tissue characteristics, such as macromolecular content, as well as by the magnetic field strength of the MRI system, so that it has rather low specificity for a particular tissue state or pathology.

2.3. T_2 Measurements

To describe T_2, the so-called transverse relaxation time, we again consider the tissue magnetization vector induced by the external magnetic field of the MRI magnet. After this vector is reoriented perpendicular to the external field by the above-mentioned electromagnetic pulse, not only does it regain length in the direction of that field as described above in the context of T_1, but it also loses its component perpendicular to the magnetic field. This again occurs according to an exponential law, with a time constant called T_2. See Figure 2. As in the case of T_1, by using particular MRI methods, the T_2 value of tissue regions can be mapped and measured. T_2 also increases with increasing hydration. T_2 relaxation is driven by a greater range of molecular processes than is T_1 relaxation, so T_2 relaxation is faster and T_2 values are shorter than T_1 values. T_2 is particularly sensitive to the rate of motion of water molecules, so that freely moving fluid, such as in CSF or urine, tends to have a long T_2. In contrast, for tissue in which the water is less mobile, the T_2 value is much shorter. This is seen, for example, in dense tissue. T_2 values for typical tissues are in the range of 50-500 milliseconds, and approximately 50 milliseconds in articular cartilage (Lusse *et al.*, 2000). Its value has been found to increase in cartilage as one moves progressively from the subchondral bone to the synovial surface (Mosher *et al.*, 2000). While T_2-weighted images have proved useful for detecting focal lesions in cartilage (Bredella *et al.*, 1999), the many factors that influence T_2 have limited the ability of T_2 maps to probe specific biochemical changes.

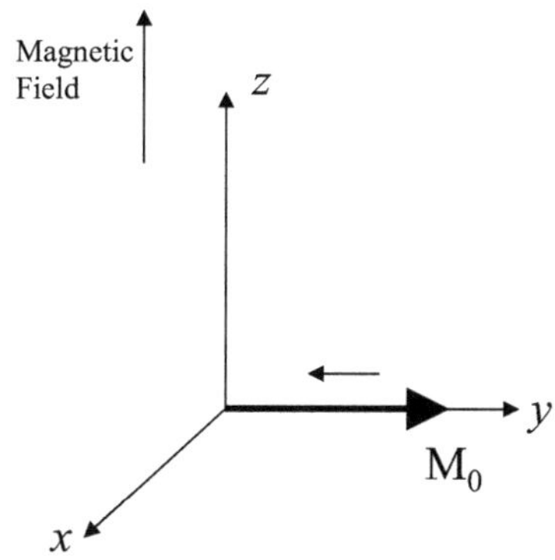

Figure 2: T_2 relaxation. After the magnetization vector is tipped, loss occurs perpendicular to the magnetic field according to $M_\perp(t)=M_0(0)\exp(-t/T_2)$.

2.4 Diffusion Measurements

Diffusion is a process by which molecules move randomly while undergoing frequent collisions with other molecules. If barriers in the form of fixed interfaces or structures, or in the form of less-mobile, typically larger, molecules, are present, then the diffusion rate of water will be accordingly slowed. Similarly, greater viscosity is associated with slower diffusion. The process of diffusion can be measured with MRI by use of special pulse sequences that are sensitive to microscopic motion, and which permit the operator to quantify the average motion of water molecules in a fixed amount of time. Initial studies suggested that the diffusion constant was increased in trypsin-degraded cartilage, and decreased in compressed cartilage compared to normal cartilage (Burstein *et al.*, 1993). Later experiments showed that increases in the diffusion constant were dependent on the enzymatic method of cartilage degradation, and suggested the change in the diffusion constant was due to microstructural changes in the cartilage matrix, rather than total GAG content (Xia *et al.*, 1995). It has been suggested that over short timescales, the diffusion constant is related to the water content of the cartilage, while over longer times, the diffusion constant is related to the structural properties of the tissue (Knauss *et al.*, 1999). Overall, the diffusion constant appears to increase in degraded as compared to intact cartilage, but there remains debate as to the biochemical basis for this effect (Mlynarik *et al.*, 2003).

2.5 Magnetization Transfer Experiments

Magnetization transfer (MT) is the name given to the technique in MRI in which the interplay between relatively bound and relatively free water is investigated. In general, the bound water is associated with macromolecules, while the free water is largely independent of macromolecules, that is, in the solution phase. The basis of the method can be understood in the following way. Bound and free water interact in a variety of ways in tissue, including by direct chemical exchange and through electromagnetic interactions between their nuclei. Thus, it stands to reason that if the magnetization of bound water is perturbed in some way, then this perturbation will be transferred to, and affect the magnetization of, free water. It is this free water which is mapped by MRI; the bound water is not directly detected. In the MT technique, the magnetization of bound water nuclei is saturated by introduction of a particular type of MRI pulse. This perturbation is transferred to free water, and the degree of that perturbation is monitored by standard MRI methods. From these MRI images, an MT value, defined as $1\text{-}M_{sat}/M_0$, where M_{sat} and M_0 are, respectively, amplitudes observed in the presence and absence of the perturbation pulse, and a magnetization transfer rate, k_m, can be calculated.

As the macromolecular content in the cartilage increases, a greater portion of the water in the tissue becomes bound to the macromolecules. The MT rate is believed to serve as a marker of collagen content in particular. This has been validated in phantoms, in degraded cartilage, and in goat knees (Laurent *et al.*, 2001).

2.6 Gadolinium Enhanced MR Measurements

Gadolinium (Gd)-enhanced MRI is used extensively throughout MRI imaging research and practice. It is based on the fact that gadolinium is paramagnetic, which results in accelerated T_1 relaxation of water molecules in the vicinity of gadolinium. Because of the manner in which MRI pulse sequences are constructed, this decrease in T_1 results in enhanced signal intensity. To make use of this fact, one of the several commercially available gadolinium-containing compounds is introduced, either by intravenous injection into an animal or human, or directly as a bath surrounding a tissue explant or other sample of interest. A Gd-containing compound may also be administered directly into the synovial space for certain cartilage studies in the intact joint. Administration of Gd causes increased brightness in the MR image in regions of the body or tissue which are accessible to the compound. In cartilage, the distribution of gadolinium in the tissue can be used to measure matrix FCD and hence proteoglycan content (Bashir *et al.*, 1996; Bashir *et al.*, 1997; Bashir *et al.*, 1999; Gillis *et al.*, 2001). This method, delayed gadolinium enhanced magnetic resonance imaging of cartilage (dGEMRIC), works as follows. The fixed charge in cartilage is associated with negatively charged groups on proteoglycans. This charge acts to repel the negatively charged gadolinium chelate, Gd-DTPA. In cartilage regions in which proteoglycans have been depleted, as in osteoarthritis, or, in *in vitro* experiments, by enzymatic digestion or other degradative interventions, there is a loss of FCD and hence less repulsion of Gd-DTPA. Therefore, these regions will have a higher local concentration of Gd-DTPA, leading to faster local T_1 relaxation, and increased image brightness. Calculations based on chemical and charge equilibrium have shown that the concentration of proteoglycans can be assessed quantitatively by this method.

2.7. Other MR Techniques in Cartilage Imaging

Other pulse sequences have been developed in an attempt to quantify proteoglycan content *in vivo* without the use of an injected contrast agent such as gadolinium. T_1 relaxation in the rotating frame of reference ($T_{1\rho}$) is another MR parameter which is sensitive to a certain range of molecular motion and can be mapped using the appropriate pulse sequence. It has been found to correlate with the proteoglycan content of degraded tissue samples (Regatte et al., 2002). Increased water content has been shown to occur early in OA, and so specific pulse sequences have been developed in connection with phantom calibration to determine the water content of cartilage (Shapiro et al., 2001).

While most clinical MR imaging experiments detect changes in the properties of tissue water, the MR technique is not limited to determining water properties. With specialized equipment and pulse sequences, it can be used to detect protons in molecules other than water, such as lactate. Indeed, lactate imaging has been used for investigating muscle metabolism (Asllani et al., 2001). Furthermore, MRI is not limited to detection of protons; other biologically relevant MRI-active nuclei include sodium, lithium, carbon and phosphorus. Phosphorus spectroscopy is a convenient method of probing cellular bioenergetics, via assessment of the concentrations of inorganic phosphate, creatine phosphate and ATP, as well as intracellular pH (Moon and Richards, 1973). Ionic sodium serves as a cation to balance part of the large negative charge of GAGs, and hence sodium images can be used to calculate the FCD, similar to the dGEMERIC method (Shapiro et al., 2002).

Note that with the MRI techniques described above, one can in principle describe the two main components of the cartilage matrix, namely proteoglycans (via Gd-DTPA enhanced imaging) and collagen (via MT), as well as water content (via T_1 and T_2). One would like to make use of these noninvasive measurements to characterize the matrix. However, for tissue properties to be determined on the basis of non-invasive MR measurements, further work is required to establish and validate correlations between MRI-derived parameters and gold-standard invasive biochemical measurements. This must be done rigorously under a wide variety of circumstances; it is by no means evident that correlations obtained in normal articular cartilage would hold in diseased tissue or in developing reparative tissue. Therefore, a central theme in our studies has been the establishment of detailed correlations among the results of MRI studies, biochemical assays, and mechanical testing.

2.8 Utility of High-Field Studies

The most common magnetic field for clinical MR instruments is 1.5 Tesla, which is roughly 30,000 times larger than the Earth's average magnetic field, and 5 times larger than the magnetic field of sunspots. However, there are a number of advantages to working at higher magnetic fields for studying tissue, usually based on the fact that the signal-to-noise ratio (SNR) is roughly proportional to field strength for biological samples. This greater SNR permits acquisition of images with improved spatial resolution, permitting greater detection and definition of structures, and more accurate area and volume measurements. Alternatively, for a given resolution and signal-to-noise ratio, images can be obtained more rapidly at higher magnetic field strengths. This is of obvious importance for patient studies, but speed of data acquisition is a significant concern even for tissue engineering studies. This is because high-resolution experiments using several MRI contrast mechanisms can require many hours of data acquisition. Another advantage of high field studies is specific to spectroscopic investigations. An NMR spectrum is comprised of a number of spectral lines, each one of which represents a nucleus in a particular chemical environment. At higher field, the separation between these spectral lines is greater, resulting in more accurate quantitation of the areas of these lines and hence of the concentrations of the metabolites they represent. In addition, similar considerations to those discussed above in the context of MRI regarding rate of data acquisition pertain to spectroscopic studies as well; at higher field, spectra of a given quality, or SNR, can be obtained more rapidly. Conversely, for a given experimental time, higher quality spectra can be acquired, leading to more accurate quantitation. However, to avoid undue expense, high field magnets usually have a small bore compared with clinical systems, limiting the size of samples which can be studied.

3. DESIGN OF AN MRI-COMPATIBLE BIOREACTOR SYSTEM

3.1 MRI-Compatibility

To make use of the capabilities of MR in tissue analysis, the bioreactor system must satisfy certain design criteria. The first and most obvious is size; tissue engineering studies are best performed at high field, and hence in relatively small-bore instruments. For example, in our laboratory, we have used a 9.4 Tesla MRI system to acquire data on bioreactor cartilage. This magnet has a bore size of only 10.4 cm, and in addition to the sample itself, the gradient coils and the radio frequency probe must fit within this diameter. The result is that the diameter of the sample to be imaged must be no greater than 2.5 cm.

A further design requirement relates to the fact that the magnetic forces due to the MRI field on magnetic materials are extremely large, so that no magnetic metals can be used in the construction of the HFBR, perfusion system, or any other component that will be within or even near the magnet. Indeed, the lore of MRI includes many examples of ferromagnetic materials being attracted into the main bore of a MRI magnet. Removal requires a winch or, if that approach fails, powering-down the magnetic field. This is no small task, as it requires quite a bit of cryogenic engineering to remove the liquid nitrogen and liquid helium cooling materials used to maintain the magnet windings at superconducting temperatures without damaging the magnet.

Even non-magnetic metals such as aluminum and titanium cause large distortions in magnetic fields. The effect of this is to distort both the geometry and the intensity of the image in the vicinity of the metal. Nonetheless, provided that this non-magnetic metal is kept a specified distance away from the sample to be imaged, such metals can be used in constructing apparatus for use inside a magnet. In fact MRI sample holders are frequently made of metal, and the signal receiving and transmitting circuits which must be located within the magnet have to date always been made of metal.

An additional requirement is the need to use construction materials that do not provide a signal to the MRI system. Because standard MRI studies image mobile protons, such as are found in water, this is generally not a problem. The situation is somewhat different for spectroscopic studies, in which moieties other than water are typically of interest. In these studies, the signal received is interpreted to indicate the presence and amount of a particular molecular species in the sample. Therefore, it must be ensured that the signal originates from the sample itself, and not from any other part of the HFBR or ancillary equipment.

There are other design requirements which are consequences of the above. The limited access available within the bore of an MRI magnet dictates extreme economy in design. Control over temperature, atmosphere, and perfusion may be regarded as the minimum for tissue experiments. Temperature control is greatly simplified in small-bore vertical magnets by the variable temperature device built in to virtually all such instruments. Atmospheric control can be achieved in a number of ways. We use a home-designed system that involves producing the appropriate gas mixture outside of the magnet, and then directing this mixture into the magnet via gas-impermeable tubing. The sample is flooded continuously with this mixture. Perfusion in the system we are using is a simple matter of a single inflow tube leading into the HFBR from an external reservoir, and a single outflow tube leading back into the reservoir. Even this simple tubing arrangement requires a great deal of care so that it will fit inside the bore of the magnet. Further, the perfusion system must permit repetitive placement into, and removal of the HFBR from, the magnet over the duration of the culture period, often several weeks, without compromising sterility.

3.2 Preparation of Chick Chondrocytes and Seeding of the Bioreactor

Our basic tissue model to date has been the chick sternum. Cells are obtained by collagenase digestion of the sterna of 16 day old chick embryos as previously described (Horton and Hassell, 1986; Gerstenfeld and Landis, 1991). Seeding of the bioreactors is illustrated schematically in Figure 3. After the embryos are decapitated, the entire breast plate is excised. Noncartilaginous tissue is dissected away, and, for most experiments, the proximal and distal thirds of the sterna are then removed and handled separately. Sterna are placed in phosphate-buffered saline (PBS) and undergo a digest with 10 mL of a 3.6 mg/mL collagenase solution for 20 minutes. Production of isolated cells is accomplished by digestion in 15 mL of collagenase with agitation for 3 hours. The suspension is centrifuged for 5 minutes at 660 g, and the pellet is re-suspended in tissue culture medium. Four dozen sterna typically yield approximately 96 million cells by this procedure. On the order of 10 to 30 million cells are inoculated into the HFBR through the side port. Perfusion

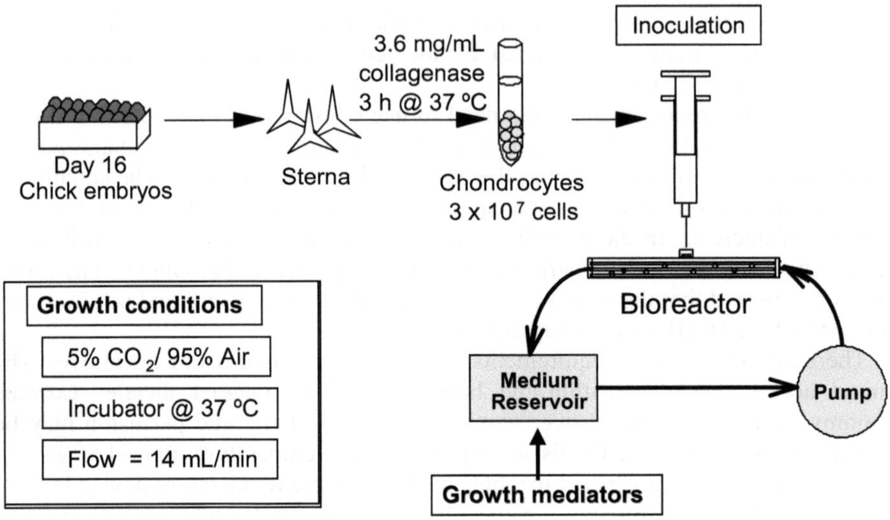

Figure 3: Preparation and seeding of the bioreactors. Sterna are isolated from day 16 chick embryos and digested in collagenase. The isolated chondrocytes are then inoculated through the side port of the bioreactor, and flow of tissue culture medium through the bioreactor is gradually increased to 14 mL/min.

flow is initiated at a rate of 5 mL/minute, and then gradually increased to 14 mL/minute over the following 10 days.

Our TCM consists of Dulbecco's modified Eagle's medium with phenol red (DMEM), augmented with heat-inactivated fetal bovine serum at 2% or 10%, and appropriate antibiotics. The TCM is sterilized through a 0.2 micron cellulose filter. Ascorbic acid is added to the TCM to a final concentration of 10 micrograms/mL upon each change of the TCM, typically every 48 hours.

The bioreactors themselves (Figure 4) are constructed from high-purity glass tubing. While we have used a variety of configurations, we most commonly use tubing of outside diameter 5 mm and inside diameter 3.5 mm. Porous polypropylene hollow fibers are inserted along the length of the bioreactor, and potted at the ends with silicon rubber adhesive. The fibers have an inside diameter of 330 microns, a pore diameter of 0.2 microns, and a wall thickness of 150 microns. Typically, six such fibers are potted within each bioreactor. After construction, the reactors are flushed with ethanol to enhance the wettability of the fibers. The perfusion system circulates TCM into the bioreactor circuit as described above. The circuit consists of, in addition to the bioreactor, a temperature-controlled 100-mL reservoir bottle, a pin compression pump, and an adequate length of size 14 gas permeable silicon tubing to permit placement of the HFBR into the MRI microscopy system. The perfusion circuit and bioreactor are filled with water and autoclaved prior to inoculation with cells. Tissue growth takes place within an incubator gassed with a mixture of 95% air and 5% CO_2.

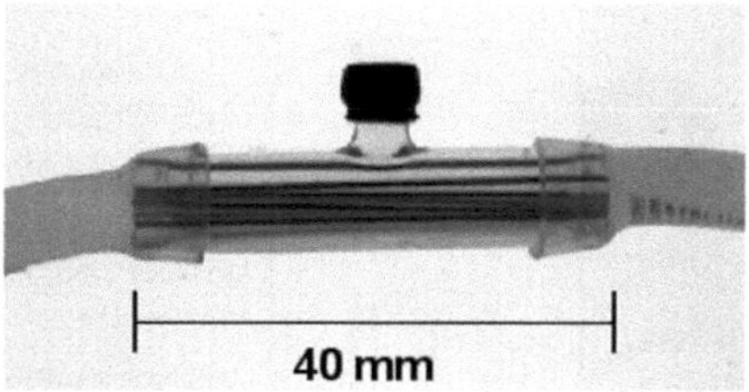

Figure 4: Silicon glass bioreactor (Petersen et al., 2000). The rubber septum, which covers the protruding side port, is used for inoculation of the chondrocytes. Figure reprinted from Magnetic Resonance in Medicine © 2000 (Wiley-Liss, Inc.).

3.2 Imaging of the Bioreactor

For MR imaging, each bioreactor is removed from the incubator and mounted into a homebuilt MRI-compatible incubation chamber that is continuously flushed with a gas mixture of 95% air and 5% CO_2. The bioreactor, with all perfusion connections intact, is then inserted into a vertical-bore DMX 400 MHz MRI system (Bruker Medizintechnik, Ettlingen, Germany) equipped with a 9.4 T magnet (Magnex Scientific, Abingdon, UK) equipped with a commercial microimaging probe capable of generating up to 100 G/cm gradients along all three spatial directions. The temperature of the bioreactor is maintained at 37° using the MRI system's built-in heater. The perfusion reservoir is maintained outside the magnet at 37° by a water bath (Figure 5). A metallic hose between the magnet and the pin compression perfusion pump and perfusion reservoir provides insulation for the entire system, and is used to maintain the desired CO_2 partial pressure throughout the system. With this design, the full range of MRI and NMR spectroscopy experiments can be performed.

4. MRI STUDIES OF HFBR-DERIVED CARTILAGE

4.1 Studies of Basic Cartilage Development

As noted above, one of the limitations of traditional chondrocyte culture studies is the dedifferentiation of chondrocytes and the generation of fibrous tissue rather than hyaline cartilage. Therefore, it is important to confirm that the cartilage grown from chick sternal cells in the HFBR develops and maintains the hyaline phenotype. The defining element of the hyaline cartilage phenotype is ongoing production of type II collagen. We found that bioreactor-derived neocartilage expresses mRNA for type II collagen as determined by dot-blot analysis throughout the growth period (Figure 6) (Potter *et al.*, 1998). To evaluate the cartilage phenotype histologically, 5-micron sections were prepared and stained with Alcian blue, a metachromatic dye that stains

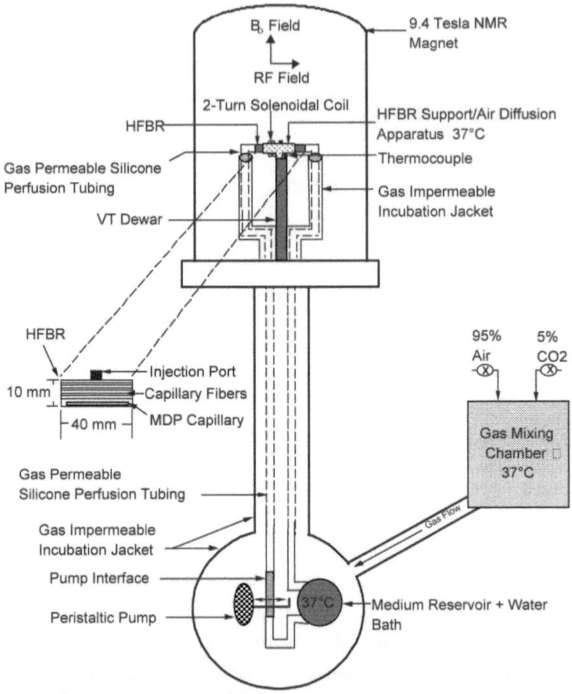

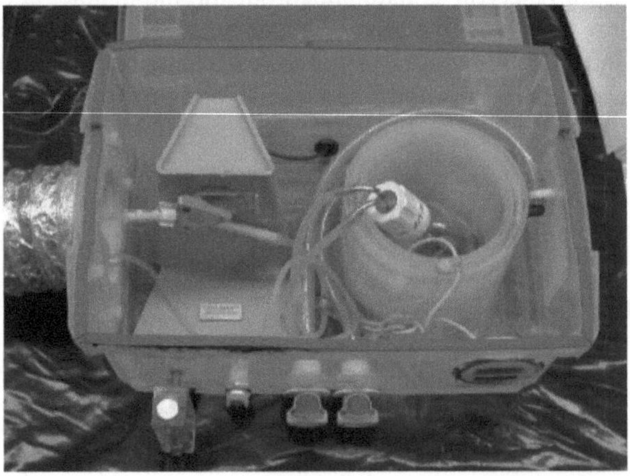

Figure 5: Imaging of the HFBR (Petersen et al., 2000). The HFBR is inserted into the magnet of the MRI system. The perfusion bottle and pin compression pump are located outside the magnet and connected to the HFBR via flexible tubing. The photograph depicts the bottle and pump. The metal vent tubing leading to the magnet is seen on the left side. Adapted from Magnetic Resonance in Medicine © 2000 (Wiley-Liss, Inc.).

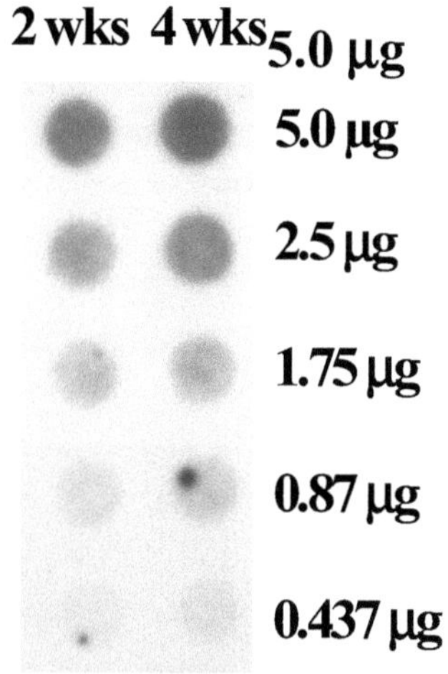

Figure 6: Dot blot analysis demonstrating expression of mRNA for Type II collagen (Potter et al., 1998). There was an increased expression of collagen II mRNA between two and four weeks of culture. Dilutions are as indicated. Figure reprinted from Matrix Biology © 1998 (Gustav Fischer Verlag).

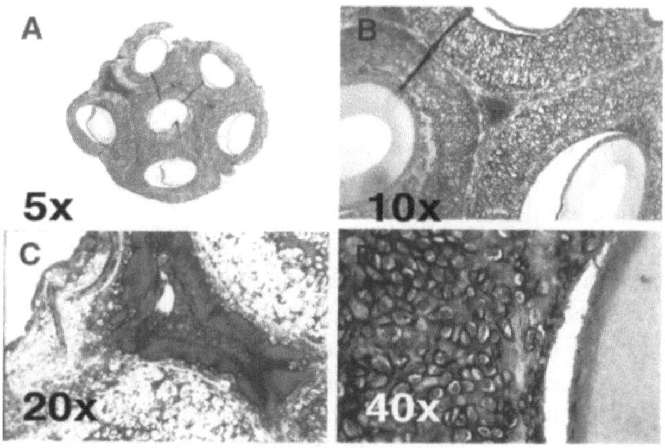

Figure 7: Alcian blue cross-sectional slices (5μ) of neocartilage grown in the HFBR (Petersen 1997). The arrangement of the six hollow fibers is clearly seen in panel A, and the growth units surrounding each fiber are evident in panels A and B. A prominent site of matrix degradation was seen at the periphery of the bioreactor (light staining), as was a proteoglycan deposit (dark staining), as shown in panel C. The chondrocytes in their lacunae are clearly visualized at higher magnification, as shown in panel D. Figure reprinted from International Journal of Imaging Systems and Technology © 1997 (John Wiley & Sons, Inc.).

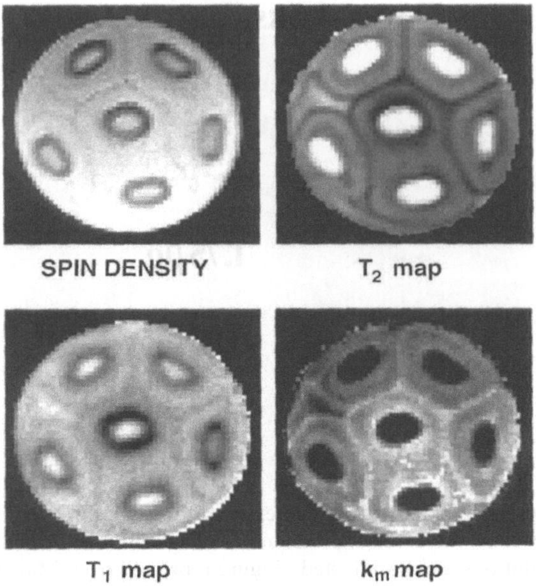

Figure 8: MR parameter maps of bioreactor-derived cartilage (Petersen et al., 1997). These MR maps of the spin density, T_2, T_1, and k_m correspond to the histological sections seen in Figure 7. Figure reprinted from International Journal of Imaging Systems and Technology © 1997 (John Wiley & Sons, Inc.).

purple for proteoglycan molecules. The Alcian blue stained sections of bioreactor-grown neocartilage revealed dense purple staining and rounded chondrocytes surrounded by their characteristic lacunae (Petersen *et al.*, 1997) (Figure 7), as expected for hyaline cartilage.

The histologic features of the tissue after 4 weeks of growth were consistent with evaluation of the tissue by MRI. This was demonstrated by comparing Alcian blue stained sections, as shown in Figure 7, to corresponding maps of MRI parameters, as shown in Figure 8. Histology from this particular sample demonstrated that discrete growth units surround each hollow fiber (Petersen *et al.*, 1997) often bordered by a stromal layer. Note that the absence of such a stromal layer between the central and lowermost of the fibers is evident both in the histologic section and in the MR images. A particularly dense proteoglycan deposit is readily visualized in both the section and the MRI images at approximately 10 o'clock. Note in particular that different MRI contrast parameters highlight these features differently. Limited contrast is seen in the proton density and T_1 images, while the T_2 image shows structural details very well. The k_m image highlights the stromal layers in particular, and is relatively insensitive to the proteoglycan deposit. These findings correspond to the known relative uniformity of water density in biological tissues, to the often limited sensitivity of T_1, as compared with T_2, in delineating tissue features, and

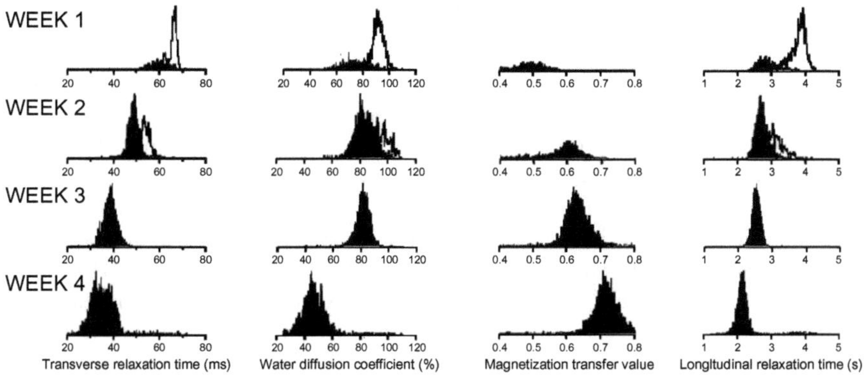

Figure 9. Histograms demonstrating changes in MR parameters during four weeks of tissue growth (Potter et al., 1998). The parameters quantified are transverse relaxation, water diffusion, magnetization transfer and longitudinal relaxation. The solid histograms represent the values obtained from the water in the neocartilage, and the open histograms represent the values obtained for the water in the fiber lumen and the extracapillary space. Figure reprinted from Matrix Biology © 1998 (Gustav Fischer Verlag).

finally to the sensitivity of magnetization transfer to the collagen component of cartilage.

The development over time of neocartilage in bioreactors was evaluated serially by both Alcian blue staining and MR imaging. Tissue growth was initiated around the hollow fibers. After two weeks of growth, the tissue nearly filled the extracapillary space, and by three weeks the tissue appeared as an opaque mass. No further gross morphological changes were visible between weeks three and four. Consistent with these observations, there was an increase in the wet and dry weight of the tissue, and a decrease in hydration, between weeks 2 and 4. Matrix deposition continued throughout the four weeks of tissue growth, with increasing GAG and collagen contents (Potter *etb* ., 1998).

These findings are consistent with those obtained using MRI microscopy (Potter *etb* ., 1998; Potter *etb* , 1998). The MT images demonstrated heterogeneity in the collagen distribution in the tissue; such spatially localized analysis is not readily available through standard biochemical assays. The transverse and longitudinal relaxation times were also observed to decrease during the four-week period, consistent with the biochemical observation of decreasing hydration as the tissue matures. There was also a substantial drop in the water diffusion coefficient between weeks 3 and 4 of growth, suggesting that the movement of water was becoming limited. This presumably reflects increasing macromolecular content. Histograms demonstrating the changes that occur in the MR maps are shown in Figure 9.

4.2 *Modification of Tissue Growth and Correlations Between MRI and Biochemical Assays*

The cartilage growth in the HFBR can be modified by introducing biologically active compounds. This is useful both for evaluation of the effect of an agent on tissue growth, as well as for extending the range and applicability of correlations between MRI measures and biochemical assays.

As examples of this paradigm, developing cartilage in the bioreactor was treated with retinoic acid and interleukin-1, and with increased doses of ascorbic acid. Retinoic acid has been shown to inhibit cartilage matrix synthesis (Horton *et al.*, 1987) and promote matrix degradation by a mechanism which is independent of matrix metalloproteinases. Consistent with this, treatment of cartilage in the HFBR with retinoic acid throughout week four of a four-week growth period led to a reduction in tissue wet and dry weights, and resulted in formation of tissue which was nearly devoid of GAG. Interleukin-1 is an inflammatory cytokine that leads to matrix degradation and suppresses chondrocyte differentiation (Lotz, 2001). It is thought to be of central importance in the early stages of osteoarthritis. Bioreactors treated with interleukin-1 throughout the final week of a four week growth period produced cartilage that was substantially different from that obtained from the control bioreactors. We observed trends towards decreased chondrocyte proliferation and an increase in proteoglycan production on a per cell basis, as well as enlargement of cell lacunae (Potter *et al.*, 2000). Ascorbic acid is known to

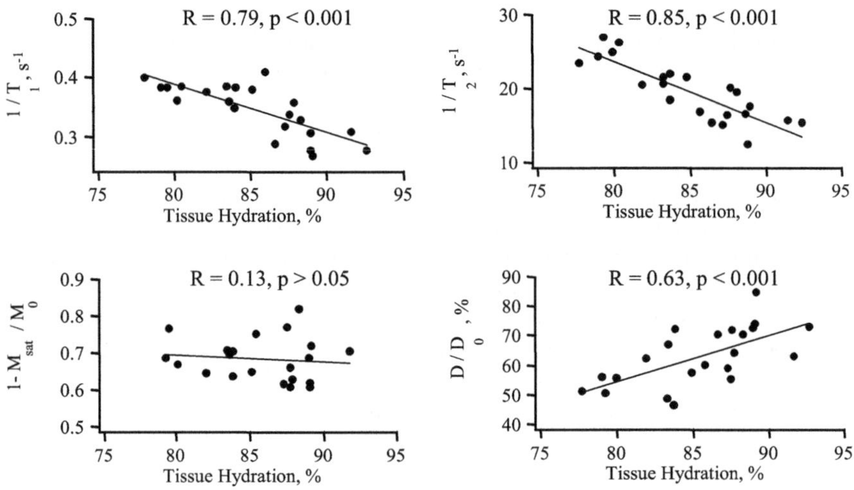

Figure 10: Correlations between MR-derived quantities and tissue hydration (Potter et al., 2000). Each data point represents a separate bioreactor. The hydration of the tissue was found to correlate with the T_1 and T_2 relaxation times and the diffusion coefficient, but not with the magnetization transfer rate constant ($1\text{-}M_{sat}/M_0$). Figure reprinted from Arthritis and Rheumatism © 2000 (American College of Rheumatology).

upregulate collagen synthesis (Clark *et al.*, 2002), and it has been suggested that high intake of vitamin C may slow the progression of osteoarthritic disease, (McAlindon *et al.*, 1996). In the HFBR system, high ascorbic acid dosing accelerated the formation of cartilage and increased the collagen content of the mature tissue.

There is a great deal of current interest in determining the biochemical composition of the extracellular matrix using cartilage MR techniques. Relationships between MRI-derived parameters and hydration (Figure 10) and biochemically-determined matrix composition (Figure 11) were derived for control bioreactors and bioreactors exposed to pharmacologic agents. The MRI parameters that were examined include the longitudinal relaxation rate ($1/T_1$), the transverse relaxation rate ($1/T_2$), the magnetization transfer parameter (MT), and the water diffusion coefficient (D). It was found that there was an inverse correlation between longitudinal and transverse relaxation rates and tissue hydration. These findings are consistent with the general understanding of the relationship between T_1 and T_2 relaxation and tissue hydration. Further consistent with our understanding of MRI parameters, the water diffusion coefficient was directly correlated with tissue hydration, and the MT value was unrelated to tissue hydration. There was a direct correlation between the GAG content and both longitudinal and transverse relaxation rates. The GAG content was not related to the MT constant, but it was inversely proportional to the diffusion coefficient. The collagen content was directly related to the longitudinal and transverse relaxation rates and the MT constant. There was no relationship between collagen content and diffusion constant.

4.3 Correlations Between MRI, Biochemical, and Biomechanical Outcome Measures

While there is a substantial literature relating MR parameters to the biochemical composition of cartilage matrix, and relating biochemical to biomechanical properties of the matrix, there is less work relating MR-derived properties of tissue to biomechanical properties. In order to assess the functional characteristics of tissue *in vivo*, it is this relationship which must be established

To pursue this, bioreactor-derived neocartilage was sampled at 1.5, 2.8 and 4.2 weeks of growth (Chen *et al.*, 2003) under control conditions. Increasing culture time led to an increase in tissue cross-sectional area, GAG content as a function of dry weight, equilibrium modulus, and transverse relaxation rate. Furthermore, it was noted that in mature bioreactors, there was a decrease in GAG content from the inflow side of the bioreactor toward the outflow side.

Treatment of cartilage with chondroitinase leads to loss of matrix proteoglycans (Yamada *et al.*, 2001) and serves as a model for *in vivo* cartilage degradation. Accordingly, an additional group of chondrocyte-seeded HFBRs was allowed to develop for 10 days with control medium, treated for 14 days with chondroitinase-ABC treated medium, and then allowed to grow for an additional 7 days in control medium. There was a decrease in tissue cross-sectional area, GAG content, FCD, dynamic modulus and equilibrium modulus in the chondroitinase treated bioreactors as compared with controls.

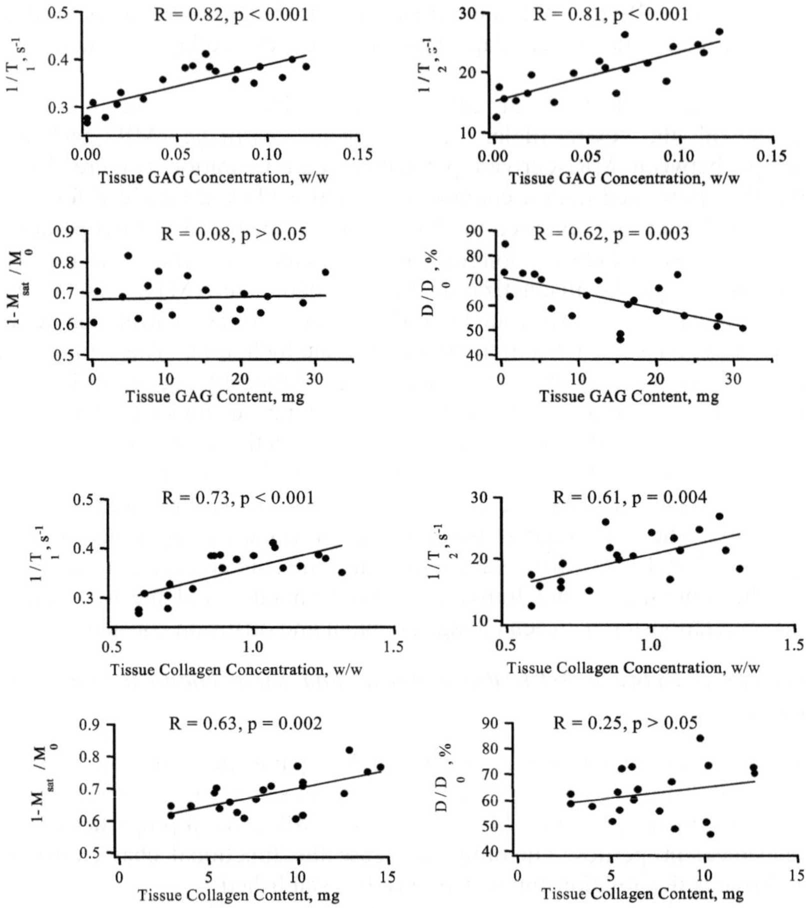

Figure 11: Correlation between MR and biochemical properties (Potter et al., 2000). Each data point represents a separate bioreactor. GAG content was correlated with the relaxation times and the diffusion coefficient, but not with MT. The collagen content was correlated with the relaxation times and MT, but not with the diffusion coefficient. MT appears to be a relatively specific marker for collagen content. Figure reprinted from Arthritis and Rheumatism © 2000 (American College of Rheumatology).

Corresponding tissue regions were used to correlate MRI, biochemical and biomechanical properties (Figure 12). Consistent with observations in tissue explants, the MRI-derived FCD measured by the dGEMRIC method was correlated with the GAG content, but not the hydroxyproline content. Further, the GAG content was directly correlated with dynamic modulus. Consistent with both of these observations, the FCD measured by dGEMERIC was correlated with the dynamic modulus of the tissue. This correlation between the functional properties of bioengineered tissue and non-invasive MRI measurements is important in

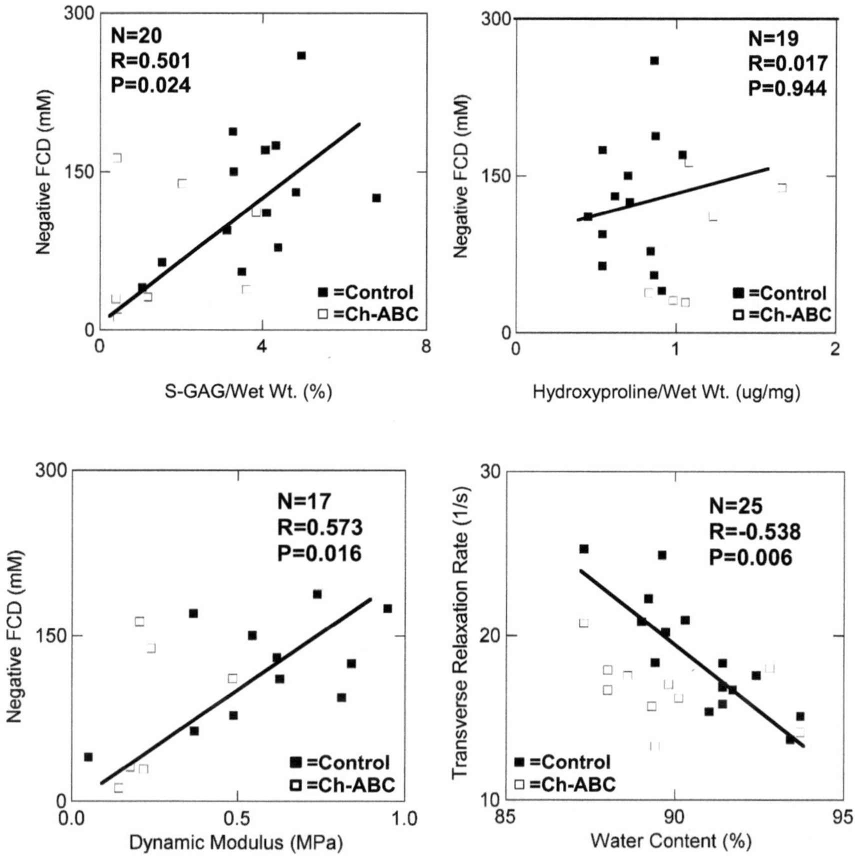

Figure 12: Relationship between biochemical, biomechanical and MRI-derived parameters (Chen et al., 2003). Each data point represents a separate bioreactor. MRI-derived FCD correlates well with biochemically determined S-GAG (top left). This is consistent with the usual interpretation of the dGEMRIC method. Lack of correlation between MRI-determined FCD and hydroxyproline, which is proportional to collagen content, is consistent with the usual interpretation of these experiments (top right). A correlation is seen between MRI-determined FCD and dynamic compressive modulus (lower left), demonstrating the ability of this non-invasive MRI measure to predict biomechanical properties. The correlation between the transverse relaxation rate, $1/T_2$, and water content (lower right), demonstrates the ability of MRI to predict biochemical parameters. Figure reprinted from Arthritis and Rheumatism © 2003 (American College of Rheumatology).

establishing the potential use of MRI to monitor cartilage repair and chondrocyte transplantation *in situ*.

5. BIOENERGETIC STUDIES OF DEVELOPING CARTILAGE

5.1 Overview of the Bioenergetics of Developing Cartilage

Provision of oxygen is of metabolic importance in both engineered and *in situ* cartilage, in spite of the fact that chondrocytes derive energy primarily from anaerobic metabolism and have an oxygen consumption only 2-5% that of the typical cell. However, appropriate oxygen tension is still important for developing and maintaining cartilage matrix. In engineered tissue anaerobic conditions lead to suppression of matrix production (Obradovic *et al.*, 1999). In osteoarthritis, there is diminished mitochondrial enzymatic activity (Maneiro *et al.*, 2003). Hence it may be expected that there is decreased O_2 consumption in this setting, and potentially decreased ability to produce high energy phosphate compounds.

5.2 Phosphorus Spectroscopy of HFBR-Derived Neocartilage

NMR spectroscopic studies utilizing the ^{31}P nucleus permit the measurement of intracellular concentrations of high-energy phosphate compounds such as ATP and phosphocreatine, as well as intracellular inorganic phosphate and pH. These techniques therefore permit the study of many bioenergetic properties of cartilage in the bioreactor. Our initial examination of this developing cartilage by ^{31}P spectroscopy had several goals. First, we wished to ascertain the visibility of phosphorus-containing metabolites. We then sought to validate the long-term bioenergetic stability of the construct. Finally, we were interested in evaluating the spin-lattice relaxation times of the inorganic phosphate resonance to look at potential immobilization related to the mineralization process. These investigations were carried out using serial ^{31}P NMR measurements of phosphorus metabolites (Petersen *et al.*, 2000). While ^{31}P spectroscopy of cartilage is generally extremely difficult due to the low cellularity of the tissue and resultant low NMR signal, the high cellularity of the HFBR tissue and the stability of our experimental arrangement permitted us to make these measurements. Typical ^{31}P spectra are shown in Figure 13. The spectral lines corresponding to ATP and inorganic phosphate are clearly visible, and permit the quantification of these metabolites. Further, the spectral separation between inorganic phosphate and the methylenediphosphonic acid (MDP) external standard allowed us to determine the pH of the tissue. From analysis of these types of spectra, chondrocytes obtained from both the proximal and distal sternum were found to be metabolically stable over a four week period, while producing abundant matrix. The relaxation times of the visible metabolites were also measured. In neocartilage which was produced by chondrocytes cultured from the proximal sternum there was a substantial increase in the T_1 relaxation time of inorganic phosphate between week 1 and week 4. No such increase in T_1 was seen in tissue derived from distal sternal chondrocytes. This suggests that in the proximal, but not the distal sternum, the inorganic phosphate was becoming immobilized. Consistent with this, we note that in the living animal, only the proximal sternum ossifies. We applied von Kossa staining to detect calcium phosphate deposition in the neocartilage. This staining was entirely

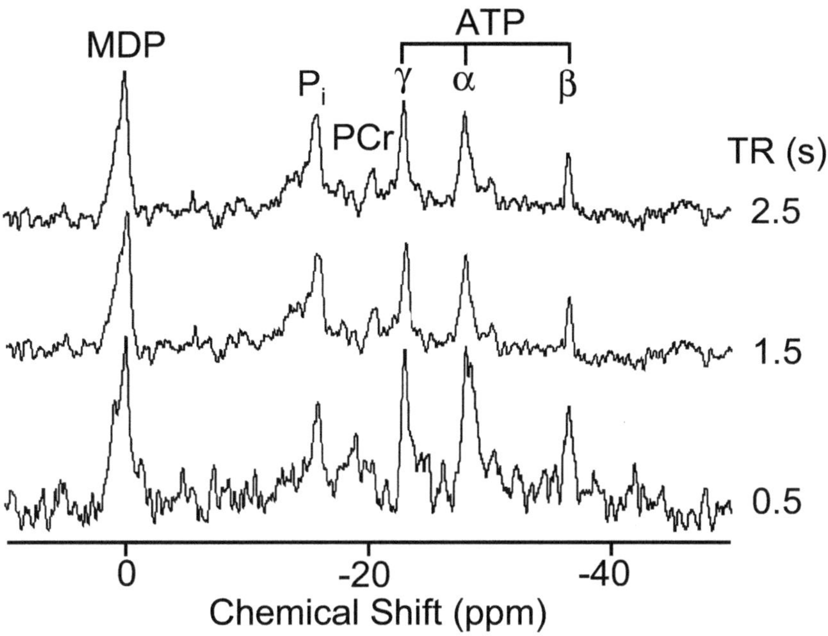

Figure 13: ^{31}P spectrum of neocartilage obtained after 4 weeks of development in the HFBR (Petersen et al., 2000). The spectra were obtained using three different TR's in order to obtain T1 values by the progressive saturation technique. Statistically indistinguishable spectra were obtained at 1 week of development (data not shown), demonstrating metabolic stability of the system. Figure reprinted from Magnetic Resonance in Medicine © 2000 (Wiley-Liss, Inc.).

negative, suggesting that the potential immobilization of inorganic phosphate seen in proximal sternum-derived bioreactor tissue may indicate a premineralization state of the tissue.

5.3 Lactate Imaging of Developing Cartilage

Proper development of cartilage requires not only the delivery of necessary metabolic substrates such as glucose and oxygen, but also the removal of metabolic products such as lactate. We used double quantum-edited chemical shift imaging to measure the distribution of lactate in the bioreactor (Velan *et al.*, 1998). This technique permits the suppression of the large (110 M) water proton signal, so that the much smaller (~10 mM) lactate proton signal can be detected. Using a single-fiber bioreactor we found that substantial amounts of lactate were being produced, and that the lactate concentration was higher in the tissue than in the extracapillary space surrounding the tissue. As expected, the amount of lactate in the return tube was negligible. Typical results are seen in Figure 14. A standard water proton image is shown for reference, with the cross-hairs indicating the region from which

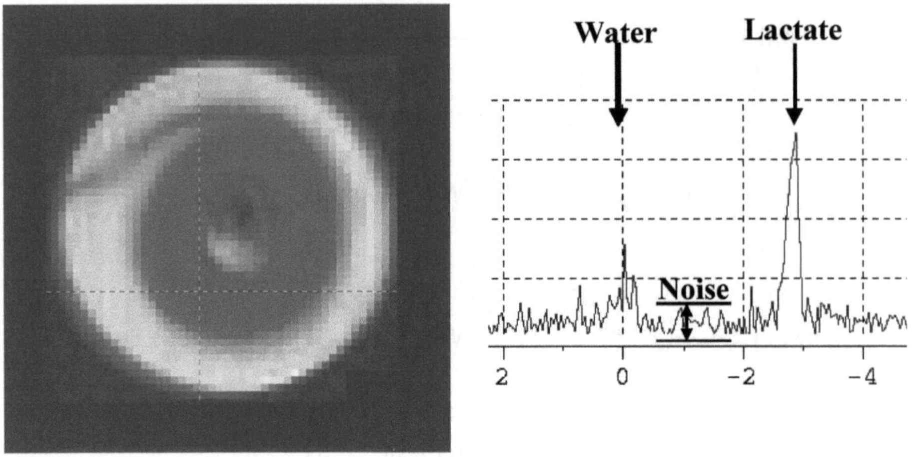

Figure 14: Lactate spectrum obtained from cartilage within a hollow fiber bioreactor (Velan et al., 1998). A reference image of the developing cartilage is shown on the left. A double-quantum edited water-suppressed proton spectrum obtained from within the neocartilage at the location indicated.

the lactate spectrum was obtained using a double quantum-edited chemical shift imaging sequence.

5.4 Oxygen Distribution in the HFBR Measured by Electron Paramagnetic Resonance

There are a limited number of methods for measuring oxygen tension in tissue. These include microelectrodes (Bartels *et al.*, 1998), Pd-coporphyrin phosphorescence (Haselgrove *et al.*, 1993) and the Warburg technique (Otte, 1972), which calculates pressure and volume changes in a closed system. In MR, the relaxation rates of compounds containing ^{19}F is a function of oxygen concentration, so that mapping these rates in the presence of fluorine-containing probes (Williams *et al.*, 1997) permits measurement of oxygen concentration. Electron paramagnetic resonance (EPR) has also been used to directly create maps of oxygen tension, with the addition of an incorporated spin label. A particulate label can be used to measure oxygen partial pressure, whereas a soluble label can be used to measure dissolved oxygen concentration.

We developed an approach (Velan *et al.*, 2000) using EPR spectroscopic imaging with a soluble probe molecule. We acquired an EPR spectrum for each voxel in an image data set, and subsequent analysis permitted the oxygen concentration in each voxel to be determined. To achieve this, we made use of two types of information available from EPR spectra: the local resonance amplitude (peak height), and the local value of spin density (peak area). With appropriate noise filtration, the ratio of these two values gives local linewidth. The process of

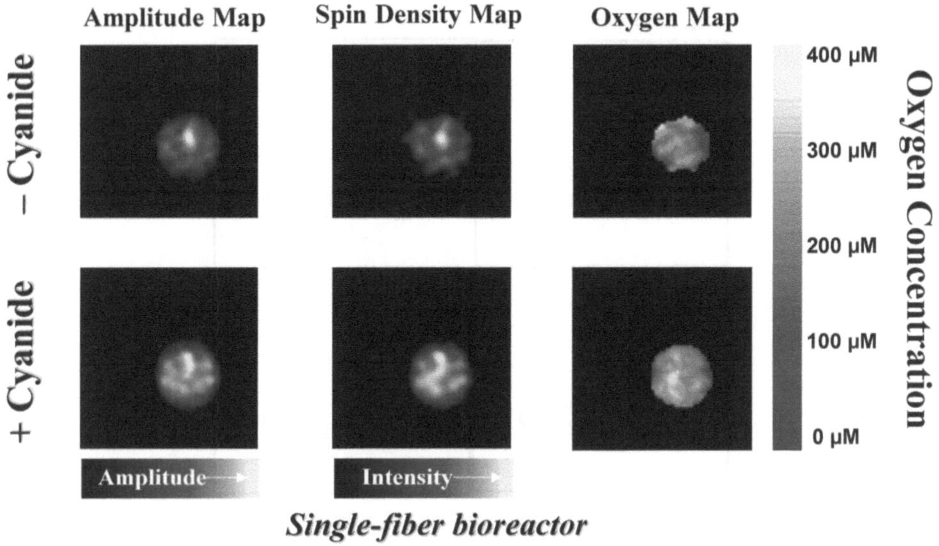

Figure 15: Images obtained by EPR oxygen mapping of a single fiber bioreactor (Ellis et al., 2001). The bioreactor was studied pre- and post- administration of cyanide. The oxygen map was computed using the amplitude map (derived from the height of the absorption line) and the spin density map (derived from the area of the absorption signal). The oxygen map indicates a 40% increase in oxygen concentration after perfusion with cyanide, reflecting the accumulation of oxygen. Figure reprinted Magnetic Resonance in Medicine © 2001 (Wiley-Liss, Inc.).

linewidth broadening by oxygen is well-understood in EPR spectroscopy, so that this local linewidth map could readily be converted into a map of oxygen concentration by use of a calibration curve. We have called our method EPROM, for EPR oxygen mapping.

EPROM was applied to cartilage growing within a single-fiber bioreactor (Ellis *et al.*, 2001). After determining the rate of perfusion of a nitroxide probe through the bioreactor, an oxygen map of the bioreactor was created (Figure 15). The oxygen concentration was then measured after provision of cyanide, which inhibits cytochrome c in the mitochondrial respiratory chain and would therefore be expected to lead to increased oxygen tension in this closed system. Indeed, this was observed. After perfusion of the bioreactor with 10 mM cyanide, the average oxygen concentration increased from 225 μM to 310 μM. These results suggest tha the EPROM technique is indeed sensitive to changes in oxygen concentrations in biological samples, and can be applied to engineered cartilage tissue.

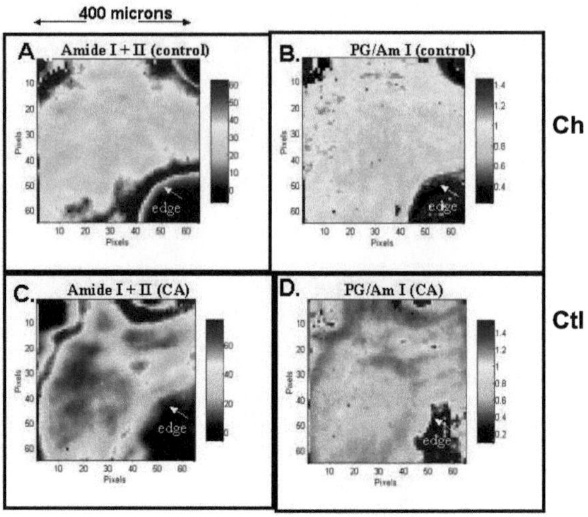

Figure 16: FT-IRIS images of control and Ch-ABC treated (CA) HFBR neocartilage. Resonance ratios, which can be interpreted in terms of matrix characteristics as described in the text, are displayed.

6. FOURIER TRANSFORM INFRARED IMAGING SPECTROSCOPY OF CHONDROITINASE-TREATED ENGINEERED CARTILAGE

While the emphasis in this Review is on magnetic resonance studies of bioreactor-derived engineered cartilage, the tissue is also suitable for evaluation using other methodologies. One powerful technique for detecting early degenerative changes in cartilage is FTIR imaging spectroscopy (FT-IRIS) (Camacho *et al.*, 2001; Potter *et al.*, 2001), which is sensitive to subtle molecular alterations in matrix. This technique has particular significance because of its extension from an *in vitro* modality to an arthroscopic device (West *et al.*, 2004), designed to provide the capability for detection of early OA in patients.

We used FT-IRIS to evaluate chondroitinase-ABC (Ch-ABC) treated HFBR-derived neocartilage as a model of cartilage degeneration (Camacho *et al.*, 2003). Biochemical analysis of Ch-ABC treated tissue demonstrated a significant reduction in chondroitin sulfate content without a statistically significant change in collagen content. Amide I and amide II absorbances were monitored in the 1590-1720 and 1480-1490 cm^{-1} infrared regions respectively. Images based on the integrated area of these regions were created to determine type II collagen quantity and distribution. The proteoglycan (PG) absorbance that arises from the sugar moiety was monitored in the 980 – 1145 cm^{-1} region. The ratio of the integrated area of this absorbance to the area of the amide I collagen absorbance, PG/Am-I, was used to obtain relative quantity and distribution of the PG component. The ratio of the amide I to amide II absorbance, Am-I/Am-II, was utilized as an indicator of collagen quality. See

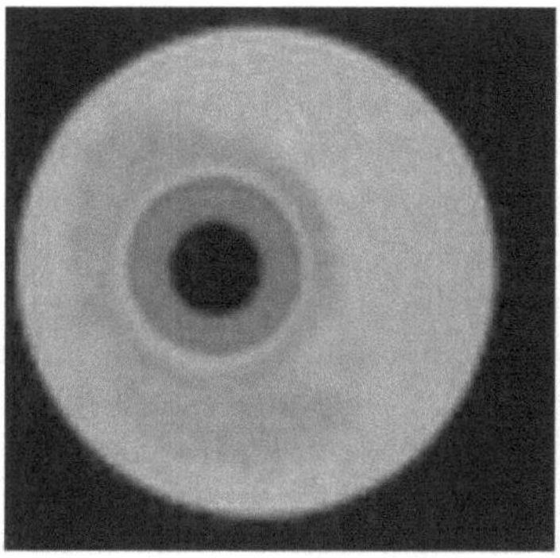

Figure 17: Typical MT-weighted image of bioreactor cartilage derived from bovine cells.

Figure 16. Quantitation of FT-IRIS images (n=4 for controls, n=4 for Ch-ABC treated) demonstrated a reduction in PG concentration in the Ch-ABC treated tissue as compared to controls (0.92 $\pm$ 0.13 vs 0.97 $\pm$ 0.12, $p < 0.001$), as well as reduced Am-I/Am-II (1.35 $\pm$ 0.13 vs 1.6 $\pm$ 0.0, $p < 0.01$). This alteration in collagen integrity as evaluated by FT-IRIS in the absence of a biochemically-detectable difference in overall collagen concentration may reflect changes in the collagen environment arising from altered proteoglycan content.

These results indicate that FT-IRIS is sensitive to molecular alterations in the matrix of engineered cartilage. We note further that the correlation of FT-IRIS measurements with appropriate MRI studies will potentially strengthen the utility of both techniques.

7. GROWTH OF MAMALLIAN CELLS IN BIOREACTORS

While the studies that have been outlined above were all performed using cartilage that developed from chick chondrocytes, mammalian tissue is of particular interest. Further experiments have been performed with chondrocytes isolated from newborn calves. The condyles and the patellar-femoral groove were dissected away from the subchondral bone, diced, and digested with pronase followed by collagenase. The isolated chondrocytes were then inoculated into single-fiber bioreactors in order to produce radially symmetric tissue. Cartilage grown from bovine chondrocytes was examined 2 and 4 weeks after inoculation into the bioreactors. Good cartilage growth was observed by MR (Figure 17). MR-derived parameters were similar to those previously obtained for cartilage grown from chick chondrocytes.

8. CONCLUSIONS AND FURTHER DEVELOPMENTS

The ability of MRI to make detailed non-invasive biophysical measurements on engineered tissue has numerous ramifications for optimizing growth protocols and for assessing the effects of biochemical and physical interventions. The HFBR described here permits exposure of the developing neocartilage to growth factors and other agents, and permits full control over substrate composition, dissolved O_2 and CO_2 concentrations, and temperature. While conditions pertaining to *in situ* development of cartilage from cells in an animal or human will differ in important ways from bioreactor conditions, these *in vitro* studies will nevertheless be of value in defining appropriate conditions for growth. Although chondrocytes are an obvious point of departure for such studies, other approaches may be envisioned, including utilization of bone marrow stromal cells which can be driven towards the chondrocyte phenotype. Exploration of these concepts in realistic *in vitro* systems may represent an important avenue towards developing therapeutic protocols for osteoarthritis. Furthermore, growth of high-quality cartilage in the bioreactor may result in a source of tissue for transplantation.

While the capabilities of the MR compatible bioreactor are substantial, the need for the entire system to be MR compatible does cause significant constraints on bioreactor design. An important element which has not yet been incorporated into our studies is application of mechanical stimuli to the developing cartilage tissue; it is well-known that cartilage is responsive to such stimuli, and that mechanical input is in fact required for proper cartilage development. Designing an MRI-compatible bioreactor which can accommodate mechanical energy input, such as ultrasound (Zhang *et al.*, 2002, Zhang *et al.*, 2003) or hydrostatic pressure, is a significant engineering challenge. For introduction of ultrasound energy, the bioreactor must exhibit the required transparency to this form of energy. The hydrostatic pressures of relevance to cartilage studies are significant, so that the requirement for non-metallic construction presents a challenge. One of the current design constraints, that of small bioreactor size to permit high-field, high-resolution studies, is becoming more relaxed with the increased penetration into the marketplace of MRI systems with field strengths of 7 Tesla and greater, high-performance gradient sets, and clear bores of 20 cm or more.

Finally, regardless of the specifics of eventual clinical cartilage repair and regeneration procedures, the ability to monitor tissue quality is of clear importance. While arthroscopic biopsies permit assessment of the biochemical and histologic state of the tissue, it is clearly more desirable to utilize noninvasive assessment methods. Although MRI is becoming increasingly accepted as a noninvasive tool for the measurement of cartilage thickness and volume and for delineation of localized pathology, the ability of MRI to noninvasively assess cartilage quality, as opposed to just anatomy, is currently a topic of great interest. Most investigations of the correlations between MR-derived parameters and biochemistry have utilized enzymatically degraded cartilage. However, to fully evaluate potential therapies using MRI it is important to understand the properties of actively regenerating tissue. The ability to monitor tissue growth in the MRI-compatible hollow fiber bioreactor as described here provides this opportunity.

9. ACKNOWLEDGEMENTS

It is a pleasure to acknowledge the following for their contributions to the research described: Daisy Bang, Nancy Pleshko Camacho, Chih-Tung Chen, Scott J. Ellis, Kenneth W. Fishbein, Walter E. Horton, Jr., Periannan Kuppusamy, Eric W. McFarland, Erik F. Petersen, Kimberlee Potter, Peter A. Torzilli, and S. Sendhil Velan. We are especially grateful to Dr. McFarland for introducing us to the HFBR.

10. REFERENCES

Asllani I, Shankland E, Pratum T, Kushmerick M. 2001. Double quantum filtered 1H NMR spectroscopy enables quantification of lactate in muscle. J Magn Reson 152:195-202.

Bartels EM, Fairbank JC, Winlove CP, Urban JP. 1998. Oxygen and lactate concentrations measured in vivo in the intervertebral discs of patients with scoliosis and back pain. Spine 23:1-7; discussion 8.

Bashir A, Gray ML, Burstein D. 1996. Gd-DTPA2- as a measure of cartilage degradation. Magn Reson Med 36:665-73.

Bashir A, Gray ML, Boutin RD, Burstein D. 1997. Glycosaminoglycan in articular cartilage:in vivo assessment with delayed GdDTPA2--enhanced MR imaging. Radiology 205:551-8.

Bashir A, Gray ML, Hartke J, Burstein D. 1999. Nondestructive imaging of human cartilage glycosaminoglycan concentration by MRI. Magn Reson Med 41:857-65.

Bischoff HA, Roos EM. 2003. Effectiveness and safety of strengthening, aerobic, and coordination exercises for patients with osteoarthritis. Curr Opin Rheumatol 15:141-144.

Bredella MA, Tirman PF, Peterfy CG, Zarlingo M, Feller JF, Bost FW, Belzer JP, Wischer TK, Genant HK. 1999. Accuracy of T_2-weighted fast spin-echo MR imaging with fat saturation in detecting cartilage defects in the knee:comparison with arthroscopy in 130 patients. AJR Am J Roentgenol 172:1073-80.

Brittberg M, Lindahl A, Nilsson A, Ohlsson C, Isaksson O, Peterson L. 1994. Treatment of deep cartilage defects in the knee with autologous chondrocyte transplantation. N Engl J Med 331:889-895.

Burstein D, Gray ML, Hartman AL, Gipe R, Foy BD. 1993. Diffusion of small solutes in cartilage as measured by nuclear magnetic resonance NMR spectroscopy and imaging. J Orthop Res 11:465-478.

Burstein D, Bashir A, Gray ML. 2000. MRI techniques in early stages of cartilage disease. Invest Radiol 35:622-638.

Burstein D, Gray M. 2003. New MRI techniques for imaging cartilage. J Bone Joint Surg Am. 85-A:Suppl 2, 70-77.

Camacho NP, West P, Torzilli PA, Mendelsohn R. 2001. FTIR microscopic imaging of collagen and proteoglycan in bovine cartilage. Biopolymers 62:1-8.

Camacho NP, Fishbein KW, Horton WE, Spencer RG. 2003. Fourier transform infrared imaging spectroscopy FT-IR of chondroitinase-treated engineered cartilage. 49th Annual Meeting of the Orthopedic Research Society, New Orleans, LA.

Chen CT, Fishbein KW, Torzilli PA, Hilger A, Spencer RG, Horton WE, Jr. 2003. Matrix fixed-charge density as determined by magnetic resonance microscopy of bioreactor-derived hyaline cartilage correlates with biochemical and biomechanical properties. Arthritis Rheum 48:1047-1056.

Clark AG, Rohrbaugh AL, Otterness I, Kraus VB. 2002. The effects of ascorbic acid on cartilage metabolism in guinea pig articular cartilage explants. Matrix Biol 21:175-184.

Darling EM, Athanasiou KA. 2003. Articular cartilage bioreactors and bioprocesses. Tissue Eng 9:9-26.

Ellis SJ, Velayutham M, Velan SS, Petersen EF, Zweier JL, Kuppusamy P, Spencer RG. 2001. EPR oxygen mapping EPROM of engineered cartilage grown in a hollow- fiber bioreactor. Magn Reson Med 46:819-826.

Evelhoch JL, Gillies RJ, Karczmar GS, Koutcher JA, Maxwell RJ, Nalcioglu O, Raghunand N, Ronen SM, Ross BD, Swartz HM. 2000. Applications of magnetic resonance in model systems:cancer therapeutics. Neoplasia 2:152-165.

Gerstenfeld LC, Landis WJ. 1991. Gene expression and extracellular matrix ultrastructure of a mineralizing chondrocyte cell culture system. J Cell Biol 112:501-513.

Gillies RJ, Galons JP, McGovern KA, Scherer PG, Lien YH, Job C, Ratcliff R, Chapa F, Cerdan S, Dale BE. 1993. Design and application of NMR-compatible bioreactor circuits for extended perfusion of high-density mammalian cell cultures. NMR Biomed 6:95-104.

Gillis A, Bashir A, McKeon B, Scheller A, Gray ML, Burstein D. 2001. Magnetic resonance imaging of relative glycosaminoglycan distribution in patients with autologous chondrocyte transplants. Invest Radiol 36:743-748.

Gray ML, Burstein D, Xia Y. 2001. Biochemical and functional imaging of articular cartilage. Semin Musculoskelet Radiol 5:329-343.

Haselgrove JC, Shapiro IM, Silverton SF. 1993. Computer modeling of the oxygen supply and demand of cells of the avian growth cartilage. Am J Physiol 265:C497-506.

Horton W, Hassell JR. 1986. Independence of cell shape and loss of cartilage matrix production during retinoic acid treatment of cultured chondrocytes. Dev Biol 115:392-397.

Horton WE, Yamada Y, Hassell JR. 1987. Retinoic acid rapidly reduces cartilage matrix synthesis by altering gene transcription in chondrocytes. Dev Biol 123:508-516.

Hunziker EB. 2002. Articular cartilage repair:basic science and clinical progress. A review of the current status and prospects. Osteoarthritis Cartilage 10:432-463.

Knauss R, Schiller J, Fleischer G, Karger J, Arnold K. 1999. Self-diffusion of water in cartilage and cartilage components as studied by pulsed field gradient NMR. Magn Reson Med 41:285-292.

Kuritzky L, Weaver A. 2003. Advances in rheumatology:coxibs and beyond. J Pain Symptom Manage 25:S6-20.

Laurent D, Wasvary J, Yin J, Rudin M, Pellas TC, O'Byrne E. 2001. Quantitative and qualitative assessment of articular cartilage in the goat knee with magnetization transfer imaging. Magn Reson Imaging 19:1279-1286.

Lotz M. 2001. Cytokines in cartilage injury and repair. Clin Orthop S108-115.

Lusse S, Claassen H, Gehrke T, Hassenpflug J, Schunke M, Heller M, Gluer CC. 2000. Evaluation of water content by spatially resolved transverse relaxation times of human articular cartilage. Magn Reson Imaging 18:423-430.

Macdonald JM, Grillo M, Schmidlin O, Tajiri DT, James TL. 1998. NMR spectroscopy and MRI investigation of a potential bioartificial liver. NMR Biomed 11:55-66.

MacDonald JM, Wolfe SP, Roy-Chowdhury I, Kubota H, Reid LM. 2001. Effect of flow configuration and membrane characteristics on membrane fouling in a novel multicoaxial hollow-fiber bioartificial liver. Ann N Y Acad Sci 944:334-343.

Maneiro E, Martin MA, de Andres MC, Lopez-Armada MJ, Fernandez-Sueiro JL, del Hoyo P, Galdo F, Arenas J, Blanco FJ, Lema B, Lombardi VR, Lagares R, Cacabelos R, Miguel-Hidalgo JJ, Mancino G, Poccia F, Placido R, Ron-Corzo A, Julve J, Goyanes VJ, Costas E. 2003. Mitochondrial respiratory activity is altered in osteoarthritic human articular chondrocytes. Arthritis Rheum 48:700-708.

McAlindon TE, Jacques P, Zhang Y, Hannan MT, Aliabadi P, Weissman B, Rush D, Levy D, Felson DT. 1996. Do antioxidant micronutrients protect against the development and progression of knee osteoarthritis? Arthritis Rheum 39:648-656.

Mlynarik V, Sulzbacher I, Bittsansky M, Fuiko R, Trattnig S. 2003. Investigation of apparent diffusion constant as an indicator of early degenerative disease in articular cartilage. J Magn Reson Imaging 17:440-444.

Moon RB, Richards JH. 1973. Determination of intracellular pH by ^{31}P magnetic resonance. J Biol Chem 248:7276-7278.

Mosher TJ, Dardzinski BJ, Smith MB 2000 Human articular cartilage:influence of aging and early symptomatic degeneration on the spatial variation of T_2--preliminary findings at 3 T. Radiology 214:259-266.

Neves AA, Medcalf N, Brindle K. 2003. Functional assessment of tissue-engineered meniscal cartilage by magnetic resonance imaging and spectroscopy. Tissue Eng 9:51-62.

Obradovic B, Carrier RL, Vunjak-Novakovic G, Freed LE. 1999. Gas exchange is essential for bioreactor cultivation of tissue engineered cartilage. Biotechnol Bioeng 63:197-205.

Otte P. 1972. Biology of articular cartilage with special reference to transplantation. Z Orthop Ihre Grenzgeb 110:677-85.

Petersen E, Potter K, Butler J, Fishbein KW, Horton W, Spencer RGS, McFarland EW. 1997. Bioreactor and probe system for magnetic resonance microimaging and spectroscopy of chondrocytes and neocartilage. International Journal of Imaging Systems and Technology 8:285-292.

Petersen EF, Fishbein KW, McFarland EW, Spencer RG. 2000. 31P NMR spectroscopy of developing cartilage produced from chick chondrocytes in a hollow-fiber bioreactor. Magn Reson Med 44:367-372.

Pilatus U, Ackerstaff E, Artemov D, Mori N, Gillies RJ, Bhujwalla ZM. 2000. Imaging prostate cancer invasion with multi-nuclear magnetic resonance methods:the metabolic Boyden chamber. Neoplasia 2:273-279.

Potter K, Butler JJ, Adams C, Fishbein KW, McFarland EW, Horton WE, Spencer RG. 1998. Cartilage formation in a hollow fiber bioreactor studied by proton magnetic resonance microscopy. Matrix Biol 17:513-523.

Potter K, Fishbein KW, Horton WE, Spencer RG. 1998. Morphometric analysis of cartilage grown in a hollow fiber bioreactor using NMR microscopy. In: P Blümler, B Blümlich, R Botto, E Fukushima, editors. Spatially Resolved Magnetic Resonance Weinheim, Wiley–VCH Press: pp 363-371.

Potter K, Butler JJ, Horton WE, Spencer RG. 2000. Response of engineered cartilage tissue to biochemical agents as studied by proton magnetic resonance microscopy. Arthritis Rheum 43:1580-1590.

Potter K, Kidder LH, Levin IW, Lewis EN, Spencer RG. 2001. Imaging of collagen and proteoglycan in cartilage sections using Fourier transform infrared spectral imaging. Arthritis Rheum 44:846-855.

Regatte RR, Akella SV, Borthakur A, Kneeland JB, Reddy R. 2002. Proteoglycan depletion-induced changes in transverse relaxation maps of cartilage:comparison of T_2 and $T_1\rho$. Acad Radiol 9:1388-1394.

Risbud MV, Sittinger M. 2002. Tissue engineering:advances in in vitro cartilage generation. Trends Biotechnol 20:351-356.

Schnabel M, Marlovits S, Eckhoff G, Fichtel I, Gotzen L, Vecsei V, Schlegel J. 2002. Dedifferentiation-associated changes in morphology and gene expression in primary human articular chondrocytes in cell culture. Osteoarthritis Cartilage 10:62-70.

Shapiro EM, Borthakur A, Kaufman JH, Leigh JS, Reddy R. 2001. Water distribution patterns inside bovine articular cartilage as visualized by 1H magnetic resonance imaging. Osteoarthritis Cartilage 9:533-538.

Shapiro EM, Borthakur A, Gougoutas A, Reddy R. 2002. 23Na MRI accurately measures fixed charge density in articular cartilage. Magn Reson Med 47:284-291.

Velan SS, Peterson E, Fishbein KW, Spencer RGS. 1998. Mapping the distribution of lactate in cartilage grown from chondrocytes in a hollow fiber bioreactor. ISMRM Workshop on Magnetic Resonance of Connective Tissues and Biomaterials, Philadelphia, PA.

Velan SS, Spencer RG, Zweier JL, Kuppusamy P. 2000. Electron paramagnetic resonance oxygen mapping EPROM:direct visualization of oxygen concentration in tissue. Magn Reson Med 43:804-809.

West PA, Bostrom MPG, Torzilli PA, Camacho NP. 2004. Fourier transform infrared spectral analysis of degenerative cartilage: An infrared fiber optic probe and imaging study. Applied Spectroscopy 58:376-381.

Williams SNO, Callies RM, Brindle KM. 1997. Mapping of oxygen tension and cell distribution in a hollow-fiber bioreactor using magnetic resonance imaging. Biotechnol Bioeng 56:56-61.

Wolfe SP, Hsu E, Reid LM, Macdonald JM. 2002. A novel multi-coaxial hollow fiber bioreactor for adherent cell types: Part 1:hydrodynamic studies. Biotechnol Bioeng 77:83-90.

Xia Y, Farquhar T, Burton-Wurster N, Vernier-Singer M, Lust G, Jelinski LW. 1995. Self-diffusion monitors degraded cartilage. Arch Biochem Biophys 323:323-328.

Yamada K, Tanabe S, Ueno H, Oinuma A, Takahashi T, Miyauchi S, Shigeno S, Hirose T, Miyahara K, Sato M. 2001. Investigation of the short-term effect of chemonucleolysis with chondroitinase ABC. J Vet Med Sci 63:521-525.

Zhang ZJ, Huckle J, Francomano CA, Spencer RG. 2002. The influence of pulsed low-intensity ultrasound on matrix production of chondrocytes at different stages of differentiation:an explant study. Ultrasound Med Biol 28:1547-1553.

Zhang ZJ, Huckle J, Francomano CA, Spencer RG. 2003. The effects of pulsed low intensity ultrasound on chondrocyte viability, proliferation, gene expression and matrix production. Ultrasound Med Biol 29:1645-1651

CHAPTER 7

MECHANICAL CONDITIONING OF CELL-SEEDED CONSTRUCTS FOR SOFT TISSUE REPAIR - ARE OPTIMISATION STRATEGIES POSSIBLE?

D.L. BADER AND D.A. LEE

Department of Engineering and Interdisciplinary Research Centre in Biomedical Materials, Queen Mary University of London, London, UK

1. INTRODUCTION

Tissue Engineering is a major focus of biotechnological research today, with the expectation that this type of technique will ultimately transform medical practice (Nerem, 2000). The most ambitious tissue engineering schemes assume that specific tissues and organs will be restored, in a multi stage fabrication procedure. For example, cells derived from the patient may be processed to increase the total number available, seeded into a suitable three-dimensional resorbable scaffold and the resultant construct may be further processed *in vitro* using specially designed bioreactor systems to induce the elaboration of neo-tissue prior to implantation (Bader and Lee, 2000).

The importance of mechanical stimulation in maintaining the integrity of healthy load-bearing soft tissues is widely accepted, both for metabolism and homeostasis. As an example, during normal activity, the articular cartilage of the major load bearing joints is exposed to complex loading, which is predominantly compressive in nature. It is well known that this mechanical environment is essential for the lubrication and nutrition of chondrocytes embedded in the extracellular matrix (ECM) of this avascular tissue. Previous studies have reported that static compression of explanted cartilage causes a down-regulation of matrix synthesis, while cyclic compression may induce an up-regulation of matrix synthesis dependent on the dynamic frequency applied (Sah et al., 1989; Kim et al., 1994). Accordingly, the manner in which cells respond to different mechanical environments has attracted a wide interest, with studies investigating the role of various intracellular signalling pathways associated with mechanotransduction. In

J.B. Chaudhuri & M. Al-Rubeai (eds.), Bioreactors for Tissue Engineering, 165 - 192.

addition, there are an increasing number of authors who have proposed *in vitro* mechanical conditioning strategies for cell-seeded scaffolds as an essential feature for the long term functionality of tissue engineered implants. This requires the development of suitable bioreactors, incorporating mechanical loading modules for use in a controlled biological environment. These are designed to provide appropriate mechanical conditioning regimes recognised to stimulate the formation of a functional neo-tissue, thereby improving the efficacy and efficiency of production of tissue engineered implants. In order to successfully develop and optimise bioreactor systems for mechanical conditioning of cell-seeded constructs, it is necessary to understand the complex interplay between the nature of the loading regime applied, the scaffold material and the cells seeded within the scaffold.

This chapter describes studies conducted by the authors and others, aimed at developing mechanical loading strategies for use in functional tissue engineering of soft tissue defects. These studies primarily address issues related to the nature of the loading regimes adopted, the sourcing of cells for tissue engineering and the nature of the scaffold material.

2. THE IMPORTANCE OF MECHANOTRANSDUCTION

The cells in many load bearing soft tissues are sparsely populated, often occupying less than 10% of the tissue volume, but are necessary for the synthesis, assembly, maintenance and degradation of the ECM. The cells are therefore crucial to the structural integrity and function of the tissue. It is known that the cells in all such tissues, whether chondrocytes in cartilage, fibrochondrocytes in menisci or tenocytes and fibroblasts in tendon and ligament respectively, are able to alter their metabolic activity in response to applied load. Both the level of the applied strain and the dynamic frequency are known to be important in eliciting a specific response. These processes are believed to be major factors in determining cellular activity in soft tissues, such as articular cartilage. Indeed *in vivo* studies have shown that the mechanical environment will influence the structure and function of articular cartilage. Typically, moderate exercise leads to an increase in cartilage proteoglycans, whereas immobilisation of joints leads to a reversible release of proteoglycans (Kiviranta et al., 1988; Saamanen et al., 1990).

The mechanisms by which cells detect and respond to the mechanical environment are termed mechanotransduction pathways and are both complex and poorly understood. For example, mechanotransduction events in articular cartilage may be resolved into extracellular components including cell deformation, hydrostatic pressures, fluid flow and streaming potentials, followed by intracellular signalling events, such as intracellular calcium fluxes, cAMP production and cytoskeletal alterations, which finally lead to altered effector cell response. It is unclear whether mechanotransduction is a universal process or whether various metabolic parameters are influenced by distinct and uncoupled intracellular signalling pathways. A better understanding of the relationship between initial application of load and effector cell response is necessary for the appreciation of metabolic control in normal and pathological tissues and will be of use in the development of tissue engineered repair systems. This goal is complicated by the need to investigate processes at the cellular level. However, this approach has been facilitated in the last decade by the considerable developments in molecular biology.

Two fundamental approaches have been adopted to examine the response of chondrocytes to biomechanical stimuli *in vitro*. One involves the use of cartilage explants in which the chondrocytes will be associated with ECM similar to that *in situ* (Gray et al., 1989; Sah et al., 1989). The alternative involves model systems incorporating isolated chondrocytes which may be maintained in various cultures systems, including pellet cultures, suspension cultures or monolayer cultures of either high or low cell densities, or the cells may be seeded within a 3-D construct, typically comprising a hydrogel or a porous scaffold (Buschmann et al., 1995; Kisiday et al., 2002; Freeman et al., 1994; Lee and Bader, 1997).

Explant systems have the disadvantage of being unsuitable to examine individual extracellular components of mechanotransduction, such as cell deformation, due to the inherent coupling of mechanical and physicochemical processes in the ECM. Moreover, while the use of cartilage explants is appropriate for studies investigating fundamental mechanotransduction events associated with normal turnover and pathology, model systems used for bioreactor development for tissue engineering should reflect the nature of the cell-seeded constructs typically used in such therapeutic strategies. Accordingly, a wide variety of model systems have been adopted by researchers working within this research area (Table 1).

Key variables associated with the application of mechanical conditioning to cell seeded 3-D scaffolds for tissue engineering include the following:

- The nature of the loading regimen applied
- The source of the cells
- The composition of the scaffold

As a consequence, much of the data reported in the literature provides phenomenological observation, specific to a particular model system, which may not be easily translated to other model systems. Only by conducting systematic studies, involving the alteration of one variable at a time, can underlying mechanistic parameters be defined leading to the derivation of predictive strategies suitable for the optimisation of bioreactor systems. The following sections describe the model systems adopted by the authors for their studies and further focus on specific studies aimed at elucidating the influence of biomechanical loading regimens, cell sub-populations and scaffold materials on the efficacy of mechanical conditioning strategies.

Table 1. Summary of Selected Studies Illustrating the Range of In Vitro Dynamic Loading Regimens Used to Enhance Chondrocyte Metabolism

Source	*Cell Source*	*Sample/Scaffold Material*	*Passaged Y/N*	*Loading Conditions*
Present authors	Bovine (Adult)	Agarose	N	15%, 0.3, 1 and 3Hz, 48 hr
Bonassar et al., 2001	Bovine (Calf)	Cartilage discs	N	2%, 0.1 Hz, 48 hrs
Kisiday et al., 2002	Bovine (Calf)	Agarose and self assembling peptide gel	N	2.5 – 3%, 1 Hz, 1 hr on / 1 hr off, d9-d12 or 1 hr on / 7 hr off, d27 –d34
Buschmann et al., 1995	Bovine (Calf)	Agarose	N	0.001 Hz to 1 Hz, 10 hr
Buschmann et al., 1999	Bovine (Calf)	Explants	N	0.001 Hz, 23 hr, day 4 & day 6 0.1 Hz, 23 hr, d3 & d5
Sah et al, 1989; Li et al., 2001	Bovine (Calf)	Explants	N	2.4%, 0.01 Hz (200 kPa); 2 hr on/ 2hr off, 23 hr, 0.88-1 mm compression
Mauck et al., 2000	Bovine (Calf)	Agarose	N	10%, 1 Hz, 3 x 1 hr intermittent/day, 5 days/week for 4 weeks
Millward-Sadler et al., 1999	Human normal or OA	Monolayer	Y	0.33 Hz, 20 min
Guilak et al., 1994; Fermor et al., 2001	Bovine (Calf) / porcine	Explants	N	0.1, 0.5 MPa, 0.5 Hz, 24 hr
Altman et al., 2002	Human Bone marrow stromal cells	Silk fibre matrices	Y	0.0167 Hz, 2 mm deformation
Martin et al., 2000	Bovine (Calf)	Polyglycolic acid	N	Compressive deformation for samples 1.3 – 3 mm thickness
Hunter et al., 2002	Bovine (Calf)	Collagen I	N	Oscillatory 25$\pm$4 %, 1 Hz, 24 hr

3. MECHANICAL CONDITIONING OF MODEL SYSTEMS

3.1 Loading Systems

Conventional *in vitro* mechanical tests require the soft tissues to be kept in a moist environment and are either performed on a Universal materials test facility or on a specially designed test system. However, if tissue explants or cell-seeded biomaterial constructs are to be tested, viability must be maintained in an environment similar to a conventional CO_2 incubator, whilst the system is subjected to either static or dynamic loading. A wide variety of "one-off" research systems have been described in the literature, which exhibit a diversity of features associated with the application of mechanical loading and the maintenance of a sterile environment suitable for cell culture. A smaller number of commercial systems are available although, for a variety of reasons, few have gained widespread acceptance by the research community. In this chapter data will be presented from the authors' research using a commercially available system, the Flexercell system, and two in-house systems, designed for the application of dynamic tensile and compressive strains. These systems are described below.

Flexercell system: The Flexercell system is a commercially available strain unit (Flexercell, Flexcell Intl. Corp., McKeesport, USA), in which cells are attached to a flexible substrate at the base of culture plates. The essential design feature is based on radial and circumferential stretching of a circular silicone rubber membrane by the application of controlled negative air pressure directed towards causing the membrane to dome downwards. The strain profiles for the system have been well documented (Gilbert et al., 1994; Grymes and Sawyer, 1997). The strains are defined as non-uniform when the membrane is allowed to deform freely or biaxial, when a loading post is placed underneath the membrane thereby inducing radial strains. Typical studies have used this system to apply cyclic strain to tendon cells cultured in monolayer and examine effects, in conjunction with growth factors, on the stimulation of DNA synthesis (Banes et al., 1995).

Tensile cell strain system: An alternative cell straining system was developed by the authors (Cacou et al. 2000), designed to apply uniaxial strains either to cells attached to flexible membranes, or to cells seeded in 3-D constructs or to tissue explants. The cyclic strain is applied via a linear stepper motor and the loads as well as the strain on each specimen are recorded, enabling the material properties of the specimens to be determined during testing. A photograph of this Cell Straining unit is presented in Figure 1d, which also incorporates appropriate dumbell specimens (Figure 1a), grips (Figure 1b) and individual Perspex chambers (Figure 1c).

Compressive cell strain system: A separate cell straining apparatus (Zwick Testing Machines Ltd, Leominster, UK) was also developed by the authors to apply static and dynamic compression to biomaterial constructs seeded with chondrocytes (Lee and Bader, 1997). This apparatus, as shown schematically in Figure 2, consists of a conventional loading frame with an hydraulic actuator-controlled vertical assembly, which enters a tissue culture incubator (Heraeus Instruments, Brentwood,

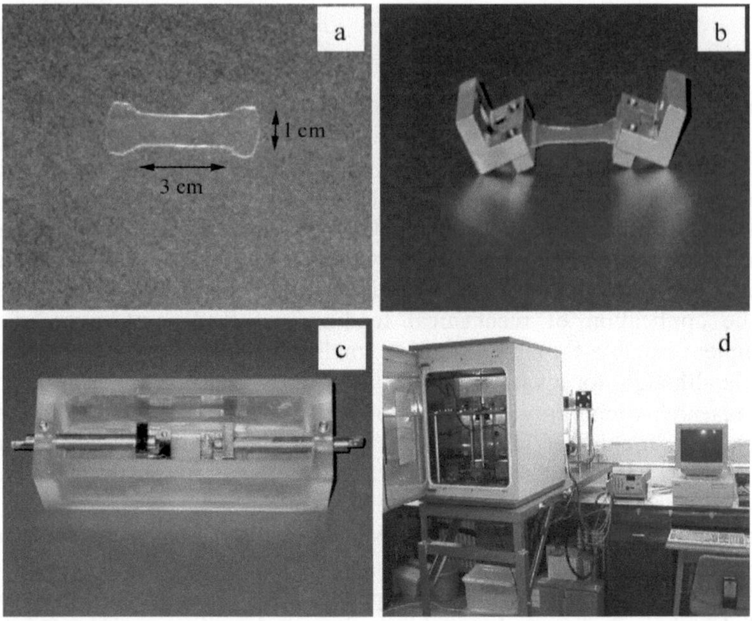

Figure 1: Components within the system used for the application of uniaxial tensile strain to cell seeded membranes. (a) Dumbell specimen indicating dimensions held within grips (b); (c) individual Perspex chambers with grips and membranes in position; (d) Six Perspex chambers housed within the tensile Cell Straining system.

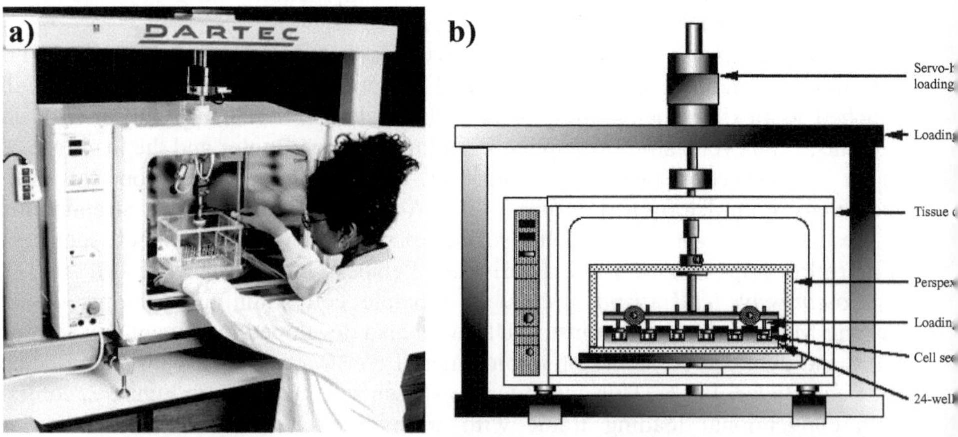

Figure 2: Photograph (a) and schematic representation (b) of the Compressive cell strain system.

U.K.). The assembly is connected to a central rod which is attached to a mounting plate located within a perspex box. The box, in turn, is placed on a circular platten fixed to the base of the loading frame. The mounting plate holds 24 loading pins, half of which are unconstrained to move vertically in harmony with the loading assembly. Each loading pin incorporates an 11 mm circular perspex indenter, which applies compressive strain to samples located within separate wells of a tissue culture plate.

3.2 Cell Seeded Model Systems

Tensile studies: Many studies by the authors and others, investigating fibroblast and tenocyte mechanotransduction, have employed a simple 2-D model system comprising cells seeded onto elastomeric silicone membranes (Banes et al., 1995; Berry et al., 2003). The application of tensile strain to the membrane is transferred to the cells via cell adhesion molecules. Accordingly the system may be further adapted by the grafting of known ligands for integrins, such as fibronectin or collagen, onto the membrane. Cellular strain levels may not be equivalent to the membrane strain, as they are determined by the precise number and location of adhesion sites, which may vary from cell to cell. Moreover, under the influence of tensile strain, the cells are known to reorganise their focal adhesion sites in order to minimise the magnitude of the resultant cellular strain.

In the studies presented in this chapter two fibroblast sources were employed, isolated respectively from neonatal human dermis (HuFFs) or adult human dermis (1BR3). The cells were seeded at a density of 1×10^4 cm^{-2} onto type I collagen-grafted silicone membranes either as 35 mm diameter membranes within 6 well plates (Flexcell Corporation, McKeesport, USA) or as dumbell-shaped specimens comprising a gauge length of 20 mm and width of 10 mm (Figure 1b). The cells were cultured for 5 days with media changes every 2 days prior to the application of dynamic tensile strain, which was achieved using the Flexercell system or the tensile cell strain system. Both systems were programmed to apply a nominal strain of 5% at 1Hz for 24 hours using a sawtooth waveform. Cell seeded control membranes were maintained unstrained during the 24 hour conditioning period.

Compression studies: Numerous studies have examined chondrocyte mechanotransduction using a well characterised experimental system consisting of isolated chondrocytes embedded within a 3D agarose construct. Although recognised as inherently non-physiological, the authors and others have shown that the agarose/chondrocyte model system exhibits several suitable features for the study of fundamental mechanotransduction processes and the development of conditioning strategies for tissue engineering. The agarose system is known to maintain chondrocytic phenotype over extended culture periods (Benya and Shaffer, 1982; Aydelotte et al, 1990; Hauselmann et al., 1994), whilst enabling the application of cell deformation through gross compression of the cell-agarose construct (Lee et al., 2000). The model system therefore enables the role of cell deformation in cartilage mechanotransduction to be examined in isolation from factors associated with compression of the charged extracellular matrix (Buschmann et al., 1995; Lee et al., 1998a, 2000; Mauk et al., 2000). Moreover the agarose system is relatively simple and reproducible and has been characterised in mechanical terms (Knight et al., 1996). The model provides a relevant insight into mechanotransduction within tissue engineered constructs as results indicate important differences between the potential signalling pathways within cartilage and cell seeded scaffolds (Lee et al., 2000).

Although compression of cell-agarose constructs will initiate transient changes in hydrostatic pressure and fluid flow, which have both been implicated in mechanotransduction, it is widely believed that cell deformation is the primary mediator. The effect of loading on cell morphology and deformation during the application of physiological levels of compressive strain, typically in the order of 5 to 20%, has been studied extensively (Knight 1997). Various factors have been shown to influence the level of cell deformation and hence the ability of the cells to respond to gross compressive strain. These include the presence and mechanical properties of elaborated extracellular matrix, the modulus of the scaffold relative to that of the cells and matrix and the loading regime and viscoelastic properties of the scaffold (Knight et al., 1998, 2001; Bader et al., 2002).

The production of chondrocyte seeded agarose constructs has been previously detailed (Lee and Bader, 1995, 1997). To review briefly, chondrocytes are isolated by sequential enzyme digestion using pronase and collagenase and the resultant chondrocyte suspension is added to an equal volume of 6% agarose (type VII, Sigma, Poole, U.K.) to give a final concentration of 4 x 10^6 cells.mL^{-1} in 3% agarose. Cylindrical core constructs, approximately 5 mm in diameter and 5 mm in height are prepared and cultured in 1ml of DMEM+ 20% FCS at 37°C in 5% CO_2 for 24 hr to allow equilibration to culture conditions prior to the application of dynamic compression. The control electronics of the compressive cell strain system is set to produce a crosshead movement equivalent to a compressive strain ranging from 0% to 15%. For the dynamic strain conditions, a sinusoidal waveform was used a frequencies of 0.3Hz, 1Hz and 3Hz, although some studies employed the single frequency of 1Hz.

In the majority of studies described below the chondrocytes were derived from cartilage dissected from the proximal surface of the metacarpalphalangeal joint of 18 month old steers. However, in addition, other studies utilised equine or human chondrocytes.

Outcome measures: The chapter will concentrate on a selected number of metabolic markers to examine the effects of a range of mechanical conditioning regimens. Of particular relevance for functional tissue engineering studies is the ability of the cells to proliferate, which is necessary to allow re-population of the defect site and for the cells to synthesis a functional neo-tissue, comprising key ECM molecules. Full details of the methodologies used are described fully in the literature (Lee & Bader, 1997; Lee et al., 1998a) and are outlined briefly hereafter.

[^{3}H]-thymidine incorporation into newly synthesised DNA, which occurs during s-phase of the cell cycle provides an appropriate measure of cell proliferation. Typical experiments involve incubation of the cell seeded constructs in medium supplemented with 1 μCi.mL^{-1} [^{3}H]-thymidine for 24-48 hours. Following solubilisation of the constructs incorporated [^{3}H]-thymidine was measured using trichloroacetic acid precipitation.

For the tensile studies, collagen production in response to mechanical conditioning was assayed using [^{3}H]-proline incorporation, using the sirius red precipitation method to ensure selectivity for newly synthesis collagen (Lee et al., 1998c). Typical experiments involve incubation of the cell seeded constructs in medium supplemented with 10 μCi.mL^{-1} [^{3}H]-proline for 24-48 hours. For compression studies the rate of proteoglycan synthesis was assessed by measuring alterations in the levels of sulphated glycosaminoglycan using an established method (Farndale et al., 1982). Total GAG synthesised during the culture period was determined as described previously (Lee and Bader, 1997). In other studies, the incorporation of $^{35}SO_4$ into newly synthesised proteoglycan was determined using the alcian blue precipitation method (Lee et al., 1998a), after incubation of the constructs in medium supplemented with 10 μCi.mL^{-1} $^{35}SO_4$ for 24-48 hours. Total DNA, determined using the Hoescht 33258 method (Rao & Otto, 1992), was used as a baseline for cell proliferation, collagen synthesis and proteoglycan synthesis. In some studies, nitrite was measured in the medium of cultured cells using the Griess reaction (Lee et al., 1998b).

For all loading experiments, GAG or collagen synthesis and [^{3}H]-thymidine incorporation values in experimentally strained systems were normalised to unstrained control values.

4. FACTORS AFFECTING MECHANICAL CONDITIONING OF CELL-SEEDED CONSTRUCTS

4.1 Loading Regimens

Continuous Dynamic Compression for 48h – Full Depth Bovine Chondrocytes

An important aspect of any prolonged culturing period is the maintenance of cell viability. During the 48 hour culture period in the compressive cell strain system, the chondrocyte viability remained above 95% within both unstrained and dynamically strained constructs (Lee and Bader, 1997). Thus any differences in metabolism could not be attributed to alterations in chondrocyte viability. The findings for GAG synthesis values, normalised to unstrained control levels, are presented in Figure 3a. It can be seen that at the low frequency (0.3Hz) dynamic strain inhibited GAG synthesis, while a frequency of 1Hz induced a significant stimulation of GAG synthesis. At the higher frequency of 3Hz GAG synthesis returned to unloaded control levels. The normalised data for tritiated thymidine incorporation are presented in Figure 3b. It is evident that constructs subjected to dynamic strain at all three frequencies exhibited significant stimulation of [^{3}H]-thymidine uptake in comparison with unstrained control values. However, there was no statistically significant difference in the level of stimulation between the three dynamic strain regimes. Although all loading regimens yielded an inhibition in protein synthesis, as measured using proline incorporation, the analysis of data revealed an association between the frequency rate and the level of inhibition (data not presented, Lee and Bader, 1997).

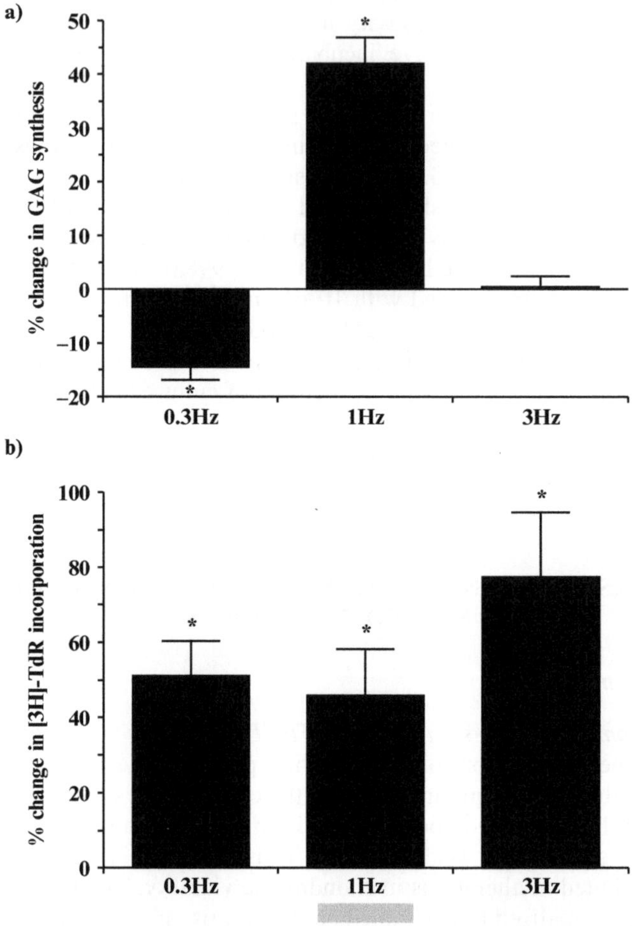

Figure 3: GAG synthesis (a) and [^{3}H]-thymidine incorporation (b) by full depth bovine chondrocytes embedded in agarose constructs and subjected 15% dynamic compressive strain amplitude at various frequencies for 48 hr. The values are presented as % change from unstrained control levels. Each value represents the mean and standard error of at least 16 replicates from at least two separate experiments. Unpaired student's t-test results indicate differences from control values as follows: * = $p \leq 0.05$.

This study represented the first in a series by the authors to demonstrate that the metabolic response of chondrocyte/agarose model is influenced by the application of dynamic compression. In particular, it indicated that chondrocytes pooled from the full depth of cartilage and subjected to 15% dynamic strain amplitude for 48 hours exhibited frequency dependent alterations in GAG synthesis, when compared to unstrained control values. Although other studies have suggested an increase in

GAG synthesis with frequency, few had examined the response to frequencies exceeding 1Hz. The findings of the authors' study suggested that at the strain level of 15% there is a window of dynamic frequencies, which can induce GAG synthesis, with a lower cut-off frequency between 0.3 and 1 Hz and an upper cut-off between 1 and 3 Hz. This cut-off frequency may be associated with a frequency dependent initiation of marked fluid flow in the periphery of the specimen, which will vary with the dynamic stiffness of the system. This frequency will be dependent on the strain level, the mechanical properties of the matrix and the size and shape of the individual specimens. Indeed it has been established that increasing frequency will increase the rate of fluid flow but will reduce the effective width of the peripheral ring in which it occurs (Sah et al., 1989). It is conceivable, therefore, that the influence of the central core may mask stimulatory effects within the peripheral ring at the 3Hz frequency. The findings also implied that each metabolic parameter may be influenced by dynamic strain regimens in a distinct manner, implying that the associated signalling mechanisms are uncoupled (Lee and Bader, 1997).

Continuous Dynamic Compression for 1.5h - 48h – Full Depth Bovine Chondrocytes
Several studies associated with different cell types have indicated that dynamic loading for shorter periods than 48 hours may be sufficient to stimulate cellular activity (Rubin and Lanyon, 1992; Fermor et al., 2001; Robling et al., 2002). This prompted an investigation by the authors to examine the temporal response of full depth chondrocytes seeded in agarose constructs to continuous compression regimens. Six different periods of compressive strain were examined ranging from 1.5 to 48 hours with corresponding recovery periods of 46.5 hours to 0 hour. During each strained period, the constructs were subjected to 15 % dynamic compressive strain at 1 Hz.

Dynamic continuous compression resulted in significant increases in both $^{35}SO_4$ and [^{3}H]-thymidine incorporation (Figure 4). However there were clear differences in the optimal profiles over the 48 hour culture period. For example, the maximum $^{35}SO_4$ incorporation occurred after at least 12 hours of continuous compression. By contrast, [^{3}H]-thymidine was maximal after 1.5 hours of continuous compression followed by 46.5 hour recovery (labelled C 1.5 in Figure 4). Longer bursts of cyclic compression appeared to inhibit the proliferative response (Figure 4b). Nitrite release was inhibited by dynamic compression by values ranging from 18% and 43%, although the magnitude appeared to be independent of the compression period (data not shown).

The authors postulate that dynamic compressive strain acts as a competence and/or progression factor for DNA synthesis in chondrocytes. Once the cell has been stimulated to enter the cell cycle it will progress through the complete cycle, which may take several days. Thus stimulation of cell proliferation within the time scale of this study is a unique event, as opposed to repeated stimulation, which is required to upregulate proteoglycan synthesis long term. It is feasible therefore that the frequency of dynamic strain does not affect the response, as it appears that a finite number of cycles of dynamic strain are sufficient for stimulation. This suggests that the cells are temporally processing the mechanical stimulus.

(a)

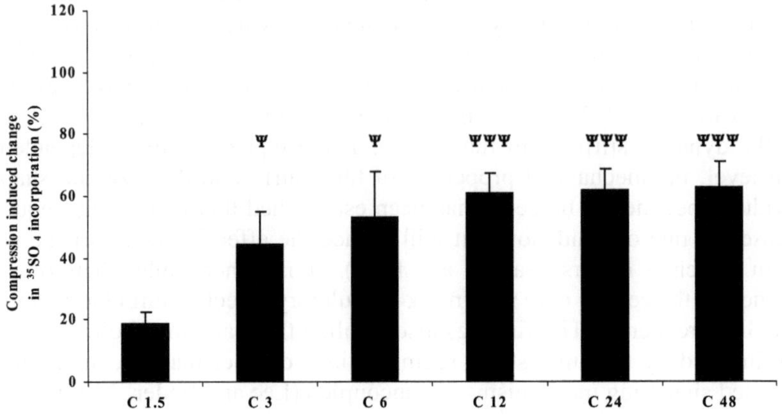

(b)

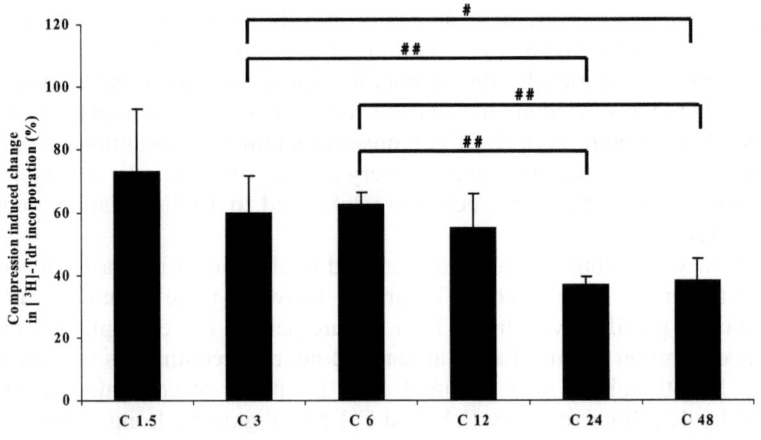

Figure 4: The percentage change from unstrained control values for $^{35}SO_4$ incorporation (a) and [^{3}H]-thymidine incorporation (b), by chondrocytes seeded in 3 % agarose and subjected to different periods of continuous compression of 15 % dynamic strain at a frequency of 1 Hz. Error bars represent the mean and SEM of 24 replicates from two separate experiments. Unpaired Student's t-test results indicate significant differences between strained and unstrained control groups (ψ) and between strain regimes (#) as follows: ψ/#= $p<0.05$, ψψ/##=$p<0.01$, ψψψ/###=$p<0.001$. All other comparisons were not significant ($p>0.05$).

Such a study demonstrates that continuous loading, applied in a dynamic manner can temporally modulate the metabolism of chondrocytes seeded in agarose gel. These results will have important implications in the optimisation strategies for both temporal and spatial processing of functional neo-tissues.

Continuous Dynamic Tension for 24h – HuFFs and 1BR3 Fibroblasts
Both the neonatal (HuFF) and adult (1BR3) fibroblasts were viable following the dynamic tensile conditioning period and also exhibited a flattened and elongated fibroblast morphology. It was noted that both HuFFs and 1BR3s exhibited cell alignment in response to the uniaxial strain profile, in an oblique direction to that of the uniaxial strain (Berry et al., 2003). This alignment was also apparent with the non-uniform strain profile, in the circumferential direction, towards the periphery of the membrane, an area in which the strain can be regarded as being uniaxial in nature. By contrast, no alignment was evident in response to the biaxial strain profile.

The mean values of [^{3}H]-proline incorporation for unloaded specimens were 236 cpm/μgDNA and 162 cpm/μgDNA for HuFF and 1BR3 cells, respectively (Berry et al., 2003). The corresponding normalised data for strained specimens are shown in Figure 5a. The three strain profiles produced a range of responses for the two cell types. For HuFF cells, there was a small increase in proline incorporation in response to both non-uniform and uniaxial strain, although the differences were not statistically significant. Biaxial strain, however, significantly inhibited the proline incorporation. There was minimal change in the proline incorporation for 1BR3 cells conditioned in a non-uniform strain field. By contrast, both biaxial and uniaxial strains produced an inhibitory effect, the latter being statistically significant.

The mean values of [3H]-thymidine incorporation for unloaded specimens were 881 cpm/μgDNA and 520 cpm/μgDNA for HuFF and 1BR3 cells, respectively. The corresponding normalised data are shown in Figure 5b. It is evident that the three strain profiles produced differing effects on the two fibroblast types. HuFF cells were significantly stimulated in response to uniaxial strain, but largely unaffected by either non-uniform or biaxial strain profiles. Similarly, uniaxial strain stimulated 1BR3 cells to produce an enhanced thymidine incorporation when compared to control values, although the differences were not statistically significant. By contrast, both non-uniform and biaxial strains inhibited thymidine incorporation compared to control values, the differences being statistically significant.

The study demonstrated that the exact nature of the response to dynamic tensile conditioning is highly dependent on both strain profile and cell source. In addition, it is important to clearly define the magnitude and strain profile perceived by the cells and this should be equivalent to that experienced *in situ* for fibroblasts and tenocytes (Screen et al., 2003). These are important issues in developing tensile conditioning strategies for cell-seeded constructs intended to repair load bearing tissues, such as ligaments and tendons.

Continuous Dynamic Compression for 48h – Surface and Deep Zone Bovine Chondrocytes
Articular cartilage is an heterogeneous tissue exhibiting characteristic differences in both cell morphology and activity and ECM from its surface to the calcified tissue adjacent to the subchondral bone. This prompted the authors to examine the metabolic response of bovine cells from different zones of cartilage to continuous

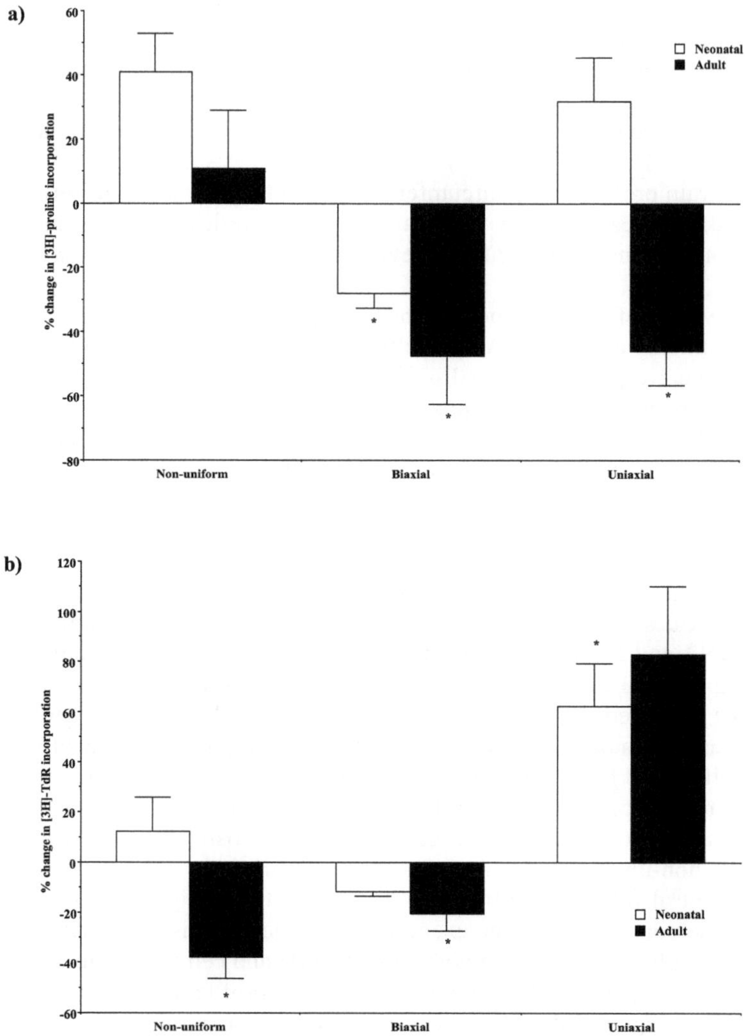

Figure 5: [^{3}H]-proline incorporation (a) and [^{3}H]-thymidine incorporation (b) by neonatal (white) and adult (black) human fibroblasts in monolayer in response to non-uniform, biaxial and uniaxial strain at a nominal amplitude of 5%. The values are presented as % change from unstrained control levels. Each value represents the mean and SE of 12 replicates. Unpaired Student's t-test results indicated differences from the control values as follows: * = $p \leq 0.05$.

dynamic compression (Lee et al., 1998a). Thus slices of cartilage, approximately 45-60 μm in thickness, representing the uppermost 15-20% of the total uncalcified tissue depth were removed from the proximal joint surface. The cells isolated from these slices were termed "superficial cells". Cells from the residual uncalcified cartilage tissue were isolated and termed "deep cells". The cells were seeded in 3% agarose constructs as previously described. Superficial and deep cell-containing constructs were cultured under a continuous compressive amplitude of 15% at 1Hz for 48 hours.

Normalised GAG synthesis data are presented in Figure 6a. It can be seen that all frequencies of dynamic strain produced an inhibition in GAG synthesis by superficial cells, the differences being statistically significant at 0.3 Hz and 3 Hz. With reference to the GAG synthesis by deep cells, 0.3 Hz produced a significant reduction, and 1 Hz induced a highly significant stimulation of 50% (Figure 6a). All three dynamic frequencies induced a significant increase in [^{3}H]-thymidine incorporation by superficial cells (Figure 6b). There was no statistically significant difference in the level of stimulation between the three dynamic strain regimens. By contrast, [^{3}H]-thymidine incorporation by deep cells was not greatly influenced by the application of compressive strain. The only significant results was the increase associated with the 0.3 Hz dynamic regime when compared to unstrained control values (Figure 6b).

Intrinsic differences in GAG synthesis and the proliferative response of chondrocyte sub-populations are revealed on application of mechanical compression. Data from this study demonstrated that the control of GAG synthesis and proliferation in response to dynamic compression are not merely uncoupled, but occur in different sub-populations of chondrocytes within the full depth cell isolate (Lee et al., 1998a). This conclusion raises the possibility of the involvement of distinct intracellular mechanotransduction mediators. One possible mediator is nitric oxide, which is known to influence both GAG synthesis and proliferation in chondrocytes, and may be modulated by physical stimuli. This has been examined by the authors in recent papers (Lee *et al.*, 1998b, 2000; Chowdhury *et al.*, 2001). Both studies indicated a down regulation of nitric oxide production, and associated PGE_2, by dynamic compression. These results have important implications both in the conditioning of cell-seeded constructs within a bioreactor and in the development of novel strategies for the treatment of degenerative cartilage disorders.

4.2 Cell Source

As previously mentioned, the use of isolated cells seeded in 3D scaffolds can be justified with reference to tissue engineered cartilage repair systems and the development of associated bioreactor systems. As such, these models should accord with the basic features of established or novel cell-based repair procedures which, for articular cartilage, typically involve the implantation of autologous chondrocytes into defect sites with or without associated 3-D scaffold (Bentley and Greer, 1971; Grande et al., 1989; Vancanti et al., 1991; Brittberg et al., 1994; Paige et al., 1996; Kawamura et al., 1998).

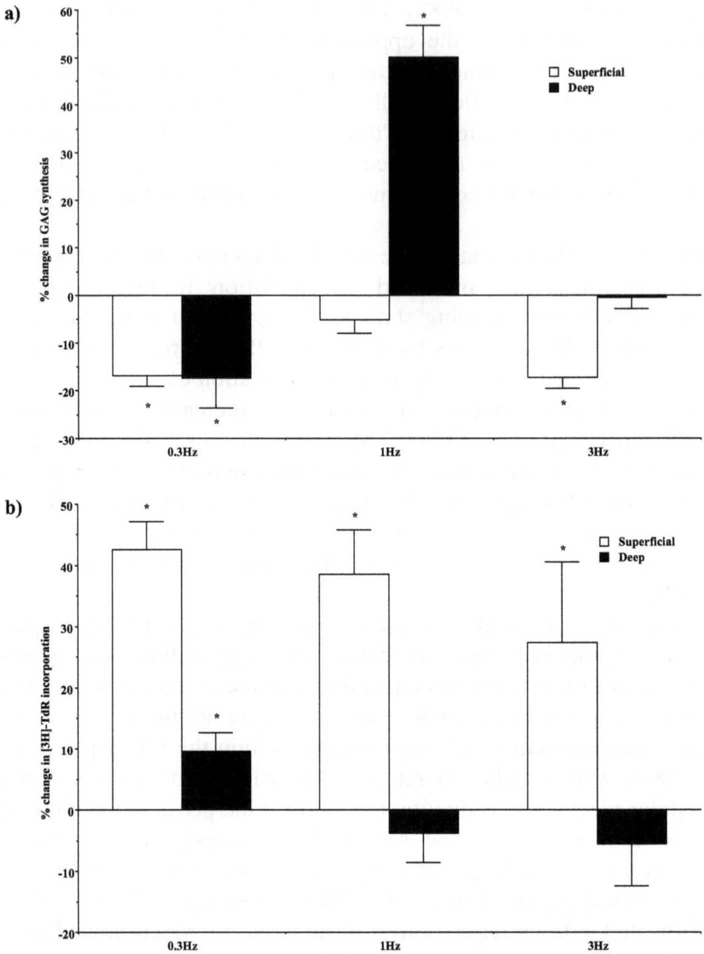

Figure 6: GAG synthesis (a) and [^{3}H]-thymidine incorporation (b) by superficial (white) and deep (black) bovine chondrocytes embedded in agarose constructs and subjected to 15% dynamic compressive strain amplitude at various frequencies for 48 hours. The values are presented as % change from unstrained control levels. Each value represents the mean and standard error of at least 12 replicates. Unpaired student's t-test results indicate differences from control values as follows: * = $p<0.05$.

The cells are typically isolated from a small tissue biopsy, removed from a low load-bearing site at the periphery of the cartilaginous joint surface, which may not conform anatomically or physiologically to the load bearing site of the traumatic or pathology-induced lesion (Brittberg et al., 1994). Moreover, the loading history of the donor and recipient sites will be different, which may influence the ability of the cells to respond appropriately to the application of mechanical conditioning. Using a small biopsy minimises potentially detrimental effects at the donor site, but limits the number of available cells. Thus the isolated cells must be cultured *in vitro* in order to increase cell number. Typically, a ten fold expansion of cells is required in approximately four weeks to ensure the practicality and financial viability of the Autologous Chondrocyte Implantation (ACI) technique (Brittberg et al., 1994). The induction of cell proliferation during expansion will necessarily increase the telomeric age of the cells which may influence the response of the cells to mechanical stimulation, as was demonstrated earlier in this chapter for neonatal (HuFFs) and adult dermal (1BR3) fibroblasts.

Most repair systems use monolayer culture techniques to expand the chondrocytes prior to implantation. While these methods are extremely effective at inducing cell proliferation, chondrocytes are known to dedifferentiate and adopt a fibroblastic phenotype when cultured in monolayer and this process is only slowly reversible (Mayne et al., 1976; Benya et al., 1978; Benya and Shaffer, 1982; deHaart et al., 1999). The dedifferentiation process is believed to be mediated by the formation of actin stress fibres which occurs when the cells spread on an adhesion-permitting substrate (Mayne et al., 1976; Benya et al., 1978). Culture systems which maintain the chondrocyte in a rounded morphology, such as agarose and alginate, prevent the formation of stress fibres and therefore induce retention of the chondrocytic phenotype or enhance the re-expression of chondrocytic phenotype by monolayer expanded chondrocytes (Takigawa et al., 1984; Brown and Benya, 1988; Benya et al., 1988; Newman and Watt, 1988; Loty et al., 1995; Guo et al., 1989; Hauselmann et al., 1992, 1994; Bonaventure et al., 1994; Binette et al., 1998). More recent studies have demonstrated that the provision of specified differentiation factors, such as FGF and TGF-β, either during expansion or subsequent culture in 3-D, can influence the re-expression of the chondrogenic phenotype (Yaeger et al., 1997; Lemare et al., 1998; deHaart et al. 1999; Jakob et al., 2001).

Accordingly a number of factors, related to the choice of cells, should be addressed by researchers developing appropriate bioreactor systems for cartilage tissue engineering. These factors lead to a series of questions, which may be summarised as follows:

- Do chondrocytes isolated from a low load-bearing site respond to mechanical conditioning in a manner which matches that of cells derived from high load-bearing regions?
- Does phenotypic modulation and acceleration of telomeric aging, which occur during monolayer expansion, influence the response of the cells to mechanical conditioning when compared to freshly isolated cells?
- Can these processes be influenced by factors, such as TGF- β, which are known to enhance the re-expression of the chondrogenic phenotype by monolayer expanded chondrocytes?

Analysis of the literature with reference to these issues is not particularly enlightening. Although there are numerous studies using model systems involving chondrocytes seeded within 3-D constructs, which employ freshly isolated or passaged cells, the presence of compounding variables such as tissue age, scaffold material and load regimen renders definitive conclusions, as opposed to simple phenomenological observation, difficult to sustain. A summary of the compounding variables associated with studies from selected key researchers has already been presented (Table 1). A number of studies by the authors have, however, attempted to address these issues in a systematic fashion.

The Effects of Tissue Location on the Response of Full Depth Bovine Chondrocytes
A recently published study involved the application of dynamic compression to chondrocyte-seeded agarose constructs containing cells isolated from either the femoral condyle and the patella groove of the equine knee joint (Wiseman et al., 2003). These tissue locations were selected to represent a highly loaded area and a low-loaded area respectively, as identified during normal articulation of the joint (Athanasiou et al., 1991). Accordingly these anatomic regions represent, respectively, the lesion and donor sites for typical tissue engineering strategies. The data from this study revealed differences between cells isolated from the two selected tissue locations, in terms of both the absolute levels of proteoglycan synthesis and cell proliferation and also changes induced by dynamic compression. Interestingly, nitric oxide, a key mediator of chondrocyte mechanotransduction was inhibited by the application of loading for cells isolated from both anatomic regions for all equine samples tested. However the heterogeneous response may have been compounded by the use of tissue of different ages and unknown exercise history, a factor which is highly relevant for the translation of tissue engineered cartilage repair systems from the laboratory to clinical use.

The Effects of Passaging on the Response of Full Depth Bovine Chondrocytes
A further study by the authors investigated the influence of passage in monolayer on the response of bovine articular chondrocytes to the application of dynamic compression. It revealed a clear relationship between passage number and mechanical sensitivity. The cells were either seeded directly into 3-D agarose constructs, to determine cell phenotype and the ability of the cells to respond to mechanical conditioning, or were passaged up to four times at weekly intervals prior to seeding in 3-D constructs. The ability of the cells to re-express a chondrocytic phenotype was assessed using immunolocalisation to detect collagen types I and II, which are key markers of fibroblastic and chondrocytic phenotypes, respectively (Figure 7). The data indicate that cells at passage 2 and beyond expressed a fibroblastic phenotype even when cultured in alginate beads, as evidenced by the presence of type I collagen and the absence of type II collagen staining. On application of dynamic compression, cells at early passages (P1 and P2) exhibited a stimulation of both proteoglycan synthesis and cell proliferation, as illustrated in Figures 8a and b, respectively. However at passages 3 and 4, the reverse was evident with dynamic compression acting to inhibit these factors, which are essential for the re-population of defect sites and the formation of a cartilaginous neo-tissue and are, therefore key for the success of tissue engineered cartilage repair systems. Thus the

application of mechanical conditioning may be considered beneficial during early passage, but detrimental at later passages.

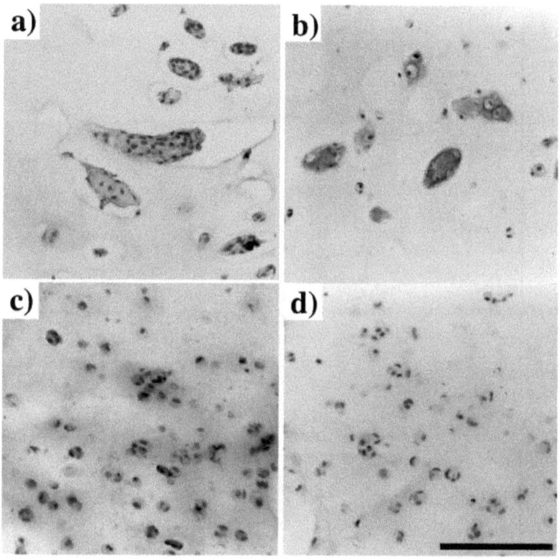

Figure 7: Micrographs representing immunolocalisation of bovine chondrocytes cultured in alginate beads for 14 days. The chondrocytes were seeded into alginate directly after isolation (a, b) or were passaged four times in monolayer prior to seeding in alginate (c, d). Immunolocalisation was performed for type I collagen (a, c) and type II collagen (b, d). All micrographs are at the same magnification, scale bar = 100 μm.

The Effects of Conditioning Medium on the Response of Full Depth Adult Human Chondrocytes

It is tempting to speculate that the modulation in response of the monolayer-expanded cells to mechanical conditioning is determined by the underlying phenotype of the cells, which is, in turn, determined by the expansion process. In order to establish a causal link, however, it is necessary to develop an experimental protocol in which cell phenotype can be influenced without an associated alteration in the telomeric age of the cells. A possible mechanism is by the application of defined chondrogenic factors to chondrocytes which have been subject to identical monolayer expansion protocols. A recent study, as yet unpublished by the authors, has utilised monolayer-expanded human chondrocytes surplus to requirement for clinical ACI repair procedure (expanded and supplied by Verigen AG, Leverkusen, Germany). The cells were seeded into agarose constructs and subjected to dynamic compression during incubation in either standard medium comprising DMEM + 20% FCS, or a defined chondrogenic medium comprising DMEM + ITS + 10 $ng.mL^{-1}$ TGF-β. This defined medium has been demonstrated, previously, to

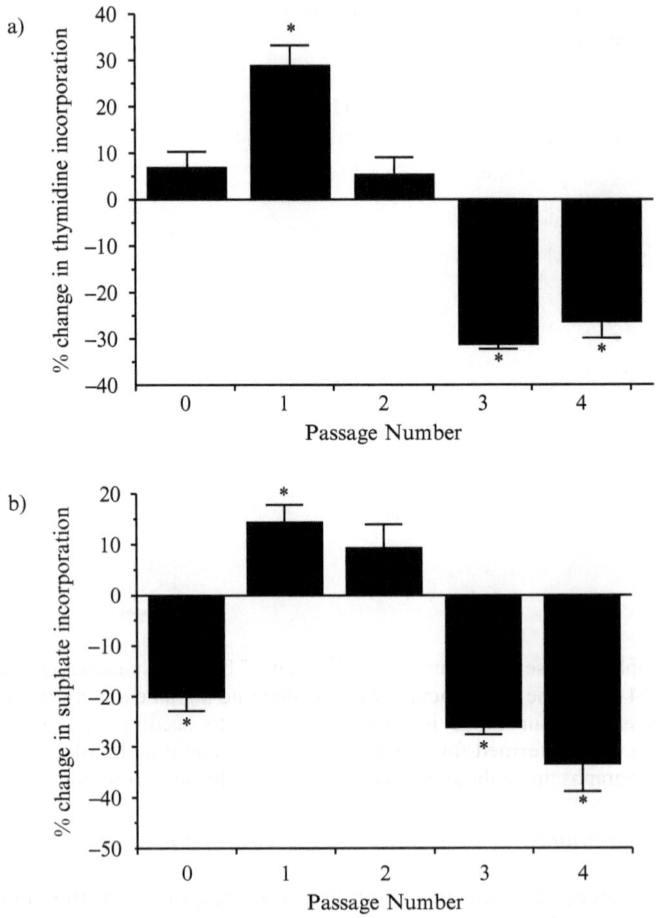

Figure 8: Sulphate incorporation (a) and [^{3}H]-thymidine incorporation (b) by freshly isolated bovine chondrocytes (designated passage 0) and chondrocyte passaged in monolayer between one and four times. The cells were subsequently embedded in agarose constructs and subjected to 15% dynamic compressive strain amplitude at 1 Hz for 24 hours. The values are presented as % change from unstrained control levels. Each value represents the mean and standard error of at least 12 replicates. Unpaired student's t-test results indicate differences from control values as follows: * = $p<0.05$.

enhance the re-expression of a chondrocytic phenotype by human chondrocytes following expansion in monolayer (Jakob et al., 2001). Absolute levels of both proteoglycan synthesis and cell proliferation were elevated during incubation in DMEM + ITS + 10 ng.mL^{-1} TGF-β compared to DMEM + 20% FCS. Importantly, during incubation in DMEM + 20% FCS, dynamic compression failed to stimulate either proteoglycan synthesis or cell proliferation, similar to the response demonstrated for monolayer-expanded bovine chondrocytes reported above (Figure 9). However, during incubation in DMEM + ITS + 10 ng.mL^{-1} TGF-β, both proteoglycan synthesis and cell proliferation were significantly stimulated by the application of dynamic compression (Figure 9). These data provide further evidence for a link between the expression of a chondrocytic phenotype and a beneficial response to dynamic compression. In addition this study demonstrates the importance of interactions between biophysical and biochemical stimuli, an understanding of which is essential for the successful development of conditioning strategies during bioreactor processing of tissue engineered constructs.

4.3 Scaffold Materials

Although the use of agarose has proved a suitable model systems to examine the metabolic effects of dynamic compression, several studies have employed other scaffold materials. For example, alginate, polyglycolic acid and protein scaffolds, such as fibrin and collagen, have all been used in mechanical conditioning studies (Table 1). It might be supposed that the cell-scaffold interactions in agarose gels are substantially different from those in protein scaffolds (Hunter and Leveston, 2002). In the former the cells preferentially bind to the synthesised pericellular ECM as it is deposited, while in protein scaffolds, cell adhesion molecules such as integrins, will enable direct interactions between the cell and the scaffold. Cell receptor binding is well known to alter mechanical behaviour, and may also alter the manner in which cells respond to mechanical conditioning. Indeed in a recent study involving chondrocytes seeded in fibrin glue, results suggested that early sustained oscillatory compression, inhibited both cell proliferation and matrix accumulation (Hunter et al., 2002). These results are in marked contrast to those found in other systems (Lee and Bader, 1997; Sah et al., 1989).

An alternative series of scaffolds proposed for cartilage tissue engineering is based on poly(ethylene glycol) (PEG) hydrogels (Bryant and Anseth, 2001). These synthetic gels provide the advantage of greater control over the scaffold degradation and the macroscopic gel properties, while photopolymerization allows in situ gelation with control, both temporally and spatially, over the polymerization reaction. Previous work has reported that photocrosslinkable hydrogels formulated from PEG macromers provide a synthetic matrix which maintains chondrocyte viability and promotes deposition of extracellular matrix rich in proteoglycans and type II collagen (Elisseeff et al., 2000). A recent, as yet unpublished, study by the authors investigated whether changes in the hydrogel crosslinking density influence the metabolic response of chondrocytes to continuous dynamic compression of 15% at a frequency of 1 Hz for 48 hours. The bovine chondrocytes were seeded into two PEG dimethacrylate (PEGDM) gels crosslinked with final concentrations of 10% and 20% (w/w).

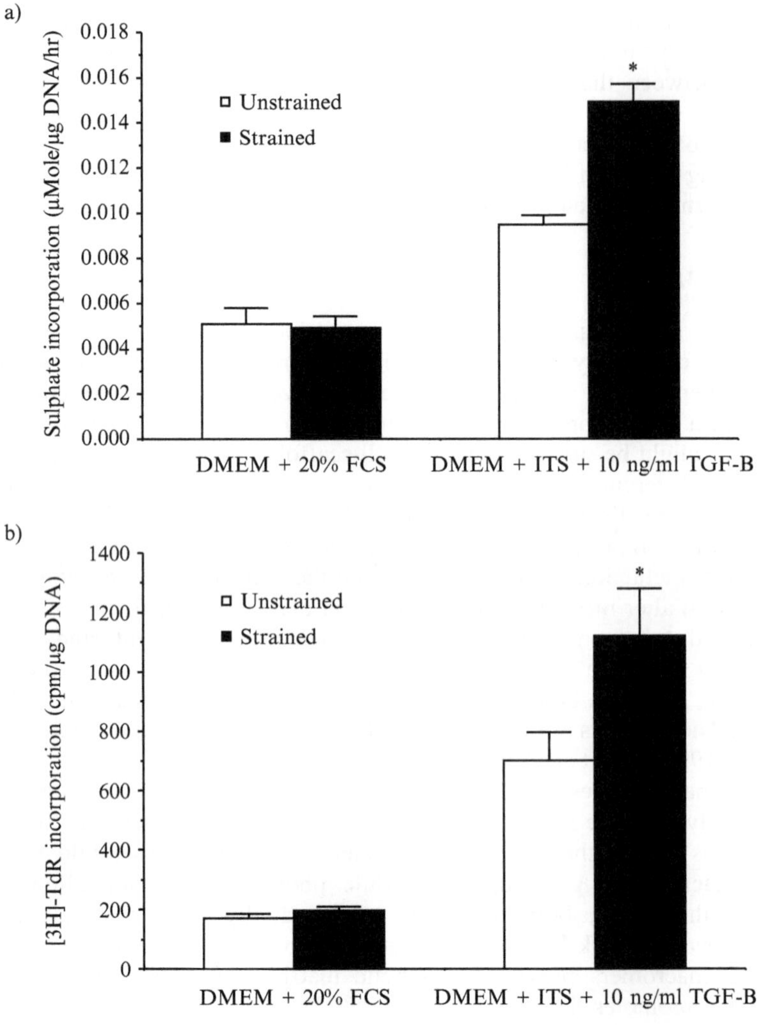

Figure 9: Sulphate incorporation (a) and [^{3}H]-thymidine incorporation (b) by human monolayer-expanded chondrocytes embedded in agarose constructs and subjected to 15% dynamic compressive strain amplitude at 1 Hz for 48 hours (black) or remained unstrained (white). The constructs were maintained in DMEM + 20% FCS or a defined medium comprising DMEM + ITS + 10 ng.mL^{-1} TGF-β. Each value represents the mean and standard error of at least 12 replicates. Unpaired student's t-test results indicate differences from control values as follows: * = $p<0.05$.

An increase in crosslinking density resulted in an inhibition in cell proliferation and proteoglycan (PG) synthesis. The normalised data for both sulphate and thymidene incorporation and nitrite production are presented in Figure 10. Dynamic compression only marginally influenced GAG synthesis in the 10% gel, although for the 20% gel, there was a marked decrease in PG production. Cell proliferation was inhibited in both crosslinked gels but, particularly, in the highly crosslinked gel. By contrast, nitrite release was slightly increased as a result of dynamic stimulation. The trends in the metabolic response to dynamic stimulation of these PEGDM gels are in marked contrast to those found for 3% agarose constructs seeded with bovine chondrocytes (Figure 10). It is clear that the interactions between cells and scaffold materials must be well characterised before successful strategies for mechanical conditioning can be adopted during processing within a bioreactor.

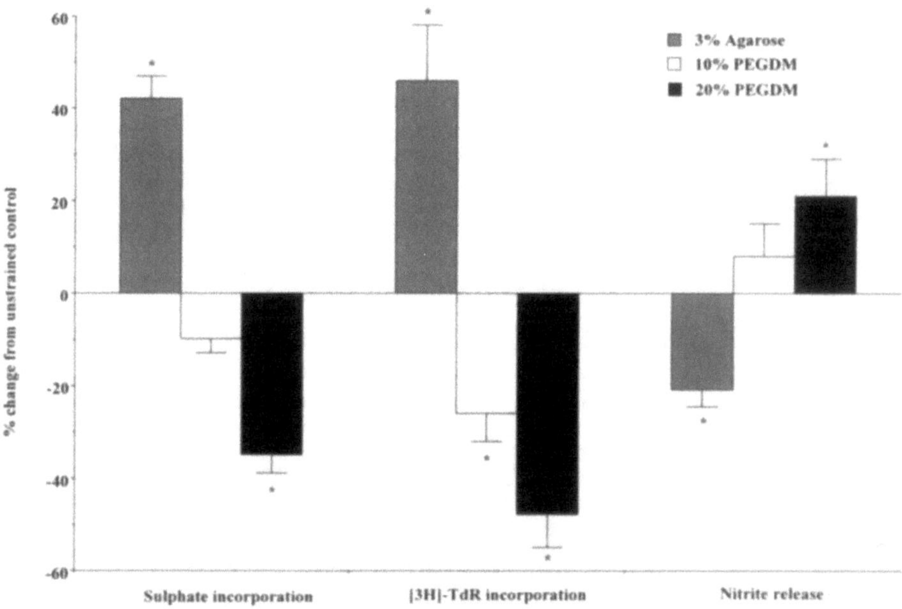

Figure 10: Sulphate incorporation (a) and [^{3}H]-thymidine incorporation (b) by full depth bovine chondrocytes embedded in 3% agarose constructs (grey) or constructs comprising 10% (white) or 20% (black) PEGDM. All constructs were subjected to 15% dynamic compressive strain amplitude at 1 Hz for 48 hours. The values are presented as % change from unstrained control levels. Each value represents the mean and standard error of at least 12 replicates. Unpaired student's t-test results indicate differences from control values as follows: * = $p<0.05$.

5. FINAL COMMENTS

This chapter has described a series of studies, all of which demonstrate the potential of dynamic mechanical conditioning in functional tissue engineering of soft tissue defects.
They also, however, confirm that the precise metabolic response of cell-seeded constructs will depend on many factors including:

- the temporal nature and the type of loading modality
- the cell sub-population in terms of age, location, differentiation state etc.
- the nature and type of the scaffold material and its interaction with the cell population

The studies also highlight the complex interplay between stimuli and metabolic parameters, which appear to be distinctive and uncoupled. Nonetheless provided suitable monitoring systems are available within a bioreactor environment, there remains a possibility of fine tuning the mechanical stimulation to elicit a specific cellular response during the *in vitro* conditioning period. The adoption of this systematic approach, involving the alteration of one variable at a time, can ultimately result in the definition of underlying mechanistic parameters leading to the derivation of predictive strategies suitable for the optimisation of neo-tissue formation within bioreactor systems.

6. ACKNOWLEDGEMENTS

The authors acknowledge the invaluable assistance of Dr Julia Shelton, Dr Tina Chowdhury, Dr Stephanie Bryant, Dr Mike Wiseman and other colleagues, who have contributed to much of the experimental work described within this chapter. We also have received generous financial support from the EPSRC, BBSRC, Smith and Nephew Group Research, Verigen AG and the European Community for a Vth Framework Project Grant ("IMBIOTOR").

7. REFERENCES

Altman GH, Lu HH, Horan RL, Calabro T, Ryder D, Kaplan DL, Stark P, Martin I, Richmond JC, Vunjak-Novakovic G. 2002. Advanced bioreactor with controlled application of multi-dimensional strain for tissue engineering. J Biomech Eng 124:742-749.

Athanasiou KA, Rosenwasser MP, Buckwalter JA, Malinin TI, Mow VC. 1991. Interspecies comparisons of in situ intrinsic mechanical properties of distal femoral cartilage. J Orthop Res 9:330-340.

Aydelotte MB, Schumacher BL and Kuettner KE. 1990. In: Maroudas A, Kuettner KE eds. Methods in Cartilage Research. London: Academic Press. pp 90-92.

Bachrach NM, Valhmu WB, Stazzone E, Ratcliffe A, Lai WM, Mow VC. 1995. Changes in proteoglycan synthesis of chondrocytes in articular cartilage are associated with the time-dependent changes in their mechanical environment. J Biomech 28:1561-1569.

Bader DL, Lee DA. 2000. Structure-properties of soft tissues: articular cartilage In: Elices M. ed. Structural Biological Materials: Design and Structure Property Relationships. Oxford: Pergamon. pp 73-104.

Bader DL, Ohashi T, Knight MM, Lee DA, Sato M. 2002. Deformation properties of articular chondrocytes: a critique of three separate techniques. Biorheol 39:69-78.

Banes AJ, Gilbert J, Taylor D, Monbureau O. 1985. A new vacuum-operated stress-providing instrument that applies static or variable cyclic tension or compression to cells in vitro. J Cell Sci 75:35-42.

Banes AJ, Tsuzaki M, Hu PQ, Brigman B, Brown T, Almekinders L, Lawrence WT, Fischer T. 1995. PDGF-BB, IGF-1 and mechanical load stimulate DNA synthesis in avian tendon fibroblasts in vitro. J Biomechanics 28:1505-1514.

Banes AJ, Link GW, Gilbert JW, Tay RTS, Monbureau O. 1990. Culturing cells in a mechanically active environment. Am Biotechnol Lab 8:12-22.

Bentley G, Greer RB. 1971. Homotransplantation of isolated epiphyseal and articular cartilage chondrocytes into joint surfaces of rabbits. Nature 230:184-197.

Benya PD, Padilla SR, Nimni ME. 1978. Independent regulation of collagen types of chondrocytes during the loss of differentiated function in culture. Cell 15:1313-1321.

Benya PD, Brown PD, Padilla SR. 1988. Microfilament modification by dihydrocytochalasin B causes retinoic acid-modified chondrocytes to reexpress the differentiated collagen phenotype without a change in shape. J Biol Chem 106:161-170.

Benya PD, Shaffer JD. 1982. Dedifferentiated chondrocytes reexpress the differentiated collagen phentotype when cultured in agarose gels. Cell 30:215-224.

Berry CC, Cacou C, Lee DA, Bader DL, Shelton JC. 2003. Dermal fibroblast respond to mechanical conditioning in a strain profile dependent manner. Biorheol 40:337-345.

Binette F, McQuaid MP, Haudenschild DR, Yaeger PC, McPherson JM, Tubo R. 1998. Expression of a stable articular cartilage phenotype without evidence of hypertrophy by human articular chondrocytes in vitro. J Orthop Res 16:207-216.

Bishop JE, Michell JJ, Absher PM, Baldor L, Geller HA, Woodcock-Michell J, Hamblin MJ, Vacek P, Low RB. 1993. Cyclic mechanical deformation stimulates human lung fibroblast proliferation and autocrine growth factor activity. Am J Respir Cell Mol Biol 9:126-33.

Bonassar LJ, Grodzinsky AJ, Frank EH, Davila SG, Bhaktav NR and Trippel SB. 2001. The effect of dynamic compression on the response of articular cartilage to IGF-1. J Orthop Res 19:11-17.

Bonaventure KM, Kadhom N, Cohen-Solan L, Ng KH, Bouguisgnon J, Lasselin C and Freisinger P. 1994. Re-expression of cartilage-specific genes by de-differentiated human articular chondrocytes cultured in alginate beads. Exp Cell Res 212:97-104.

Brittberg M, Lindahl A, Nilsson A, Ohlsson C, Isaksson O and Peterson L. 1994. Treatment of deep cartilage defects in the knee with autologous chondrocyte transplantation. New Eng J Med 331:889-895 .

Bryant SJ, Anseth KS. 2001. Hydrogel properties influence ECM production by chondrocyte photoencapsulated in polyethylene glycol. hydrogels. J Biomed Mater Res 59:63-72.

Brown PD, Benya PD. 1988. Alterations in chondrocyte cytoskeletal architecture during phenotypic modulation by retinoic acid and dihydrocytochalasin B-induced reexpression. J Cell Biol 106:171-179.

Buschmann MD, Gluzband YA, Grodzinsky AJ, Hunziker EB. 1995. Mechanical compression modulates matrix biosynthesis in chondrocyte/agarose culture. J Cell Science 108:1497-1508 .

Buschmann MD, Kim YJ, Wong M, Frank E, Hunziker EB, Grodzinsky AJ. 1999. Stimulation of aggrecan synthesis in cartilage explants by cyclic loading is localized to regions of high interstitial fluid flow. Arch Biochem Biophys 366:1-7.

Cacou C, Palmer, D, Lee DA, Bader DL, Shelton JC. 2000. A system for monitoring the response of uniaxial strain on cell seeded collagen gels. Medical Engineering and Physics 22:327-333.

Chowdhury TT, Bader DL, Lee DA. 2001. Dynamic compression inhibits the synthesis of nitric oxide and PGE2 by IL-1B stimulated chondrocytes cultured in agarose constructs. Biochem Biophys Res Comm 285:1168-74.

Elisseeff J, McIntosh W, Anseth K, Riley S, Ragan P, Langer R. 2000. Photoencapsulation of chondrocytes in polyethylene oxide - based semi-interpenetrating networks. J Biomed Mater Res 51:164-171.

Farndale RW, Buttle DJ, Barrett AJ. 1986. Improved quantitation and discrimination of sulfated glycosaminoglycans by use of dimethylmethylene blue. Biochim Biophys Act 883:173-177.

Fermor B, Weinberg JB, Pisetsky DS, Misukonis MA, Banes AJ, Guilak F. 2001. The effects of static and intermittent compression on nitric oxide production in articular cartilage explants. J Orthop Res 19:729-737.

Freeman PM, Natarajan RN, Kimura JH, Andriacchi TP. 1994. Chondrocytes respond mechanically to compressive loads. J Orthop Res 12:311-320.

Gilbert JA, Weinhold PS, Banes AJ, Link GW, Jones GL. 1994. Strain profiles for circular cell culture plates containing flexible surfaces employed to mechanical deform cells in vitro. J Biomechanics 27:1169-1177.

Grande DA, Pitman MI, Peterson L, Menche D, Klein M. 1989. The repair of experimentally produced defects in articular cartilage by autologous chondrocyte transplantation. J Orthop Res 7:208-218.

Gray ML, Pizzanelli AM, Lee RC, Grodzinsky AJ, Swann DA. 1989. Kinetics of the chondrocyte biosynthetic response to compressive load and release Biochim Biophys Acta 991:415-425.

Grymes RA, Sawyer CA. 1997. Novel culture morphology resulting from applied mechanical strain. In Vitro Cell Dev Biol Anim 33:392-397.

Guilak F, Meyer BC, Ratcliffe A, Mow VC. 1994. The effects of matrix compression on proteoglycan metabolism in articular cartilage explants. Osteoarthritis Cartilage 2:91-101.

Guilak F, Sah R, Setton LA 1997. In: Mow VC, Hayes WC eds. Basic Orthopaedic Biomechanics. Philadelphia: Lippincott-Raven Publishers. pp 179-207.

Guilak F, Butler DL, Goldstein SA. 2001. Functional tissue engineering: the role of biomechanics in articular cartilage repair. Clin Orthop Relat R 391:S295-S305 Suppl. S.

Guo J, Jordan GW, MacCallum DK. 1989. Culture and growth characteristics of chondrocytes encapsulated in alginate beads. Conn Tiss Res 19:277-297.

deHaart M, Marijnissen WJCM, van Osch GJVN, Verhaar JAN. 1999. Optimization of chondrocyte expansion in culture: effects of TGF-β, bFGF and L-ascorbic acid on bovine articular chondrocytes. Acta Orthop Scand 70:55-61.

Hauselmann HJ, Fernandes RJ, Mok SS, Schmid TM, Block JA, Aydelotte MB, Kuettner KE, Thonar EJMA. 1994. Phenotypic stability of bovine articular chondrocytes after long-term culture in alginate beads. J Cell Sci 107:17-27.

Hauselmann HJ, Aydelotte MB, Schumacher BL, Kuettner KE, Gillelis SH, Thonar EJMA. 1992. Synthesis and turnover of proteoglycan by human and bovine articular chondrocytes cultured in alginate beads. Matrix 12:116-129.

Heath CA, Magari SR. 1996. Mechanical factors affecting cartilage regeneration in vitro. Biotechnol Bioeng 50:430-437.

Hunter CJ, Imler SM, Malaviya P, Nerem RM, Levenston ME. 2002. Mechanical compression alters gene expression and extracellular matrix synthesis by chondrocytes cultured in collagen I gels. Biomaterials 23: 1249-59.

Hunter CJ, Mouw JK, Levenston ME. 2002. Oscillatory compression stimulates matrix production in an in vitro repair model. Trans Orthop Res Soc 27:136.

Hunter CJ, Levenston ME. 2002. Native/engineered cartilage adhesion varies with scaffold material and does not correlate to gross biochemical content. Trans Orthop Res Soc 27:479.

Jakob M, Demarteau O, Schafer D, Hintermann B, Dick W, Heberer M, Martin I. 2001. Specific growth factors during the expansion and redifferentiation of adult human articular chondrocytes enhance chondrogenesis and cartilaginous tissue formation in vitro. J Cell Biochem 81:368-77.

Jones GL. 1994. Strain profiles for circular cell culture plates containing flexible surfaces employed to mechanical deform cells in vitro. J Biomechanics 27:1169-1177.

Kawamura S, Wakitani S, Kimura T, Maeda A, Caplan AI, Shino K, Ochi T. 1998. Articular cartilage repair, rabbit experiments with a collagen gel-biomatrix and chondrocytes cultured in it. Acta Orthop Scand 69:56-62.

Kim YJ, Sah RLY, Grodzinsky AJ, Plaas AHK, Sandy JD. 1994. Mechanical regulation of cartilage biosynthetic behaviour: physical stimuli. Arch Biochem Biophys 311:1-12.

Kisiday J, Jin M, Grodzinsky AJ. 2002. Effects of dynamic compressive loading duty cycle on in vitro conditioning of chondrocyte seeded peptide and agarose scaffolds. Trans Orthop Res Soc 27:216.

Kiviranta I, Tammi M, Jurvelin J, Saamanen A-M, Heliminen HJ. 1988. Moderate running exercise augments glycosaminoglycans and thickness of articular cartilage in knee joint of young beagle dogs. J Orthop Res 6:188-95.

Knight MM, Lee DA, Bader DL. 1996. Distribution of chondrocyte deformation in compressed agarose gel using confocal microscopy. J Cell Eng 1:97-102.

Knight MM. 1997. Deformation of isolated articular chondrocytes cultured in agarose constructs. PhD Thesis, University of London.

Knight MM, Ghori SA, Lee DA, Bader DL. 1998a. Measurement of the deformation of isolated chondrocytes in agarose subjected to cyclic compression. J Med Eng Phys 20:684-688.

Knight MM, Lee DA, Bader DL. 1988b. The influence of elaborated pericellular matrix on the deformation of isolated articular chondrocytes cultured in agarose. Biochim Biophys Act 1405:67-77.

Knight MM, Bravenboer JVDB, Lee DA, van Osch GJVM, Weinans H, Bader DL. 2002. Cell and nucleus deformation in compressed chondrocyte-alginate constructs: temporal changes and calculation of cell modulus. Biochim Biophys Act 1570:1-8.

Lee DA, Bader DL. 1995. The development and characterization of an in vitro system to study strain-induced cell deformation in isolated chondrocytes. In Vitro Cell Dev-Am. 31:828-835.

Lee DA, Bader DL. 1997. Compressive strains at physiological frequencies influence the metabolism of chondrocytes seeded in agarose. J Orthop Res 15:181-188.

Lee DA, Noguchi T, Knight MM, O'Donnell L, Bentley G, Bader DL 1998a. Response of chondrocyte sub-populations cultured within unloaded and loaded agarose. J Orthop Res 16:726-733.

Lee DA, Frean S, Lees P, Bader DL. 1998b. Dynamic mechanical compression influences nitric oxide production by articular chondrocytes seeded in agarose. Biochem Biophys Res Comm 251:580-585.

Lee DA, Assoku E, Doyle V. 1998c. A specific quantitative assay for collagen synthesis by cells seeded in collagen-based biomaterials using sirius red F3B precipitation. J Mat Sci: Materials in Medicine 9:47 – 51.

Lee DA, Noguchi T, Frean SP, Lees P, Bader DL. 2000. The influence of mechanical loading on isolated chondrocytes seeded in agarose constructs. Biorheology 37:149-161.

Lee DA Knight MM, Bolton JF, Idowu BD, Kayser MV, Bader DL et al 2000. Chondrocyte deformation within compressed agarose constructs at the cellular and sub-cellular levels. J Biomechanics 33:81-95.

Lemare F, Steinberg N, Le Griel C, Demignot S, Adolphe M. 1998. Dedifferentiated chondrocytes cultured in alginate beads: restoration of the differentiated phenotype and of the metabolic response to interleukin-1 beta. J Cell Physiol 176:303-313.

Li KW, Williamson AK, Wang AS, Sah RL. 2001. Growth responses of cartilage to static and dynamic compression. Clin Orthop 391:S34-48.

Loty S, Forest N, Boulekbache H, Sautier J. 1995. Cytochalasin D induces changes in cell shape and promotes in vitro chondrogenesis: a morphological study. Biol Cell 83:149-161 .

Martin I, Obradovic B, Treppo S, Grodzinsky AJ, Langer R, Freed LE, Vunjak-Novakovic G. 2000. Modulation of the mechanical properties of tissue engineered cartilage. Biorheology 37:141-147.

Mauck RL, Soltz MA, Wang CC, Wong DD, Chao PH, Valhmu WB, Hung CT, Ateshian GA. 2000. Functional tissue engineering of articular cartilage through dynamic loading of chondrocyte-seeded agarose gels. J Biomech Eng 122, 252-260.

Mayne R, Vail MS, Mayne PM, Miller EJ. 1976. Changes in the type of collagen synthesised as clones of chick chondrocytes grow and eventually lose division capacity. Proc Nat Acad Sci USA 73:1674-1678.

Millward-Sadler SJ, Wright MO, Lee H, Nishida K, Caldwell H, Nuki G, Salter DM. 1999. Integrin-regulated secretion of interleukin 4: a novel pathway of mechanotransduction in human articular chondrocytes. J Cell Biol 145:183-189.

Nerem, RM. 2000. Tissue engineering: confronting the transplant crisis. Proc Instn Mech Engrs Part H 214:95-99.

Newman P, Watt FM. 1988. Influence of cytochalasin D-induced changes in cell shape on proteoglycan synthesis by cultured articular chondrocytes. ExpCell Res 178:192-210.

Paige KT, Cima LG, Yaremchuk MJ, Schloo B, Vacanti JP, Vacanti CA. 1996. De novo cartilage generation using calcium alginate-chondrocyte constructs. Plastic Reconstruct Surg 97:168-178 .

Rao J, Otto WR. 1992. Fluorimetric DNA assay for cell growth estimation. Anal Biochem 207:186-192 .

Robling AG, Hinant FM, Burr DB, CH Turner CH. 2002. Improved bone structure and strength after long-term mechanical loading is greatest if loading is separated into short bouts. J Bone Miner Res 17:1545-1554.

Rubin CT, Lanyon LE. 1992. Regulation of bone formation by applied dynamic loads. J Bone Joint Surg 66:397-402.

Saamanen A-M, Tammi M, Jurvelin J, Kiviranta I, Helminen HJ. 1990. Proteoglycan alteration following immobilisation and remobilisation in the articular cartilage of young canine joint. J Orthop Res 8:863-873.

Sah RLY, Kim YJ, Doong JY, Grodzinsky AJ, Plaas AH, Sandy JD. 1989. Biosynthetic response of cartilage explants to dynamic compression. J Orthop Res 7:619-636.

Screen HRC, Lee DA, Bader DL, Shelton JC. 2003. Development of a technique to determine strains in tendons using the cell nuclei. Biorheology 40:361-368.

Takigawa M, Takano T, Shirai E, Suzuki F. 1984. Cytoskeleton and differentiation: effects of cytochalasin B and colchicine on expression of the differentiated phenotype of rabbit costal chondrocytes in culture. Cell Diff 14:197-204.

Urban JPG. 1994. The chondrocyte: a cell under pressure. Brit J Rheumatol 33:901-908.

Vacanti CA, Langer R, Schloo B, Vacanti JP. 1991. Synthetic biodegradable polymers seeded with chondrocytes provide a template for new cartilage formation in vivo. J Plast Reconstruct Surg 87:753-759.

Vunjak-Novakovic G, Martin I, Obradovic B, Treppo S, Grodzinsky AJ, Langer R, Freed LE. 1999. Bioreactor cultivation conditions modulate the composition and mechanical properties of tissue engineered cartilage J Orthop Res 17, 130-138.

Wiseman M, Henson F, Lee DA, Bader DL. 2003. Dynamic compressive strain inhibits nitric oxide synthesis by equine chondrocytes isolated from different areas of the cartilage surface. Equine Vet J 35:451-456.

Yaeger PC, Masi TL, deOrtiz JL, Binette F, Tubo R, McPherson JM. 1997. Synergistic action of transforming growth factor-beta and insulin-like growth factor-I induces expression of type II collagen and aggrecan genes in adult human articular chondrocytes. Exp Cell Res 237:318-325.

CHAPTER 8

MECHANICAL BIOREACTORS FOR BONE TISSUE ENGINEERING

S.H. CARTMELL AND A.J. EL HAJ

Centre for Science and Technology in Medicine, University of Keele, Stoke-on-Trent, U.K.

1. INTRODUCTION

Static *in vitro* culture for cell monolayer and small explants has been regularly employed for many decades. These protocols have provided adequate nutrients and oxygenation to the cells or tissue by incubation in a temperature controlled and CO_2/bicarbonate buffered environment. Scale up to increase the number of cells or size of a cell seeded scaffold leads to limitations in the diffusion of nutrients and waste products throughout the tissue or engineered construct. This often allows the formation of a construct that has cell viability and proliferation at the periphery but a cell seeded scaffold core that is necrotic. Bioreactors can control the environment for tissue engineering and conditioning such that these diffusion limitations are reduced. The biochemical environment in a bioreactor can be controlled by allowing the transport of nutrients, (e.g. glucose and dissolved oxygen) to the cells and the removal of degradation products away from the cells seeded throughout the construct. Factors such as pH, growth factors and other cell signalling molecules available to the construct can also be monitored. Figure 1 demonstrates some of these criteria that is needed to be considered when designing a bioreactor for bone tissue engineering.

J.B. Chaudhuri & M. Al-Rubeai (eds.), Bioreactors for Tissue Engineering, 193-208.

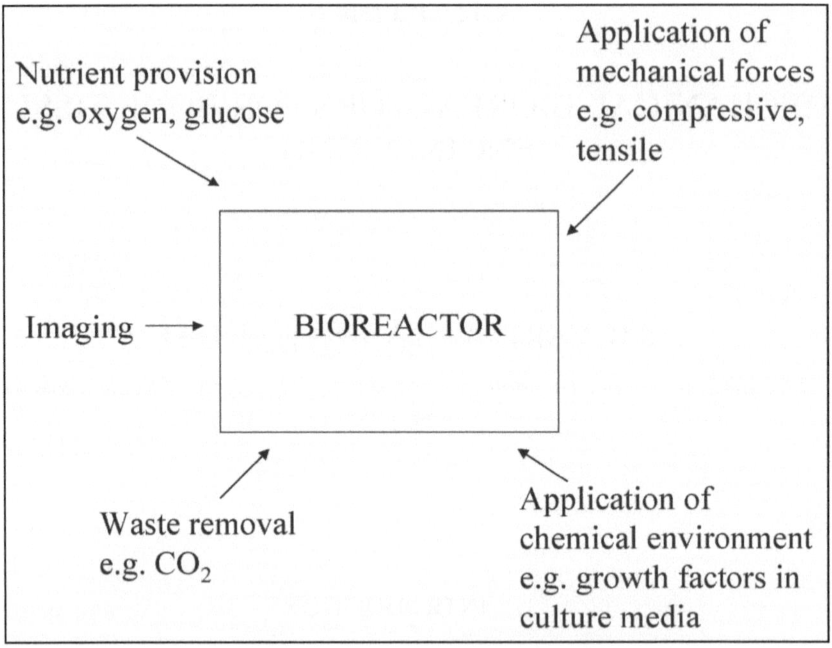

Figure 1: Schematic of the various parameters considered when designing a bioreactor for bone tissue engineering.

Developing large-scale cell bioreactors for scale up has been widely applied in the cell recombinant technology processing industry. In these reactors, much work has been carried out into improving cell nutrient supply and viability when culturing large numbers of cells in mass culture. These circulation or mixing strategies improve mass transfer within the cultures but are not specifically designed to provide physiologically relevant levels of strain or mechanical mixing. These types of production units may meet the requirements for cell proliferation required as a source of high cell numbers for stem cell therapies without delivery vehicles or implant tissue production. In contrast for tissue engineering where the ultimate aim is an 'off the shelf' tissue, bioreactors aim to meet two requirements, firstly, to improve mass transfer as described above potentially through mixing or perfusion strategies and secondly, to apply physiologically relevant loads to the tissues which enables constructs to be conditioned according to the implantation environment and matrix production to be accelerated *in vitro*, thus reducing production time. This is particularly relevant to bone tissue engineering where load bearing is a major requirement for any replacement tissue.

Studies *in vivo* have demonstrated that the biomechanical profile on the skeleton is made up of a number of components. Long bones such as the femur are not entirely straight but curvilinear in nature, therefore the application of force induces both localised compression and bending (Rubin and Lanyon, 1982). Bending induces both compression and tension forces on the opposing surfaces of a long bone with the neutral axis between the two found to experience rotational force (Jones *et al.*, 1995). The nature of the applied load is important in determining any possible response, e.g. dynamic loading of approximately 1000 □icrostrain *in vivo* is known to induce bone formation (Rubin and Lanyon, 1984; Rubin and Lanyon, 1985; Turner *et al.*, 1994a) which is absent when bones are subjected to static loading (Turner *et al.*, 1994b; Lanyon and Rubin, 1984). The major parameters which define the effectiveness of a particular loading regime in stimulation of an osteogenic response are the strain magnitude, strain rate and frequency (Lanyon and Rubin, 1984; Rubin and Lanyon, 1984; Rubin and Lanyon, 1985; Rubin & McLeod, 1994; Turner 1998). The duration of the period of mechanical loading is also known to have an effect, but extended periods of loading are known to have a diminishing effect upon load induction (Rubin and Lanyon, 1984), as the bone formation response becomes saturated (Turner, 1998). The sensitivity of various bone cell types to minute changes in perceived strain is very high with deformations of 0.1% (1000 □icrostrain) sufficient to induce a response *in vivo* (Rubin & Lanyon, 1984) and at considerably lower strains *in vitro* (Thomas and El Haj, 1996; Salter *et al.*, 1997, Walker, 1999). As not all of the cells within the bone experience the peak strain, then any response must be co-ordinated by intercellular signalling mechanisms.

In bone fracture repair, the type and magnitude of loading (compressive, tensile, hydrostatic pressure and strain) seem to control which type of tissue will be formed from stem cells sources such as bone marrow in the callus tissue, i.e. fibrous tissue, fibrocartilagenous tissues or intramembranous bone (Claes and Heigele, 1999). High strains (5-15% deformation) have been proposed to induce connective tissue or fibrocartilage, wheras low strains (0.15-0.3%) of compression of tension have been proposed to induce intramembranous bone formation.

With this in mind, many investigators have studied the effects of differing components of load, shear, tensile and compressive forces on bone cells in 2D and organ culture. The biomechanical environment can be controlled in many ways. Many different stages of bone cell differentiation are responsive to these types of mechanical forces. Cell activity (such as proliferation rate, differentiation or matrix production) has been shown to be influenced by the application of such forces. Shear stress can be applied to the cells via altering the flow rate and pattern of the culture media in the bioreactor. It is also possible to deliver tensile or compressive forces to cellular constructs in a bioreactor via a number of strategies. In this chapter, we outline the different methods of applying mechanical forces to cells using bioreactors and also describe the effects that these forces have on cell activity.

2. MECHANICAL FORCE APPLICATION VIA SHEAR STRESS FROM FLUID FLOW IN MONOLAYER

Methodologies have been optimised to apply shear stress stimulation to monolayer cultures. These systems have differed in the pattern of fluid flow that is applied (i.e.

unidirectional, bi-directional, pulsatile and/or oscillatory) and have also differed in their mode of shear stimulus application to cells. These include the parallel plate flow chamber in which fluid is 'forced' across the cell monolayer by a pressure gradient. These parallel plate flow chambers have been widely used to study the effects of fluid flow on bone cells. Such a system utilises a monolayer of cells grown on polylysine coated glass slides, which forms the bottom of the parallel plate chamber (Klein-Nulend *et al.*, 1997) modified from the original design by Frangos and associates (Frangos *et al.*, 1985). Culture media is circulated through the chamber at adjustable and determinable rates allowing the study of the effects of varying the flow's rate, duration and frequency on the bone cells. The flow of fluid in this manner induces two different stimuli – shear stress and streaming electrical potentials (Salzstein *et al.*, 1987; Salzstein and Pollack, 1987). Both osteocytes and osteoblasts have been studied using this system.

Mechanical loading of osteocytes, the terminal differentiation stage of bone cells, using the parallel flow chamber has been shown to induce bone related gene expression and protein production. Primary osteocytes that were isolated from chick calvariae have been shown to respond to a 1hour pulsating fluid flow by upregulating prostaglandin E_2 and I_2 production (Klein-Nulend *et al.*, 1995a). Prostaglandins upregulate osteoblast proliferation and bone matrix formation and are essential in the transfer of mechanical stimuli to bone formation. An even earlier response to shear stress is seen in chick osteocytes by the elevation of levels of nitric oxide production in response to fluid flow (Klein-Nulend *et al.,* 1995b). Comparisons between cell types using a pulsating fluid flow (0.5±0.02Pa, 5Hz, 0.4Pa/sec) showed that osteocytes and calvarial derived less mature bone cells but not periosteal fibroblasts derived from dental tissues upregulate NO release 2-3 fold. This effect was at a maximum after 5 minutes and then levelled off. Nitric oxide is a short lived free radical that has been shown to inhibit osteoclastic activity and stimulate osteoblast proliferation (Sikavitsas *et al.*, 2001) and may be involved in the signalling from one cell to another along the bone cell lineage.

Mature osteoblasts have also been shown to be responsive to mechanical forces applied in the form of fluid flow. Using the parallel plate chamber, it has been shown that primary murine osteoblasts respond to pulsatile fluid flow (5 Hz, 0.7±0.03 Pa, 12.2 Pa/sec) by upregulating prostaglandin E_2 and I_2 production (Klein-Nulend *et al.*, 1997). Primary human osteoblasts have shown a rapid response in upregulation of nitric oxide in response to pulsatile fluid flow at these same conditions (Klein-Nulend *et al.*, 1998). The same research group have shown that these responses were due to the shear stress application of fluid flow rather than the streaming electrical potential (Bakker *et al.*, 2001). This was accomplished by comparing the effects of low (0.2cm^3/s, 3Hz shear stress 0.4±0.12Pa) medium (0.33cm^3/s, 5Hz shear stress 0.6±0.27Pa) and high (0.63cm^3/s, 9Hz shear stress 1.2±0.37Pa) fluid flow rates on primary murine osteoblasts. In some of the low pusatile fluid flow rate experiments, 2.8% dextran was added to the media to allow approximately the same shear stress to be applied to the cells as in the high pulsatile fluid flow samples but at the low flow rate. Upregulation of nitric oxide and prostaglandin production was seen by the osteoblasts exposed to the higher stresses (low flow rate with dextran and high flow rate) in comparison to the low flow rate (no dextran) and medium flow rates indicating that this response was shear stress dependant.

A nine and twenty fold increase in prostaglandin E_2 production was seen when rat calvarial osteoblasts were exposed to fluid flow shear stresses of 6 dyn/cm^2 and 30 dyn/cm^2 respectively (Reich and Frangos, 1991). These experiments utilised a flow chamber designed by Frangos and associates (Frangos *et al.*, 1985; Frangos *et al.*, 1988) and marketed by Cytodyne (San Diego, CA) similar to the parallel plate chamber design. The same steady state flow conditions (6 dyn/cm^2) was applied to rat calvarial osteoblasts for a 12 hour time period and it was seen that the cells sustained elevated production of nitric oxide throughout the force application (Johnson *et al.*, 1996). Expression of c-Fos and COX-2 was increased by MC3T3-E1 murine osteoblast-like cells in response to fluid shear (Pavalko *et al.*, 1998). McAllister and Frangos have also demonstrated using this same set-up that rat calvarial osteoblasts respond to steady state fluid flow by upregulating nitric oxide production by two distinct pathways – a G-protein and calcium dependant phase sensitive to flow transients, and a G-protein and a calcium-independent pathway stimulated by sustained flow (McAllister and Frangos, 1999). A burst in nitric oxide production (8.2 nmol/mg of protein/hour) was seen initially in response to the fluid flow with a lower sustained nitric oxide production (2.2 nmol/mg of protein/hour). Using these systems it has been possible to further understand the nature of flow induced signalling in these cells.

Jacobs and associates have extended experiments to look at the effects of fluid flow on osteoblasts using a parallel plate flow chamber modified from the Frangos and associates design (Frangos *et al.*, 1985) to accept quartz glass microscope slides required for fluorescent imaging techniques (Jacobs *et al.*, 1998). The chamber is made up of a polycarbonate manifold, rubber gasket and a microscope slide which is all held together by a vacuum pressure. A hydraulic actuator and a motor is also connected to the system to allow the fluid flow to be oscillated over the monolayer of cells rather than steady or pulsatile flow. Oscillating flow was chosen for study, as physiologically, bones are loaded repetitively rather than in a sinusoidal manner. The osteoblasts have been shown to respond to oscillating flow in a very different manner than to steady or pulsatile flow (Jacobs *et al.*, 1998). Fluid flow was applied to human fetal osteoblasts (hFOB 1.19) in either a steady state (resulting in a wall shear stress of 2Nm^{-2}), pulsating state (0 to 2Nm^{-2}) or an oscillating state (±2Nm^{-2}) at 0.5, 1 or 2 Hz. It was found that oscillating fluid flow was a much less potent stimulator of bone cells than either steady or pulsating fluid flow.

Oscillating fluid flow (±2Nm^{-2}, 1Hz) also results in an increase in the levels of intracellular calcium (You *et al.*, 2001, Allen *et al.*, 2000, Hung *et al.*, 1995). Further downstream, mRNA levels of osteopontin were seen to be elevated. It has also been demonstrated that oscillating fluid flow inhibits the production of the mRNA of several adhesion proteins such as TNF-α-induced intercellular adhesion molecule-1 by UMR106 cells experiencing stresses in the range of 1.9 to 9.3 dyn/cm^2 (Kurokouchi *et al.*, 2001).

A parallel fluid flow chamber was also utilised in experiments described by Nauman and associates (Nauman *et al.,* 2001) to investigate effects on primary rat stromal cells. An acute response to pulsatile fluid flow (Shear stress: 0.6±0.5Pa, 3Hz) by a fourfold production in prostaglandin E_2 production was seen from primary rat stromal cells. The application of low (Shear stress: 0.06±0.05Pa, 0.3Hz), high (Shear stress: 0.6±0.5Pa, 3Hz) pulsatile fluid flow and static culture showed no significant difference in cell proliferation or matrix mineralization after 7 days.

Other systems have been designed to induce fluid flow in different monolayer culture systems other than the parallel plate flow chamber. Ogata has performed experiments on fluid flow over MC3T3-E1 osteoblasts by shaking the culture dishes on a shaker, horizontally with an amplitude of 3cm at a rate of 60 shakes/min for 1 or 5 minute time periods (Ogata, 2000). The samples were stimulated in this manner nine times and it was found that an upregulation of egr-1 mRNA and tyrosine phosphorylation enhancement of many proteins including ERK2 and Shc occurred as a result.

Rotating disk apparatus has been used to study the effect of fluid flow on bone cells in monolayer (Horikawa *et al*., 2000). A fluid shear stress of approximately 1Pa (~10dynes/cm^2) induced MC3T3-E1 osteoblasts to change from a polygonal to a spindle shape after 1 hour of exposure. The control cells remained polygonal and the stressed cells exhibited an increase in the number of microvilli. A more prolonged stress over 1 hour caused cell shrinkage and a decrease in intracellular calcium levels suggesting that these conditions may be causing cellular damage.

Different kinds of mechanical stress have also been applied to MC3T3-E1 cells such as centrifugation (Fitzgerald and Hughes-Fulford, 1996) and vibration (Tjandrawinata *et al*, 1997), which result in similar cell responses such as small changes in c-Fos and osteocalcin mRNA expression. It has been shown that mild centrifugation (287 x *g*) stimulated a ten-fold increase in c-Fos and prostaglandin E_2 production by MC3T3-E1 osteoblasts (Fitzgerald and Hughes-Fulford, 1999). In order to apply this force, confluent cultures were centrifuged at 1600 RPM for 5 minutes. The centrifuge had specially designed trunnions to hold the culture plates in place and the strain was experienced primarily through the apical-basal axis.

The studies described in this section indicate that flow conditions generated *in vitro* have the potential to be a powerful tool that could be used in the manufacture of a tissue engineered bone construct. It is clear that when bone cells experience a given force within the correct parameters they respond by upregulating bone-related proteins and genes which ultimately could lead to tissue engineered constructs that either have more strength than their non-mechanically stimulated counterparts or are created in a shorter period of time. Thus mechanical stimuli can act as an agonist in a similar fashion to more expensive growth factor approaches potentially in a more specific and targeted manner. The analysis of the mechanical forces applied by fluid flow to cells in monolayer is important as it gives us a simplified indicator of response of bone cells to stresses, however, it is necessary to perform similar studies in more complex and larger 3D environment potentially more analogous to the environment *in vivo*. The difficulties lie in the interpretation of the strain profile across a complex porous construct.

3. MECHANICAL FORCE APPLICATION VIA SHEAR STRESS FROM FLUID FLOW IN 3D

Fluid flow in 3D has been performed using bioreactors such as spinner flasks, rotating wall vessels and perfusion systems (Sikavitsas *et al.*, 2002; Gooch *et al.*, 2001; Rucci *et al.*, 2002; Qiu *et al.*, 1998; Goldstein *et al.*, 2001; Van Den Dolder *et al.*, 2003; Shelton *et al.*, 1992). Spinner flasks consist of a container full of culture media where several cell seeded constructs are held in position by a vertical wire suspended from the top of the flask. A magnetic stirrer bar is rotated at the bottom of the flask at a typical speed of 50 rpm. This environment improves nutrient diffusion and promotes cell proliferation throughout the constructs in comparison to static conditions, however, the shear forces that act on the constructs are heterogeneous.

Antigravity effects on cells such as osteoblasts have been studied by NASA using rotating wall vessel bioreactors. The constructs are cultured together in the same media and continuously 'tumble' in a rotating container. The shear force experienced by the constructs is reduced in the rotating wall vessel in comparison to the spinner flask, as the culture media is gradually introduced to the centre of the vessel via a porous tube. Rotating bioreactors similar to this are manufactured by Synthecon Inc where culture of a variety of tissue engineered constructs have been performed (Unsworth *et al.*, 1998; Duray *et al.*, 1997; Qiu *et al.*, 2001; Qiu *et al.*, 1999; Granet *et al.*, 1998).

Perfusion bioreactors have been studied by a variety of research groups for different purposes such as bone tissue engineering, (Rucci *et al.*, 2002; Qiu *et al.*, 1998; Goldstein *et al.*, 2001; Bancroft *et al.*, 2002; Cartmell *et al.*, 2003; Porter *et al.*, 2003). Typically, this type of bioreactor consists of a scaffold with an individual culture media reservoir specifically for that one construct. This allows independent control over the biochemical environment to each construct. A porous scaffold is seeded with cells and the culture media is perfused through the construct either transversely or axially. The flow rate of the culture media is obviously important as is the morphology and porosity of the construct. Research into varying the rate of the perfusing media has shown that for an 80% porous construct seeded with 2 million osteoblasts, optimum flow rates were below 0.1ml/min (Porter *et al.*, 2001). This flow rate produced the highest proliferation of cells. However, a flow rate of 2ml/min, although produced constructs with significantly less cell proliferation, upregulated production of bone-related genes such as osteopontin and osteonectin. It may be that the increased shear stress applied to the cells from the higher flow rate induced this upregulation.

Glowacki and associates have shown that GPIa murine stromal cells seeded onto porous collagen scaffolds and perfused with culture media at 1.3ml/min for 7 and 16 days, showed increased viability in comparison to their static counterparts (Glowacki *et al.*, 1998). The perfusion system consisted of a horizontal glass column

(1cm diameter, 10cm length) in which the collagen/cell construct resided, a peristaltic pump and a media reservoir. The media was in a volume of 30ml and was changed every 7 days. The same research group used this system to study murine osteosarcoma cells seeded onto 3D collagen sponges after perfusing at 1.3ml/min for 21 days (Mueller *et al.*, 1999). DNA content was doubled and bone-related gene expression was significantly increased in the perfused samples in comparison to the static controls. In a similar experiment, bovine chondrocytes were seeded onto collagen sponges and perfused at 0.33ml/min for 15 days (Mizuno *et al.*, 2001). Significant upregulation in the production of S-GAG and aggrecan and collagen type II gene expression was seen with perfused samples relative to the static controls.

Goldstein and associates have analysed the effect of media convection on rat stromal cell seeded PLGA foams by using static culture, a spinner flask, a rotary vessel and a perfusion system (Goldstein *et al.*, 2001). It was found that convection did not have a significant effect on cell proliferation but the rotary vessel and flow system produced a more uniform distribution of cells throughout the constructs. Alkaline phosphatase production was increased after 7 days in both the flow system and the spinner flask but osteocalcin production after 14 days of culture was unaffected by the different bioreactor conditions utilised.

Botchwey and associates have shown that 7 day culture of SaOS-2 cells seeded onto a buoyant PLGA microcarrier material significantly upregulate alkaline phosphatase production and alzarin red staining when cultured in a synthecon rotating bioreactor in comparison to static controls (Botchwey *et al.*, 2001). Numerical model simulations and particle motion analysis showed that the constructs established a near circulary trajectory in the fluid medium with a terminal velocity of 98mm/s whilst avoiding bioreactor wall collisions. The maximum shear stress seen by the cells was estimated to be 3.9 dynes/cm^2.

Computer modelling of the flow parameters of the media used in each of these bioreactors to understand the forces experienced by these cellular constructs is important but complicated. Williams and associates have begun to model steady state momentum and mass transport in a concentric cylinder bioreactor using computational fluid dynamics (Williams *et al.*, 2002). It was found that in their system, the shear stress distribution ranged from 1.5 to 12 dyne/cm^2 and varied little with relative number or placement within the bioreactor. Approximately 80% of the construct surface was found to be exposed to shear stresses between 1.5 and 4 dyne/cm^2. Characterisation of flow through a porous media has also been attempted by Kohles and associates (Kohles *et al.*, 2001). The anisotropic properties of trabecular bone (the porous media in question with this particular experiment) was demonstrated and the effect this had on the permeability of direct fluid perfusion was shown.

Further characterisation of the fluid flow through the porous constructs created in these bioreactors for bone tissue engineering is needed if we are to fully understand the effects of environmental factors such as the mechanical forces exposed to the cells and the scaffold. As well as the potential of shear stress application to the cells that are seeded onto a perfused 3D porous scaffold, many research groups have been performing experiments that look at the effects of simultaneous perfusion with other different types of mechanical force application. These include tensile and axial compression forces and will be discussed in more detail in the following sections.

4. APPLICATION OF TENSILE MECHANICAL FORCE ON BONE CELLS IN MONOLAYER

One of the most widely used model systems for application of tensile or compressive forces has been the stretchable membrane substrate mechanism that has been utilised in a number of (subtly) different ways. The way in which cells perceive the strains depends upon the type of stretch applied to the substrate. Two modes may be defined, namely uniaxial stretching and biaxial stretching. It has been suggested that uniaxial strains most accurately model the types of loads experienced in the periosteum *in vivo* (Duncan and Turner, 1995). It has also been argued that the majority of tension and compressive forces perceived by bone cells *in vivo* are uniaxial in nature and as such biaxial strain may not be an ideal mechanism to model the effects of strain *in vitro*. However, biaxial substrate strain mechanisms have also been widely used for a number of years to model similar mechanical load responses. For example, the stimulation of osteopontin expression has been shown to result from exposure of osteoblastic cells to both uniaxial (Owan *et al.*, 1997) and biaxial strains (Walker *et al.*, 2000) of similar magnitudes.

Examples of such uniaxial mechanisms include the stretchable collagen ‘ribbon’ type substrate models that have been utilised to transfer cyclical uniaxial stretch to cells cultured on them (Yeh and Rodan, 1984). Biaxial strain models have been developed from culture dish deformation models, initially of non-elastic materials (Somjen *et al.*, 1980), and later utilised expandable membrane bases with elastic properties. These membranous substrates are cyclically deformed as a result of defined pattern of pneumatic pressure and subsequent pressure release (Brighton *et al.*, 1991; Duncan and Hruska 1994; Walker *et al.*, 2000). This model of strain has been widely used in the study of bone cell responses to strain, with load responses in alkaline phosphatase (Thomas and El Haj, 1996) and other markers of the late load response including osteopontin and osteocalcin (Walker *et al.*, 2000). However, cells cultured on this membrane are cyclically strained in a non-homogeneous fashion as strains increase from zero at the edge of the membrane to a maximal strain at the centre of the circular membrane.

Alternative *in vitro* models have been developed to apply homogeneous strains to monolayer cultures (Jones *et al.*, 1991; Jones *et al.*, 1994; Owan *et al.*, 1997; Peake *et al.*, 2000). These models are based around the 4-point bending principle that had been previously used to apply defined strains *in vivo*. The basis of this loading model comes from the fact that cells located between the two central loading fulcra are homogeneously loaded due to the constant curve induced between these two points by this mode of substrate deformation. The nature of this curve not only permits the application of uniform and homogeneous strains to cells located therein, but also allows these strains to be quantified (mathematically) in a simple manner.

Such 4-point bending models have been used to induce similar transcriptional and other load responses as have been characterised in alternative *in vivo* and *in vitro* studies. For example, load induced prostaglandin release (Zaman *et al.*, 1997) and also induction of osteopontin gene expression (Owan *et al.*, 1997) have been observed following loading of osteoblastic cells in this manner. One of the advantages of the four-point bending system is that a large range of strains can be applied to cells with the application of an identical deformation. Owan and associates have used this fact to 'separate' the effects of displacement (deformation) from strain magnitude in an effort to evaluate the role of shear stress in the load response of osteoblasts (Owan *et al.*, 1997).

In addition to these widely used mechanical loading systems, a further, more specialised laser based loading system has been devised to apply mechanical load at a much smaller scale, with loading possible not only to individual cells, but also to particular regions of the cell. This optical gradient light trap 'optical tweezer' system was used to apply loads to individual primary human osteoblasts, resulting in an increase in intracellular calcium concentration resulting (Walker *et al.*, 1999). A less elaborate method of loading was utilised to stimulate and characterise an identical type of calcium response in osteoclasts, with 'loading' consisting simply of touching the surface of the cell with a glass pipette (Xia & Ferrier, 1995). An alternative and more widely used load model utilises osmotically induced mechanical (membrane) stress to stimulate load inducible signalling pathways in osteoblasts (Tsai *et al.,* 1999).

5. MECHANICAL FORCE APPLICATION VIA AXIAL COMPRESSION OF CONSTRUCT

A number of systems have been designed which enable mechanical compression to be applied to a 3D explant or a cell seeded construct in culture. However, the majority have focused on other tissue types such as cartilage or vascular systems with less work focusing on bone (Altman *et al.,* 2002; Ochoa and Vacanti, 2002; Pei *et al.,* 2002). Essentially, the predominant design relies on a piston, which can cyclically compress a construct whilst enabling perfusion of the media through the tissue. The method by which compressive forces are applied to the piston can vary from pneumatic to mechanical stepper motors in nature but in most of the systems the strain rate and frequency can be controlled over long durations in culture. More direct compression models include the seeding of bone cells onto deformable macroporous gelatine beads that are placed into columns and subsequently subjected to cyclical compressive forces (Shelton & El Haj, 1992). Such compression loading of periosteal cells within this model was shown to result in increased RNA synthesis (Shelton and El Haj, 1992).

Tensile forces have also been applied to osteoblast seeded scaffolds cultured *in-vitro* (Yang *et al.*, 2002). A tensile strain of 1000 microstrain was applied to primary human osteoblast seeded porous PLLA scaffolds for 30 minutes daily at 1hz. These conditions showed an upregulation in alkaline phosphatase production by the cellular constructs in comparison to the controls with no load. Addition of a calcium channel agonist incorporated into the PLLA scaffold further elevated the alkaline phosphatase production with the load present.

Further work is clearly needed to investigate the potential for mechanical conditioning using compression and tension in a 3D bioreactor. This is needed with a better understanding of the strain profiles obtained as a result of modelling of the constructs within the system.

6. MECHANICAL FORCE APPLICATION VIA MAGNETIC PARTICLE TECHNOLOGY

New ways of trying to deliver mechanical forces to cells in a bioreactor are currently being researched. One of these ways includes the development of magnetic particle technology for applying forces directly to the cells whilst cultivating them in a perfusion bioreactor (Dobson *et al.*, 2002; Cartmell *et al.*, 2002). This method of applying mechanical forces uses magnetic particles, of varying sizes, from nanometres to microns in diameter. These particles are coated with a variety of proteins such as RGD (Arg-Gly-Asp), or more specific proteins that target, for example, calcium channel receptors, and attached via these proteins to the cell membrane (see figure 2).

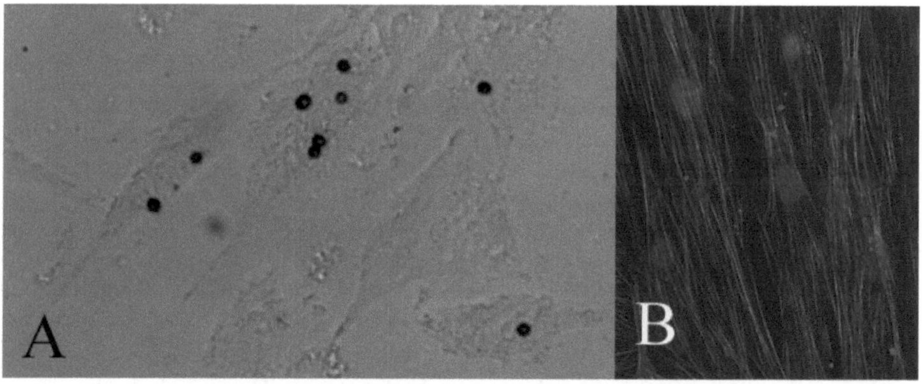

Figure 2: (A) Light microscopy image of primary human osteoblasts with attached 4.5μm ferromagnetic chromium dioxide particles (B) Human osteoblast-like MG63 cells cultured in monolayer for 7 days with magnetic particles: actin filament staining (green), nuclei staining (blue) and magnetic particles (pink).

When an oscillating magnetic field is then applied to the cell/particle, the particle undergoes a translational and rotational movement in response to the field. This in turn, applies a stretch and torque directly to the cell, in the order of 10 piconewtons. Translational stretches and torques have been applied to individual cells in this manner in the short-term (less than 24 hours) and an upregulation of Ca^{2+} influx into cells and significant alterations in the cytoskeletal network such as actin filament stiffening have been reported as a result of such strains (Bausch *et al.*, 2001; Glogauer *et al.*, 1997; Wu *et al.*, 1999). Long-term studies utilising this technique have been performed for bone tissue engineering purposes (Cartmell *et al.*, 2002). A 21 day culture of primary human bone cells in monolayer, with RGD coated 4.5μm chromium dioxide particles attached, underwent a 1Hz cyclical magnetic field for 1 hour daily. These conditions showed an upregulation in osteopontin gene production and early production of mineralized matrix in the experimental groups in comparison to the controls employed (Cartmell *et al.*, 2002). The use of magnetic particle technology is now being applied in 3D by cultivating the cell/magnetic particles on porous PLLA scaffolds in a perfusion bioreactor. Preliminary studies, in the authors' laboratory, have shown an upregulation in osteocalcin and osteopontin gene production after 7 days, in cells with particles attached that were exposed to a magnetic field, compared to the control group which had no magnetic field exposure. The bioreactor set-up used for this experiment is as shown in figure 3.

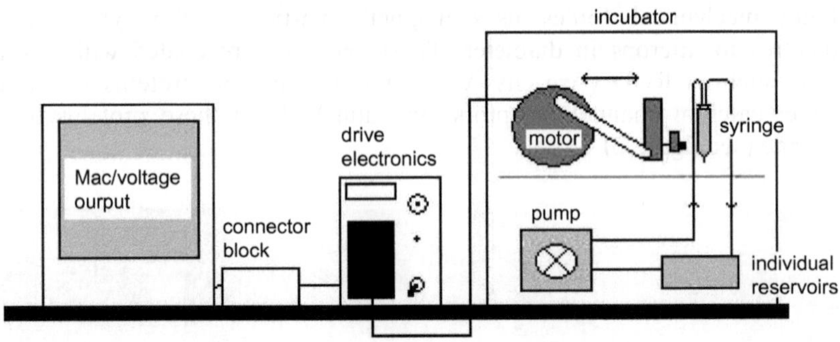

Figure 3: Schematic of 3D perfusion and magnetic force bioreactor. Computer software specifically designed for the bioreactor, controls the motor inside an incubator. The motor has a permanent NdFeB magnet attached to it that moves to and from (at a frequency and period duration controlled by the computer) the cell/magnetic particle/3D scaffold sample. The twelve samples that are held in individual syringes are perfused with culture media simultaneously.

Using this type of bioreactor has several advantages. These include the use of mechanically weak scaffolds as a cell growth substrate (as comparable strain application using axial compression demands substantial scaffold mechanical strength) and the use of a scaffold that has differing dimensions (as axial compression requires uniform dimensions to ensure equally distributed strain across all parts of the construct). It also has the advantage of applying different magnitudes of forces throughout a tissue engineered construct by either altering the size of the attached particles or the number of particles per cell and seeding these cells at different parts of the construct.

The development of bioreactors for bone tissue engineering is an ongoing research area. Much more knowledge needs to be gained regarding the characterisation of fluid flow within the bioreactor, the full effect of mechanotransduction on the bioengineered construct and the scaffold type that is optimum for the creation of functional implants. The manufacture of such bone tissue engineered constructs using mechanical conditioning in combination with altering chemical environments and gene therapy techniques could potentially provide improved implant performance. The goal of creating a biological replacement tissue for osteo defects is still the aim that many researchers aspire to and the use of mechanical bioreactors constructively aids this goal.

7. REFERENCES

Allen FD, Hung CT, Pollack SR, Brighton CT. 2000. Serum modulates the intracellular calcium response of primary cultured bone cells to shear flow. J Biomech 33:1585-1591.

Altman GH, Lu HH, Horan RL, Calabro T, Ryder D, Kaplan DL, Stark P, Martin I, Richmond JC, Vunjak-Novakovic G. 2002. Advanced bioreactor with controlled application of multi-dimensional strain for tissue engineering. J Biomech Eng 1246:742-749.

Bakker AD, Soejima K, Klein-Nulend J, Burger EH. 2001. The production of nitric oxide and prostaglandin e2 by primary bone cells is shear stress dependant. J Biomech 34:671-677.

Bancroft GN, Sikavitsas VI, Van Den Dolder J, Sheffield TL, Ambrose CG, Jansen JA, Mikos AG. 2002. Fluid flow increases mineralized matrix deposition in 3D perfusion culture of marrow stromal osteoblasts in a dose-dependent manner. Proc Natl Acad Sci USA 9920:12600-12605.

Bausch AR, Hellerer U, Essler M, Aepfelbacher M, Sackmann E. 2001. Rapid stiffening of integrin receptor-actin linkages in endothelial cells stimulated with thrombin: a magnetic bead microrheology study. Biophys J 80:2649-2657.

Botchwey EA, Pollack SR, Levine EM, Laurencin CT. 2001. Bone tissue engineering in a rotating bioreactor using a microcarrier matrix system. J Biomed Mater Res 55:242-253.

Brighton CT, Strafford B, Gross SB, Leatherwood DF, Williams JL, Pollack SR. 1991. The proliferative and synthetic response of isolated calvarial bone cells of rats to cyclic biaxial mechanical strain. J Bone Joint Surg [Am] 73:320-331.

Cartmell SH, Dobson J, Verschueren SB, El Haj AJ. 2002. Development of magnetic particle techniques for long term culture of bone cells with intermittent mechanical activation. IEEE Transactions on NanoBioscience 12:92-97.

Cartmell SH, Porter BD, Garcia AJ, Guldberg RE. 2003. Effects of medium perfusion rate on cell-seeded three-dimensional bone constructs in vitro 9:1197-1203.

Claes LE, Heigele CA. 1999. Magnitudes of local stress and strain along bony surfaces predict the course and type of fracture healing. J Biomech 32(3):255-266.

Cowin SC, Moss-Salentijn L, Moss ML. 1991. Candidates for the mechanosensory system in bone. J Biochem Eng 113:191-197.

Cowin SC, Weinbaum S, Zeng Y. 1995. A Case For bone canaliculi as the anatomical site of strain generated potentials. J Biomech 28:1281-1296.

Dobson J, Keramane A, El Haj AJ. 2002. Theory and applications of a magnetic force bioreactor Europ Cells Mater 4(2):42-44.

Duncan RL, Hruska KA. 1994. Chronic intermittent loading alters mechanosensitive channel characteristics in osteoblast-like cells. Am J Physiol 267:F909-F916.

Duncan RL, Turner CH. 1995. Mechanotransduction and the functional response of bone to mechanical strain. Calcif Tissue Int 57:344-358.

Duray PH, Hatfill SJ, Pellis NR. 1997. Tissue culture in microgravity. Science and Medicine May/Jun:45-55.

Fitzferald J, Hughes-Fulford M. 1996. Gravitational loading of a simulated launch alters mrna expression in osteoblasts Exp Cell Res 228:168-171.

Fitzferald J, Hughes-Fulford M. 1999. Mechanically induced c-fos expression is mediated by cAMP in MC3T3-E1 osteoblasts. FASEB J 13 553-557.

Frangos JA, Eskin SG, McIntire LV, Ives CL. 1985. Flow effects on prostaglandin production by cultured human endothelial cells Science 227:1477-1479.

Frangos JA, McIntire LV, Eskin SG. 1988. Shear stress induced stimulation of mammalian cell metabolism. Biotechnol Bioeng 32:1053-1060.

Glogauer M, Arora P, Yao G, Sokholov I, Ferrier J, McCulloch CA. 1997. Calcium ions and tyrosine phosphorylation interact coordinately with actin to regulate cytoprotective responses to stretching. J Cell Sci 110(Pt 1):11-21.

Glowacki J Mizuno S, Greenberger JS. 1998. Perfusion enhances functions of bone marrow stromal cells in three-dimensional culture. Cell Transplan 7(3):319-326.

Goldstein AS, Juarez TM, Helmke CD, Gustin MC, Mikos AG. 2001. Effect of convection on osteoblastic cell growth and function in biodegradable polymer foam scaffolds. Biomaterials 22(11) 1279-1288.

Gooch KJ, Kwon JH, Blunk T, Langer R, Freed LE, Vunjak-Novakovic G. 2001. Effects of mixing intensity on tissue-engineered cartilage. Biotechnol Bioeng 72(4): 402-407.

Granet C, Laroche N, Vico L, Alexandre C, Lafage Profust MH. 1998. Rotating-wall vessels promising bioreactors for osteoblastic cell culture; comparison with other 3D conditions. Med Biol Eng Comput 36:513-519.

Horikawa A, Okada K, Sato K, Sato M. 2000. Morphological changes in osteoblastic cells MC3T3-E1 due to fluid shear stress: cellular damage by prolonged application of fluid shear stress. Tohoku J Exp Med 191:127-137.

Hung CT, Pollack SR, Reilly TM, Brighton CT. 1995. Real-time calcium response of cultured bone cells to fluid flow. Clin Orth Rel Res 313:256-269.

Jacobs CR, Yellowley CE, Davis BR, Zhou Z, Cimbala JM, Donahue HJ. 1998. Differential effect of steady versus oscillating flow on bone cells. J Biomech 31:969-976.

Johnson DL, McAllisteer TN, Frangos JA. 1996. Fluid flow stimulates rapid and continuous release of nitric oxide in osteoblasts. Am J Physiol 271 (Endocrinol Metab 34): E205-E208.

Jones DB, Nolte H, Scholubbergs JG, Turner E, Veltel D. 1991. Biochemical signal transduction of mechanical strain in osteoblast-like cells. Biomaterials 12:101-110.

Jones DB, Leivseth G, Sawada Y, Van Der Sloten J, Bingman D. 1994. Application of homogeneous defined strains to cell cultures. In Lyall F, El Haj AJ eds. Biomechanics and Cells. New York: Cambridge University Press.

Jones D, Leivseth G, Tenbosch J. 1995. Mechano-reception in osteoblast-like cells. Biochem Cell Biol 73:525-534.

Klein-Nulend J, Van Der Plas A, Semeins CM, Ajubi NE, Frangos JA, Nijweide PJ, Burger EH. 1995a. Sensitivity of osteocytes to biomechanical stress in vitro. FASEB 9:441-445.

Klein-Nulend J, Semeins CM, Ajubi NE, Nijweide PJ, Burger EH. 1995b. Pulsating fluid flow increases nitric oxide NO synthesis by osteocytes but not periosteal fibroblasts--correlation with prostaglandin upregulation. Biochem Biophys Res Commun 217:640-648.

Klein-Nulend J, Burger EH, Semeins CM, Raisz LG, Pilbeam CC. 1997. Pulsating fluid flow stimulates prostaglandin release and inducible prostaglandin G/H synthase mrna expression in primary mouse bone cells. J Bone Min Res 12(1):45-51.

Klein-Nulend J, Helfrich MH, Sterck JGH, MacPherson H, Joldersma M, Ralston SH, Semeins CM, Burger EH. 1998. Nitric oxide response to shear stress by human bone cell cultures is endothelial nitric oxide synthase dependant. Biochem Biophys Res Commun 250:108-114.

Kohles SS, Roberts JB, Upton ML, Wilson CG, Bonassar LJ, Schlichting AL. 2001. Direct perfusion measurements of cancellous bone anisotropic permeability, J Biomech 34(9):1197-1202.

Kurokouchi K, Jacobs CR, Donahue HJ. 2001. Oscillating fluid flow inhibits TNF-α-induced NF-κB activation via an IκB kinase pathway in osteoblast-like UMR106 cells. J Biol Chem 276(16):13499-13504.

Lanyon LE, Rubin CT. 1984. Static vs dynamic loads as an influence on bone remodelling. J Biomech 17:897-905.

McAllister TM, Frangos JA, 1999. Steady and transient fluid shear stress stimulate NO release in osteoblasts through distinct biochemical pathways. J Bone Min Res 14(6):930-936.

Mizuno S, Allemann F, Glowacki J, 2001. Effects of medium perfusion on matrix production by bovine chondrocytes in three-dimensional collagen sponges. J Biomed Mat Res 56:368-375.

Mueller SM, Mizuno S, Gerstenfeld LC, Glowacki J. 1999. Medium perfusion enhances osteogenesis by murine osteosarcoma cells in three-dimensional collagen sponges. J Bone Min Res 14(12):2118-2126.

Nauman EA, Satcher RL, Keaveny TM, Halloran BP, Bikle DD. 2001. Osteoblasts respond to pulsatile fluid flow with short-term increases in PGE2 but no change in mineralization. J Appl Physiol 90:1849-1854.

Ochoa ER, Vacanti JP. 2002. An overview of the pathology and approaches to tissue engineering. Ann N Y Acad Sci Dec 979:10-26.

Ogata T. 2000. Fluid flow-induced tyrosine phosphorylation and participation of growth factor signaling pathway in osteoblast-like cells. J Cell Biochem 76:529-538.

Owan I, Burr DB, Turner CH, Qiu J, Tu Y, Onyia JE, Duncan RL. 1997. Mechanotransduction in bone: osteoblasts are more responsive to fluid forces than mechanical strain. Am J Physiol 273:C810-C815.

Pavalko FM, Chen NX, Turner CH, Burr DB, Atkinson S, Hsieh Y-H, Qiu J, Duncan RL. 1998. Fluid shear-induced mechanical signaling in MC3T3-E1 osteoblasts requires cytoskelton-integrin interactions. Am J Physiol 275 (Cell Physiol 44): C1591-C1601.

Peake M, Yang Y, Cooling L, El Haj AJ. 2000. Identifying cellular mechanotransduction pathways for use in designing mechano-active scaffolds for tissue engineering. Proc Mechanotransduction 2000:177-184.

Pei M, Solchaga LA, Seidel J, Zeng L, Vunjak-Novakovic G, Caplan AI, Freed LE. 2002. Bioreactors mediate the effectiveness of tissue engineering scaffolds. FASEB J 16(12): 1691-1694.

Porter B, Cartmell S, Guldberg R. 2001. Design of a 3D perfused cell culture system to evaluate bone regeneration technologies. In: Transactions of the 47th Annual Meeting of the Orthopaedic Research Society 25th-28th February San Fransisco CA USA.

Porter BD, Zauel R, Cartmell SH, Stockman HW, Fyhrie D, Guldberg R. 2003. 3D computational modeling of media flow through scaffolds in a perfusion bioreactor. In: Transactions of 49th Annual Meeting of the Orthopaedic Research Society 2nd-5th February New Orleans LA, USA.

Qiu Q, Ducheyne P, Gao H, Ayyaswamy P. 1998. Formation and differentiation of three-dimensional rat marrow stromal cell culture on microcarriers in a rotating-wall vessel. Tissue Eng 4(1):19-34.

Qiu QQ, Ducheyne P, Ayyasawamy PS. 1999. Fabrication characterization and evaluation of bioceramic hollow microspheres used as microcarriers for 3D bone tissue formation in rotating bioreactors. Biomaterials 20:989-1001.

Qiu QQ, Ducheyne P, Ayyaswamy PS. 2001. 3D bone tissue engineered with bioactive microspheres in simulated microgravity. In Vitro Cell Dev Biol Anim 37:157-165.

Reich KM, Frangos JA. 1991. Effect of flow on prostaglandin E_2 and inositol trisphosphate levels in osteoblasts. Am J Physiol 261 (Cell Physiol 30):C428-C432.

Rubin CT, Lanyon LE. 1982. Limb mechanics as a function of speed and gait: a study of functional strains in the radius and tibia of horse and dog. J Exp Biol 101:187-211.

Rubin CT, Lanyon LE. 1984. Regulation of bone formation by applied dynamic loads. J Bone Joint Surg Am 66:397-402.

Rubin CT, Lanyon LE 1985. Regulation of bone mass by mechanical strain magnitude. Calcif Tissue Int 37:411-417.

Rubin CT, Mcleod KJ. 1994. Promotion of bony ingrowth by frequency-specific low-amplitude mechanical strain. Clin Orthop 165-174.

Rucci N, Migliaccio S, Zani BM, Taranta A, Teti A. 2002. Characterization of the osteoblast-like cell phenotype under microgravity conditions in the NASA-approved rotating wall vessel bioreactor RWV. J Cell Biochem 85(1):167-79.

Salter DM, Robb JE, Wright MO. 1997. Electrophysiological responses of human bone cells to mechanical stimulation: evidence for specific integrin function in mechanotransduction. J Bone Miner Res 12:1133-1141.

Salzstein RA, Pollack SR, Mak AFT, Petrov N. 1987. Electromechanical potentials in cortical bone – a continuum approach. J Biomech 20:261-270.

Salzstein RA, Pollack SR. 1987. Electromechanical potentials in cortical bone – experimental analysis. J Biochem 20:271-280.

Shelton RM, El Haj AJ. 1992. A novel microcarrier bead model to investigate bone cell responses to mechanical compression in vitro. J of Bone and Mineral Res 7(supp 2):S403-S405.

Sikavitsas VI, Temenoff JS, Mikos AG. 2001. Biomaterials and bone mechanotransduction. Biomaterials 22:2581-2593.

Sikavitsas VI, Bancroft GN, Mikos AG. 2002. Formation of three-dimensional cell/polymer constructs for bone tissue engineering in a spinner flask and a rotating wall vessel bioreactor. J Biomed Mater Res 62(1):136-148.

Somjen D, Binderman I, Berger E, Harrell A. 1980. Bone remodelling induced by physical stress is prostaglandin E2 mediated. Biochim Biophys Acta 627(1):91-100.

Thomas GP, El Haj AJ. 1996. Bone marrow stromal cells are load responsive in vitro. Calcif Tissue Int 58:101-108.

Tjandrawinata RR, Vincent VL, Hughes-Fulford M. 1997. Vibrational force alters mRNA expression in osteoblasts. FASEB J 11:493-497.

Tsai JA, Larsson O, Kindmark H. 1999. Spontaneous and stimulated transients in cytoplasmic free Ca2+ in normal human osteoblast-like cells: aspects of their regulation. Biochem Biophys Res Commun 263:206-212.

Turner CH, Forwood MR, Rho JY, Yoshikawa T. 1994a. Mechanical loading thresholds for lamellar and woven bone formation. J Bone Miner Res 9:87-97 .

Turner CH, Forwood MR, Otter MW. 1994b. Mechanotransduction in bone: do bone cells act as sensors of fluid flow. FASEB J 8:875-878.

Turner CH. 1998. Three rules for bone adaptation to mechanical stimuli. Bone 23:399-407.

Unsworth BR, Lelkes PI. 1998. Growing tissues in microgravity. Nature Medicine 4:901-907.

Van Den Dolder J, Bancroft GN, Sikavitsas VI, Spauwen PH, Jansen JA, Mikos AG. 2003. Flow perfusion culture of marrow stromal osteoblasts in titanium fiber mesh. J Biomed Mater Res 64A(2): 235-241.

Walker LM. 1999. The effect of mechanical and hormonal stimuli on femur derived osteoblasts; intracellular calcium fluxes and calcium channels. PhD Thesis, University of Birmingham, UK.

Walker LM, Holm A, Cooling L, Maxwell L, Oberg A, Sundqvist T, El Haj AJ. 1999. Mechanical manipulation of bone and cartilage cells with 'optical tweezers'. FEBS Lett 459:39-42.

Walker LM, Publicover SJ, Preston MR, Said Ahmed MA, El Haj AJ. 2000. Calcium-channel activation and matrix protein upregulation in bone cells in response to mechanical strain. J Cell Biochem 79:648-661.

Williams KA, Saini S, Wick TM. 2002. Computational fluid dynamics modeling of steady-state momentum and mass transport in a bioreactor for cartilage tissue engineering. Biotechnol Prog 18(5):951-963.

Wu Z, Wong K, Glogauer M, Ellen RP, McCulloch CA. 1999. Regulation of stretch-activated intracellular calcium transients by actin filaments. Biochem Biophys Res Commun 261:419-425.

Xia SL, Ferrier J. 1995. Calcium signal induced by mechanical perturbation of osteoclasts. J Cell Physiol 163:493-501.

Yang Y, Magnay J, El Haj AJ. 2002. Development of a mechano responsive scaffold for tissue engineering. Biomaterials 23:2119-2126.

Yeh CK, Rodan GA. 1984. Tensile forces enhance prostaglandin E synthesis in osteoblastic cells grown on collagen ribbons. Calcif Tissue Int 36(Suppl 1): S67-S71.

You J, Reilly GC, Zhen X, Yellowley CE, Chen Q, Donahue HJ, Jacobs CR. 2001. Osteopontin gene regulation by oscllatory fluid flow via intracellular calcium mobilization and activation of mitogen-activated protein kinase in MC3T3-E1 osteoblasts. J Biol Chem 276(16):13365-13371.

Zaman G, Suswillo RF, Cheng MZ, Tavares IA, Lanyon LE. 1997. Early responses to dynamic strain change and prostaglandins in bone-derived cells in culture. J Bone Miner Res 12:769-777.

CHAPTER 9

A DYNAMIC STRAINING BIOREACTOR FOR COLLAGEN-BASED TISSUE ENGINEERING

Y. SHI AND I. VESELY

Saban Research Institute, Children's Hospital at Los Angeles, Los Angeles, USA

1. INTRODUCTION

Collagen-based tissue constructs for vascular, orthopedic, and other applications have been studied extensively for many years (L'Heureux et al., 1993; Seliktar et al., 2000; Shi et al., 2002; Shi and Vesely, 2004; Shi and Vesely, 2003; Tranquillo et al., 1996; Weinberg and Bell, 1986). These models involve first mixing soluble, fibrillar collagen with the appropriate cells, serum and medium. After the collagen-cell mixture is neutralized and brought up to 37°C, soluble collagen reassembles into fibrils and a gel is created. Cells become entrapped within the collagen gel and begin to interact with the collagen fibrils. These cells reorganize the surrounding collagen matrix, contract it and exclude water. In many ways, this in-vitro contraction mimics wound healing *in vivo* (Grinnell, 1994). When the gel is mechanically constrained, the collagen fibrils align in the direction of constraint, and a highly aligned, compacted collagenous construct can thus be fabricated. The collagen fibrils can therefore be manipulated into useful structures using the principle of directed collagen gel shrinkage.

So far, inappropriate mechanical properties have been one of the main limitations of most collagen-based tissue equivalents (Girton et al., 1999; Kanda and Matsuda, 1994; Kanda et al., 1993a; Kanda et al., 1993b; Kim et al., 1999; Seliktar, et al., 2000; Tranquillo et al., 1992). Various methods for improving these constructs have been developed and evaluated. They include mandrel compaction (L'Heureux, et al., 1993), magnetic prealignment (Tranquillo, et al., 1996), glycation etc (Girton, et al., 1999).

Mechanical stimuli are known to exert a variety of effects on smooth muscle cells (SMCs) in culture, increasing proliferation, matrix production and expression of phenotype-specific proteins and growth factors (Keeley and Bartoszewicz, 1995;

J.B. Chaudhuri & M. Al-Rubeai (eds.), Bioreactors for Tissue Engineering, 209 - 219.

O'Callaghan and Williams, 2000; Seliktar et al., 2001; Stanley et al., 2000; Williams, 1998; Wilson et al., 1995). It is also known that fibroblasts can generate substantial traction forces in culture that are, in part, regulated by external mechanical forces (Akhouayri et al., 1999; Brown et al., 1998; Brown et al., 1996; Fredberg et al., 1997; Petroll et al., 1993; Shelburne and Pandy, 1997). The contractile forces generated by fibroblasts in three-dimensional collagen substrates have been studied for many years (Arora et al., 1999; Brown, et al., 1998; Brown et al., 2002). The observation that external mechanical forces can regulate both fibroblast contractile forces and matrix production suggested to us that mechanically stimulating the constructs could improve their microstructure and enhance their mechanics. The enhanced cell growth and extracellular matrix (ECM) production, as well as the improved structural integrity of collagen constructs, indeed appear to be related directly to the application of cyclic strain. Our application of a dynamic loading bioreactor is for the fabrication of tissue engineered mitral valve chordae. These "one dimensional" structures need to be loaded only longitudinally, requiring a relatively simple dynamic bioreactor.

Dynamic mechanical conditioning systems being used by most groups make use of compressed air. We were interested in developing a much simpler, very inexpensive system, and thus focused on a cam-based approach. We also used stereolithography (SLA) for the fabrication of the dynamic bioreactor. SLA is typically used for rapid prototyping, not for fabrication of the final product. We found that the resin from which SLA components are fabricated withstands autoclaving, is not toxic to cells (once thoroughly washed), and is very cost effective. It can thus be used for large-scale product manufacturing and may help offset some of the huge scale-up costs that have plagued the tissue engineering industry.

2. DYNAMIC STRAIN BIOREACTOR

Pro-Engineer CAD/CAM software (Tri Cad/Cam Systems, Inc., Phoenix, Arizona) was used for the design and SLA was used for most of the fabrication (Figure 1). The dynamic strain bioreactor is a relatively simple device, consisting of a plastic replica of a standard 100 mm diameter Petri dish with stretching rods passing through the sidewalls. These pull rods have cam followers at their distal ends and connect to the collagen constructs at their proximal, inner ends. The culture dish slowly rotates, driven by a motor, and the cam followers are pulled against a static cam by a rubber band. The cam follower consists of a bead that rolls, following the cam as the culture dish rotates. Displacement of the cam follower causes the pull rods to move in and out, subjecting the collagen constructs to cyclic mechanical strain. The displacement waveform of the cam is a circle perturbed by a simple sine function (Figure 2). The culture dish is firmly attached to the shaft of a DC Gear-head motor powered by a 0-25 V DC bench-top power supply (both from Jameco Electronics, Belmont, CA). Because the entire apparatus, except the power supply, is placed in the incubator, the motor was enclosed within a sealed SLA box to protect it from humidity. The holes in the wall of the culture dish that accommodated the stretching rods were sealed with a mixture of vacuum grease and neomycin.

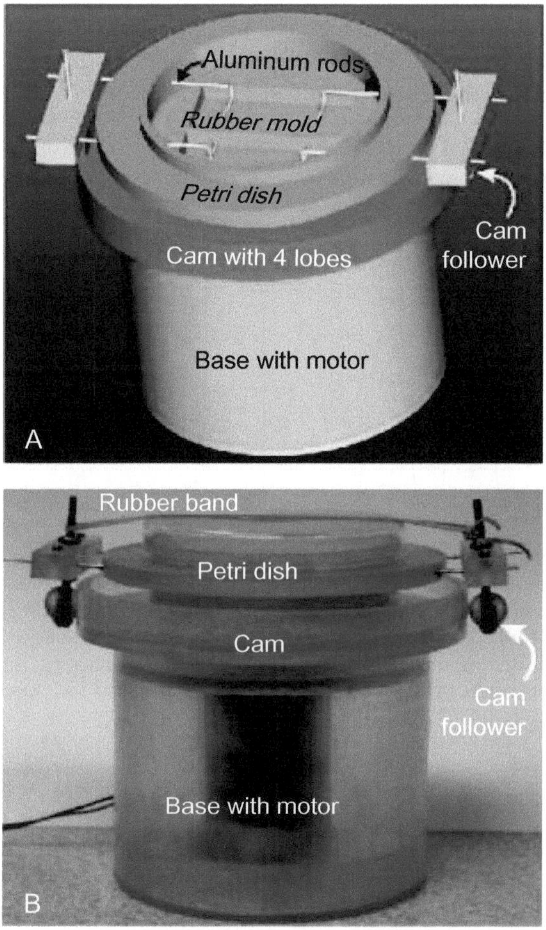

Figure 1: Engineering drawing and photo of the dynamic loading bioreactor. Image A is a 3-D Pro-Engineer rendering of the bioreactor design. Image B is a photo of the stereolithography (SLA) product. Note the configuration of the pull rods and the simple cam followers. Although SLA is typically used to fabricate prototypes or mock-ups of devices, we found that the resin material is suitable for long-term use in the finished bioreactor. Note the presence of the motor within the sealed SLA box, protecting it from humidity. The resin culture dish was covered by a standard 100 mm diameter Petri dish lid, and a rubber band was used to pull the cam followers together against the cam. The stretching rods were made from stainless steel and passed through holes in the wall of the culture dish. Contamination of the cultured tissues was prevented by sealing the gap around the pull-rods with a mixture of vacuum grease and neomycin and by making the wall of the culture dish relatively thick to minimize particle transport during cyclic sliding of the pull rods.

Prior to use, the complete silicone-rubber-lined SLA culture dish with stretching rods was steam sterilized at 121 psi for 15 minutes. The collagen-cell suspension was then pipetted into the culture dish and incubated at 37°C for 2 weeks under static conditions to enable the gel to set and undergo cell-mediated compaction. Once initial tendon-like structures formed, the motor was turned on and the constructs cultured dynamically for up to 6 weeks. The constructs were stretched up to 10% strain, at a rate of 1.33 Hz. The system was stopped periodically only to change the culture media. All of the constructs were cultured for up to 8 weeks. After culture, constructs were examined for mechanical properties, microstructure, and cell and collagen content.

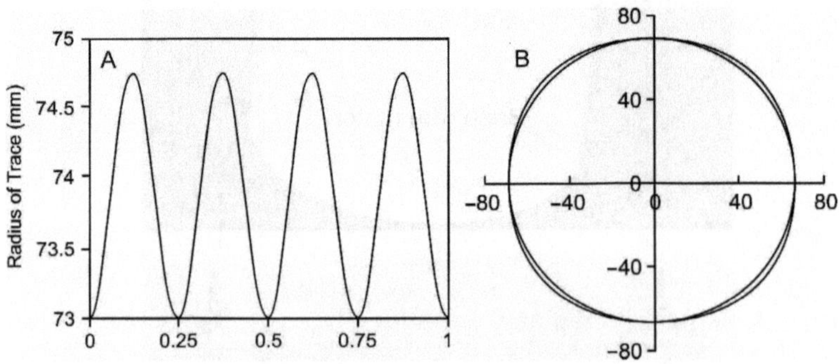

Figure 2: The cam profile consisted of four sinusoidal waves (A) mapped onto a circle, thus creating the four lobes of the cam (B). For every 360 degrees of rotation, the collagen constructs inside the culture dish were stretched four times.

3. PARTS OF DYNAMIC STRAIN BIOREACTOR

The cyclic device has four main parts: motor housing, culture dish, cam and cam follower.

3.1 Motor Housing

The motor housing was composed of a motor box (Figure 3) and a base (Figure 4). Figure 3 shows illustration of Pro-E version of motor box. The shaft of the motor (Figure 5) passed through the hole in the middle of the motor box. The motor was connected to the platform of the motor box by screws. A hole in the sidewall of the motor box allowed the electric wire of the motor to pass. The edge of the wall of the motor box had a small slot (Figure 3) that fitted with a ridge of the cam to prevent the cam from spinning while acted upon by the cam followers. The motor housing also acted as the support for the cyclic stretching device.

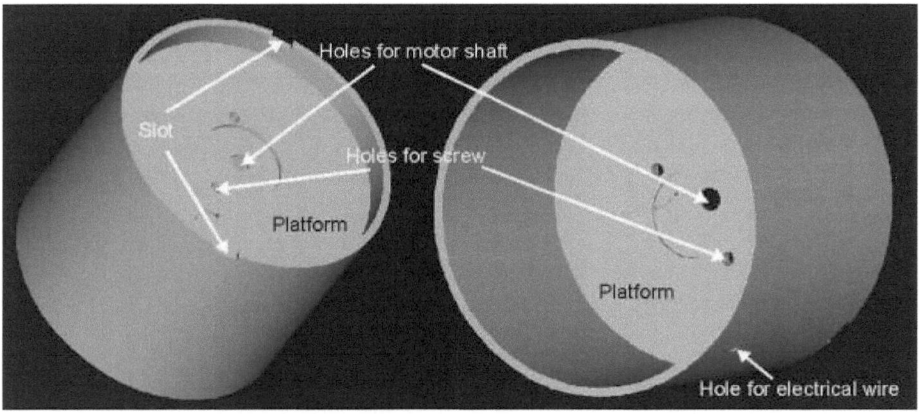

Figure 3: Illustration of Pro-E version of motor box. Left is topside view of the motor box and right is bottom-side view of the motor box.

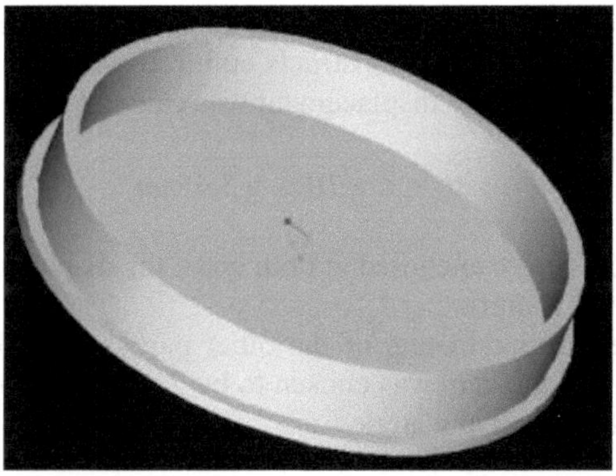

Figure 4: Illustration of Pro-E version of base. After the motor was put into the motor box, the base was fitted into the bottom of the box. The motor housing was sealed to prevent motor from humidity

3.2 Cam

Figure 2 shows the theoretical base of the cam, the four sinusoidal waves (A) correspond to the four lobes of the cam (B).

Figure 5: Photo of DC gear head motor.

3.2.1 Calculations for Cam Geometry

For consistency, the length of the collagen constructs in the dynamic straining bioreactor was the same as for the constructs cultured statically (L = 36 mm). If the desired strain was 10%, then the displacement was calculated as follows:

$$\Delta L = L \times 10\% = 3.6\ mm \tag{1}$$

Since the constructs were anchored at both ends, the displacement at each end was half this amount (1.8 mm).

To match the design parameters of the other parts (motor box, Petri dish), the diameter of the circle of the cam was chosen to be 73 mm (R1 = 73 mm). Therefore, the diameter of the 4 lobes at any angle was:

$$R2 = R1 + \Delta L/2 + \Delta L/2 \times \sin[4 \times (v\pi/180 - \pi/8)] \tag{2}$$
$$(v = 0, 1, 2, \ldots\ldots 360)$$

The sinusoidal curve of the cam was obtained according to Equation 2 (Figure 2A) and mapped onto a circle (Figure 2B). For every 360 degrees of rotation, the collagen constructs inside the culture dish were stretched four times.

The displacement of the cam follower caused the pull rods to move in and out, and subjected the collagen constructs to cyclic mechanical strain. Figure 6 shows the pro-E version of the cam. Note the ridge at the bottom of the cam. This ridge fitted in the slot in the wall of the motor box, and prevented the cam from spinning while the dish was spun.

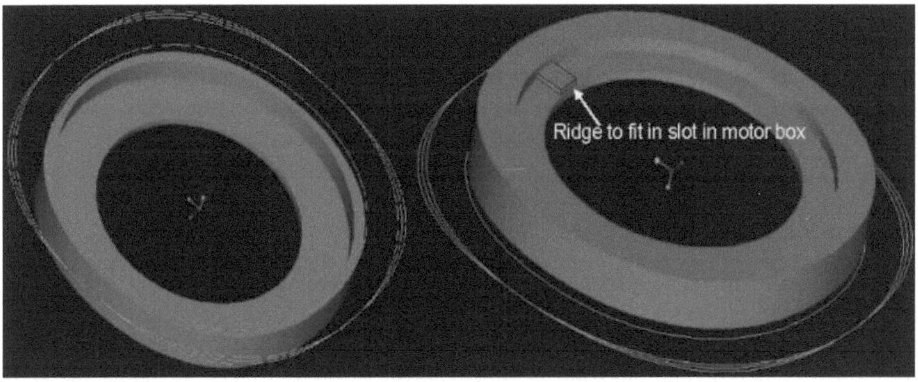

Figure 6: Pro-E illustration of cam. Left is topside view of cam and right is bottom-side view of the cam. The four lobes in the cam correspond to the sinusoidal curve.

3.3 Culture Dish

The standard 100 mm diameter Petri dish was replicated in SLA to enable the stretching rods to pass through the annular ring of culture dish. The culture dish had a small slot at the bottom, into which the shaft of DC gear head motor was inserted (Figure 7).

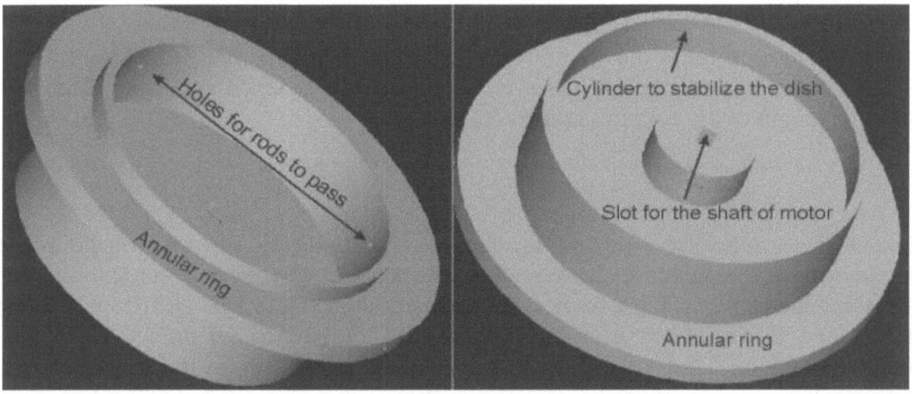

Figure 7: Illustration of Pro-E version of culture dish. Left is topside view of culture dish and right is bottom-side view of the dish.

The cylinder at the bottom of the culture dish contacted with the platform of the motor box. This prevented the culture dish from wobbling when it spun. Lubricant was put between the cylinder and platform to reduce friction. Four holes were drilled through the annular ring of the culture dish to allow the pulling rods to move in and out, subjecting the constructs to cyclic strain. Contamination of the cultured tissues was prevented by sealing the gap around the pull-rods with a mixture of vacuum grease and neomycin and by making the annular ring of the culture dish relatively thick to minimize particle transport during cyclic sliding of the pull rods. Therefore, the culture dish could remain sterile during long-term culture.

3.4 Cam Follower

Figure 8 shows the image of the cam follower. The collagen constructs attached to the ends of the pull rods in the culture dish. The pull rods passed through the holes in the annular ring of the culture dish to the holder block. The use of two joined pull rods prevented the rods, and the attached constructs, from twisting when the rods moved in and out. The holder block also acted as the cam follower, by way of a spinning bead that followed the cam as the culture dish rotated. The collagen constructs inside the culture dish were stretched by the reciprocating movement of the rods when the beads rotated along the cam.

Rectangular chambers were constructed from silicone rubber in the 100 mm SLA culture dish similar to those used in the static loading bioreactor (Figure 9). First, silicone rubber (Silicones, Inc, NC) was poured into the Petri dish to a height of 5 mm. Blocks (60 ×18 ×18mm) were then placed in the dish and more silicone rubber was poured into the dish to a further height of 6 mm, and allowed to set overnight. The blocks were then removed, leaving a rectangular well. Strips of glass fiber (Millipore, Ireland) were wrapped around the ends of stainless steel rods to act as anchors for the constructs. These rods passed through the holes in the annular ring of the culture dish and connected to the cam followers. The top view of the culture dish and the cam follower is shown in Figure 9. It is assembled before steam sterilization. The distance between the two ends of the pull rods was fixed at 36 mm. This was also the length of the collagen constructs.

4. PRIMARY CELL CULTURE

Neonatal rat aortic SMCs and bovine chordae fibroblasts (BCF) were isolated by the explant method outlined by Oakes and coworkers. Segments of aorta and chordae were incubated with 2 ml of type II collagenase (2 mg/ml in DMEM/F12 (1:1) medium; Life Technologies, Grand Island, NY) for 10 minutes at 37°C to remove the endothelium. The explants were then washed with PBS several times, minced into small pieces, transferred onto a sterile Petri dish and incubated in 3 ml of equal amounts of DMEM and F12, supplemented with 10% fetal bovine serum (Life Technologies, Grand Island, NY) at 37°C for 1 week to establish the primary culture. After reaching confluence, the cells were detached by trypsinization with 1 ml of 0.05% fresh trypsin containing 0.2% EDTA (Gibco Laboratories, Grand Island, NY), suspended in the medium and centrifuged at 1500 rpm. The cell pellet obtained was resuspended, counted and seeded in a 100 ml Petri dish. Cells were stained with trypan blue, and a hemocytometer was used to determine the cell density and viability. Culture medium was changed twice a week.

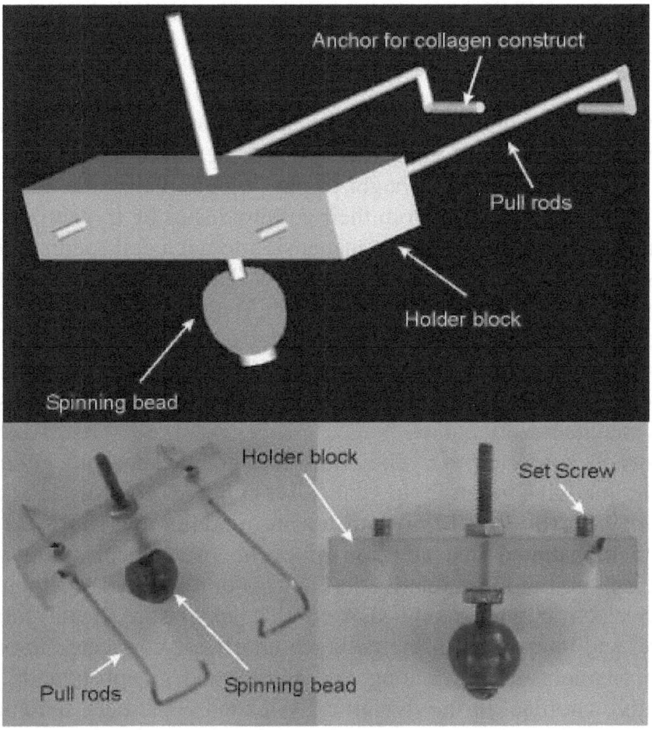

Figure 8: Pro-E illustration (top) and SLA prototype of the cam follower (bottom).

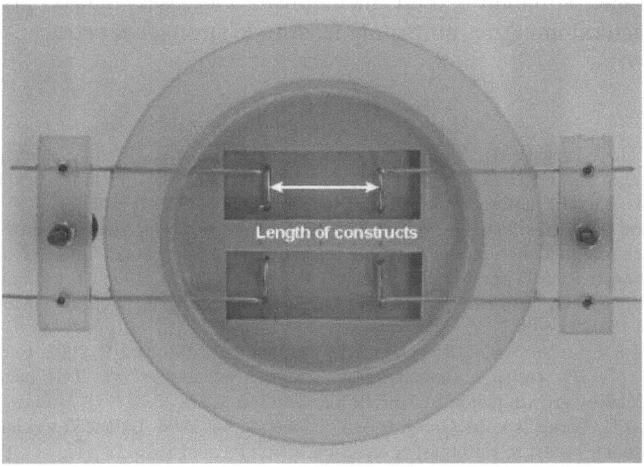

Figure 9: Top view of the culture dish and the cam follower, assembled prior to sterilization.

5. DIRECTED COLLAGEN GEL SHRINKAGE

Fetal bovine serum and penicillin-streptomycin (pen-strep) were thawed and added to the medium (5 × DMEM/F12) to obtain a solution of 20% serum and 100 units/ml pen-strep. Sterile acid-soluble type I collagen (BD Biosciences, Bedford, MA; Rat tail; 3.94 mg/ml, 0.02 N acetic acid) was added to get an initial concentration of 2.0 mg/ml, and the suspension was brought to physiologic pH by the addition of 0.1 N NaOH. The cells were detached from their culture dishes by trypsinization, counted, centrifuged and added to the collagen suspension to get a cell seeding density of 1.0 million cells/ml. All mixing was done on ice. The collagen-cell suspension was pipetted into the wells and incubated at 37°C. Within several hours, a gel formed and attached to the microporous holders at the ends of the wells. These holders prevented longitudinal contraction and allowed shrinkage to occur only transverse to the long axis of the wells. The constructs were cultured for 2 to 8 weeks and culture medium was changed every 2 days.

The constructs subjected to cyclic strain had greater mechanical strength and a higher elastic modulus relative to statically cultured controls. Failure strength increased by 195% (to 3.25 MPa) and tensile modulus by 188% (to 18.7 MPa). DNA assay demonstrated that addition of cyclic mechanical strain resulted in a 155% increase in cell number relative to controls during the 8-week culture period. Transmission electron microscopy showed that an elastin sheath surrounded the collagen core, and that fibrils in the collagen core were connected to each other by proteoglycan filaments (50 μm long and 6 μm wide, cuprolinic blue staining), typical of native chordae. In the dynamically loaded constructs, the collagen fibers were longer, more aligned and compacted, and the elastin sheath was 1-2 μm thicker than in statically cultured constructs. This has demonstrated that mechanical stimulation offers a means of producing a denser matrix with more cells. Increasing cell content and stimulating cell metabolism increases matrix production and improves structural integrity, ultimately leading to stronger constructs.

6. REFERENCES

Akhouayri O, Lafage-Proust MH, Rattner A, Laroche N, Caillot-Augusseau A, Alexandre C, Vico L. 1999. Effects of static or dynamic mechanical stresses on osteoblast phenotype expression in three-dimensional contractile collagen gels. J Cell Biochem 76:217-230.

Arora PD, Narani N, McCulloch CA. 1999. The compliance of collagen gels regulates transforming growth factor-beta induction of alpha-smooth muscle actin in fibroblasts. Am J Pathol 154:871-882.

Brown RA, Prajapati R, McGrouther DA, Yannas IV, Eastwood M. 1998. Tensional homeostasis in dermal fibroblasts: mechanical responses to mechanical loading in three-dimensional substrates. J Cell Physiol 175:323-332.

Brown RA, Sethi KK, Gwanmesia I, Raemdonck D, Eastwood M, Mudera V. 2002. Enhanced fibroblast contraction of 3D collagen lattices and integrin expression by TGF-beta1 and -beta3: mechanoregulatory growth factors? Exp Cell Res 274:310-322.

Brown RA, Talas G, Porter RA, McGrouther DA, Eastwood M. 1996. Balanced mechanical forces and microtubule contribution to fibroblast contraction. J Cell Physiol 169:439-447.

Fredberg JJ, Inouye D, Miller B, Nathan M, Jafari S, Raboudi SH, Butler JP, Shore SA. 1997. Airway smooth muscle tidal stretches and dynamically determined contractile states. Am J Respir Crit Care Med 156:1752-9.

Girton TS, Oegema TR, Tranquillo RT. 1999. Exploiting glycation to stiffen and strengthen tissue equivalents for tissue engineering. J Biomed Mater Res 46:87-92.

Grinnell F. 1994. Fibroblasts myofibroblasts and wound contraction. Journal of Cell Biology 124:401-404.

Kanda K, Matsuda T. 1994. In vitro reconstruction of hybrid arterial media with molecular and cellular orientations. Cell Transplant 3:537-545.

Kanda K, Matsuda T, Oka T. 1993a. In vitro reconstruction of hybrid vascular tissue. Hierarchic and oriented cell layers Asaio J 39:M561-565.

Kanda K, Matsuda T, Oka T. 1993b. Mechanical stress induced cellular orientation and phenotypic modulation of 3-D cultured smooth muscle cells. Asaio J 39:M686-690.

Keeley FW, Bartoszewicz LA. 1995. Elastin in systemic and pulmonary hypertension. Ciba Found Symp 192:259-273; discussion 273-278.

Kim BS, Nikolovski J, Bonadio J, Mooney DJ. 1999. Cyclic mechanical strain regulates the development of engineered smooth muscle tissue. Nat Biotechnol 17:979-83.

L'Heureux N, Germain L, Labbe R, Auger FA. 1993. In vitro construction of a human blood vessel from cultured vascular cells: a morphologic study. J Vasc Surg 17:499-509.

O'Callaghan CJ, Williams B. 2000. Mechanical strain-induced extracellular matrix production by human vascular smooth muscle cells: role of TGF-beta1. Hypertension 36:319-324.

Petroll WM, Cavanagh HD, Barry P, Andrews P, Jester JV. 1993. Quantitative analysis of stress fiber orientation during corneal wound contraction. J Cell Sci 104, Pt 2:353-363.

Seliktar D, Black RA, Vito RP, Nerem RM. 2000. Dynamic mechanical conditioning of collagen-gel blood vessel constructs induces remodeling in vitro. Ann Biomed Eng 28:351-362.

Seliktar D, Nerem RM, Galis ZS. 2001. The role of matrix metalloproteinase-2 in the remodeling of cell-seeded vascular constructs subjected to cyclic strain. Ann Biomed Eng 29:923-934.

Shelburne KB, Pandy MG. 1997. A musculoskeletal model of the knee for evaluating ligament forces during isometric contractions. J Biomech 30:163-176.

Shi Y, Ramamurthi A, Vesely I. 2002. Towards tissue engineering of a composite aortic valve. Biomed Sci Instrum 38:35-40.

Shi Y, Vesely I. 2004. Characterization of statically loaded tissue-engineered mitral valve chordae tendineae. Journal of Biomedical Materials Research 69A:26-39.

Shi Y, Vesely I. 2003. Fabrication of tissue engineered mitral valve chordae using directed collagen gel shrinkage. Tissue Engineering 96:1233-1242.

Stanley AG, Patel H, Knight AL, Williams B. 2000. Mechanical strain-induced human vascular matrix synthesis: the role of angiotensin II. J Renin Angiotensin Aldosterone Syst 1:32-35.

Tranquillo RT, Durrani MA, Moon AG. 1992. Tissue engineering science: consequences of cell traction force. Cytotechnology 10:225-250.

Tranquillo RT, Girton TS, Bromberek BA, Triebes TG, Mooradian DL. 1996. Magnetically orientated tissue-equivalent tubes: application to a circumferentially orientated media-equivalent. Biomaterials 17:349-357.

Weinberg CB, Bell E. 1986. A blood vessel model constructed from collagen and cultured vascular cells. Science 231:397-400.

Williams B. 1998. Mechanical influences on vascular smooth muscle cell function. J Hypertens 16:1921-1929.

Wilson E, Sudhir K, Ives HE. 1995. Mechanical strain of rat vascular smooth muscle cells is sensed by specific extracellular matrix/integrin interactions. J Clin Invest 96:2364-2372.

CHAPTER 10

BIOREACTORS FOR LIGAMENT ENGINEERING

B. J. AINSWORTH AND J. B. CHAUDHURI

Department of Chemical Engineering, University of Bath, United Kingdom

1. INTRODUCTION

Much of the interest in engineering replacement ligaments and tendons has focused on the Anterior Cruciate Ligament (ACL), as it is frequently ruptured (over 200000 cases in the USA per annum (Weitzel et al., 2002). Currently the default option when reconstructing a ruptured ACL is to surgically reconstruct it with an autograft of either hamstring tendon or patellar tendon (Dopirak et al., 2004; Fu et al., 1999; Fu et al., 2000; Spindler et al., 2004). Two recent long term studies of soccer players with ruptured ACLs (Lohmander et al., 2004; Von Porat et al., 2004) suggest that the soccer players are at high risk of developing osteoarthritis regardless of surgical reconstruction. Indeed there is little evidence to suggest a benefit of current ACL reconstruction techniques (Feller, 2004). Cadaveric models suggest that these problems might be caused by failure to replicate the anatomy of the ACL (Yagi et al., 2002; Yamamoto et al., 2004).

In some patients neither hamstring tendon nor patellar tendon is suitable for harvesting and use as an autograft. For this population several options are available including: allografts, which remain controversial, due to the possibility of infection and high cost (Barber, 2003; Johnson, 2003; McGuire, 2003), and the Leeds-Keio polyester ligament which has poor long term outcomes (Murray and Macnicol, 2004), possibly because the tissue inside the ligament is still immature 60 months postoperatively (Nomura et al., 2005). A major objective of ligament engineering is to produce a better replacement ACL.

Tissue engineering a replacement ligament offers a new approach to solve these problems. Using a bioreactor, stem cells from the patient can be differentiated into ligament cells (Altman et al., 2002a) with good manufacturing practice minimising the risks of infection. Biodegradable scaffolds such as modified silk (Chen et al., 2003), collagen gels (Altman et al., 2002a) and polylactide-co-glycolide 10:90 (Cooper et al., 2005), can be used and their properties optimised for ACL

J.B. Chaudhuri & M. Al-Rubeai (eds.), Bioreactors for Tissue Engineering, 221- 233.

reconstruction. The main roles of the bioreactor in ligament tissue engineering are to allow a 3D construct to be mechanically manipulated (either with static or cyclical strain), and to maintain the physiochemical environment (e.g. pH, concentrations of O^2, CO^2, and temperature). This review will discuss bioreactors for ligament tissue engineering; the bioreactors might also be used to further understanding of ligaments. First, the basic anatomy of ligaments will be described, then a consideration of how knowledge of biomechanics influences bioreactor design, and finally descriptions of current bioreactors and possibilities for future bioreactors.

2. LIGAMENT ANATOMY

2.1. ACL Appearance and Location

The Anterior Cruciate Ligament (ACL) is a helical intraarticular ligament named for its attachment to anterior intercondylar area of the tibia. From this attachment it extends upwards, backwards and laterally (Figure 1) past the Posterior Cruciate Ligament (PCL). Together with the PCL it forms the main bond between the femur and tibia (Snell, 2000).

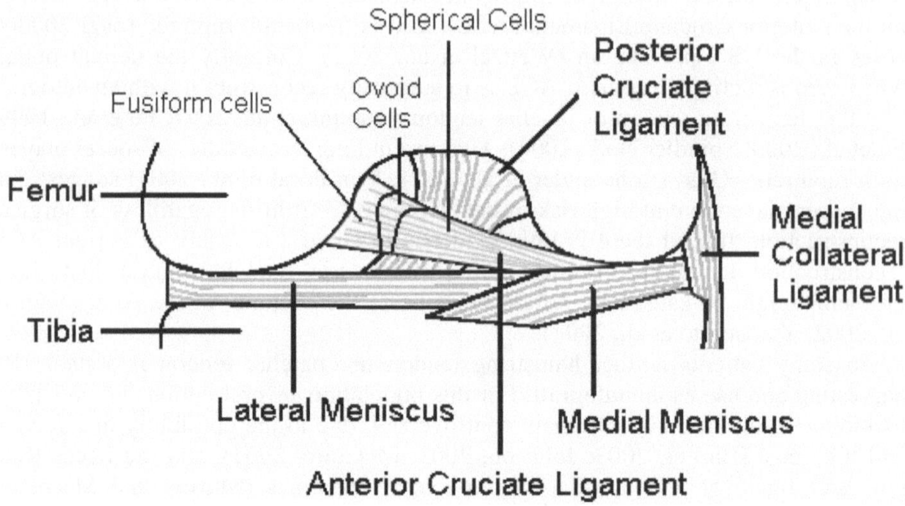

Figure 1. Anterior view of the tissues within the right knee and cell distribution of the ACL (after Murray and Spector (1999) and Snell (2000))

The ACL can be divided into three zones based on the shape of the cells within the ligament (Murray and Spector, 1999). As you move along the ligament from the femur to the tibia the cells start out looking like typical fibroblasts – spindle shaped, the cells become more rounded as the tibia is approached becoming first ovoid and then spherical (see Figure 1). This differentiation to a more fibrocartilaginous phenotype is caused by the ligament being compressed during movement of the knee (Benjamin and Ralphs, 1998).

2.2. Structure of Ligaments and Tendons

The structure of the ACL will be described by moving from a description of the arrangement of collagen molecules at the smallest scale moving up through larger scales. This description is relevant to both ligaments and tendons, it is based on the work of Clark and Sidles (1990), it agrees with the descriptions given by Benjamin and Ralphs (1998) and Silver et al. (2003), although the three use different terminology. This structural similarity suggests that useful analogies can be made between tendons and ligaments in the absence of data on one or other tissue.

Figure 2 shows the five levels of organisation found by Clark and Sidles (1990). Collagen molecules are arranged into fibrils, with the molecules arranged in rows. These fibrils are from 50-300nm in diameter. Fibrils themselves are tightly packed to form bundles known as fibre bundles. These fibre bundles are surrounded by fibrous capsules that are largest in the fibrocartilaginous regions near the insertion of the ligament into the bone, becoming more compact and elongated further away from the insertion. The fibrous capsules contain cells arranged in rows running parallel to the fibre bundles.

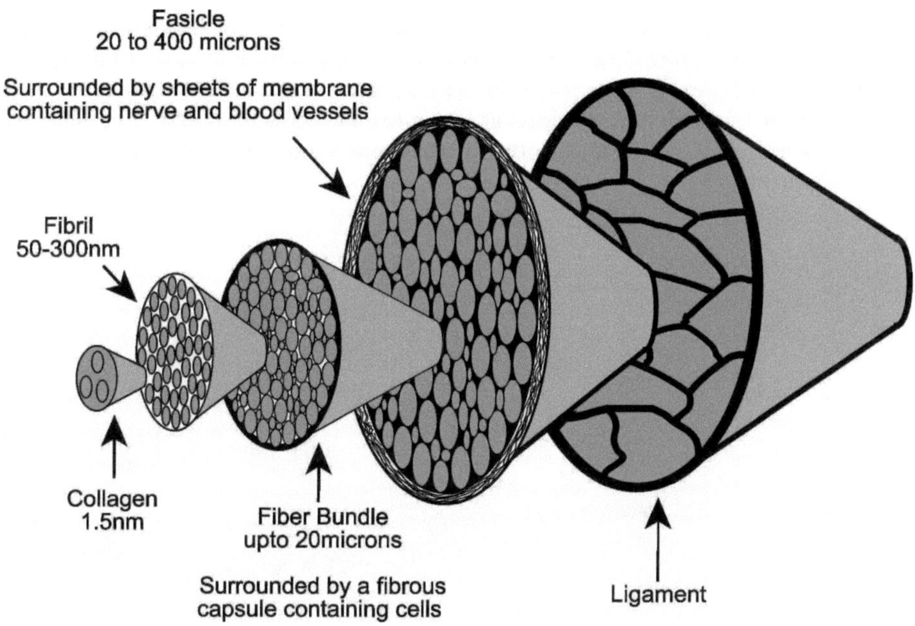

Figure 2. Structure of Ligaments (after Kastelic et al. (1978))

Fibre bundles are packed together to form fascicles, which have a diameter of 20 to 400 μm. The fascicles are separated by a membranous septae, the number of membranes varies within the ligament, normally there are several layers separated by empty space, but occasionally fascicles are only separated by one layer of membrane or there is no membrane separating the fascicles. The membranes are separated by empty space, within the membranous septae lie the blood and nerve vessels. The divisions between fascicles become more prominent closer to the insertion of the ligament into the bone as the membranes separating them become more numerous.

3. BIOMECHANICS

The ACL has two biomechanical roles: it is the primary restraint on anterior tibial displacement and is a secondary restraint of axial rotation (Smith et al., 1993). This section will explore the biomechanics of ACL to consider what mechanical properties should be replicated by a tissue engineered ACL and the mechanical forces that act upon the cells.

3.1. Viscoelasticity

Ligaments are viscoelastic materials; the simplest model of this behaviour is to imagine a spring and damper in series (Figure 3a). The three phenomena most associated with viscoelasticity are: creep - a ligament held a constant force will get extend with time (Figure 3b); stress relaxation - a ligament held at constant length will progressively exert less force on its restraint (Figure 3c); and strain rate dependence, if ligaments are strained at low rates by increasing force then they are absorb less energy, fail at a lower ultimate load and are less stiff than at high strain rates (Figure 3d).

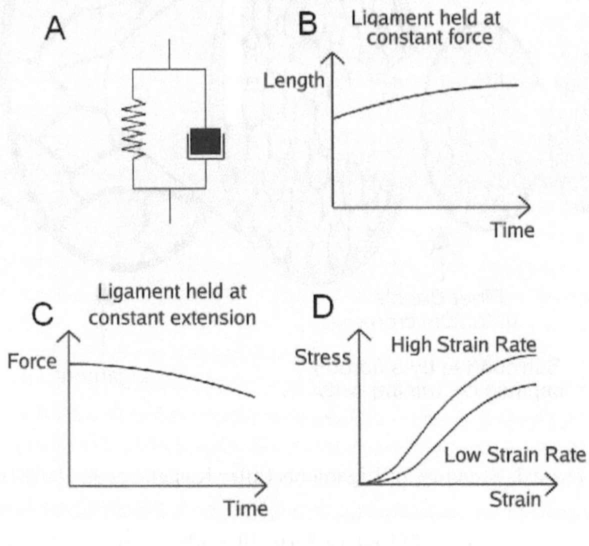

Figure 3. A Basic model of viscoelasticity B Stress relation C Creep D Viscoelastic response to changes in strain rate

3.2. Biomechanics of the ACL

The accepted standards for ACL grafts using current surgical techniques are: ultimate load at failure of 1730 N, linear stiffness of 182 N/mm (Noyes and Grood, 1976) and energy absorbed of 12.8N-m (Woo and Adams, 1990). However the material and structural properties of the ACL vary from individual to individual, with age and gender being important factors, for example: ACLs from younger donors (22-35 years old) had a mean ultimate load of 2160N, mean linear stiffness of 242 N/mm and mean energy absorbed of 11.6N-m, middle aged donors (40-50 years) had ACLs with a mean ultimate load of 1530N, mean linear stiffness of 220 N/mm and mean energy absorbed of 6.1N-m and old donors (60-97 years) had ACLs with a mean ultimate load of 658N, mean linear stiffness of 180 N/mm and mean energy absorbed of 1.8N-m (Woo et al., 1991). It is not clear whether the accepted values are the best values for a replacement ACL. It might be preferable to match the strength of an engineered ligament to the strength of the joint. Alternatively, other biomechanical properties might be more important to emulate, such behaviour when cyclically loaded (Butler et al., 2000).

One major complication when tissue engineering a replacement ACL will be replicating the different functions and behaviours of the anteromedial and posterolateral bundles. The response of the posterolateral bundle to anterior tibial loading varies with the angle of knee flexion, whereas the anteromedial bundle shows a very constant response to anterior tibial loading at knee flexions between 0° and 90° (Sakane et al., 1997); it has been hypothesised that the anteromedial bundle plays a role in guiding the ACL (Takai et al., 1993). The in-situ force in the posterolateral bundle is larger than that in the anteromedial bundle between 0° and 45° (Sakane et al., 1997), the angles of flexion when the ACL is the primary restraint of anterior tibial loading. Most ACL reconstructions have focused on replicating the anteromedial bundle, but recent work in cadavers has shown that replicating both bundles produces biomechanical results closer to a normal knee (Yagi et al., 2002; Yamamoto et al., 2004).

3.3. Forces Acting Upon the Cells

The obvious approach to tissue engineering a ligament is to replicate the forces that occur in vivo. The recent work of Screen et al. (2003) provides this evidence for tendons (ligaments are likely to behave almost identically due to their structural similarities). They used cell nuclei as markers to estimate the strain field within a fibre bundle by comparing the relative positions of nuclei next to the same fibre bundle. They estimated the displacement of adjacent fibre bundles by comparing the position of cell nuclei either side of a fibre bundle. They found that the strain within a fibre bundle was small only reaching ~1% when the fascicle was strained 8%, on the other hand the displacement between fibre bundles was much larger reaching ~4% when the fascicle was strained 8%. Figure 4 summaries the current model of how the fibre bundles within a tendon respond to strain.

As strain increases the signals and proteins produced by fibroblasts change (Agarwal et al., 2003; Howard et al., 1998; Huh et al., 2003). This change raises questions about the magnitude of strain normally used in experiments to determine the responses of fibroblasts to strain, as these strains are normally much larger than the strains that occur in vivo (Screen et al., 2003).

Important questions for future research are to consider the effect of shear upon the fibroblasts within the ligament, how this shear interacts with physiological levels of strain and design of bioreactors for these experiments. If shear is important then bioreactors and biomaterials will have to be carefully designed in conjunction to replicate the mechanical forces acting upon cells.

The fibrocartilage section of the ACL is formed by simultaneous tension and compression (Carter et al., 1998). This compression is generated in vivo by loading around a pulley (Benjamin and Ralphs, 1998; Wren et al., 2000). The rapid adaptation of ligament and tendon to compression suggests an engineered ligament will develop the appropriate fibrocartilage once transplanted.

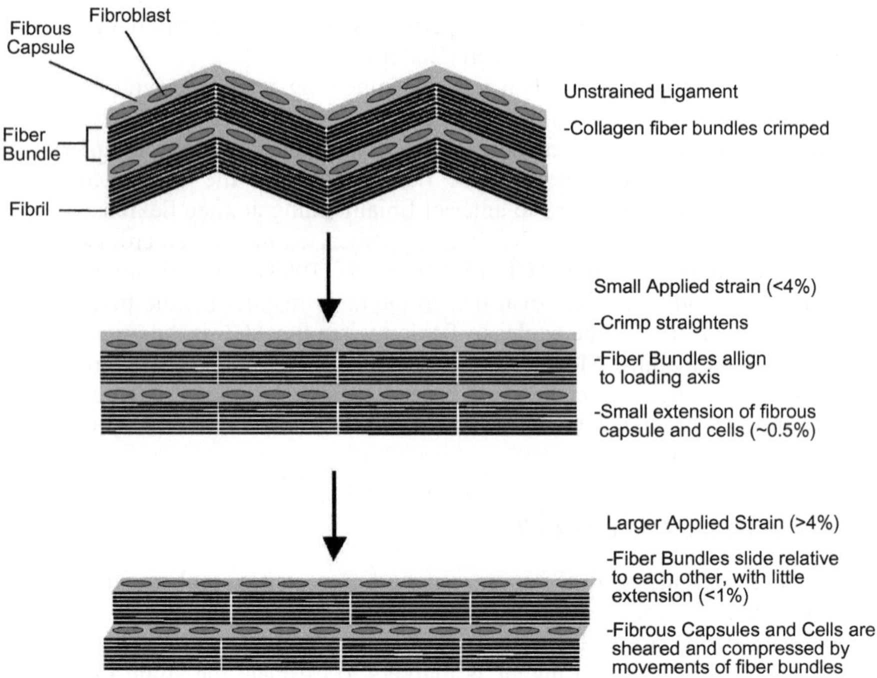

Figure 4. Effects of Tensile Strain upon Fibre Bundles, Fibroblasts and Fibrous Capsules (after Screen et al. (2003)).

3.4. Cellular Responses to Tensile Strain

Much more is known about the effect of 2D cyclical tensile strain upon ligament cells than shear. Cyclical strain increases gene expression of collagens I and III by

ACL fibroblasts, Transforming Growth Factor β1 (TGF-β1) release is a key messenger in this response as the response can be significantly reduced by anti-TGF-β1 antibodies (Hsieh et al., 2000; Kim et al., 2002). Tenascin-C, an ECM protein that reduces cells adherence to the ECM, is also released in response to cyclical stretching (Chiquet et al., 2003).

Ligament fibroblasts are also sensitive to the magnitude of strain. (Howard et al., 1998) looked at the effect of different strains upon periodontal ligament fibroblasts (PDLF), they found that 5% strains increased collagen I and fibronectin production, decreasing tropoelastin production relative to static controls, whilst 10% strains had similar effects on fibronectin and tropoelastin, but had no effect on collagen I relative to static controls.

4. BIOREACTORS

Current ligament engineering bioreactors focus using cyclical stretching upon a cell seeded biomaterial, the interaction between biomaterial and bioreactor is important (Pei et al., 2002), but poorly investigated for ligament tissue engineering. (Lee et al., 2005) have recently found that stretching cells in the same orientation as the nanofibres of their biomaterial are aligned increases collagen synthesis when compared to stretching perpendicularly to the nanofibre orientation. The cells aligned themselves with the nanofibres regardless of the orientation of stretching.

4.1. Current Bioreactors

The simplest bioreactors are those that use static strain to mechanically stimulate cells. These consist of two poles in a petri or pyrex dish, and biomaterial placed on the dish that contracts to from a three dimensional scaffold suspended between in the two posts. The first of these used a collagen gel as the biomaterial and after twelve weeks the construct had an ultimate tensile strength of 0.14 MPa. Blocking collagen crosslinking with BAPN suggested that much of this increase in ultimate tensile strength was due to collagen crosslinking (Huang et al., 1993). A recent experiment used tendon fibroblasts seeded in monolayer on lumican in culture dishes coated with SYLGARD to allow the tendon monolayer to peel off the dish and contract. This showed that, after four weeks, the tendon fibroblasts in vitro could form constructs with ultimate tensile strength of 2 MPa, and morphology resembling embryonic tendon (Calve et al., 2004). It is not clear if it would be possible to scale this technique up to form large ligaments like the ACL.

Several simple bioreactors have been designed that allow cell seeded collagen gels to be cyclically stretched (Cho et al., 2004; Garvin et al., 2003; Langelier et al., 1999; Peperzak et al., 2004; Yahia et al., 1991). The cyclical strain bioreactor designed by Langelier et al. (1999) allowed the maturation of ACL fibroblast seeded collagen gels into ligament-like construct, with dimensions similar to human ACL. After 10 days, the cells had orientated parallel to the direction of strain.

Immunohistology showed increased levels of collagen I and chondroitin sulphate compared to statically stretched controls (Goulet et al., 2000).

Garvin et al. (2003) moulded a small diameter, tendon fibroblast seeded collagen gel between two posts on a rubber bottomed Tissue Train plate. The plate and the gel could be cyclically stretched by a vacuum below the rubber plate. This process resulted in engineered tendons that were composed of a single fascicle, complete with an epitenon-like layer. The fibroblasts within the engineered tendons expressed similar levels of collagens I and III, aggrecan, fibronectin, and prolyl hydroxylase to native tendon, and slightly more tenascin-C, but much less collagen XII. The ultimate tensile strength of their cyclically stretched constructs averaged 327.65kPa after 7 days; compared to 112.2kPa for unloaded controls.

Altman et al. (2002b) designed a much more sophisticated reactor that allowed them to cyclically stretch and rotate their construct, and to pump media both through the construct and around it. The reactor can also control the physicochemical environment including partial pressures of O^2, CO^2, and N^2. This bioreactor allowed them to demonstrate that combined cyclical tensile strain (10%) and cyclical rotational strain (90°, 25%) at 0.0167 Hz applied to a collagen gel seeded with mesenchymal progenitor cells could transform the gel into a ligament like construct with a helical organisation reminiscent of the ACL and the size of goat ACL. The mesenchymal progenitor cells had apparently differentiated into ligament fibroblasts based on production RNA for tenacsin-C, and collagens I and III (Altman et al., 2002a). The reactor allowed them to simultaneously differentiate into ligament-like cells and mature them into a ligament-like construct which had a helical organisation reminiscent of the ACL, although information on the mechanical properties of the constructs has not been published.

All the bioreactors discussed so far have focused on trying to make a weak biomaterial stronger. An alternative approach to tissue engineering ligament is to use a strong biomaterial that replicates some or all of the material properties of the ACL. In this case the bioreactor might be used to: seed cells, mature a cell seeded biomaterial, test the mechanical properties of a cell seeded biomaterial or model the mechanical and physicochemical environment a cell seeded biomaterial might be exposed to in vivo. Chen et al. (2003) used the same bioreactor as Altman et al. (2002a) to determine that modifying silk by adding RGD peptides significantly increased the expression of collagen RNA when the silk was seeded with Bone Marrow Stromal cells and stretched and rotated with the same procedure as the collagen gels above.

There needs to be better established outcome measures for evaluating the performance of bioreactors. At the moment the published work is largely limited to looking at histology and RNA for extracellular matrix (ECM) components, with an implicit assumption that producing more RNA is better. However, RNA does not necessarily reflect protein synthesis and proteins in the ECM can be broken down by enzymes that appear to be synthesised in response to mechanical force (Archambault et al., 2002). There is also the problem that overproduction of collagen is associated with fibrosis and scar tissue which is mechanically weaker than normal tissue (Frank et al., 1985; Prockop and Kivirikko, 1995). As ligament is a mechanical tissue the outcome measures should include comprehensive mechanical testing (Calve et al.,

2004). There has been a lack of attention paid to the ground substance between the fibre bundles, although it plays a vital role in the mechanical strength of ligaments (Yamamoto et al., 1999).

4.2. Mass Transfer

Mass transfer, the diffusion of nutrients to cells through biomaterials, is a major limitation in tissue engineering (Yang et al., 2001), as the time taken for nutrient diffusion is proportional to the square of the distance travelled. In vivo the blood vessels surrounding the fascicles (Clark and Sidles, 1990) ensure that cells are no further than 200μm from a blood supply. In vitro there is no vasculature, so ensuring sufficient mass transfer is a major problem: assemblies of cell aggregates larger than 1mm diameter invariably develop necrotic cores in conventional bioreactors due to insufficient oxygen (Unsworth and Lelkes, 2000); the axial diameter of ACL averages 4.7 mm in women and 5.6 mm in men (Anderson et al., 2001). There are two approaches that could be used to solve this.

One option would be to rely upon the rapid neovascularisation that occurs in vivo as part of an in vivo tissue engineering strategy. When tendons are transplanted as part of surgical ACL reconstruction neovascularisation starts within 20 days of surgery and is complete after 3 months (Fu et al., 1999).

Alternatively, there are several ways a bioreactor could be designed to improve mass transfer in vitro: the design of Altman et al. (2002b) perfuses media through the biomaterial, and around the biomaterial, reducing the distance media has to diffuse through the boundary layer; mechanical stimulation might be used to create negative pressure within the construct, sucking fresh media in Langberg et al. (1999); or the vasculature could be artificially replicated with biodegradable hollow fibres (see Chapter 1 this volume).

4.3. Non-destructive Testing

There need to be standards established to test if a specific ligament construct is suitable for implantation. This testing needs to be non-destructive so that the construct can be implanted after use. A variety of methodologies could be used. There are some non-destructive biochemical tests that could be used to monitor ECM protein production (Lindholt et al., 2001; Orum et al., 1996). An important element will be rigorous non-destructive mechanical testing of the ligament construct is necessary as the construct's main role is mechanical.

Ligaments are repeatedly loaded in vivo and as viscoelastic materials they will lengthen each time they are loaded, therefore it is important to measure their performance when cyclically loaded (Butler et al., 2000). This can be achieved in a cyclical tension bioreactor by monitoring the load applied to the construct and the resulting extension, such bioreactors have already been built (Langelier et al., 1999). This could be further improved by measuring the cross sectional area (Woo et al., 1990), allowing material properties of the construct to be determined.

4.4. Bioreactor Operation

Very little is known about the ideal mechanical and physicochemical environment for the operation of a bioreactor for ligament engineering. Some clues might be available from 2D culture: low oxygen tension promotes collagen synthesis and proliferation of ACL fibroblasts in cell culture (Fermor et al., 1998); with PDLF the magnitude of strain affects collagen synthesis (Howard et al., 1998) and levels of signalling molecules (Agarwal et al., 2003). There might be useful analogies to be made with studies of the effects of exercise and rest upon the ligaments in vivo (Cabaud et al., 1980; Sakuma et al., 1992). However these results might depend upon systems not present in a 3D bioreactor.

5. CONCLUSIONS

Tissue engineering offers a promising alternative to allografts and conventional artificial ligaments. Biomechanical studies are beginning to clarify the requirements for successful surgical ligament replacement (Woo et al., 2004). There are several strategies ligament tissue engineering could choose, such as engineering a ligament in vitro, or relying upon some maturation in vivo. Bioreactors have important roles to play in all of these strategies as: platforms for simulating in vivo conditions, non-destructively testing implants, or maturing cell seeded scaffolds into constructs ready for transplantation. Very little is known about the best conditions for the operation bioreactors, including the possible effects of improving mass transfer and how to optimise mechanical loading regimes. So far it has been shown that: tissue resembling embryonic tendon can be grown in vitro (Calve et al., 2004), cyclical stretching appears to result in stronger constructs with ECM more closely resembling natural ligament (Garvin et al., 2003; Goulet et al., 2000), and that cells can be differentiated into ligament fibroblasts inside a collagen gel that develops a helical organisation similar to ACL (Altman et al., 2002a). Much more research is necessary before a replacement ACL can be engineered.

6. REFERENCES

Agarwal S, Long P, Seyedain A, Piesco N, Shree A, Gassner R. 2003. A central role for the nuclear factor-kappaB pathway in anti-inflammatory and proinflammatory actions of mechanical strain. Faseb J 17(8):899-901.

Altman GH, Horan RL, Martin I, Farhadi J, Stark PR, Volloch V, Richmond JC, Vunjak-Novakovic G, Kaplan DL. 2002a. Cell differentiation by mechanical stress. Faseb J 16(2):270-2.

Altman GH, Lu HH, Horan RL, Calabro T, Ryder D, Kaplan DL, Stark P, Martin I, Richmond JC, Vunjak-Novakovic G. 2002b. Advanced bioreactor with controlled application of multi-dimensional strain for tissue engineering. J Biomech Eng 124(6):742-9.

Anderson AF, Dome DC, Gautam S, Awh MH, Rennirt GW. 2001. Correlation of anthropometric measurements, strength, anterior cruciate ligament size, and intercondylar notch characteristics to sex differences in anterior cruciate ligament tear rates. Am J Sports Med 29(1):58-66.

Archambault JM, Elfervig-Wall MK, Tsuzaki M, Herzog W, Banes AJ. 2002. Rabbit tendon cells produce MMP-3 in response to fluid flow without significant calcium transients. J Biomech 35(3):303-9.
Barber FA. 2003. Should allografts be used for routine anterior cruciate ligament reconstructions? Arthroscopy 19(4):421.
Benjamin M, Ralphs JR. 1998. Fibrocartilage in tendons and ligaments--an adaptation to compressive load. J Anat 193 (Pt 4):481-94.
Butler DL, Goldstein SA, Guilak F. 2000. Functional tissue engineering: the role of biomechanics. J Biomech Eng 122(6):570-5.
Cabaud HE, Chatty A, Gildengorin V, Feltman RJ. 1980. Exercise effects on the strength of the rat anterior cruciate ligament. Am J Sports Med 8(2):79-86.
Calve S, Dennis RG, Kosnik PE, 2nd, Baar K, Grosh K, Arruda EM. 2004. Engineering of functional tendon. Tissue Eng 10(5-6):755-61.
Carter DR, Beaupre GS, Giori NJ, Helms JA. 1998. Mechanobiology of skeletal regeneration. Clin Orthop(355 Suppl):S41-55.
Chen J, Altman GH, Karageorgiou V, Horan R, Collette A, Volloch V, Colabro T, Kaplan DL. 2003. Human bone marrow stromal cell and ligament fibroblast responses on RGD-modified silk fibers. J Biomed Mater Res A 67(2):559-70.
Chiquet M, Renedo AS, Huber F, Fluck M. 2003. How do fibroblasts translate mechanical signals into changes in extracellular matrix production? Matrix Biol 22(1):73-80.
Cho SW, Kim IK, Lim SH, Kim DI, Kang SW, Kim SH, Kim YH, Lee EY, Choi CY, Kim BS. 2004. Smooth muscle-like tissues engineered with bone marrow stromal cells. Biomaterials 25(15):2979-86.
Clark JM, Sidles JA. 1990. The interrelation of fiber bundles in the anterior cruciate ligament. J Orthop Res 8(2):180-8.
Cooper JA, Lu HH, Ko FK, Freeman JW, Laurencin CT. 2005. Fiber-based tissue-engineered scaffold for ligament replacement: design considerations and in vitro evaluation. Biomaterials 26(13):1523-32.
Dopirak RM, Adamany DC, Steensen RN. 2004. A comparison of autogenous patellar tendon and hamstring tendon grafts for anterior cruciate ligament reconstruction. Orthopedics 27(8):837-42; quiz 843-4.
Feller J. 2004. Anterior cruciate ligament rupture: is osteoarthritis inevitable? Br J Sports Med 38(4):383-4.
Fermor B, Urban J, Murray D, Pocock A, Lim E, Francis M, Gage J. 1998. Proliferation and collagen synthesis of human anterior cruciate ligament cells in vitro: effects of ascorbate-2-phosphate, dexamethasone and oxygen tension. Cell Biol Int 22(9-10):635-40.
Frank C, Amiel D, Woo SL, Akeson W. 1985. Normal ligament properties and ligament healing. Clin Orthop(196):15-25.
Fu FH, Bennett CH, Lattermann C, Ma CB. 1999. Current trends in anterior cruciate ligament reconstruction. Part 1: Biology and biomechanics of reconstruction. Am J Sports Med 27(6):821-30.
Fu FH, Bennett CH, Ma CB, Menetrey J, Lattermann C. 2000. Current trends in anterior cruciate ligament reconstruction. Part II. Operative procedures and clinical correlations. Am J Sports Med 28(1):124-30.
Garvin J, Qi J, Maloney M, Banes AJ. 2003. Novel system for engineering bioartificial tendons and application of mechanical load. Tissue Eng 9(5):967-79.
Goulet F, Rancourt D, Cloutier R, Germain L, Poole AR, Auger FA. 2000. Tendons and ligaments. In: Lanza RP, Langer R, Vacanti J, editors. Principles of Tissue Engineering. 2nd ed: Academic Press.
Howard PS, Kucich U, Taliwal R, Korostoff JM. 1998. Mechanical forces alter extracellular matrix synthesis by human periodontal ligament fibroblasts. J Periodontal Res 33(8):500-8.
Hsieh AH, Tsai CM, Ma QJ, Lin T, Banes AJ, Villarreal FJ, Akeson WH, Sung KL. 2000. Time-dependent increases in type-III collagen gene expression in medical collateral ligament fibroblasts under cyclic strains. J Orthop Res 18(2):220-7.
Huang D, Chang TR, Aggarwal A, Lee RC, Ehrlich HP. 1993. Mechanisms and dynamics of mechanical strengthening in ligament-equivalent fibroblast-populated collagen matrices. Ann Biomed Eng 21(3):289-305.
Huh YH, Kim SH, Kim SJ, Chun JS. 2003. Differentiation status-dependent regulation of cyclooxygenase-2 expression and prostaglandin E2 production by epidermal growth factor via mitogen-activated protein kinase in articular chondrocytes. J Biol Chem 278(11):9691-7.

Johnson DH. 2003. Should allografts be used for routine anterior cruciate ligament reconstructions? No, allografts should not be used for routine ACL reconstruction. Arthroscopy 19(4):424-5.

Kastelic J, Galeski A, Baer E. 1978. The multicomposite structure of tendon. Connect Tissue Res 6(1):11-23.

Kim SG, Akaike T, Sasagaw T, Atomi Y, Kurosawa H. 2002. Gene expression of type I and type III collagen by mechanical stretch in anterior cruciate ligament cells. Cell Struct Funct 27(3):139-44.

Langberg H, Skovgaard D, Bulow J, Kjaer M. 1999. Negative interstitial pressure in the peritendinous region during exercise. J Appl Physiol 87(3):999-1002.

Langelier E, Rancourt D, Bouchard S, Lord C, Stevens PP, Germain L, Auger FA. 1999. Cyclic traction machine for long-term culture of fibroblast-populated collagen gels. Ann Biomed Eng 27(1):67-72.

Lee CH, Shin HJ, Cho IH, Kang YM, Kim IA, Park KD, Shin JW. 2005. Nanofiber alignment and direction of mechanical strain affect the ECM production of human ACL fibroblast. Biomaterials 26(11):1261-70.

Lindholt JS, Heickendorff L, Vammen S, Fasting H, Henneberg EW. 2001. Five-year results of elastin and collagen markers as predictive tools in the management of small abdominal aortic aneurysms. Eur J Vasc Endovasc Surg 21(3):235-40.

Lohmander LS, Ostenberg A, Englund M, Roos H. 2004. High prevalence of knee osteoarthritis, pain, and functional limitations in female soccer players twelve years after anterior cruciate ligament injury. Arthritis Rheum 50(10):3145-52.

McGuire DA. 2003. Should allografts be used for routine anterior cruciate ligament reconstructions? Yes, allografts should be used in routine ACL reconstruction. Arthroscopy 19(4):421-4.

Murray AW, Macnicol MF. 2004. 10-16 year results of Leeds-Keio anterior cruciate ligament reconstruction. Knee 11(1):9-14.

Murray MM, Spector M. 1999. Fibroblast distribution in the anteromedial bundle of the human anterior cruciate ligament: the presence of alpha-smooth muscle actin-positive cells. J Orthop Res 17(1):18-27.

Nomura E, Inoue M, Sugiura H. 2005. Histological evaluation of medial patellofemoral ligament reconstructed using the Leeds-Keio ligament prosthesis. Biomaterials 26(15):2663-70.

Noyes FR, Grood ES. 1976. The strength of the anterior cruciate ligament in humans and Rhesus monkeys. J Bone Joint Surg Am 58(8):1074-82.

Orum O, Hansen M, Jensen CH, Sorensen HA, Jensen LB, Horslev-Petersen K, Teisner B. 1996. Procollagen type I N-terminal propeptide (PINP) as an indicator of type I collagen metabolism: ELISA development, reference interval, and hypovitaminosis D induced hyperparathyroidism. Bone 19(2):157-63.

Pei M, Solchaga LA, Seidel J, Zeng L, Vunjak-Novakovic G, Caplan AI, Freed LE. 2002. Bioreactors mediate the effectiveness of tissue engineering scaffolds. Faseb J 16(12):1691-4.

Peperzak KA, Gilbert TW, Wang JH. 2004. A multi-station dynamic-culture force monitor system to study cell mechanobiology. Med Eng Phys 26(4):355-8.

Prockop DJ, Kivirikko KI. 1995. Collagens: molecular biology, diseases, and potentials for therapy. Annu Rev Biochem 64:403-34.

Sakane M, Fox RJ, Woo SL, Livesay GA, Li G, Fu FH. 1997. In situ forces in the anterior cruciate ligament and its bundles in response to anterior tibial loads. J Orthop Res 15(2):285-93.

Sakuma K, Mizuta H, Takagi K, Takashima K. 1992. Effects of enforced exercise on biomechanical properties of the anterior cruciate ligament of bipedal rats. Nippon Seikeigeka Gakkai Zasshi 66(11):1146-55.

Screen HR, Lee DA, Bader DL, Shelton JC. 2003. Development of a technique to determine strains in tendons using the cell nuclei. Biorheology 40(1-3):361-8.

Silver FH, Freeman JW, Seehra GP. 2003. Collagen self-assembly and the development of tendon mechanical properties. J Biomech 36(10):1529-53.

Smith BA, Livesay GA, Woo SL. 1993. Biology and biomechanics of the anterior cruciate ligament. Clin Sports Med 12(4):637-70.

Snell RS. 2000. Clinical anatomy for medical students: Lippincott Williams & Wilkins.

Spindler KP, Kuhn JE, Freedman KB, Matthews CE, Dittus RS, Harrell FE, Jr. 2004. Anterior cruciate ligament reconstruction autograft choice: bone-tendon-bone versus hamstring: does it really matter? A systematic review. Am J Sports Med 32(8):1986-95.

Takai S, Woo SL, Livesay GA, Adams DJ, Fu FH. 1993. Determination of the in situ loads on the human anterior cruciate ligament. J Orthop Res 11(5):686-95.

Von Porat A, Roos EM, Roos H. 2004. High prevalence of osteoarthritis 14 years after an anterior cruciate ligament tear in male soccer players: a study of radiographic and patient relevant outcomes. Br J Sports Med 38(3):263.

Weitzel PP, Richmond JC, Altman GH, Calabro T, Kaplan DL. 2002. Future direction of the treatment of ACL ruptures. Orthop Clin North Am 33(4):653-61.

Woo SL, Adams DJ. 1990. The tensile properties of human anterior cruciate ligament (ACL) and ACL graft tissues. In: Daniel D, Akeson W, O'Connor J, editors. Knee Ligaments: Structure, Function, Injury, and Repair: Lippincott Williams and Wilkins.

Woo SL, Danto MI, Ohland KJ, Lee TQ, Newton PO. 1990. The use of a laser micrometer system to determine the cross-sectional shape and area of ligaments: a comparative study with two existing methods. J Biomech Eng 112(4):426-31.

Woo SL, Hollis JM, Adams DJ, Lyon RM, Takai S. 1991. Tensile properties of the human femur-anterior cruciate ligament-tibia complex. The effects of specimen age and orientation. Am J Sports Med 19(3):217-25.

Woo SL, Thomas M, Chan Saw SS. 2004. Contribution of biomechanics, orthopaedics and rehabilitation: the past present and future. Surgeon 2(3):125-36.

Wren TA, Beaupre GS, Carter DR. 2000. Mechanobiology of tendon adaptation to compressive loading through fibrocartilaginous metaplasia. J Rehabil Res Dev 37(2):135-43.

Yagi M, Wong EK, Kanamori A, Debski RE, Fu FH, Woo SL. 2002. Biomechanical analysis of an anatomic anterior cruciate ligament reconstruction. Am J Sports Med 30(5):660-6.

Yahia LH, Desrosiers EA, Rivard CH. 1991. A computer-controlled apparatus for in vitro mechanical stimulation and characterization of ligaments. Biomed Mater Eng 1(4):215-22.

Yamamoto E, Hayashi K, Yamamoto N. 1999. Mechanical properties of collagen fascicles from stress-shielded patellar tendons in the rabbit. Clin Biomech (Bristol, Avon) 14(6):418-25.

Yamamoto Y, Hsu WH, Woo SL, Van Scyoc AH, Takakura Y, Debski RE. 2004. Knee stability and graft function after anterior cruciate ligament reconstruction: a comparison of a lateral and an anatomical femoral tunnel placement. Am J Sports Med 32(8):1825-32.

Yang S, Leong KF, Du Z, Chua CK. 2001. The design of scaffolds for use in tissue engineering. Part I. Traditional factors. Tissue Eng 7(6):679-89.

CHAPTER 11

BIOMECHANICAL CONSIDERATIONS FOR TISSUE ENGINEERED HEART VALVE BIOREACTORS

M.S. SACKS[1], G.C. ENGELMAYR JR.[1], D.K. HILDEBRAND[1] AND J.E. MAYER, JR.[2]

[1]*McGowan Institute for Regenerative Medicine, Department of Bioengineering, University of Pittsburgh, Pittsburgh, USA*

[2]*Department of Cardiovascular Surgery, Children's Hospital Boston, Harvard Medical School, Boston, USA*

1. INTRODUCTION

1.1 Tissue Engineered Heart Valves

First performed successfully in 1960, surgical replacement of diseased human heart valves by mechanical and tissue valve substitutes is now commonplace and enhances survival and quality of life for many patients. Tissue heart valve substitutes are now used in approximately 40% of the estimated 75,000 domestic and 275,000 worldwide valve replacements done annually (Schoen 2001). Tissue valves have the advantage of low rates of thromboembolic complications without chronic anticoagulation therapy; however, they suffer high rates of late structural dysfunction owing to tissue degradation (Schoen and Levy 1999; Turina et al. 1993). While much effort in currently underway to improve current aortic valve (AV) bioprostheses, in the long-term new technologies will have to be developed.

Thus, a non-obstructive, non-thrombogenic tissue valve substitute that lasts the lifetime of the patient, provides ongoing remodeling and repair of cumulative injury and potentially grows with the recipient is an important goal. Such a valve would be particularly useful in the treatment of the approximately 20,000 children with congenital heart disease born in the United States each year and those with valvular disease. In this population the anticoagulation required with mechanical valves is particularly dangerous and tissue valve substitutes undergo accelerated calcification.

J.B. Chaudhuri & M. Al-Rubeai (eds.), Bioreactors for Tissue Engineering, 235 - 267.

Thus, a fundamental problem inherent to the use of existing mechanical and biological prostheses in the pediatric population is their failure to grow, repair and remodel.

Tissue engineering (TE) is an approach that attempts to combine engineering principles with the biological sciences to produce viable structures for replacement of diseased or deficient native structures. In the case of the TEHV, the scaffold must also have a design and mechanical characteristics that allow valve-like function during in-vivo "maturation" of the TE construct. The basic properties necessary for a bioabsorbable scaffolding material are non-toxicity (including degradation products), biocompatibility, resorbability, strength, and processability. Besides the proper material properties, a TE scaffold needs to be engineered to possess the proper geometry for tissue formation. Proper engineering and construction techniques are expected to meet many of the other necessary design properties such as geometry, density, compliance, hemodynamics etc. while most of the biological activity of the TEHV should be fulfilled by the growing cells. The polymer scaffold temporarily provides the biomechanical structural characteristics for the replacement "tissue" until the cells produce their own extracellular matrix which will ultimately provide the structural integrity and biomechanical profile for the replacement "tissue." During this process of tissue development, the scaffold will be gradually degraded eventually leaving no foreign materials within the replaced tissues.

Work by the Mayer group (Hoerstrup et al. 2000a; Hoerstrup et al. 1998; Hoerstrup et al. 1999; Sodian et al. 1999) have focused on the creation of cardiovascular structures using this tissue engineering approach. They reasoned that the creation of a tissue engineered structure from autologous cells could potentially offer several advantages over the currently available prosthetic, bioprosthetic, or homograft devices used to replace cardiac valves and arteries. These structures would be living, viable structures which could retain the normal biological mechanisms for repair and development, thereby leading to greater durability. A TE autologous heart valve could be completely biocompatible with a reduced risk of infection and thrombus formation.

Of particular interest in the pediatric population, a viable TE structure could potentially grow with the patient, eliminating the need for multiple reoperations. In addition, TE structures may prove to be less costly than currently available prosthetic devices, and there may be fewer supply limitations compared to homografts. They have demonstrated the feasibility of creating pulmonary valve leaflets and segments of main pulmonary artery. Both structures have functioned for periods of up to four months in growing lambs and have demonstrated short term durability, growth, and lack of thrombogenicity. The gross and microscopic characteristics of these TE structures approximate those of normal structures (Hoerstrup et al. 2000a). Stock (2000) had also demonstrated formation of pulmonary artery tissue with histology very similar to native tissue.

Despite this encouraging progress, significant problems have arisen in our previous work in creating TE pulmonary valves (PV) and valved conduits, and critical questions of function and pathobiology remain. In-vivo related issues that we have identified is that TE PV replacements have developed moderate amounts of central valvular regurgitation, despite the in vivo evolution of the TE valve tissue into a tri-layered structure which resembles native valve tissue. We thus have only very limited information on the extent to which a TEHV duplicates the biomechanical function and structure of the native PV cusp, and the events and mechanisms that guide the remodeling process. There is thus a need for a detailed understanding of the micromechanical events that take place during in-vitro incubation and subsequent in-vivo remodeling.

1.2 General Biomechanical Considerations For Bioreactor Design And In-Vitro Tissue Development

Recent studies have shown that incubation under pulsitile loading upregulates extracellular matrix (ECM) deposition. However, it is unknown what the resultant changes in mechanical properties were, and to what extent these changes are due to the degrading scaffold and the increased ECM mass. Particular difficulties are encountered when attempting to quantify the mechanical properties of engineering scaffolds in the early phases of incubations due to their very low mechanical stiffness.

Bioreactors have been developed for the dynamic mechanical stimulation of tissue engineered cardiovascular constructs, including vascular grafts (Niklason et al. 1999; Seliktar et al. 2000), myocardial patches (Sodian et al. 2001), and heart valves (Hoerstrup et al. 2000a; Zeltinger et al. 2001). These devices typically rely on pulsatile flow to generate a complex biomechanical environment resembling *in vivo* conditions, and have been demonstrated to promote both the development of mechanical strength (Hoerstrup et al. 2000b; Niklason et al. 1999; Seliktar et al. 2000), and the modulation of cellular function (Seliktar et al. 2001) within these tissue engineered constructs. While these devices have shown promise in the development of functional tissue replacements (Hoerstrup et al. 2000b; Niklason et al. 1999), they present several drawbacks when applied to the study of fundamental biomechanical phenomena, including: small sample capacity, anatomical sample geometry, and coupled mechanical stimuli.

Few bioreactors have been designed for the explicit purpose of studying the individual effect of specific modes of mechanical stimulation on engineered cardiovascular tissues, such that candidate scaffolds and constructs can be systematically evaluated (Cacou et al. 2000; Jockenhoevel et al. 2002; Kim and Mooney 2000; Mitchell et al. 2001). Such devices should be designed to provide a user-defined, simple mode of mechanical stimulation, offer a sufficient sample capacity for statistically significant comparisons at multiple time points, and accommodate simple sample geometries amenable to subsequent mechanical testing. As is the case for most tissue engineering bioreactors, these devices should allow for the culture of engineered tissue samples under sterile, physiological (37 °C and 5% CO_2) conditions for the duration of the experiment, and should be relatively easily to clean and maintain.

In our laboratory, we are interested in studying the effects of dynamic mechanical stimulation on the development of TEHV. Our foci are both on the evaluation of candidate scaffold materials and to conduct mechanistic studies in order to elucidate fundamental biomechanical phenomena. Moreover, our long-term goal is that the development of a rational basis for tissue engineering design may eventually be derived from such studies, allowing tissue structure and function to be accurately predicted from cell, scaffold, media, and environmental conditions.

As alluded to above, utmost care must be taken to control mechanical conditions when conducting mechanistic studies on the behavior of candidate tissue engineering scaffolds and constructs. If we aspire to incorporate mechanical and biological test results as parameters in the design of tissue engineered constructs, the individual effect of simple modes of mechanical stimulation must be accurately quantified and modeled in the absence of potentially confounding effects. Likewise, the definitive role of a particular mode of mechanical stimulation cannot easily be resolved in a system composed of coupled mechanical stimuli. These stringent control requirements preclude the immediate use of pulse duplicator or flow loop bioreactors for such discriminating studies.

1.3 Focus of This Chapter

The present chapter reviews our laboratory's recent approaches for 1) mechanical assessment of TEHV biomaterials during in-vitro development, 2) design and evaluation of a novel flexural bioreactor, and 3) biomechanics of native heart valve tissues (for future assessment of mature TEHV tissues). Further, it is speculated that once optimal conditions for tissue development are established, there will be a need to design precision pulsatile bioreactors that can subject TEHV to precise hemodynamic conditions. In the final section, we thus present a novel pulsitile heart valve bioreactor that can subject TEHV to highly controlled pressure and flow conditions, both within and outside the functional physiological range.

2. FLEXURAL MECHANICAL PROPERTIES OF TEHV BIOMATERIALS

Flexure mechanical testing is a highly sensitive method to assess the mechanical properties of TEHV biomaterials, which generally have very low mechanical stiffness. For example, in a recent study Perry et al. (2001) fabricated tri-leaflet valve constructs (TEHV) from P4HB coated PGA sheet, and seeded the TEHV using peripheral blood-derived mesenchymal cells. The cell-polymer construct was allowed to rotate in suspension for two weeks in a rotating bioreactor before being transported to a pulse-duplicating bioreactor for an additional two weeks. Using the flexure mechanical testing methods described previously (Gloeckner et al. 1999), flexural mechanical properties were assessed at zero and four weeks. Results clearly indicated a marked reduction in flexural stiffness with incubation time, as well as non-linearities in the moment-curvature relation. This latter finding suggested that non-linear soft tissue properties may become more prominent as the polymer scaffold degrades and the deposited soft tissue mass increases. In the following section, we demonstrate the effects of cell seeding and static vs. dynamic conditioning on the resulting flexural mechanical properties using a functional TEHV model subjected to in-vitro mechanical conditioning.

2.2 Methods

2.2.1 Valve Construction, Seeding and Incubation

TEHV scaffolds consisted of a non-woven mesh of polyglycolic acid (PGA) fibers (Albany International Research, Mansfield, MA) dip-coated with poly-4-hydroxybutyrate (P4HB) (Tepha, Inc., Cambridge, MA). Sheets of the PGA/P4HB were cut and formed to create a tri-leaflet valve approximately 20 mm in diameter. Cell seeding methods have been previously described (Hoerstrup et al. 2000a). Briefly, carotid artery vessels were harvested from Dover lambs of various ages supplied by a local farm. In a laminar flow hood the tissue were minced into 1- to 2-mm pieces, serially washed with phosphate buffered saline (Gibco BRL-Life Technologies) supplemented with 1% antibiotic solution (Irvine Scientific) and evenly distributed onto 15 mm x 60 mm tissue culture plates (Becton-Dickinson). After the addition of tissue culture media (Dulbelcco's modified Eagle medium supplemented with 10% fetal bovine serum with 1% antibiotic solution), the dishes were incubated for 8 to 10 weeks at 37°C with 5% CO_2. The medium was changed every 7 days. The cells migrate off from the tissue explants and form a mixed cell population of endothelial, smooth muscle cells, and fibroblasts with a ratio of 1:20:13.

After the mixed cell population is grown to confluence, the cells were labeled with a fluorescent acetylated low-density lipoprotein (LDL) marker that is selectively taken into endothelial cells via the scavenger pathway. After a 24-hour incubation period the cells are separated into two cell groups (LDL-positive and LDL-negative) using a fluorescent activated cell sorter. The two cell populations were separately passaged so as to obtain sufficient numbers of cells for cell seeding.

Each valve was initially seeded with $\sim 1 \times 10^7$ smooth muscle cells and fibroblasts, and then one week later the scaffold-SMC construct will be seeded with 2×10^6 LDL positive endothelial cells. Valves were cultured using 4 day static/14 day dynamic (18 day total) in-vitro protocol established by the Mayer lab (Hoerstrup et al. 2000a). Upon removal from the bioreactor, each valve was placed in a plastic container, filled with media and sealed, then placed in ice slush and shipped overnight.

A second group of two valves were prepared and incubated identically to the seeded valves described above, but were not seeded will cells. This allowed separation of the effects of scaffold degradation and cell/ECM-deposition on the flexural mechanical properties. A final third group of valves were prepared up to albumin coating and media immersion, and immediately shipped. This last group served as a control to establish baseline mechanical properties of the preprocessed (i.e. up to incubation) scaffold. For each group, two valves were prepared, resulting in a total of six leaflets per group. All valves were shipped in tissue culture media in a sealed container in ice slush via overnight courier.

2.2.2 Flexural mechanical testing

Upon arrival, each TEHV was cut along its length and opened, and the leaflets dissected free from the conduit using micro-scissors. From each leaflet, a single 2 mm x 15 mm circumferentially oriented specimen was prepared (Figure 1), using parallel-spaced razor blades with 2 mm spacers. Next, along one edge approximately fifteen 0.5 mm diameter graphite markers were attached with cyanoacrylate (Figure 2). A small metal sleeve was glued ~2 mm from one end and used to hold the specimen in the tester (Figure 2). Based on techniques established for valvular tissues (Billiar and Sacks 2000), thickness measurements were made in the center of the belly region using a non-rotating gauge (resolution 12.5 μm) that applied a small tare load (~100g). Five measurements were made along the length of each specimen and the mean value utilized for subsequent calculations.

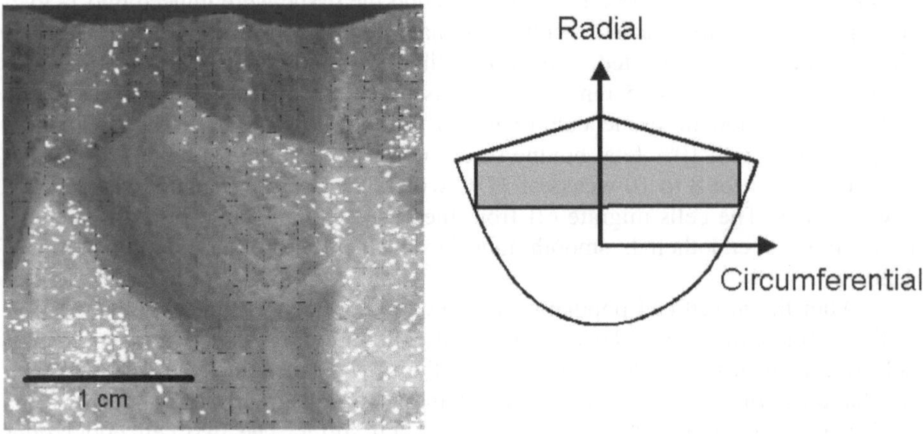

Figure 1. a) A TEHV leaflet attached to the vessel conduit after 18 days of incubation. b) A schematic of where the flexural mechanical testing specimens were removed.

Each specimen was subjected to three-point bending using a specially designed flexure testing apparatus for soft biomaterials, previously described in (Gloeckner et al. 1999). Briefly, test specimens are mounted between two posts approximately 10 mm apart and a long thin vertical metal wire (whose x-y position is controlled by stepper motors) is used as the central force applicator (bending bar in Figure 2). The deflection of the metal wire with load was calibrated so it was also used as the load transducer. The bending device has a spatial resolution of 30 μm/pixel and a load resolution of ±0.112 mg. The specimen was then flexed by translating the bending bar by 2-3 mm (Figure 2b) over period of ~15 seconds. To fully capture the flexural response, each specimen was first bent toward the inflow direction (with the TEHV leaflet curvature or WC), then remounted and flexed towards the outflow direction (against the TEHV leaflet curvature or AC). Video images of each flexure experiment were recorded on SVHS video tape. All specimens were tested in room temperature normal saline solution. Specimen thickness was determined from the captured reference state images and taken as the average of five equally spaced measurements between each post.

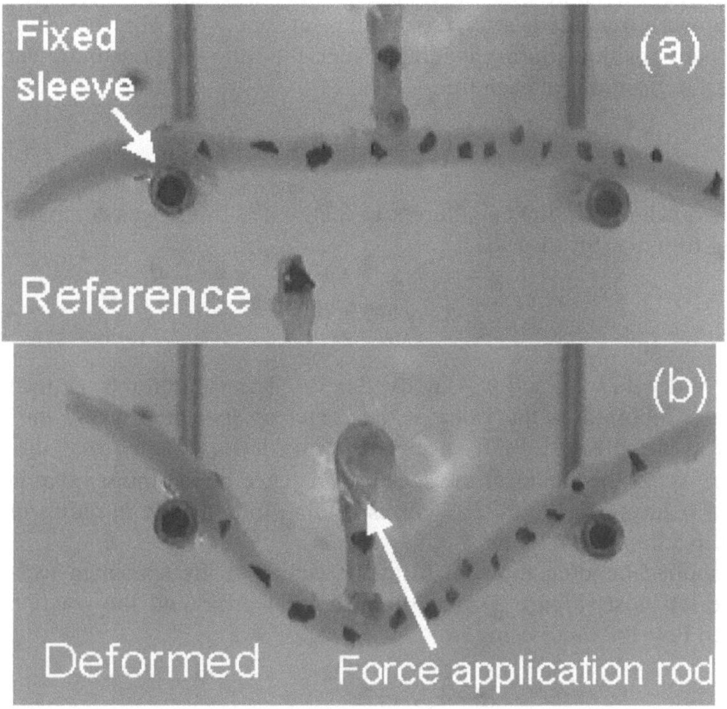

Figure 2. Photograph of a TEHV scaffold specimen subjected to three-point bending in the unloaded reference (a) and loaded deformed (b) positions.

2.2.3 Analysis Methods

We computed the specimen flexural rigidity using the generalized Bernoulli-Euler moment-curvature relation for beams undergoing arbitrary displacements (Frisch-Fay 1962)

$$M = \bar{E}I \, \Delta\kappa \tag{1}$$

where $\bar{E}$ is the effective specimen stiffness, I is the 2nd moment of inertia, $\Delta\kappa$ the change in specimen curvature from the reference state (Figures 2a, b), and M is the applied bending moment determined from the applied loads and specimen geometry. Physically, eqn. 1 is analogous to Hooke's law (i.e. $\sigma = E\,\varepsilon$, where σ is stress, E is Young's modulus, and ε is the strain) for a specimen subjected to flexure as opposed to uniaxial tension.

To compute $\Delta\kappa$, the spatial positions of the graphite markers were obtained by tracking their x-y positions from the video tapes, using video tracking software developed for biaxial mechanical testing (Sacks and Chuong 1998). The marker x-y positions were recorded at video frame rate of ~30 Hz, resulting in ~100 measurement positions during loading. Next, the following quadratic curve was fit to the marker positions at each frame.

$$y = ax^2 + bx + c \tag{2}$$

From this quadratic fit, the curvature κ at each video frame was computed using the following formula (Struik 1961):

$$\kappa = y''/(1 + (y')^2)^{3/2} \tag{3}$$

where $y'' = d^2y/dx^2 = 2a$, and $y' = dy/dx = 2ax+b$. The value for $\bar{E}$ (Eqn. 1) for each specimen is reported as the value at the center of specimen where the amount of flexure was the greatest. All specimens were deflected such that the change in curvature, $\Delta\kappa$, at the center of the specimen reached a maximum valve in the range of 0.15 to 0.20 mm^{-1} (Figure 2b), corresponding to a change in radius of curvature from flat to 5.0 mm~6.67 mm.

The applied bending moment **M** at the center of the specimen was calculated about the left post (Figure 3) from the applied load **P** and the reaction forces F_y (=P/2) and F_x (=P/2 tanθ) using

$$M = F_x y + F_y x \tag{4}$$

Note that in Figure 3b the angle used to calculate the reaction forces, θ, is determined by calculating the slope the quadratic curve (Equation 2) makes with the right post and the X axis.

Since the slope of the **M**-$\Delta\kappa$ curve represents the specimen flexural rigidity, we computed the effective stiffness $\bar{E}$ by dividing by the slope by **I**. Physically, $\bar{E}$ represents the *instantaneous effective tissue modulus* at a given level of bending (i.e. value of $\Delta\kappa$), and thus in general $\bar{E} = \bar{E}(\Delta\kappa)$. Further, due to the potential structural heterogeneity of the tissue specimens, $\bar{E}$ represents the specimen's bending stiffness only for the *current bending direction*. That is, $\bar{E}$ will generally have a different value for the same specimen when the specimen is bent in opposite directions. For the present study, the value for $\bar{E}$ is reported as the value at the center of specimen where the amount of flexure was the greatest.

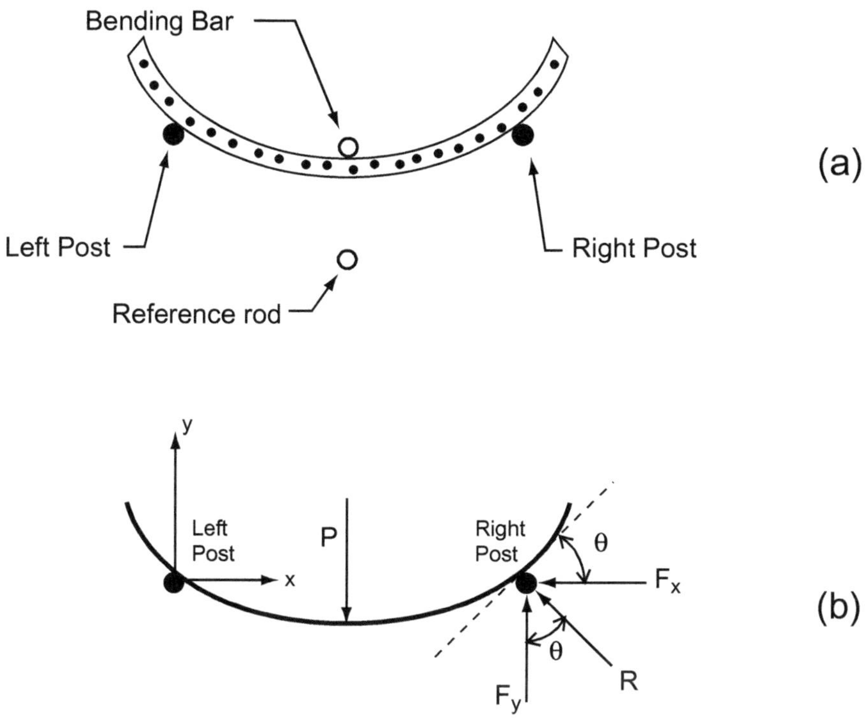

Figure 3. a) A schematic of the three-point bending technique showing the major components; b) a free-body diagram showing the definition of the x-y coordinate system and reaction forces.

2.3. Results and Discussion

2.3.1 Flexural Mechanical Properties

The pre-processed scaffold demonstrated a highly linear flexural mechanical response (i.e., a linear M/I-$\Delta\kappa$ relationship with r^2=0.98, Figure 4a). This finding was valid over the measured curvature change of -0.1 mm-1 $\leq \Delta\kappa \leq$ 0.15 mm^{-1}, corresponding to a range of curvature radii of -10.0 mm to 6.67 mm. Interestingly, there was no observable bending directional difference (Figure 4a), suggesting an effectively homogeneous transmural structure of the scaffold. The effect of 18 day dynamic incubation alone (i.e. unseeded) was to dramatically reduce the flexural rigidity, along with inducing modest directional differences (Figure 4b). More interestingly, cell seeding and subsequent ECM deposition dramatically increased the overall flexural rigidity compared to the unseeded dynamically conditioned control (Figure 4b).

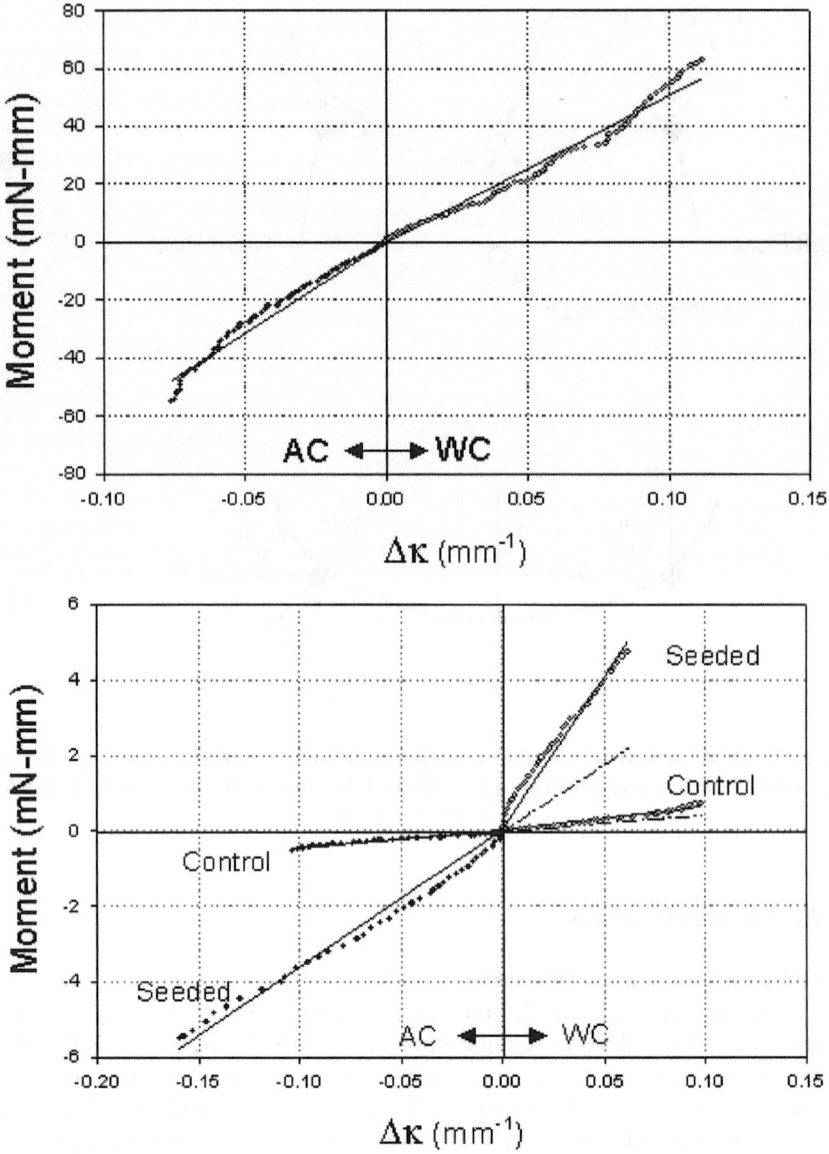

Figure 4. Measured moment-curvature (M-Δκ) response for an unmodified TEHV scaffold (PGA/P4HB) (upper), demonstrating an identical linear relation in both flexural directions. Measured moment-curvature (M-Δκ) responses for 18-day TEHV scaffold under dynamic conditions (lower). Dashed lines show the linear response in the AC direction only, underscoring that dynamic incubation induced directional differences. Moreover, seeded response demonstrated a pronounced increase in flexural rigidity.

Averaged results for $\bar{E}$ for all groups are presented in Figure 5. The pre-processed scaffold demonstrated an effective stiffness of ~1,200 kPa, with no statistically significant bending directional differences. Four day incubation (by definition static only) indicated a drop in seeded scaffolds, but no change in the control. Major decreases in $\bar{E}$ were observed in all 18 day groups. Moreover, we observed that dynamic conditioning 1) reduced the overall $\bar{E}$ more than the static groups, and 2) induced directional dependence in $\bar{E}$ the dynamic incubation groups.

These trends can be more easily visualized when examining ratios in $\bar{E}$ between seeded/control and bending direction (Figure 6). Here, the 18 day seeded leaflets were on average ~3.5 times stiffer than the unseeded controls under dynamic conditions, whereas the static incubated controls were at most 2 times stiffer (Figure 6a). Further, dynamic conditioning induced a directional dependence, with the outflow direction (against curvature or AC) exhibiting a nearly two-fold increase compared to the inflow (with curvature or WC) direction ($p<0.05$, Figure 6b).

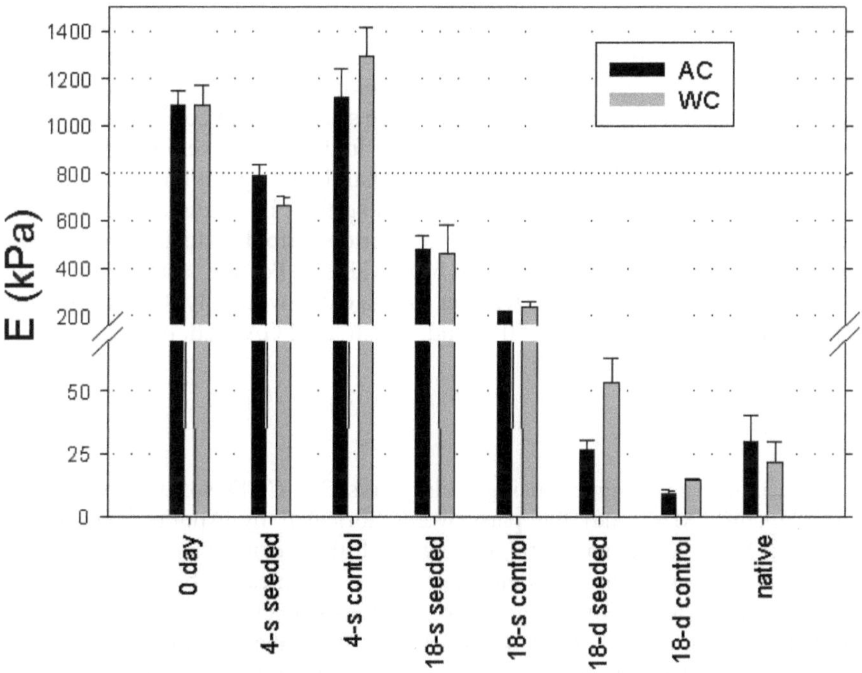

Figure 5. The measured effective moduli for all test groups, including the native pulmonary valve for comparison. Here, the major finding was that the dynamic conditioning induced both a larger decrease in effective modulus and directional differences not observed in the 18 day static groups (control or seeded). Note too the similarity in the 18 day dynamic seeded group to the native tissue.

In this study using a relatively straightforward 3-point bending mechanical testing technique, we were able to show that 1) scaffolds undergo rapid and dramatic reductions in stiffness within the 18-day test period, 2) the de-novo ECM dramatically increased scaffold stiffness compared to unseeded controls, and 3) dynamic conditioning can induce a directional dependence when utilizing pre-coated scaffolds (in this case P4HB).

3. FLEXURAL BIOREACTOR FOR HEART VALVE BIOMATERIALS

3.1. Introduction

In the above study, we demonstrated that TEHV scaffolds can undergo significant and complex changes in mechanical properties. *In vivo*, the heart valves are subjected to a unique combination of mechanical stimuli, including flexure, shear stress, and tension (Vesely and Boughner 1989). Whereas the effects of shear stress and tension on the development of engineered tissues have been explored to an extent (Jockenhoevel et al. 2002; Kim and Mooney 2000), the individual effect of dynamic flexural stimulation on TE cardiovascular constructs has been understudied to date. Because flexure represents a principal mode of deformation in heart valve leaflets (Iyengar et al. 2001), we sought to develop a device to study the individual effect of flexure on tissue engineered heart valve scaffolds and constructs.

In the current study, our goals were three-fold: 1) to develop a bioreactor to provide cyclic flexural stimulation, 2) to demonstrate the operation of the bioreactor and sterility maintenance, and 3) to evaluate the effects of unidirectional flexure on the effective stiffness of bioresorbable polymeric scaffolds which have been used extensively in the tissue engineering of heart valves. For the last goal, we used a non-woven polyglycolic acid (PGA) fiber mesh coated with poly-4-hydroxybutyrate (P4HB), and a non-woven, 50:50 blend, mesh of PGA and poly-L-lactic acid (PLLA) fibers coated with P4HB.

3.2. Bioreactor Design

The bioreactor was designed using Solidworks 3D CAD software (Solidworks Corp., Concord, MA), and the parts were fabricated by Apollo Precision, Inc. (Plymouth, MN). The structural elements of the device were machined from polysulfone; chosen for its excellent thermal and chemical stability; and abrasion-resistant acrylic; which provides good optical transparency. Neoprene gaskets were fitted between all critical joints in order to reduce the possibility of microbial contamination, and the bioreactor was assembled using 18-8 stainless steel screws (McMaster-Carr Supply Co., Cleveland, OH).

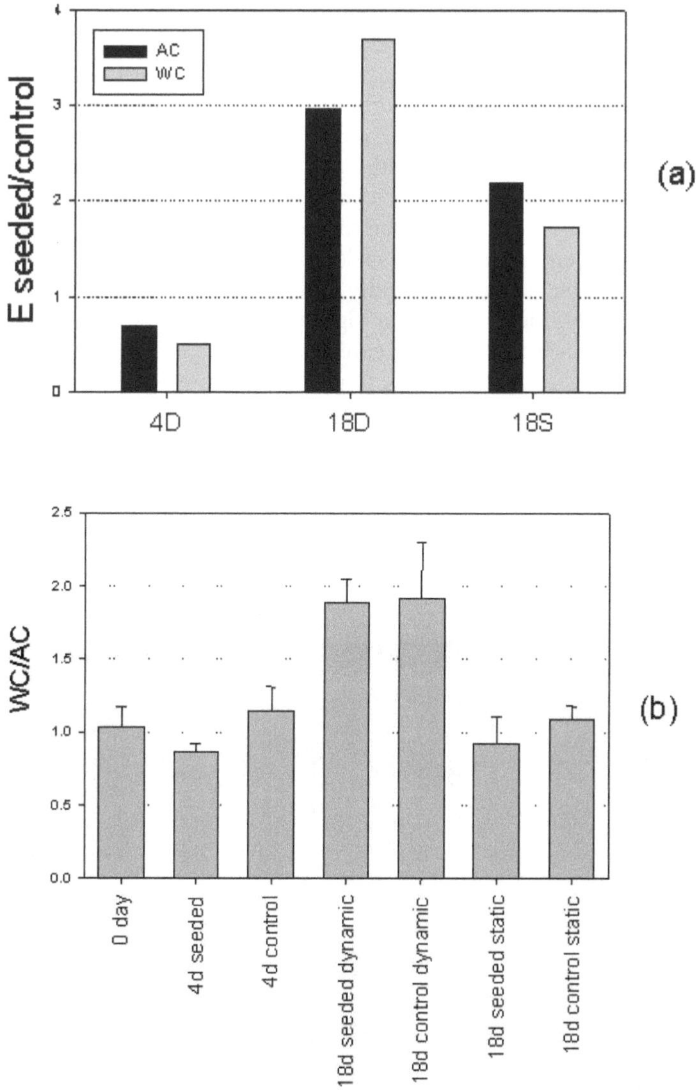

Figure 6. The effects of cell seeding and dynamic incubation are better revealed by computing the ratios with respect to (a) seeded vs. control and (b) bending direction (WC vs. AC). Under static conditions, cell seeding increased the effective stiffness by two-fold, whereas dynamic conditions increased the effective stiffness by at least three-fold. Dynamic conditioning increased directional differences by two-fold, but were the same in the seeded and control groups.

The bioreactor consists of two identical chambers (127 x 101.6 mm), each containing 6 culture wells (25.4 mm diameter, 16 mm deep). Situated within each well are four stainless steel "stationary posts" arranged orthogonally around a central channel in the floor of the culture well (1.9 mm deep) (Figure 7). The device can thus accommodate a total of 12 rectangular samples (maximum dimensions approximately 25 x 7.5 x 2 mm), with each sample being positioned between the four stationary posts, orthogonal to the central channel.

Flexural stimulation can be applied to each of the samples in the form of three-point bending by an environmentally-sealed linear actuator (UltraMotion, Mattitick, NY). The piston of the actuator is rigidly coupled to a cross-arm in the form of a T-junction. Conversely, the cross-arm is rigidly coupled to the arm of each chamber (Figure 7). Each arm bifurcates and extends into a chamber (two penetrations per chamber). Both arms terminate in six fingers through which "bending pins" can be inserted to bracket the rectangular samples in the middle. Therefore, each sample can be subjected to unidirectional or bi-directional three-point bending. Frequency, amplitude, acceleration, and deceleration profiles can be developed using Windows-based Si Programmer software (Applied Motion Products, Watsonville, CA). The structural elements of the device can be cold gas sterilized by ethylene oxide, and the entire device was designed to be operated inside of a standard humidified incubator (Figure 7).

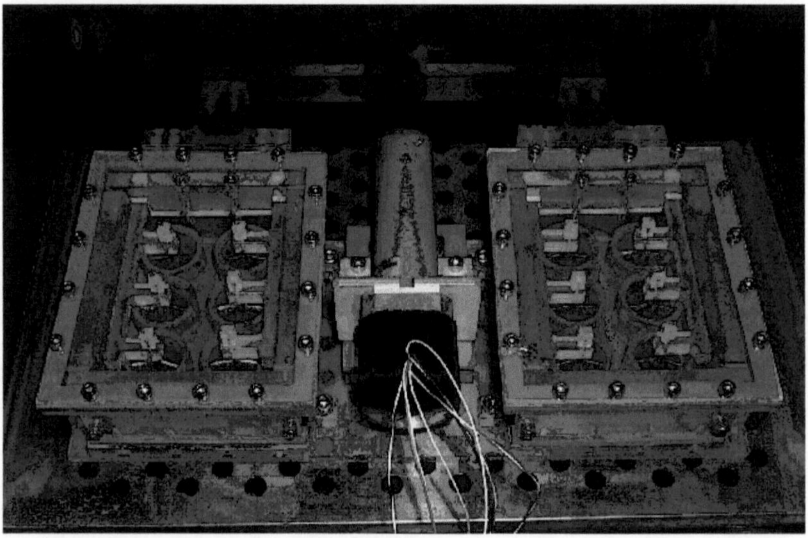

Figure 7. Bioreactor operating inside a standard cell culture incubator at 37°C and 5% CO_2. The bioreactor consists of two parallel chambers secured to a base plate and coupled via a cross-arm to a centrally positioned linear actuator.

3.3. Methods

3.3.1 Culture medium

The culture medium used in these experiments was Dulbecco's Modified Eagle's Medium with 4.5 g/L glucose and L-glutamine supplemented with 10% fetal bovine serum (BioWhittaker, Walkersville, MD). Antibiotics were excluded in order to assess the intrinsic ability of the bioreactor to maintain sterility.

3.3.2 Scaffolds

Scaffold materials consisted of a non-woven mesh of polyglycolic acid (PGA) fibers (Albany International Research, Mansfield, MA) dip-coated with poly-4-hydroxybutyrate (P4HB) (Tepha, Inc., Cambridge, MA), and a non-woven, 50:50 blend, mesh of PGA and poly-L-lactic acid (PLLA) fibers (Albany International Research, Mansfield, MA) dip-coated with P4HB. The PGA and PGA/PLLA scaffolds had an approximate fiber diameter of 0.012-0.015 mm and density of 69 mg/ml. Rectangular scaffold samples were cut to size (approximately 25 x 7.5 x 2 mm) and dipped briefly into a solution of P4HB in tetrahydrofuran (1% wt/vol), resulting in a P4HB coating following solvent evaporation. As described previously (Hoerstrup et al. 2000b), P4HB is a bioresorbable thermoplastic that allows for scaffolds to be molded into any shape. Scaffolds were cold gas sterilized with ethylene oxide prior to use.

3.3.3 Flexural mechanical evaluation

A total of 60 scaffold samples were prepared for evaluation in this study (30 x PGA/P4HB, 30 x PGA/PLLA/P4HB). Two separate runs of the bioreactor were required to test all of the samples. In the first run, PGA/P4HB scaffold samples were loaded into the culture wells of the bioreactor under aseptic conditions (n=12). Control scaffold samples were loaded into identical culture wells maintained under static conditions inside a large Petri dish (n=12). In addition, samples of the virgin scaffold material (time = 0) were retained for bending testing and scanning electron microscopy (n=6).

Approximately 6 ml of culture medium was added to each of the bioreactor and static control culture wells. The linear actuator was programmed to flex the scaffold samples at a rate of 1 cycle per second (1 Hz) in order to emulate the cardiac cycle. The center of each scaffold sample was displaced 6.35 mm in one direction, corresponding to a bending angle of approximately 62°. As in the previous study, the direction of bending is referred to henceforth as "with-flexure" (WF), and the opposing direction is referred to as "against-flexure" (AF). After loading the bioreactor, the entire device was placed inside of a humidified incubator (HERAcell, Kendro Laboratory Products, Newtown, CT) operating at 37 °C and 5% CO_2. The culture medium was not changed during the course of the run.

After 1 week (n=6) and 3 weeks (n=6) of dynamic flexure the scaffold samples were removed from the bioreactor under aseptic conditions for bending testing and scanning electron microscopy. Corresponding control samples were removed from the static culture wells at each time point. In the second bioreactor run, PGA/PLLA/P4HB scaffold samples were loaded and incubated in the same manner as the PGA/P4HB scaffolds. After 3 weeks (n=6) and 5 weeks (n=6) of dynamic flexure the PGA/PLLA/P4HB scaffold samples were removed from the bioreactor, as well as the corresponding static control samples. Flexural mechanical testing was preformed as described in section 2.

3.3.4 Sterility evaluation

The macroscopic appearance of the culture medium was checked regularly during the course of the bioreactor runs for signs of contamination (i.e., cloudiness). At each time point evaluated in the study, upon removing the scaffold sample from the bioreactor or static control culture well, the culture medium was aseptically transferred from the culture well into a sterile T-25 flask (Becton Dickinson Labware, Franklin Lakes, NJ). The culture medium was then inspected for signs of contamination using an inverted microscope (Micromaster Inverted Microscope, Fisher Scientific, Pittsburgh, PA). The culture medium was retained in the incubator at 37 °C and 5% CO_2 for one week and re-checked for signs of contamination.

3.4 Results

3.4.1 Bioreactor operation and sterility

The bioreactor demonstrated consistent operation throughout each of the two runs. Microscopic evaluation of the culture medium at each time point indicated that sterility was maintained throughout the course of each study (up to 5 weeks).

3.4.2 Effective stiffness

E_{eff} (mean ± SEM) of the virgin PGA/P4HB scaffold samples was 2406 ± 349 kPa (Figure 8). E_{eff} values were observed to decrease over time for samples incubated under both flexed and static conditions. A statistically significant decrease ($p < 0.01$) was observed in E_{eff} between the virgin sample and the flexed sample at 1 week (-43%). No significant difference was observed between the virgin material and the static control sample at 1 week, indicating that dynamic flexure has a significant effect on E_{eff} compared to static incubation conditions at 1 week. Both flexed and static samples were substantially less stiff by three weeks, however no significant difference was observed in E_{eff} between the two groups.

E_{eff} of the virgin PGA/PLLA/P4HB scaffold samples was 4724 ± 320 kPa (Figure 9), approximately twice that of the PGA/P4HB. There was no statistically significant difference in the stiffness of the static control samples between times 0, 3, and 5 weeks. There was, however, a statistically significant decrease in stiffness between the flexed and static control samples at 3 weeks (-55%) and 5 weeks (-

58%). There was not, however, any significant difference in E_{eff} between the flexed samples at 3 and 5 weeks.

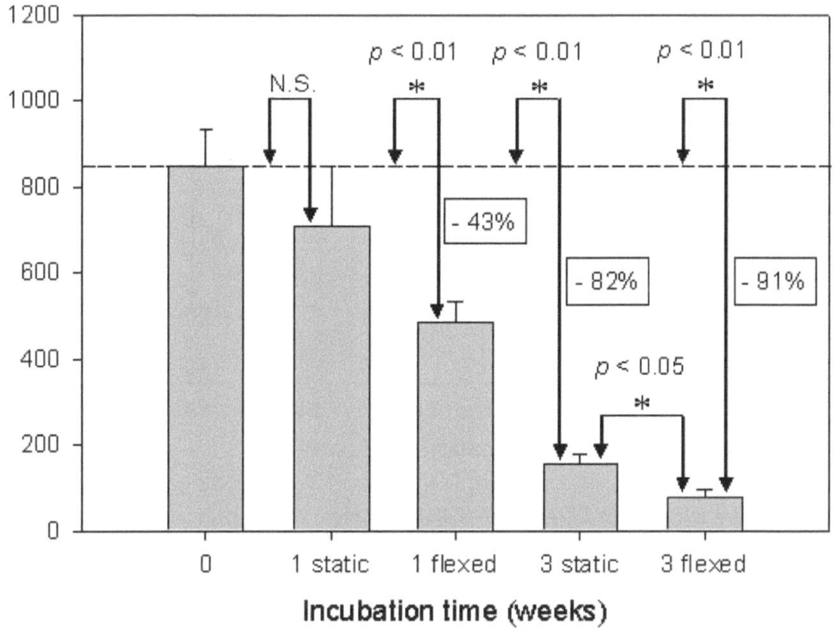

Figure 8. Effective stiffness of PGA/P4HB samples. Values presented are the mean of the measurements with-flexure (WF) and against-flexure (AF) for 6 samples. Error bars indicate standard error. Insignificant comparisons are designated by "N.S." WF and AF designations simply refer to opposing directions for the t=0 and static sample groups.

The effective stiffness ratio is calculated here as the ratio of E_{eff} measured in the "with-flexure" (WF) direction divided by that measured in the opposing "against-flexure" (AF) direction. An effective stiffness ratio of one indicates an isotropic material, whereas any significant deviation from a value of one would suggest an anisotropic material. The effective stiffness ratio of the PGA/PLLA/P4HB scaffold samples is depicted in Figure 10. Static control samples did not exhibit a statistically significant deviation from one, indicating that they were isotropic. The effective stiffness ratio of flexed samples, however, exhibited a statistically significant deviation from one (-42%) at 3 weeks and 5 weeks (-29%) ($p < 0.01$), indicating that these samples were stiffer "against-flexure." The effective stiffness ratio of flexed and static PGA/P4HB samples did not deviate significantly from one, indicating isotropic behavior (data not shown).

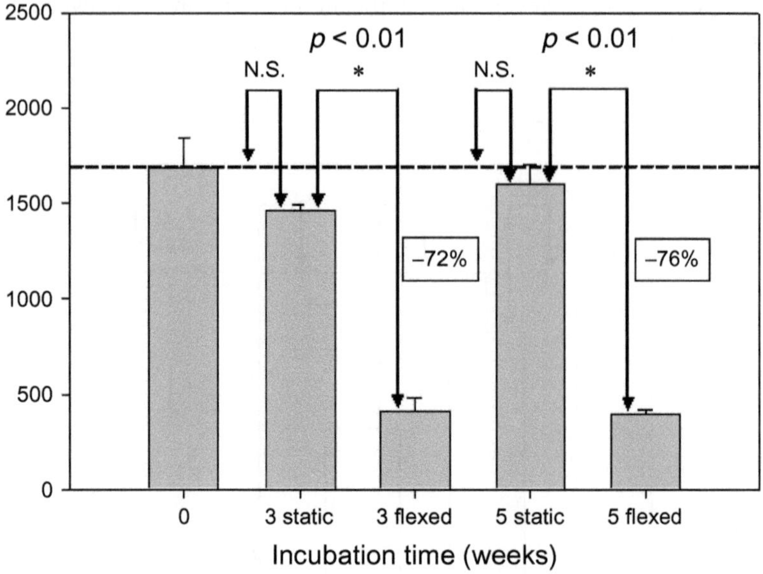

Figure 9. Effective stiffness of PGA/PLLA/P4HB samples. Values presented are the mean of the measurements WC and AC for 6 samples. Error bars indicate standard error. Insignificant comparisons are designated by "N.S."

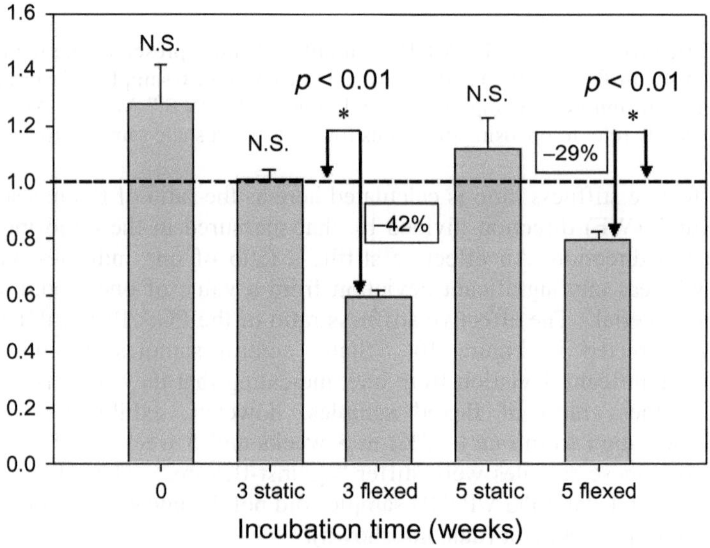

Figure 10. Effective stiffness ratio (WC/AC) for PGA/PLLA/P4HB samples. Values presented are the mean ratios (WC/AC), with error bars indicate standard error. Insignificant comparisons are designated by "N.S."

3.4.3 Scanning Electron Microscopy (SEM)

Representative SEM images are depicted for comparison (Figure 11). Several qualitative observations can be made from the images: 1) the fiber structure and orientation of the virgin scaffold materials appears to be random within each sample and across samples (data not shown), 2) the P4HB coating does not appear to be uniform over the fiber structure, but rather appears to localize in sparsely distributed aggregates (Figure 11a, 11d), 3) fiber diameter does not appear to diminish significantly over three weeks (PGA/P4HB) or five weeks (PGA/PLLA/P4HB), 4) fibers appear to fragment over the course of the incubation, and this fragmentation appears to be more pronounced in the dynamically flexed PGA/P4HB samples (Figure 11f), 5) dynamic flexure appears to induce stratification of the PGA/PLLA/P4HB scaffolds into a layer of circumferentially aligned, compressed fibers (side under cyclic compression during flexure) and a layer of random, loose fibers (side under cyclic tension during flexure) (Figure 11b).

To our knowledge, the bioreactor described herein is the first device aimed to study the individual effect of flexure on TEHV biomaterials. In these first experiments we have demonstrated the operation of the bioreactor and its ability to maintain sterility for at least 5 weeks in an incubator environment. In addition, we implemented the bioreactor to investigate the effect of dynamic flexure on the E_{eff} of two different TEHV scaffold materials, PGA/P4HB and PGA/PLLA/P4HB. These two biomaterials were chosen primarily because they are currently being evaluated as candidate scaffolds for the fabrication of TEHV (Hoerstrup et al. 2000b). Based on recent studies which evaluated the effects of constant fluid flow (Agrawal et al. 2000) and dynamic compressive loading (Thompson et al. 1996) on similar scaffold materials, we hypothesized that dynamic flexure in a non-flowing environment would augment the intrinsic degradation rate of the polymers, with concomitant losses in flexural rigidity. Moreover, because these scaffolds consist of non-woven fiber meshes, we further hypothesized that they may be more susceptible to non-destructive structural reorganization (e.g., heterogeneous fiber alignment) under conditions of dynamic stimulation than woven or non-fiber based scaffolds, potentially giving rise to unique mechanical properties.

In accordance with our hypotheses, incubation under dynamic flexure led to a trend of decreased E_{eff} compared to static conditions, and induced directional anisotropy ($p < 0.01$) in PGA/PLLA/P4HB samples at 3 and 5 weeks. SEM images of these samples hinted at some of the structural features and changes that may have been responsible for the E_{eff} results. The heterogeneity observed in the virgin scaffold fiber structure, coupled with the observation that P4HB does not uniformly coat the samples, helps explain the variability in E_{eff} measured for the virgin (t=0) replicates. Fragmentation of fibers, as opposed to diminishing fiber diameter and mass loss, appears to be a primary mechanism for loss of E_{eff} in dynamically flexed PGA/P4HB samples. Interestingly, significant fragmentation of fibers was not observed in any of the PGA/PLLA/P4HB scaffolds, indicating that PLLA fibers may shield the PGA fibers from mechanical disruption. Finally, the possible stratification of the PGA/PLLA/P4HB scaffold into structurally distinct layers under conditions of dynamic flexure could explain the directional anisotropy observed at 3 weeks.

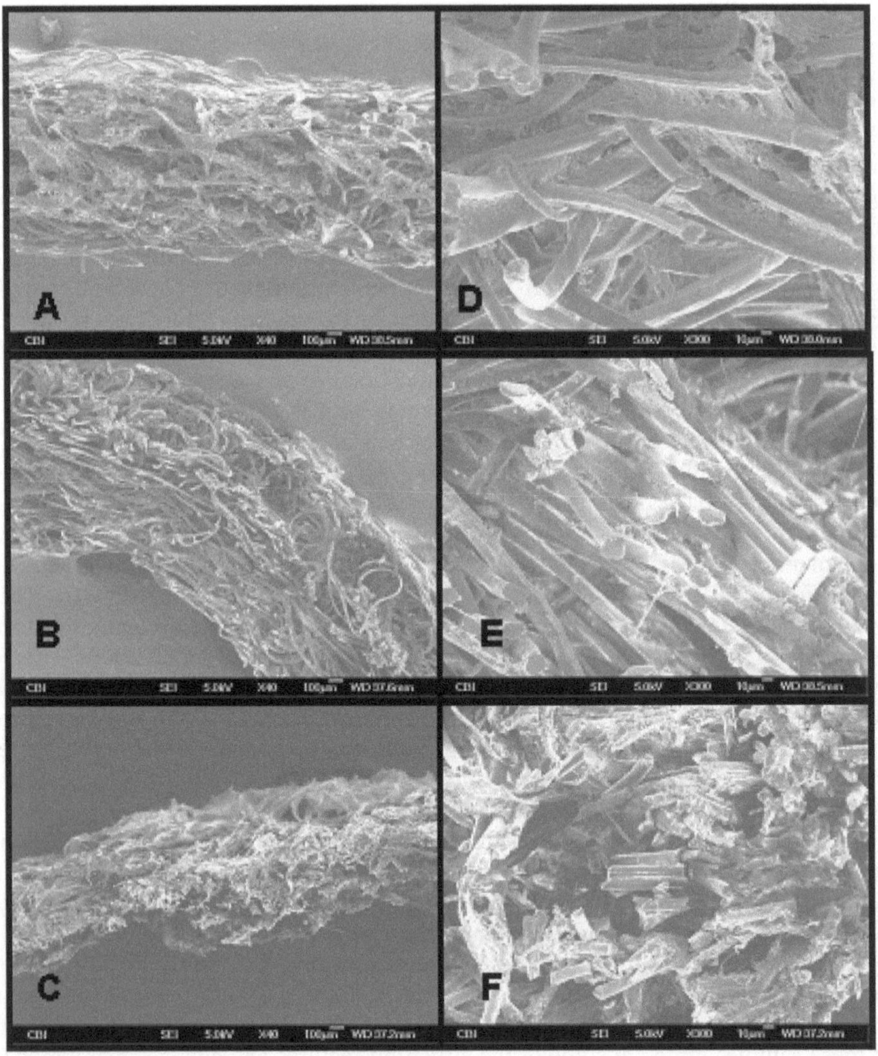

Figure 11. Scanning electron micrographs of representative 3-week static PGA/PLLA/P4HB (A, D), 3-week flexed PGA/PLLA/P4HB (B, E), and 3-week flexed PGA/P4HB (C, E) samples at 40-fold (A, B, C) and 300-fold (D, E, F) magnification. In particular, note the dramatic difference in thickness between the dynamically flexed PGA/P4HB sample (C) and the static (A) and flexed (B) PGA/PLLA/P4HB samples at three weeks. Furthermore, note the apparent stratification of the 3-week flexed PGA/PLLA/P4HB sample (C) into structurally distinct layers. The side of the scaffold undergoing cyclic compression during flexure exhibits compressed, circumferentially aligned fibers, while the side under tension exhibits a looser, more random fiber structure. Finally, the higher magnification images depict the sparse distribution of P4HB (D), as well as the cracking and fragmentation of PGA fibers under dynamic flexure (F).

Further studies are necessary in order to determine unequivocally whether the observed stratification was due to dynamic flexure, as opposed to being an artifact of the mild compression imposed by the bending pins, or due to the virgin material structure. While possible, these explanations seem unlikely in light of the fact that both sides of the scaffold would have been exposed to a slight degree of bending pin compression, as the scaffolds became stiffer in the direction "against-flexure" during dynamic incubation. Furthermore, stratification into structurally distinct layers was not observed in any of the virgin or static control samples. The possibility that dynamic flexure may induce stratification of non-woven fiber mesh scaffolds into structurally distinct layers is exciting, as this may represent a unique mechanism for promoting the development of a layered TEHV construct.

4. PLANAR BIAXIAL BIOMECHANICAL BEHAVIOR OF THE NATIVE VALVE LEAFLETS

Presumably, once mature TE valvular tissues can be formed in-vivo, functional assessment (i.e. similarity to native tissue functional biomechanics) will be come critical to evaluating their long-term success. However, the mechanics of soft tissues are complex: they exhibit a highly non-linear stress-strain relationship, undergo large deformations, complex viscoelasticity, and complex axial coupling behaviors that defy simple experiments and material models. Much of this behavior is a direct result of changes in their internal structure with strain, which involves both straightening of highly crimped collagen fibers and rotation of these fibers toward the stretch axis.

Most previous work on the mechanical properties of the native and chemically treated aortic valve has relied on uniaxial mechanical testing (Lee et al. 1984), (Lee et al. 1984), (Vesely and Noseworthy 1992). These studies demonstrate that chemical fixation of intact valves, especially under pressure, alters the mechanical properties of the cusps. Marked decreases in the extensibility are generally attributed to "locking" the collagen fibers in the uncrimped state (Broom and Christie 1982), (Christie 1992). Tests on thin tissue strips however, cannot mimic the heterogeneous multi-axial deformation fields, combined loading sequences, and native fiber kinematics found in the physiological environment. Mayne et al. (1989) and Christie et al. (1995) have performed equi-biaxial testing (i.e. equal levels of tension applied to each test axis) that overcomes many of the above limitations of uniaxial loading. However, derivation of a constitutive relationship solely from equi-biaxial test data is limited due to multiple co-linearities that confound the ability to obtain reliable, unique model parameter values (Brossollet and Vito 1995).

Billiar and Sacks (2000) generated the first complete biaxial mechanical data necessary for constitutive modeling aortic valve cusp. Their setup is shown schematically in Figure 12. Due to the small size and heterogeneous structure of the aortic valve cusp, new testing methods were developed and validated. Leaflet specimens were subjected to biaxial tests utilizing seven loading protocols to provide a range of loading states that encompass the physiological loading state. The cusps demonstrated a complex, highly anisotropic mechanical behavior, including pronounced mechanical coupling between the circumferential and radial directions. Mechanical coupling between the axes produced negative strains along the circumferential direction and/or non-monotonic stress-strain behavior in many samples subjected to equi-biaxial tension. This behavior was also noted by Mayne et al. (1989) but could not be explained. Clearly, a constitutive model is needed to truly understand the aortic cusp behavior and its implications on the mechanics of the intact valve.

The quantified fiber architecture (Sacks and Chuong 1998) and biaxial mechanical data (Billiar and Sacks 2000) suggest that a structural approach is the most suitable method for the formulation of a constitutive model for the aortic valve cusp. Our related work on native and chemically treated bovine pericardium suggests that a structural approach is both feasible and attractive for bioprosthetic heart valve biomaterials (Billiar and Sacks 2000).

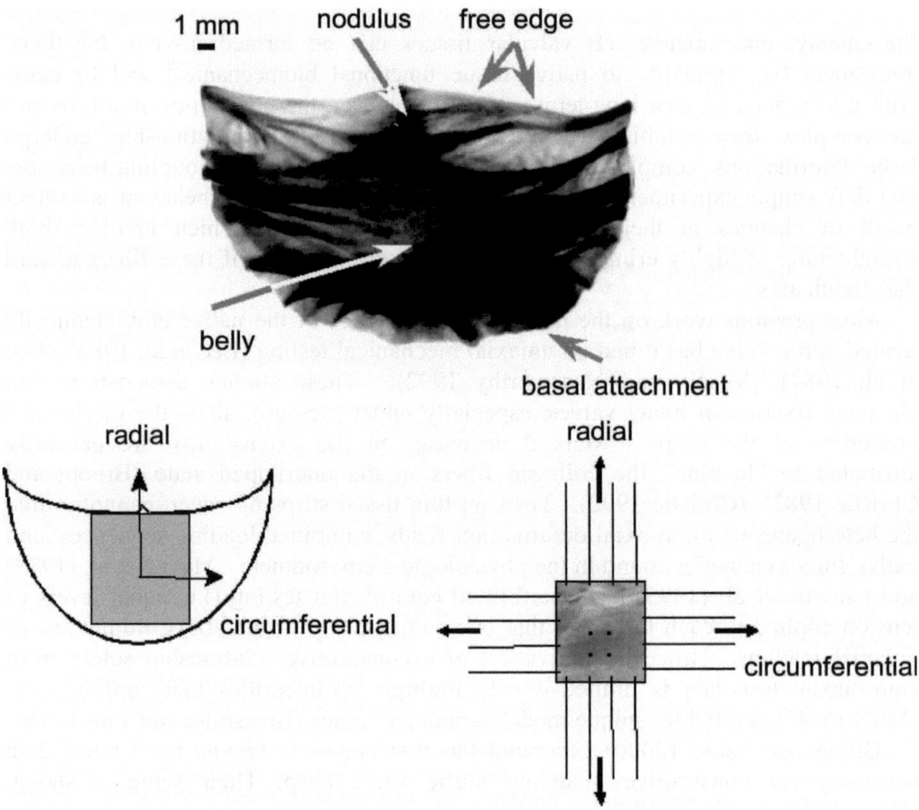

Figure 12. (a) A partially polarized image of an aortic cusp highlighting main anatomical features. (b) Schematic of the location of the biaxial test specimen taken from the aortic valve cusp, showing where the biaxial specimen was taken and how it is loaded in the biaxial testing device (Taken from (Billiar and Sacks 2000)).

Details of the model have been previously presented (Billiar and Sacks 2000), and follows the structural modeling approach by Lanir et al. (1983), (1979). In this approach, the tissues net or total strain energy is assumed to be the sum of the individual fiber strain energies, linked through appropriate tensor transformation from the fiber coordinate to the global tissue coordinates. For the aortic valve, we assume that the planar biaxial mechanical properties of the cusp can be represented as a planar array of collagen fibers. Anatomically, these fibers most closely represent the dense, highly aligned collagen fibers in the fibrosa layer. Next, the angular fiber distribution and the density of the fibers are assumed constant throughout the tissue. Based on our SALS results for the aortic valve cusp (Sacks and Chuong 1998), we utilize the fact that the angular distribution of the collagen fibers, R(θ), can be represented by a Gaussian distribution,

$$R(\theta) = \frac{1}{\sigma\sqrt{2\pi}} \exp\left[\frac{-(\theta - M)^2}{2\sigma^2}\right] \tag{5}$$

where θ is the direction with respect to the x_1 or circumferential axis (Figure 4b), σ is the standard deviation and M is the mean of the distribution. M was determined experimentally for each specimen by using the preferred fiber directions as determined by SALS (Sacks and Chuong 1998). The "effective" fiber stress-strain properties were represented using:

$$S_f = A\left[\exp(BE_f) - 1\right] \tag{6}$$

where S_f is the second Piola-Kirchhoff fiber stress, E_f is the fiber Green's strain. This formulation for the fiber stress-strain law avoids detailed descriptions of complex crimp distributions.

For valvular tissue, it is more convenient to work with membrane stresses due to considerations such as variable total and layer thickness, and heterogeneous layer structure (Billiar and Sacks 2000). Further, since the biaxial mechanical tests are run using membrane stress control using the specimen's unloaded dimensions, a Lagrangian membrane stress measure is used in the constitutive formulation. We also assume that inter-specimen variations in fiber volume fraction V_f and thickness h are negligible, so that the product hV_f can be conveniently absorbed into the material constant A. The resulting expressions for the Lagrangian membrane stresses T_{ij} are:

$$T_{11} = \int_{-\pi/2}^{\pi/2} S_f^*(E_f)R(\theta)\left(\lambda_1 \cos^2\theta + \kappa_1 \sin\theta\cos\theta\right)d\theta$$
$$T_{22} = \int_{-\pi/2}^{\pi/2} S_f^*(E_f)R(\theta)\left(\lambda_2 \sin^2\theta + \kappa_2 \sin\theta\cos\theta\right)d\theta \qquad (7)$$

where $A^*=hV_fA$ and $S_f^* = A^*\left[\exp(BE_f)-1\right]$. The parameters A^*, B, and σ were estimated by fitting Eqn. 7 to the complete biaxial data set (Billiar and Sacks 2000).

The fit of the model to the data was good despite the complexity of the mechanical response over the broad range of biaxial loading states (Figure 13). The model fit the data from all seven protocols well even though the data from the outer protocols (1 and 7, see inset in Figure 13) were not used in the parameter estimation. Using only three material parameters, the quantitative "goodness of fit" was comparable to phenomenological models of other tissues (Choi and Vito 1990), (May-Newman and Yin 1998), (Humphrey et al. 1990).

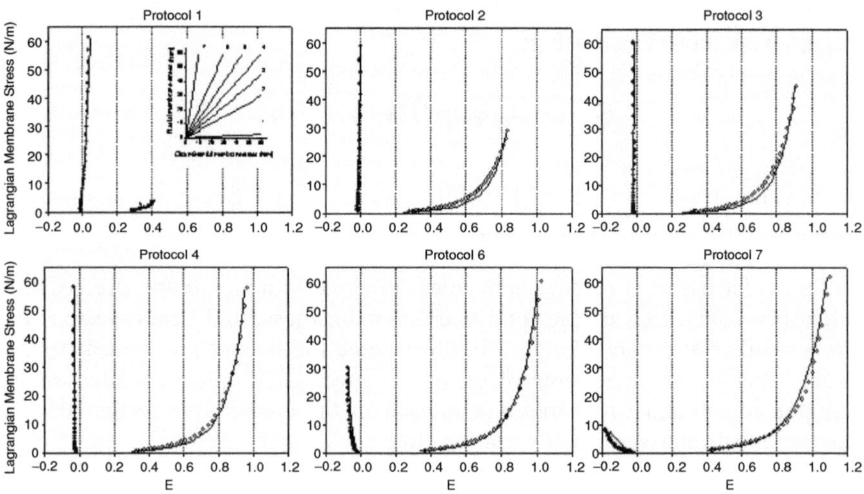

Figure 13. Stress-strain curves for six of the seven loading protocols for a 4 mmHg fixed specimen (open circles) and the simulated stress-strain curves (lines). The structural constitutive model demonstrated an excellent fit to the data, including the presence of large negative circumferential strains due to strong axial coupling. Although protocol five was not shown for clarity of presentation, it also demonstrated an equally good agreement between theory and experiment. Inset: biaxial testing protocols with showing the corresponding protocol numbers shown for reference. (Taken from (Billiar and Sacks 2000)).

Another important aspect of the structural approach is that the two distinguishing aspects of the aortic valve cusp biaxial behavior, namely the extreme mechanical anisotropy and the strong mechanical coupling between the axes, can be explained by the angular distribution of fibers. To more clearly demonstrate this effect, we generated simulations under equi-biaxial loading for a given set of A^* and B values by letting σ vary (Figure 14). These simulations indicate that the value of σ is the primary determinant of the biaxial stress-strain response, as shown for a) nearly random (σ=90°), b) moderately anisotropic (σ=35°), c) highly anisotropic, including contraction along one axis (σ=20°), and d) extremely anisotropic (σ=10°). Although we assumed a simplified tissue structure in the formulation of the model, the structural approach highlighted the importance of the angular orientation of the fibers in determining the complex anisotropic mechanical behavior of the tissue. For additional details in peforming and modeling biaxial mechanical properties of soft tissues, the interested reader is referred to (Billiar and Sacks 2000; Sacks and Sun 2003).

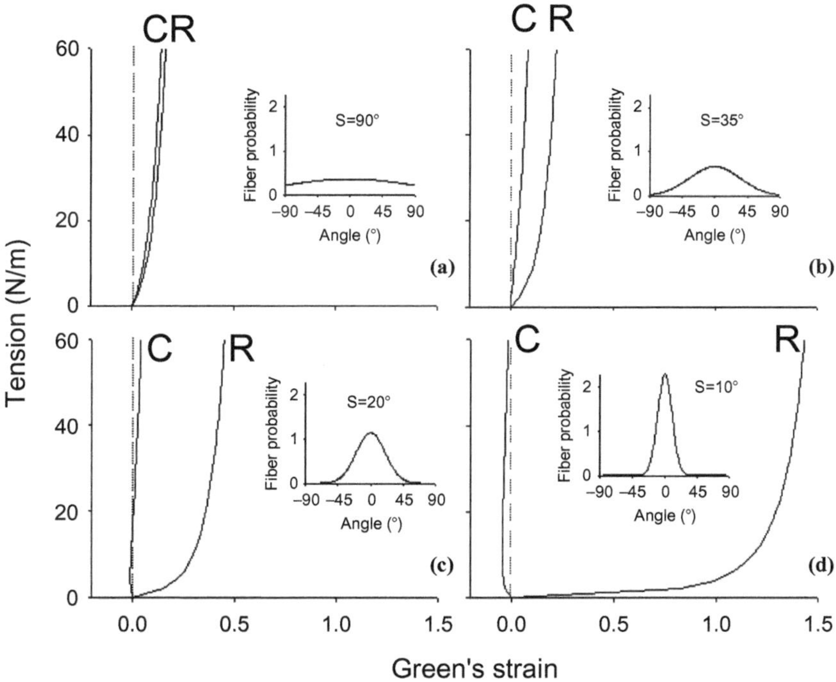

Figure 14. Simulations using the structural model of the effect of σ on equibiaxial stress-strain behavior. The insets represent the fiber probability density distribution for each σ value: a) σ =90° approximately isotropic, b) σ =35° response qualitatively similar to bovine pericardium, c) σ =20° the circumferential strains are negative at low equibiaxial tensions, and d) σ =10° the material behavior is highly anisotropic. The dotted lines indicating zero strain are included to highlight ability of the model to simulate the cross-over to negative strain observed in the pressure fixed cusps subjected to equibiaxial tension.

5. A PHYSIOLOGIC FLOW LOOP FOR BIOLOGICALLY ACTIVE HEART VALVES

5.1 Introduction

Biologically active tissue constructs as materials for use in heart valve replacements are being investigated several research groups (Dumont et al. 2002; Sodian et al. 2000; Stock et al. 2002). A major impediment to clinical use has been that these valves can not yet endure in vivo aortic pressures. It has been shown that exposing biologically active tissue constructs to dynamic pressures and flows during culture produces a stronger material (Sodian et al. 2000; Stock et al. 2002); however, to date no studies have been performed to investigate whether this results primarily from modulating pulsatile pressure, flow, or a combination of the stimuli. While the general requirements for non-sterile pulsatile mock circulatory loops to simulate the hydrodynamics of heart valves have been well established (Donovan 1975; Vandenberghe et al. 2001; Verdonck et al. 1992), these systems have primarily necessitated substantial manual intervention in order to control pressure and flow waveforms which is not ideal for a sterile incubation situation where continual user intervention could provide a source for contamination. In an effort to gain insight into the roles of pressure and flow on the development of TEHV we developed a system to study the effects of subjecting TEHV to highly controlled pressure and flow waveforms that can change during sterile incubation.

5.2 Methods

The system (Figure 15) consists of a vented atrium which passively fills a pneumatically driven ventricle with a mechanical inflow valve (On-X, MCRI) and, for purposes of hydrodynamic testing, a bioprosthetic outflow valve (Perimount, Edwards Lifesciences Corp.). The height of the atrium can be adjusted to provide different filling pressures. The waveforms are controlled via a two-element Windkessel comprised of an adjustable compliance and resistance as well as a variable driving (ventricular pressure) waveform. The ventricle has an air chamber and a fluid filled ventricular chamber separated by a thin latex bladder which deforms as air is input driving the flow. The pressure can be directly monitored with pressure transducers (19C series, Sensym ICT) (without catheters) in the air chamber, ventricular chamber, and outflow tube. Volumetric flow is measured using an ultrasonic flow probe (C series, Transonic) placed before the compliance chamber.

In order to limit user intervention and the resulting possible contamination in our system, pressure and flow are primarily controlled electronically by simultaneously modulating the driving waveform and the systemic resistance. The ventricular waveform (Figure 16) is generated by a piezoelectric pneumatic valve system (Airfit, Hoerbiger-Origa Corp.) which allows for gradual, voltage-controlled changes in air pressure, as opposed to the traditional pneumatic on/off step change, in order to better simulate the shape of the ventricular pressure waveform. The variable resistor is controlled by rotating a spiral shaped plate which is in contact with a tube network (~60 tubes, 1.3mm diameter) with the rotation provided by a high angular resolution stepping motor (C57L, Thompson Airpax). Electronic control of the device is performed by a PC with a multifunction data acquisition card (PCI 6035E, National Instruments) running custom written control software (Labview, National Instruments).

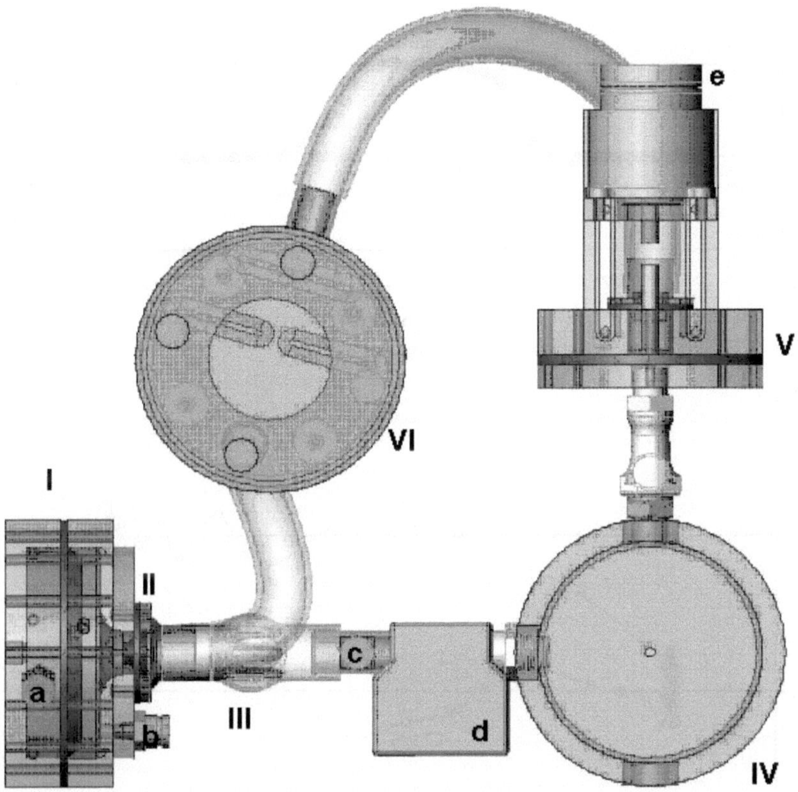

Figure 15. Schematic of the pulsatile TEHV bioreactor, showing the ventricle (I), mitral valve (II), aortic valve (III), compliance (IV), variable resistor (V), atrium (VI), pressure sensors (a-c), flow sensor (d), stepper motor (e).

The system was built from materials known to have minimal cytotoxic effects (polycarbonate, acrylic, silicone, and PVC) as well as stereolithography resin (Accura Si40, 3D Systems) for the resistor tube network and the resistor plate.

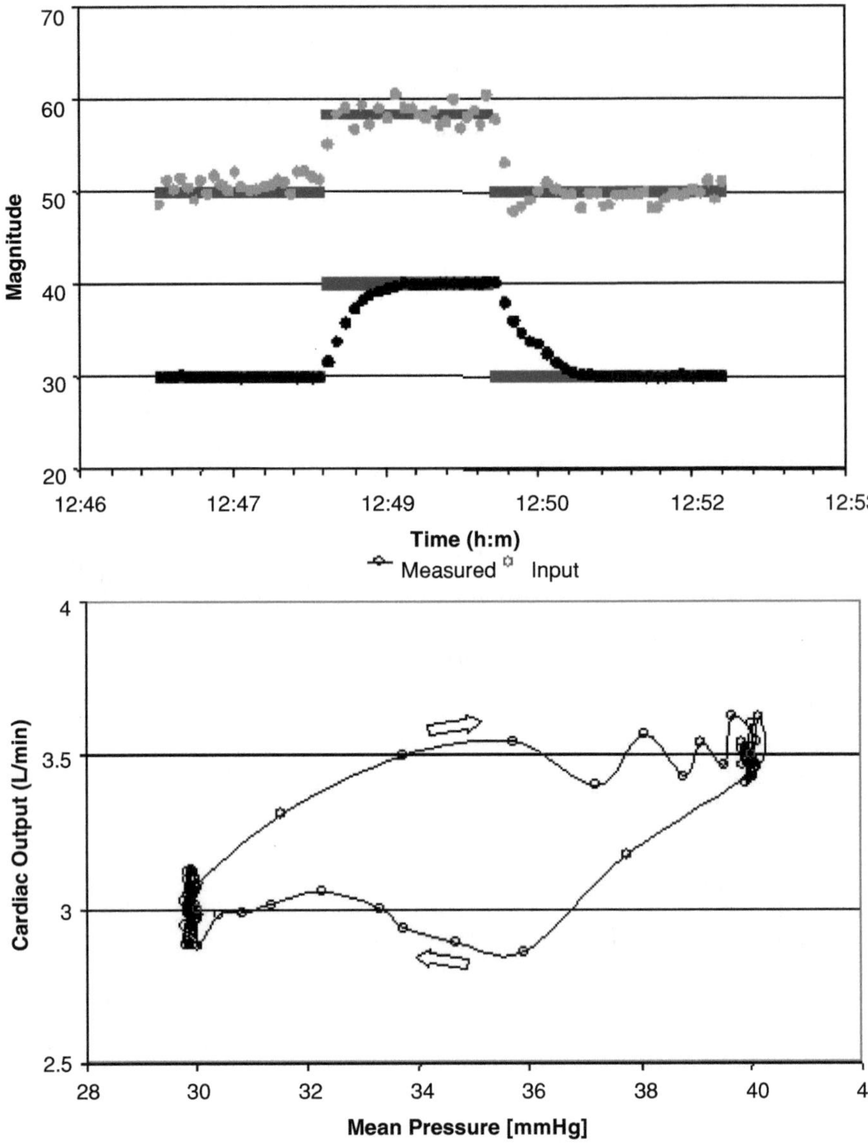

Figure 16. Stepped pressure and flow test: set point 1 at 3.0 L/min, 30 mmHg to set point 2 at 3.5 L/min, 40 mmHg. (a) Mean aortic pressure (lower closed symbols) and stroke volume (SV) (upper closed symbols) vs. time with input signals shown as solid lines. Time to reach set point was less than 60 seconds. (b) Cardiac output vs. mean aortic pressure during the test arrowS indicate the differences in paths taken between set points.

5.3 Results and Discussion

In order to first tune the system response time, a step pressure and flow test was run (Figure 16) and software control parameters were adjusted until a fast and stable response was achieved. The system response time was less than sixty seconds to go from 3.0 L/min mean aortic flow (MAF) at 30 mmHg mean aortic pressure (MAP) to 3.5 L/min at 40 mmHg. System stability was addressed by running a constant set point test of 5 L/min MAF at 30 mmHg MAP for 16 hrs (Figure 17). The results (average ± standard deviation) showed a stable reading of 5.00±0.06 L/min and 30.00±0.04 mmHg. The maximum absolute value of the residual error was 0.15 L/min for MAF and 0.21 mmHg for MAP. Resulting waveforms of the device running at pulmonary and systemic pressures and flows are shown below (Figure 18). The beat rate was fixed at 60 BPM and the MAF at 5 L/min for both cases while MAP was adjusted to be 105 mmHg for the systemic case and 27 mmHg for the pulmonary case.

TEHV tissue growth has been shown to be influenced by hydrodynamic conditions during culture, but the milieu responsible for favorable development is still not understood. A system such as the one outlined should make it possible to elucidate the effects of modulating pressure and flow on developing TEHV. The results of hydrodynamic testing indicates that the system can generate highly controlled waveforms over a wide range of pressures and flows with minimal user intervention.

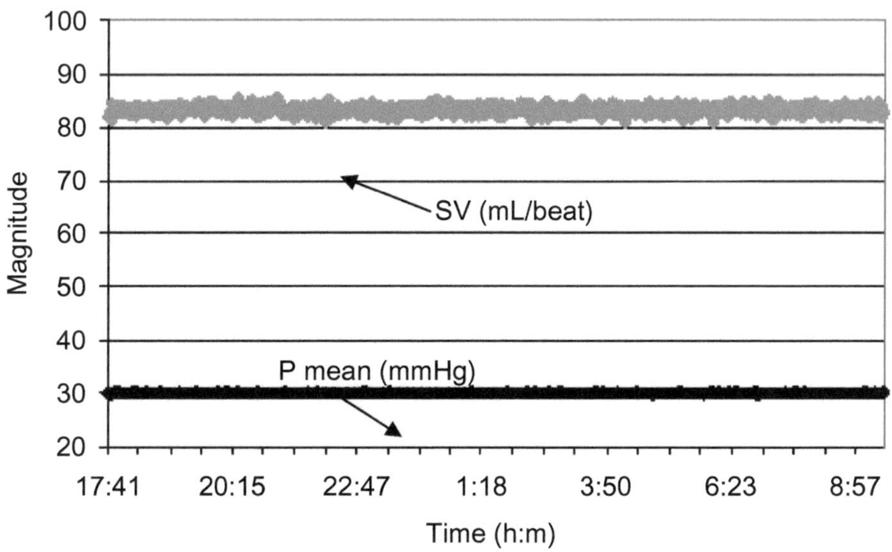

Figure 17. System stability test: 16 hr duration at set point of 5 L/min at 30 mmHg (a). The resulting average plus standard deviation was 5.00±0.06 L/min and 30.00 ±0.04 mmHg.

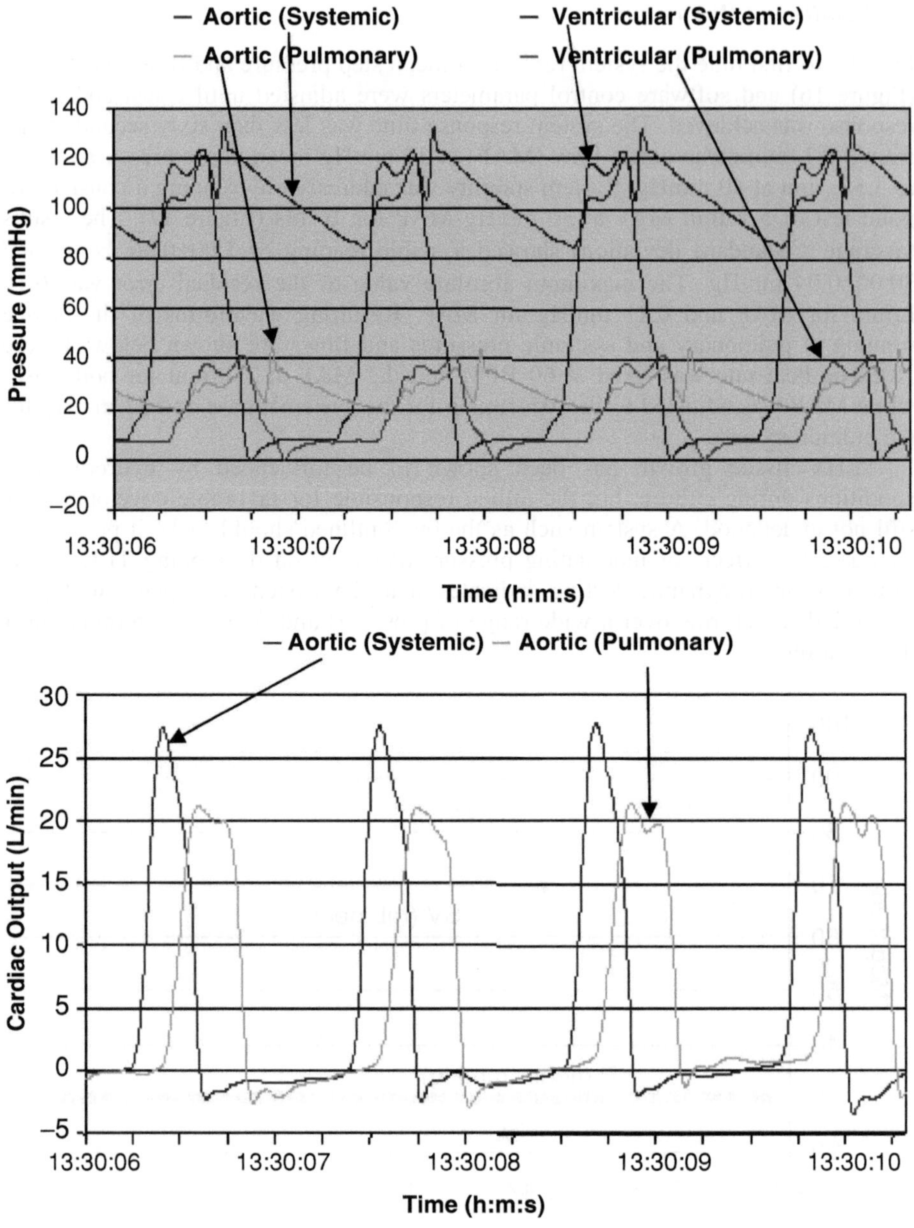

Figure 18. Acquired waveforms for the pulsatile TEHV bioreactor showing: pressure (a) and volumetric flow (b).

6. CLOSURE

Regardless of the specifics of the biomaterial design (e.g. biologically-derived collagen, de-novo synthesized collagen, collagen/polymer combination), all tissue engineered bioprostheses will have to duplicate natural tissue mechanics to some extent in order to properly simulate their in-vivo function. Biomechanics can contribute significantly to this process by utilizing fundamental continuum mechanics principals and rigorous experimental techniques to bring together many of the disparate findings in biology into a unifying theory for growth and remodeling of engineered soft biological tissues.

For example, our constitutive model was able to describe the complete measured planar biaxial stress-strain of the aortic valve cusp using a structural approach. Only three parameters were needed to fully simulate the highly anisotropic, and nonlinear in-plane biaxial mechanical behavior. Although based upon a simplified cuspal structure, the model underscored the role of the angular orientation of the fibers that completely accounted for extreme mechanical anisotropy and pronounced axial coupling. Such knowledge of the mechanics of the aortic cusp derived from this model can help lay the groundwork for the design of tissue engineered scaffolds for replacement heart valves.

Ideally, a TEHV replacement will require the geometry, mechanical properties, and cell biology closely matched to the native valve and outflow track tissues. The polymeric scaffold is meant to provide many of these features during the initial period of tissue formation. However, because the scaffold is absorbed over time, these functions must be taken over progressively by the developing tissue. Ideally, the scaffold/tissue construct will be seeded and grown *in vitro* for a period of time to achieve sufficient functionality, prior to implantation and continued growth and remodeling *in vivo*. While much progress has been made in cell biology and the chemical composition of scaffolds, comparatively little work has been performed on the tissue mechanical aspects of TEHV biomaterials. This is in part due to a lack of methods to assess and model the degrading polymer/ECM biocomposite tissue equivalent, as well as knowledge of the biomechanical behavior of the natural valves.

Computational models also needed to understand basic micromechanical phenomena of the biodegradation and ECM deposition process. Further, successful computational models can greatly aid in optimizing the incubation process by minimizing the current costly and inefficient trial-and-error approach, and in understanding the in-vivo remodeling process and ultimately the long-term fate of the TEHV. While TE offers the potential to overcome limitations of current heart valve prosthesis (especially the very limited options for the pediatric population), the bioengineering challenge is determining how to optimize biological, structural, and mechanical factors of ECM formation for in-vivo success. A major step in this long-term goal is developing a through understanding of the underlying structural and mechanical factors responsible for the initial formation of functionally optimal TEHV tissue structures. Biomechanical analyses can contribute to these critical steps in both proper assessments of the resulting tissue constructs, as well as in the design of bioreactors that provide proper mechanical signals for tissue development.

7. ACKNOWLEDGEMENTS

This research presented in this chapter was supported by the following grants from the National Institutes of Health: HL68816 (MSS) and HL97005 (JEM). MSS is an Established Investigator of the American Heart Association.

8. REFERENCES

Agrawal CM, McKinney JS, Lanctot D, Athanasiou KA. 2000. Effects of fluid flow on the in vitro degradation kinetics of biodegradable scaffolds for tissue engineering. Biomaterials 21(23):2443-52.

Billiar KL, Sacks MS. 2000. Biaxial mechanical properties of the natural and glutaraldehyde treated aortic valve cusp--Part I: Experimental results. J Biomech Eng 122(1):23-30.

Broom N, Christie GW. 1982. The Structure/Function Relationship of Fresh and Gluteraldehyde-Fixed Aortic Valve Leaflets. In: Gallucci V, editor. Cardiac Bioprosthesis. New York: Yorke Medical Books. p 477-491.

Brossollet LJ, Vito RP. 1995. An alternate formulation of blood vessel mechanics and the meaning of the in vivo property. J Biomech 28(6):679-87.

Cacou C, Palmer D, Lee DA, Bader DL, Shelton JC. 2000. A system for monitoring the response of uniaxial strain on cell seeded collagen gels. Med Eng Phys 22(5):327-33.

Choi HS, Vito RP. 1990. Two-dimensional stress-strain relationship for canine pericardium. J Biomech Eng 112(2):153-9.

Christie GW. 1992. Anatomy of aortic heart valve leaflets: the influence of glutaraldehyde fixation on function. European Journal of Cardio-Thoracic Surgery 6:S25-S33.

Donovan FM, Jr. 1975. Design of a hydraulic analog of the circulatory system for evaluating artificial hearts. Biomater Med Devices Artif Organs 3(4):439-49.

Dumont K, Yperman J, Verbeken E, Segers P, Meuris B, Vandenberghe S, Flameng W, Verdonck PR. 2002. Design of a new pulsatile bioreactor for tissue engineered aortic heart valve formation. Artif Organs 26(8):710-4.

Frisch-Fay R. 1962. Flexible bars. Washington,DC: Butterworths. 220 p.

Gloeckner DC, Billiar KL, Sacks MS. 1999. Effects of mechanical fatigue on the bending properties of the porcine bioprosthetic heart valve. Asaio J 45(1):59-63.

Hoerstrup SP, Sodian R, Daebritz S, Wang J, Bacha EA, Martin DP, Moran AM, Guleserian KJ, Sperling JS, Kaushal S and others. 2000a. Functional living trileaflet heart valves grown In vitro. Circulation 102(19 Suppl 3):III44-9.

Hoerstrup SP, Sodian R, Sperling JS, Vacanti JP, Mayer JE, Jr. 2000b. New pulsatile bioreactor for in vitro formation of tissue engineered heart valves. Tissue Eng 6(1):75-9.

Hoerstrup SP, Zund G, Schoeberlein A, Ye Q, Vogt PR, Turina MI. 1998. Fluorescence activated cell sorting: a reliable method in tissue engineering of a bioprosthetic heart valve. Ann Thorac Surg 66(5):1653-7.

Hoerstrup SP, Zund G, Ye Q, Schoeberlein A, Schmid AC, Turina MI. 1999. Tissue engineering of a bioprosthetic heart valve: stimulation of extracellular matrix assessed by hydroxyproline assay. Asaio J 45(5):397-402.

Humphrey JD, Strumpf RK, Yin FC. 1990. Determination of a constitutive relation for passive myocardium: II. Parameter estimation. J Biomech Eng 112(3):340-6.

Iyengar A, Sugimoto H, Smith DB, Sacks M. 2001. Dynamic in-vitro 3D reconstruction of heart valve leaflets using structured light projection. Annals of Biomedical Engineering 29:963-973.

Jockenhoevel S, Zund G, Hoerstrup SP, Schnell A, Turina M. 2002. Cardiovascular tissue engineering: a new laminar flow chamber for in vitro improvement of mechanical tissue properties. Asaio J 48(1):8-11.

Kim BS, Mooney DJ. 2000. Scaffolds for engineering smooth muscle under cyclic mechanical strain conditions. J Biomech Eng 122(3):210-5.

Lanir Y. 1979. A Structural Theory for the Homogeneous Biaxial Stress-Strain Relationships in Flat Collageneous Tissues. Journal of Biomechanics 12:423-436.
Lanir Y. 1983. Constitutive Equations for Fibrous Connective Tissues. journal of biomechanics 16:1-12.
Lee JM, Boughner DR, Courtman DW. 1984. The glutaraldehyde-stabilized porcine aortic valve xenograft. II. Effect of fixation with or without pressure on the tensile viscoelastic properties of the leaflet material. Journal of Biomedical Materials Research 18:79-98.
Mayne AS, Christie GW, Smaill BH, Hunter PJ, Barratt-Boyes BG. 1989. An assessment of the mechanical properties of leaflets from four second- generation porcine bioprostheses with biaxial testing techniques. J Thorac Cardiovasc Surg 98(2):170-80.
May-Newman K, Yin FC. 1998. A constitutive law for mitral valve tissue. J Biomech Eng 120(1):38-47.
Mitchell SB, Sanders JE, Garbini JL, Schuessler PK. 2001. A device to apply user-specified strains to biomaterials in culture. IEEE Trans Biomed Eng 48(2):268-73.
Niklason LE, Gao J, Abbott WM, Hirschi KK, Houser S, Marini R, Langer R. 1999. Functional arteries grown in vitro. Science 284(5413):489-93.
Sacks MS, Chuong CJ. 1998. Orthotropic mechanical properties of chemically treated bovine pericardium. Ann Biomed Eng 26(5):892-902.
Sacks MS, Sun W. 2003. Multiaxial Mechanical Behavior of Biological Materials. Annu Rev Biomed Eng.
Schoen F, Levy R. 1999. Tissue heart valves: Current challenges and future research perspectives. Journal of Biomedical Materials Research 47:439-465.
Schoen FJ. 2001. Pathology of heart valve substitution with mechanical and tissue prostheses. In: Schoen FJ, editor. Cardiovascular Pathology. New York: Livingstone.
Seliktar D, Black RA, Vito RP, Nerem RM. 2000. Dynamic mechanical conditioning of collagen-gel blood vessel constructs induces remodeling in vitro. Ann Biomed Eng 28(4):351-62.
Seliktar D, Nerem RM, Galis ZS. 2001. The role of matrix metalloproteinase-2 in the remodeling of cell-seeded vascular constructs subjected to cyclic strain. Ann Biomed Eng 29(11):923-34.
Sodian R, Hoerstrup SP, Sperling JS, Daebritz SH, Martin DP, Schoen FJ, Vacanti JP, Mayer JE, Jr. 2000. Tissue engineering of heart valves: in vitro experiences. Ann Thorac Surg 70(1):140-4.
Sodian R, Lemke T, Loebe M, Hoerstrup SP, Potapov EV, Hausmann H, Meyer R, Hetzer R. 2001. New pulsatile bioreactor for fabrication of tissue-engineered patches. J Biomed Mater Res 58(4):401-5.
Sodian R, Sperling JS, Martin DP, Stock U, Mayer JE, Jr., Vacanti JP. 1999. Tissue engineering of a trileaflet heart valve-early in vitro experiences with a combined polymer. Tissue Eng 5(5):489-94.
Stock UA, Sakamoto T, Hatsuoka S, Martin DP, Nagashima M, Moran AM, Moses MA, Khalil PN, Schoen FJ, Vacanti JP and others. 2000. Patch augmentation of the pulmonary artery with bioabsorbable polymers and autologous cell seeding [In Process Citation]. J Thorac Cardiovasc Surg 120(6):1158-67.
Stock UA, Vacanti JP, Mayer Jr JE, Wahlers T. 2002. Tissue engineering of heart valves -- current aspects. Thorac Cardiovasc Surg 50(3):184-93.
Struik DJ. 1961. Lectures on Classical Differential Geometry. New York: Dover. 232 p.
Thompson DE, Agrawal CM, Athanasiou KA. 1996. The Effects of Dynamic Compressive Loading on Biodegradable Implants of 50-50% Polylactic Acid-Polyglycolic Acid. Tissue Eng 2(1):61-74.
Turina J, Hess OM, Turina M, Krayenbuehl HP. 1993. Cardiac bioprostheses in the 1990s. Circulation 88(2):775-81.
Vandenberghe S, Segers P, Meyns B, Verdonck P. 2001. Hydrodynamic characterisation of ventricular assist devices. Int J Artif Organs 24(7):470-7.
Verdonck P, Kleven A, Verhoeven R, Angelsen B, Vandenbogaerde J. 1992. Computer-controlled in vitro model of the human left heart. Med Biol Eng Comput 30(6):656-9.
Vesely I, Boughner D. 1989. Analysis of the bending behaviour of porcine xenograft leaflets and of natural aortic valve material: bending stiffness, neutral axis and shear measurements. Journal of Biomechanics 22(6/7):655-671.
Vesely I, Noseworthy R. 1992. Micromechanics of the fibrosa and the ventricularis in aortic valve leaflets. Journal of Biomechanics 25(1):101-113.
Zeltinger J, Landeen LK, Alexander HG, Kidd ID, Sibanda B. 2001. Development and characterization of tissue-engineered aortic valves. Tissue Eng 7(1):9-22.

CHAPTER 12

DESIGN OF VASCULAR GRAFT BIOREACTORS

P.S. MCFETRIDGE[1] AND J.B. CHAUDHURI[2]

[1]*School of Chemical Engineering & Material Science, University of Oklahoma, Norman, USA*

[2]*Department of Chemical Engineering, University of Bath, Bath, UK*

1. INTRODUCTION

The use of tissue engineering has become important for the development of replacement blood vessels with prolonged patency rates (Zilla and Greisler 1999). Various groups are investigating different tissue engineering routes with the same aim of developing efficient cardiovascular grafts, particularly small diameter vesels (L'Heureux et al. 1998; Weinberg and Bell 1986); (Niklason et al. 1999) and heart valves (Shinoka et al. 1997).

An increasingly key component for tissue engineering is the use of the bioreactor for controlled tissue cultivation. This has been defined as "a system that simulates physiological environments for the creation, physical conditioning, and testing of cells, tissues, precursors, support structures, and organs in vitro" (Barron et al. 2003). The fundamental requirements for tissue engineering bioreactors are that the system delivers to the construct a homogeneous mix of nutrients, wastes and gases in a manner that minimizes the mass transfer limitations associated with static cultures. In addition, the reactor must provide the construct with a mechanical environment similar to the *in vivo* environment, into which the neo-tissue will be implanted. Finally the system must be designed with simplicity in mind, such that long term, sterile, tissue culture can be conducted. This chapter will discuss these aspects in terms of design goals with an emphasis on the pragmatic application of bioreactors to tissue engineer vascular grafts. Another key aspect of vascular tissue engineering is the choice of scaffold for cell attachment and proliferation. For further information on this topic the reader is directed to the recent review by (Schmidt and Baier 2000).

J.B. Chaudhuri & M. Al-Rubeai (eds.), Bioreactors for Tissue Engineering, 269 - 283.

2. EFFECTS OF MECHANICAL FORCES ON VASCULAR CELLS

The importance of mechanical forces on developing tissue engineered vascular grafts cannot be understated. The physical environment in which a cell persists has a profound effect on cellular behaviour, and thus vessel function (Lehoux and Tedgui 2003). Most cells persist in a state of dynamic flux where gases and diffusable molecules may interact, either indirectly or directly as ligands by binding specific receptors on the cell surface, and thus modulating cellular phenotype. Furthermore, cell–cell and cell–matrix interactions also play a critical role in modulating gene expression as the matrix and/or fluids surrounding the cells moves, expands or contracts over the cell surface (Bissell and Barcellos-Hoff 1987; Streuli 1999). Shear forces act in a variety of different ways by acting on the ECM, thus indirectly affecting the cell, by acting directly on the cell as a whole, or by stressing specific molecules integrated into the cell membrane and transmit directly via transmembrane proteins connecting to the cells interior organelles or indirectly through cell signalling cascades which lead to change in gene expression (Davies 2002).

Cell culture media are generally supplemented with foetal calf serum, and/or multiple growth factors to maintain cell viability and to retain the cells' expressed phenotype, to prevent them from de-differentiating to a less specialized state. Even with the use of these complex medias, terminally differentiated primary human cells progressively lose their specific phenotype after prolonged maintenance in static tissue culture. The use of dynamic tissue engineering principles, where mechanical forces are applied to a developing construct, may be designed to enhance phenotype retention or promote cell differentiation. It therefore necessary to emulate as closely as possible the chemical, biological and physical conditions the graft will be placed in to precondition the graft to its intended environment. This section gives an overview of the forces that can modulate the function of endothelial (EC) and vascular smooth muscle cells (VSMC), the two main cell types present in vascular tissue.

2.1 Response of Endothelial Cells to Mechanical Stress

The endothelium lining the lumenal surface of blood vessels (Figure 1) was once thought of as a simple, passive, non-thrombogenic barrier. It is now widely recognised as a dynamic participant in the active physiology of the vasculature by controlling vascular tone and haemostasis (Nerem et al. 1993; Sumpio 1993). Haemodynamic forces acting on the endothelium include: tensile stress, acting along the vessel wall due to circumferential deformations, hydrostatic pressure stress (cyclic strain), and shear stress acting along the length of the vessel due to blood flow (Sumpio 1993). All these are generated by the pulsatile action of the cardiac cycle and the resulting blood flow through the arterial system. The result of these forces being exerted on the endothelium results in a number of different responses that range from morphological changes to secretion of bioactive compounds. At a morphological level, the application of fields of force on a confluent layer of endothelial cells (Human umbilical vein endothelial cells) which under static growth conditions form a cobblestone pattern, result in elongation of the cells that align tangentially to the flow field (Ives et al. 1986). (Thoumine et al. 1995) exposed confluent bovine aortic endothelial cells (BAEC) to a range of conditions that included exposure to laminar steady shear stress of 30 dynes/cm^2 for 24 hours. Similar to the study by Ives *et al* (1986), the BAEC elongated in the direction of flow between 12 and 24 hours after the commencement of the flow regime. Not surprisingly a direct relationship was noted between the cytoskeletal proteins (f-actin, vimentin, and vinculin) with change in cell morphology. Interestingly, four other conditions also produced elongation of cells, including: altering Ca^{2+} concentrations, culture on collagen gels, floating monolayers and culturing cells on patterned surfaces (Thoumine et al. 1995). A change in morphology is therefore not necessarily a result of one condition alone, but a process in which cells adapt to each unique environment. Results by Kaiser *et al* (1997) indicate a link between mechanical shear and vascular endothelial cell apoptosis, showing that cell proliferation decreases as shear stress increases, and the lack of any stress that occurs in normal static culture conditions may increase the basal level of cell apoptosis (Kaiser et al. 1997). In addition there is a strong influence on secretion of active molecules, in some cases up regulated, in others down regulated, and

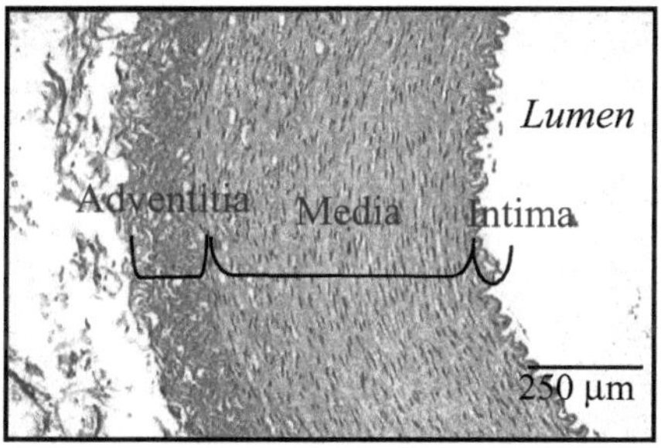

Figure 1: A haematoxylin and eosin stained section of porcine carotid artery from a 6 month old Great White pig, showing the main histological layers of the artery.

others show no change at all (Nerem et al. 1998). Importantly, the application of a specific shear regime can determine specific cell behavior. Nerem *et al* (1998) have shown HUVEC expression of VCAM-1, an adhesion molecule altered differentially between culture conditions from static to steady and pulsed flow (Nerem et al. 1998).

2.2 Response of Vascular Smooth Muscle Cells to Mechanical Stress

Vascular smooth muscle cells are a critical component of the vasculature residing in the medial layer of the vessel wall (Figure 1) and as such are directly exposed to a mechanically dynamic environment (Kim et al. 1999). If environmental cues are lost, for example when explanted vascular tissue is cultured statically, cells particularly hVSMC, revert back to a more simplistic phenotype. Like the endothelium, smooth muscle cells perform a specific function, not least to enable vessel contraction, thus providing control over peripheral blood flow. The loss of the *in vivo* phenotype is often associated with an increase in proliferation and a loss of the specialized gene expression associated with the differentiated state. A strong indicator of hVSMC dedifferentiation is the loss the ability to contract (Jones 1996). This loss of contractile ability can be identified using histological or cytological methods by localizing proteins associated with the contractile machinery of the cell, such as α-actin, within the cell cytoplasm.

It is widely acknowledged that mechanical and biochemical forces akin to the *in vivo* environment can help retain markers of cell differentiation during *in vitro* culture by modulating gene expression profiles. Platelet-derived growth factor (PDGF) is a potent mitogenic stimulus to VSMC proliferation and is likely to mediate the response of VSMC to cyclic strain. Ma *et al* (1999) demonstrated the expression of platelet-derived growth factor-B (PDGF-B) and its receptor PDGF β (PDGF β) on the cell surface in response to exposure to cyclic strain (Ma et al. 1999). However, increase in cell proliferation is often at the expense of maintaining the cells differentiated phenotype, Stegemann and Nerem (2003) have shown the interactions PDGF, transforming growth factor beta (TGF-β) and 10% strain as mechanical stimulation on graft development. In these studies PDGF increased cell proliferation and decreased expression of α-actin in 3D collagen gels, whilst reducing the compaction effects of the mechanical stimulation by producing a more open matrix compared to mechanical stimulation alone. Conversely exogenous application of TGF-β inhibited cell proliferation, increased the expression of α-actin that resulted in a compact or dense matrix (Stegemann and Nerem 2003). Birukov *et al* (1995) have shown that the application of cyclic stress on VSMC selectively induces expression of the phenotypic marker h-caldesmon, whereas omission of the stress reduces expression (Birukov et al. 1995). Interestingly Kim et al (1999) showed that rat cells cultured on fibronectin or vitronectin responded to mechanical stimulation (Flexercell unit) by an increase in proliferation, however the same cells cultured on laminin or collagen (exposed to the same conditions in serum free media) did not, but proliferation was induced when changed to serum-containing media. This was explained by the adsorption of fibronectin and vitronectin from the 10% serum-containing media onto the surface the cells were cultured on.

Kim et al (1999) also cited an increase in elastin and collagen synthesis on fibronectin-coated bonded PGA scaffolds but not on type I collagen sponges when cells were cultured in serum-free media. When cultured in 10% serum media and exposed to cyclic strain cells on the collagen sponge produced increased levels of elastin. In both cases Western blot analysis had shown that fibronectin had been adsorbed to the matrix, concluding the adsorbed proteins were responsible for the positive response of SMC when exposed to cyclic strain. Costa et al (1991) also reported an increase of elastin synthesis by bovine aortic SMC when subjected to repetitive cyclic strain (Costa et al. 1991). More recently it has been reported that mechanical stimulation of porcine SMC on a tubular PGA scaffold led to increased collagen deposition as compared to a non-stimulated control (Solan et al. 2003).

Mechanical strain of this nature can be seen as an indirect force applied to the arterial wall as a consequence of cardiovascular output. Intimal hyperplasia and medial thickening is a critical factor, particularly in vein grafts, where VSMC migrate and proliferate into the vessel lumen effectively restricting blood flow. The significant histological changes undergone in this process have been reported to be a response to mechanical stimuli and has been shown that when blood flow velocity has been reduced the occurrence of intimal hyperplasia increases (Sumpio 1993).

The importance of blood flow characteristics and vessel wall structure and mechanics are not to be underestimated in the development of a tissue engineered vascular graft. The forces mentioned above, when acting on EC and VSMC, critically modulate cell function (Davies 2002). The cells that populate blood vessels (predominantly endothelial, smooth muscle and fibroblast cell lineages) maintain a specific phenotype largely due to the direct effect of chemical and physical factors in the local environment e.g. matrix structure, mechanical properties and the fluid flow regime. Structural and dynamic effects on cell populations *in vivo* are an important factor in the prevention of graft thrombosis, a key cause of small diameter vascular graft failure.

3. VASCULAR GRAFT BIOREACTORS

The aim of graft development is to prepare a cell seeded matrix for implantation such that patterns of gene expression emulate the *in vivo* environment. Thus it is necessary to reproduce as closely as possible the conditions the graft will be placed in, such that the graft is preconditioned to its intended environment. The bioreactor is an important tool that if designed correctly can enable the application of specific biochemical and mechanical stimulation that will not only speed up development times but also produce grafting materials with superior characteristics to those currently used. This section reviews the design objectives, different approaches to bioreactor design, and integration of the bioreactor into a perfusion flow circuit for continuous, extended, cell culture.

3.1 Design Principles for Vascular Graft Bioreactors

A bioreactor has two main functions; firstly to maximise mass transfer of nutrients, and secondly as a direct method to apply force to a developing tissue construct. The forces associated with and imposed by the arterial circuit during normal haemodynamic flow on the blood vessel wall can be grouped into three major groups (1) tangential stress as a result of friction from blood flow through the vessel lumen, (2) circumferential stress, caused by cyclic changes in luminal pressure due to the pulsatile nature of large vessel blood flow, and (3) compressive stress, due to compression of the vessel wall during pressure related expansion, (Dobrin et al. 1989). Importantly, the bioreactor is not the only component, but is a component (containing the graft) that is integrated into a perfusion system and is designed specifically to allow the transition of the applied forces directly to the developing construct. The design concepts listed below are critical to achieve a functional bioreactor and process flow circuit:

- Control of the chemical environment; e.g. CO_2, PO_2, nutrients, pH
- Control of the physical environment; e.g. pulsed flow rate and frequency, and pressure
- Maintenance of sterility
- Accessibility, e.g. removal and addition of culture/growth components
- Materials compatibility, e.g. cell-media exposure to all components within the system
- Maintenance of long-term cell culture
- Reproducibility of 'standardized' reactor design
- Reproducibility of 'standardized' culture conditions
- Ability to monitor and control system and construct development
- Simplicity in design and implementation

The three main components of the perfusion circuit, the pumping apparatus, valves and tubing, and the bioreactor itself all require individual assessment to see that they fit the design criteria. This chapter will not discuss the full perfusion circuit, but will focus on our recent design of a bioreactor for a vascular graft using porcine collagen as the scaffold (McFetridge 2002).

3.2 Design of a Vascular Graft Bioreactor

3.2.1 Process Flow Circuit

Currently it is hard to design *a priori* a perfusion culture system, so development is often empirical, where incremental modifications are made as part of a problem solving regime that leads to a satisfactory result. The integrated perfusion circuit is required to operate in concert with the bioreactor to facilitate the delivery of media and its essential components to the cellular population/s within the matrix under defined pulsatile conditions (Figure 2). The complexity of design is often an inverse relationship to the usability of the system, as such compromises are often required between construct monitoring and data acquisition and the ability to assemble and operate under sterile conditions. Our previous experiences have borne this out, where earlier reactor systems that were considerably more complex in design, to allow for 'real-time' measurement of variables (pH, temperature, glucose etc), resulted in problematic operation with respect to maintenance of sterility.

In our design the bioreactor containing the scaffold is the focal point of the system (Figure 2). It is connected to the flow circuit via lumenal and ablumenal flow connectors and receives the pulsed flow generated by the pumping system (see below). The role of pump is to deliver media to the matrix with minimal disruption to the prescribed flow regime, and to deliver uniform transmatrix pressure differentials to enable uniformity of cell seeding and the delivery of nutrients/gases to the cellular populations. When designing a full bioreactor system one must

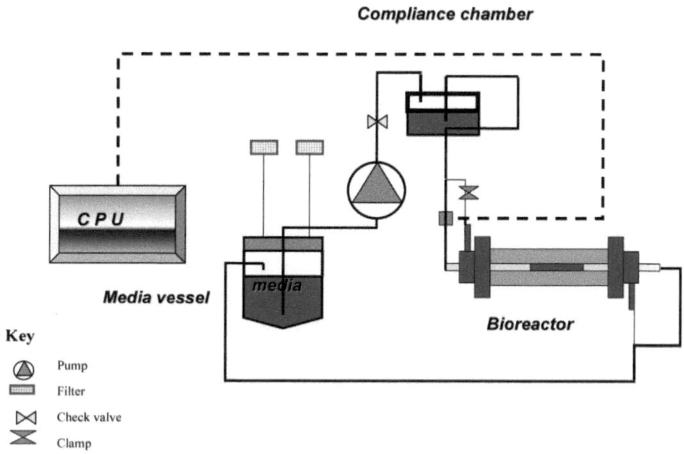

Figure 2: Schematic of process flow circuit illustrating a single reactor. The compliance chamber serves to reduce system 'noise', thus improving the flow regime.

consider the devices required for the efficient running and monitoring of the system, and the support infrastructure. The reactor variables (temperature, CO_2 and O_2 gas regulation, nutrient maintenance, pH control) must be maintained within set parameters. Whilst there are several methods of achieving this, in our case the system was run within a CO_2 incubator and thus was designed to allow transfer of the whole system from cell culture safety hood to the incubator.

A key feature of our bioreactor is the ability to pulse the media flow to the lumenal space to mimic the cardiac cycle and thus condition the EC. Figure 2 shows that media is drawn and pulsed by a peristaltic pump, from the media storage vessel through a one way check valve to prevent back flow, into a compliance chamber, designed to reduce system noise or vibration caused by the roller action of the peristaltic pump head. The figure shows that flow is then directed to the lumenal inlet port on the bioreactor, through the lumen and back to the media vessel where it is recirculated until changed. The pulsation wave form generated by this arrangement is shown in Figure 3. A separate flow circuit could also be employed to feed the SMC in the ablumenal space if two separate media were used. Alternatively, when a single media was used for both cell types, the flow was split just before the lumenal inlet and fed into the ablumenal and lumenal spaces. Immediately prior to the ablumenal inlet a clamp was used to control the flow rate into the ablumenal space of the reactor at a defined flow rate. When using single media for both cell types, care must be taken to ensure the pulse wave form or flow rates are not adversely affected.

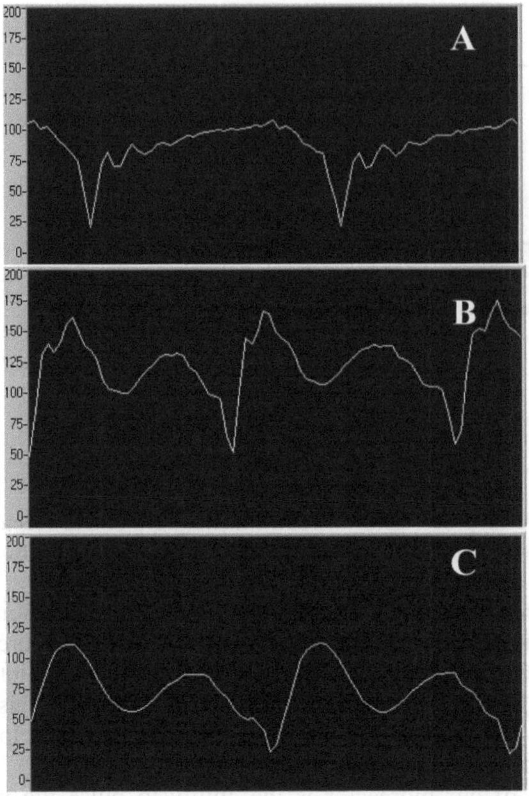

Figure 3: Pulse wave generated by the flow diagram in Figure 2. **A**, the basic pressure wave characteristics of a peristaltic pump; **B**, the addition of a compliance chamber alters both the waveform and removes much of the system noise; **C**, to prevent back flow, a one way check valve (cracking pressure of 0.33 p.s.i.) was integrated into the flow system after the peristaltic pump. Pressure readings (Y axis) are in mmHg.

A common approach is to replace the tube side media with buffered saline (Niklason et al. 1999; Niklason et al. 2001), or divert a small fraction of the single media type to the shell side prior to seeding the tube side (lumen) with endothelial cells (EC). At this later point independent circuits are required (unless using a media that different cell types such as smooth muscle cells and EC can be maintained).

An important aspect of system design that is overlooked is an appropriate method to run multiple reactors. Ideally one could run multiple, identical, units independently to ensure reliability: this, however, is likely to be prohibited by cost. Reactors could also be run in parallel, under ideal conditions with consistent pressure drops across each system. However it is difficult in practice to get identical flowrates and pressure drops through each bioreactor, and, unless care is taken, significant differences in flow rates will be observed between each system. An alternative is to use an excess of media to the cell density used and run a number of reactors in series.

3.2.2 Bioreactor Transport Considerations

Within the confines of the general design parameters two areas must be addressed: a method to apply a uniform pressure differential between the tube (lumen) and shell (albumenal) sides of the reactor; and generation and delivery of fully developed, pulsed, laminar flow to the EC in the lumen of the developing construct.

Typically, matrices cultured under static cell culture conditions are limited by mass transfer of key components, such as CO_2, O_2, growth factors, glucose etc. into the matrix material, thus limiting functional tissue development. The limit with which these components can effectively diffuse into and out of a matrix to provide the adhered cell populations with an environment conducive to cell growth and proliferation has been approximated at 100 μm (Vunjak-Novakovic 2002). As such, matrices with wall thicknesses in excess of 200 μm will be limited by this effect. Bioreactors must be designed to reduce these mass transfer limitations by applying a uniform transmatrix pressure differential to produce a homogeneous nutrient, waste, gas environment throughout the matrix wall (Figure 4). In addition, the application of convective flow may enhance the ability of cell populations to migrate into the matrix.

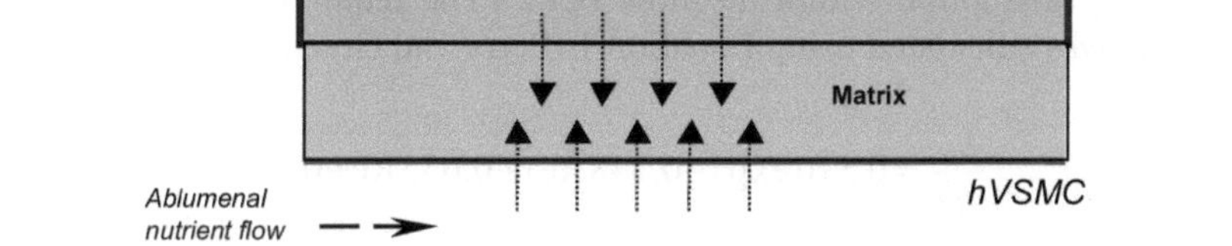

Figure 4: An illustration of perfusion conditions within the vascular graft bioreactor. ······▶ indicates cell migration into vessel, media perfusion and the transmembrane pressure differential; —➤ indicates nutrient flow direction.

It is important to consider that the diffusion characteristics across a porous scaffold will change over time (and position in the construct) as the cells grow, proliferate and deposit matrix components. The pressure drop across the scaffold will therefore change. Depending on the pressure regime the pressure drop will increase across the matrix as the pores of the matrix gradually reduce in size, with shear rates increasing as pore circumference decreases. This process is mediated by both cellular proliferation and non-specific binding of media components such as serum. However, initial matrix porosity and pressure will determine whether flow through the matrix is completely inhibited or the increased shear rates will prevent cell adhesion or protein adsorption. The latter scenario is more likely to occur in bioreactor systems that have no mechanism for maintaining uniform pressure differentials across the matrix, than a vascular perfusion circuit that has two separate flow circuits. As the construct becomes more tissue-like the ability of the bioreactor to supply the entire cell population with fresh oxygenated media is reduced.

The second major design parameter is to ensure that the bioreactor can mimic the physical environment that the developing constuct is to inhabit. In the case of vascular bioreactors, media flows through the lumen of the construct in a pulsatile fashion that simulates *in vivo* pressure and flow characteristics of blood. The reactor and support infrastructure must transfer to the matrix a physiological-like environment, which is conducive to long-term culture of human primary cell cultures.

In our design, the reactor body was fabricated from glass (8 mm ID) with stainless steel luer-type adaptors (3mm ID) to hold the tubular matrix in place (Figure 5). The stainless steel tube and luer connectors govern the flow characteristics through the matrix. At flow rates within the normal range (~140 ml/min) with arterial diameters of 3 mm ID were calculated with a velocity of 0.33 m/s. Further reduction in the diameter of the adaptors resulted in media velocities that was found to strip cells off the matrix surface even at flow rates as low as 50 ml/min. To ensure flow into the reactor was within the laminar flow range the Reynolds number (Re) was calculated using the test conditions above, which are at the high end of the flow regime for initial system testing (Best and Taylor 1991). The Reynolds number for the operating conditions above was 800.

To enhance the fluid dynamics of the reactor a modified lumenal inlet was designed that provided fully developed flow into the bioreactor, thus improving flow by minimising disruption within the flow field. The required entry length for the inlet adapter was calculated using Equation 1 (Perry and Green 1998).

$$\frac{L_{ent}}{D} = 0.370\exp(-0.148\,\mathrm{Re}) + 0.055\,\mathrm{Re} + 0.260 \qquad (1)$$

where D is the vessel diameter, Re is the Reynolds number and L_{ent} is the flow conditioning entry port length required to ensure 99% of fully developed flow into the reactor. With the use of stainless steel luer adapters with a 3.3 mm ID, under the test flow conditions described above, an entry length of 146.3 mm was required. The final design length of the lumen entry length was 250 mm to allow for variation in flow conditions during operation.

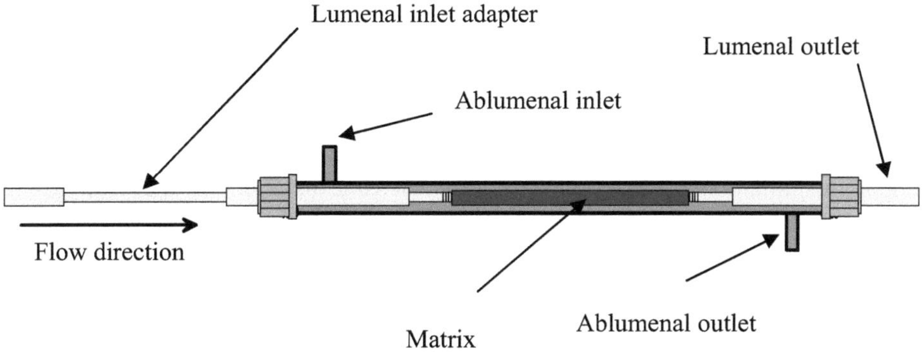

Figure 5: An illustration of the authors'bioreactor displaying inlet and outlet ports of the reactors main body. The main body had dimensions of 10 mm (OD) 8 mm (ID) x 200 mm long. The ablumenal volume was 10.1 mL.

A variety of other vascular bioreactor systems have been described in the literature (see below). The fundamental difference with our system is the ability to control both ablumenal and lumenal pressure, making it possible to manage transmembrane pressure differentials and hence provide an environment within the matrix wall that is conducive to cell growth and proliferation. In addition, minimising the ablumenal volume improves efficiency both in media and adhesion dynamics.

3.3 Development of Vascular Grafts Using Bioreactor Strategies

The first vascular grafts to be developed as living organs were grown *in vitro* without the application of mechanical forces to enhance graft development. Weinberg and Bell (1986) began these studies using collagen gels as scaffolds that could be moulded (with bovine cells embedded) into a tubular structure. Although no mechanical forces were directly applied to the graft, a fully structured vessels was created that behaved in a similar fashion to native vessels (Weinberg and Bell 1986). These vessels however required additional support by means of a Dacron mesh sleeve and extended culture times to reach acceptable burst strengths. These studies were further developed by L'Heureux *et al* (1993) with a significant difference being the use of primary human cell lineages. Although differences in the final construct were found between the Weinberg model and theirs, in terms of development and cellular orientation etc., the graft was mechanically weak; again there was no application of mechanical forces during development (L'Heureux et al. 1993). The same group later used primary human neonatal 'cell sheets' grown *in vitro* to confluence in media supplemented with 50 μg/mL of sodium ascorbate to increase extracellular matrix (ECM) formation. A number of independent 'cell sheets' where then wrapped around a mandrel to form the vessel structure. Again multiple layers of primary cells were used to form the vessels that were grown within a bioreactor, designed to provide luminal flow of culture media and construct support. Details of the flow regime were not described other than as a media supply and support mechanism for the cultured cells. This model made important advances in the technology, firstly cells sheets were used that could be layered to improve the strength of the construct (up to 2594 mmHg) well within human *in vivo* pressure extremes, and the use of a dynamic bioreactor (L'Heureux et al. 1998). An important limitation of this methodology is the time required for the construct to reach maturity. The graft requires 3 months culture in addition to cell expansion and the production of the acellular inner membrane to reach maturity.

Niklason *et al* (1999) have detailed a bioreactor used for the perfusion culture of small diameter arteries. Using a non-crosslinked PGA scaffold sewn into a tubular form, these workers cultured bovine aortic SMC for 8 weeks under pulsatile conditions using a bioreactor flow system. A compliance chamber was used to reduce high frequency vibrations and a peristaltic pump set at 165 beats/min was used to provide pulsatile radial stress (Niklason et al. 1999). Bovine SMC were found to have migrated into matrix creating a smooth lumenal surface onto which EC were seeded and subsequently cultured under perfusion conditions for 3 days. The resulting vessel displayed burst pressures well in excess of physiological conditions (8 weeks 2150 ± 709 mm Hg), compared to native saphenous vein 1680 ± 307 mm Hg. Advantages with the use of pulsatile pressure over static culture conditions were shown to be an increase in collagen deposition, though SMC density was unaffected by pulsatile pressure conditions, and a vessel histology similar to the native blood vessel compared to nonpulsed studies. Vessel contraction in response to various hormones was considered to be evidence of retained phenotype; often lost in standard culture and the EC lining the lumenal surface inhibited SMC proliferation into the vessel lumen. Implantation studies conducted with 6-month old pigs, grafts were implanted into the right saphenous artery and monitored up to 4 weeks before explantation. Results from these studies are promising with all grafts remaining patent for 2 weeks postoperatively. The pulsed xenograft remained patent for 24 days with no evidence of stenosis or dilation. The pulsed autologous vessel was patent at 4 weeks with a reduced flow rate, non-pulsed autologous vessel occluded at 3 weeks. Histological analysis of the explanted vessels displayed a highly organised structure with minimal inflammation. Importantly the effects of bioreactor conditioned (exposed to pulsatile culturing conditions) have produced positive effects on the viability of engineered vessels. Long-term implantation studies are required to fully test the effectiveness of this methodology. The application of pulsatile pressure on seeded grafts prior to implantation and the resulting benefits seen through the maintenance of cell phenotype and the up regulation of both elastin and collagen genes has lead to a significant increase in tensile strength of the engineered tissues (Kim and Mooney 1998). These distinct effects will bring the graft closer to its *in vivo* phenotype and potentially improve graft success through the first three critical months after vascular reconstructive surgery.

In a more recent publication Niklason and coworkers have focused on a variety of culture parameters to understand the mechanisms affecting graft development (Niklason et al. 2001). The effects of the degradation or presence of PGA scaffold residues on bovine SMC phenotype has been previously noted (Greisler et al. 1993), where SMC dedifferentiated to a fibroblast-like state resulting in a hyperplastic response. In this model, cultures at 5-8 weeks showed similar histology and burst strength, however the lack of elastin expression prevented native mechanical behaviour and the contractility of the graft was a fraction of native vessels. The group is currently exploring ways to induce elastin expression in order to improve mechanical compatibility (Niklason et al. 2001). Interestingly the model used by Kim *et al* (1999) (ref), confirmed elastin expression with murine cells. This effect can be attributed either to the increased mechanical stresses used or different cells types and their differential reactions to the *in vitro* environment. Both groups used PGA scaffolds, which weakens the argument that PGA residues dedifferentiate SMC under these conditions. These positive results will need to be introduced into a more stringent model where graft diameters are reduced, implanted in lower flow rate systems and adult primary human cells used to determine the effectiveness of this method.

Another bioreactor design has recently been published by (Hoerstrup et al. 2001). The experimental design involves seeding ovine cells onto a PGA scaffold, with the application of pulsed flow up to 750 ml/min. This system uses as many as 4 vessels in a parallel array. This is a more complex system where it is important to control the pressure drop between and across the 4 reactors. Unless a back pressure is applied to the downstream side of the reactors, flow rates may vary widely between the reactors.

Using an 'open bath' design with multiple constructs enclosed within a single bioreactor, Nerem and coworkers have used compressed air and a timer controlled solenoid valve to mechanically stimulate the vessel wall during development. This system was designed specifically to analyze blood vessel wall development under mechanical loading, by A cyclic distension of 10% at 60 Hz was imparted on the silicon lined collagen gel scaffolds over 4 and 8 days, and the effects of mechanical forces analysed on vessel wall development. Like other bioreactor based systems a marked difference in histological and mechanical properties compared to static cultures (Seliktar et al. 2000). Similar open bath bioreactors from the same group, using cyclic air pressure to control luminal pressure variation have been used to show improved construct histology, mechanics and phenotypic expression compared to non-mechanically stimulated constructs (Nerem 2000; Nerem and Seliktar 2001; Stegemann and Nerem 2003).

4. CONCLUSIONS

The overall aim of vascular bioreactor design and operation is the use of a perfusion system to allow adhesion, growth and proliferation of vascular cells under *in vitro* conditions that emulate the *in vivo* physiological environment. There are three key factors to be considered: design of a bioreactor that allows the uniform *in situ* seeding of specific cells to the required scaffold surfaces; use of an appropriate system to generate mechanical forces and transmit them to the cells in the bioreactor; creation of an bioreactor infrastructure to allow long-term, aseptic culture and monitoring of human primary or stem cells and the developing tissue.

Due to its relative infancy, literature describing the development of vascular bioreactors is limited with little in the way of standardized design. Such reports describe in some detail the perfusion circuits that incorporate the bioreactors, though few describe the vascular bioreactors in detail. Researchers are using unique, non-standardized systems, drawing on what little information is available. A drawback of this lack of standardisation is the inability to draw common conclusions from the different studies because of the design variation. Given that the mechanical environment is a strong determinant of gene expression, differences in design will effect gene expression and thus the performance of the graft as a whole. Clearly there is a need to define the underpinning design parameters and flow characteristics in order to facilitate comparative studies to improve design and integration methodologies.

5. REFERENCES

Barron V, Lyons E, Stenson-Cox C, McHugh PE, Pandit A. 2003. Bioreactors for cardiovascular cell and tissue growth: A review. Annals of Biomedical Engineering 31:1017-1030.

Best CH, Taylor NB. 1991. Best and Taylors physiological basis of medical practice. B. WJ, editor. Baltimore, MD 21202, USA: Williams and Wilkins.

Birukov KG, Shirinsky VP, Stepanova OV, Tkachuk VA, Hahn AWA, Resink TJ, Smirnov VN. 1995. Stretch effects phenotype and proliferation of vascular smooth muscle cells. Molecular and cellular biochemistry 144:131-139.

Bissell MJ, Barcellos-Hoff MH. 1987. The influence of extracellular matrix on gene expression: is structure the message? J Cell Sci Suppl 8:327-43.

Costa KA, Sumpio BE, Cerreta JM. 1991. Increased elastin synthesis by bovine aortic smooth muscle cells subjected to repetitive mechanical stretching. FASEB J. 5(A1069).

Davies PF. 2002. Spatial structural and genomic responces of endothelial cells to hemodynamic shear stress. Cardiovascular Pathology 11:25.

Dobrin PB, Littooy FN, Endean ED. 1989. Mechanical factors predisposing to intimal hyperplasia and medial thickening in autogenous vein grafts. Surgery 105(3):393-400.

Greisler HP, Petsikas D, Lam TM, Patel NE, Cabusao E, al. e. 1993. Kinetic of cell proliferation as a function of vascular graft material. Journal of Biomedical Materials Research 27:955-961.

Hoerstrup SP, Zund G, Sodian R, Schnell AM, Grunenfelder J, Turina MI. 2001. Tissue engineering of small caliber vascular grafts. European Journal of Cardiothoracic Surgery 20(1):164-169.

Ives CL, Eskin SG, McIntire LV. 1986. Mechanical efects on endothelial cell biology. In vitro Cell Developmental Biology 22:500-507.

Jones GE. 1996. Methods in Molecular Medicine. Walker JM, editor. Towowa: Humana Press. 545 p.

Kaiser D, Freyberg MA, Freiedl P. 1997. Lack of haemodynamic forces trigger apoptosis in vascular endothelial cells. Biochemical and Biophysical Research Communications 231:596-590.

Kim B, Mooney D. 1998. Development of biocompatible synthetic extracellular matrices for tissue engineering. Trends in Biotechnology 16(5):224-230.

Kim BS, Nikoloviski J, Bonadio J, Mooney DJ. 1999. Cyclic mechanical strain regulates the development of engineered smooth muscle tissue. Nature Biotechnology 17(October):979-983.

Lehoux S, Tedgui A. 2003. Cellular mechanics and gene expression in blood vessels. J Biomech 36(5):631-43.

L'Heureux N, Germain L, Labbe R, Auger FA. 1993. *In vitro* construction of a human blood vessel from cultured vascular cells: a morphologic study. Journal of Vascular Surgery 17:499-509.

L'Heureux N, Paquet S, Labbe R, Germain L, Auger FA. 1998. A completely biological tissue-engineered human blood vessel. FASEB Journal 12(1):47-56.
Ma Y-H, Ling S, Ives HE. 1999. Mechanical strain increases PDGF-B and PDGF-beta receptor expression in vascular smooth muscle cells. Biochemical and Biophysical Research Communications 265:606-610.
McFetridge PS. 2002. The Use of Porcine Carotid Arteries as a Matrix Material for Tissue Engineered Small Diameter Vascular Grafts [PhD]. Bath: University of Bath. 301 p.
Nerem RM. 2000. Tissue engineering a blood vessel substitute: the role of biomechanics. Yonsei Med J 41(6):735-9.
Nerem RM, Alexander WR, Chappell DC, Medford RM, Varner SE, Taylor RW. 1998. The study of the influence of flow on vascular endothelial biology. Americal Journal of Medical Science 316(3):169-175.
Nerem RM, Harrison DG, Taylor RW, Alexander WR. 1993. Haemodynamics and vascular endothelial biology. Journal of Cardiovascular Pharmacology 21 (supplement 1):S6-S10.
Nerem RM, Seliktar D. 2001. Vascular Tissue Engineering. Annual Review of Biomedical Engineering 3:225-243.
Niklason LE, Gao J, Abbott WM, Hirschi KK, Houser S, Marini R, Langer R. 1999. Functional arteries grown *in vitro*. Science 284:489-493.
Niklason LE, Gao J, Abbott WM, Klagges B, Hirschi KK, Ulubayram K, Conroy N, Jones R, Vasanawala A, Sanzgiri S and others. 2001. Morphologic and mechanical charactistics of engineered bovine arteries. Journal of Vascular Surgery 33:628-638.
Perry RH, Green DW. 1998. Perry's Chemical Engineers Handbook: McGraw-Hill International Editions, Chemical Engineering Series.
Schmidt CE, Baier JM. 2000. Acellular vascular tissues: natural biomaterials for tisue repair and tissue engineering. Biomaterials 21:2215-2231.
Seliktar D, Black RA, Vito RP, Nerem RM. 2000. Dynamic mechanical conditioning of collagen-gel blood vessel constructs induces remodeling in vitro. Ann Biomed Eng 28(4):351-62.
Shinoka T, Ma PX, Shum-Tim D, Breur CR, Cusick RA, Zund G, Langer R, Vacanti JP, Mayer JE. 1997. Tissue-engineered heart valves: autologous valve leaflet replacement, study in a lamb model. Circulation (Supplement II) 96(9):164-168.
Solan A, Mitchell S, Moses M, Niklason L. 2003. Effect of pulse rate on collagen deposition in the tissue-engineered blood vessel. Tissue Engineering 9(4):579-586.
Stegemann JP, Nerem RM. 2003. Phenotype modulation in vascular tissue engineering using biochemical and mechanical stimulation. Annals of Biomedical Engineering 31:391-402.
Streuli C. 1999. Extracellular matrix remodelling and cellular differentiation. Curr Opin Cell Biol 11(5):634-40.
Sumpio BE. 1993. Haemodynamic forces and vascular cell biology. Sumpio BE, editor. Georgetown: R.G. Landes Company.
Thoumine O, Ziegler T, Girard PR, Nerem RM. 1995. Elongation of confluent endothelial cells in culture: The importance of fields of force in the associated alterations of their cytoskeletal structure. Experimental Cell Reseaarch 219:427-441.
Vunjak-Novakovic G. 2002. Tissue engineering approach to functional myocardium. Cardiovascular Pathology 11:23.
Weinberg CB, Bell E. 1986. A blood vessel model constructed from collagen and cultured vascular cells. Science 231(4736):397-400.
Zilla PP, Greisler HP, editors. 1999. Tissue Engineering of Vascular Prosthetic Grafts. Austin: R.G. Landes Company.

CHAPTER 13

PERFUSION BIOREACTORS FOR CARDIOVASCULAR TISSUE ENGINEERING

V. KASYANOV, J.J. SISTINO, T.C. TRUSK, R.R. MARKWALD AND V. MIRONOV

Medical University of South Carolina, Charleston, USA

1. INTRODUCTION

Since the discovery of the circulation by William Harvey, it has become clear that medical research would eventually employ artificial perfusion systems for the ex vivo study of living organs. As early as the 19th century, perfusion systems were used for physiological studies. However, there were significant technical limitations including the inability to prevent infection, and the lack of adequate blood oxygenators and blood pumps.

In 1929, Alexis Carrel of the Rockefeller Institute began working on organ preservation. Together with Charles Lindbergh, who worked in Carrel's lab as a laboratory assistant, they made the major contributions to the development of organ perfusion systems. They constructed a complete pump-oxygenator apparatus for organ perfusion. Lindbergh's pump-oxygenator used compressed air, which was pulsed across cotton to maintain sterile conditions. The pulsating air oxygenated the blood as it circulated through the organ. The Lindbergh-Carrel apparatus is on display in the Smithsonian Institution of Technology in Washington D.C., and at the University of Florida where there is a collection carefully maintained by Professor Theodor Malinin. The textbook, "Culture of Organs" by Carrel and Lindbergh, which was published in 1939, has been inspiration for many generations of biomedical researchers.

During the same time period, a student and future world-renowned cardiothoracic surgeon, Michael DeBakey, designed a roller pump for use in ex vivo perfusion. Also, John Gibbon began to develop during this same time period a heart and lung machine for use as a clinical device during heart surgery. The death of a young patient in 1931 had stirred Dr. Gibbon's imagination about developing an

J.B. Chaudhuri & M. Al-Rubeai (eds.), Bioreactors for Tissue Engineering, 285 - 307.

artificial device for bypassing the heart and lungs. Many others with whom he broached the subject dissuaded him, but he continued his experiments independently. In 1935 he successfully used a prototype heart-lung bypass machine to keep a cat alive for 26 minutes. John Gibbon joined forces with Thomas Watson in 1946. Watson was an engineer and the chairman of IBM (International Business Machines). He provided the financial and technical support for Gibbon to further develop his heart-lung machine. Later, Dr. John Gibbon performed the first successful human open heart operation on May 6, 1953 using this device. Thus began a new era surgical repair of the heart utilizing cardiopulmonary bypass.

The Lindbergh-Carrel perfusion apparatus was used for a classic experiment in which a melanoma tumor was implanted in an isolated perfused rabbit thyroid gland. This experiment was performed by a young navy surgeon Judah Folkman, (Judah Folkman, personal communication). Folkman demonstrated that without an adequate vascular supply, the melonoma tumor implanted in the perfused thyroid gland stopped its growth. This was the foundation for the discovery of angiodependency of tumor growth. The eventual isolation and identification multiple angiogenic growth factors and inhibitors of angiogenesis led to the development of the concept of antitumor antiangiogenic therapy (Folkman, 1971).

Tissue engineering has reactivated interest to perfusion systems. Mechanical conditioning is a powerful stimulator for tissue growth, deposition of extracellular matrix, and tissue remodeling (Kim et al., 1999; Niklason et al., 1999, Nerem and Seliktar, 2002; Mironov et al., 2003). The first successful clinical use of tissue engineered vascular grafts stimulated an urgent need to develop clinically acceptable perfusion bioreactors (Hibino et al., 2002; Naito et al., 2003). The recent development of organ printing (Mironov et al., 2003), opens a new phase in cardiovascular tissue engineering research which will require designing perfusion bioreactors to maintain the entire heart.

The goal of this review is: i) to describe the original perfusion bioreactor for vascular tissue engineering with capacities for longitudinal stretch; ii) to discuss the special requirements for next generation of perfusion bioreactors for 3D vascularized tissue engineered myocardial tissue and the printed tissue engineered heart; and iii) to review some emerging (on different stage of development) industrial perfusion systems which could be adapted for experimental, industrial and clinical cardiovascular tissue engineering. Perfusion bioreactors for tissue engineered heart valves and clinical perfusion systems, which are used routinely in cardiothoracic surgery, are outside the scope of this review.

2. PERFUSION BIOREACTORS FOR VASCULAR TISSUE ENGINEERING

2.1 Historical Development

Tissue engineering represents a promising concept of applying the principles and methods of engineering and biology to develop functional tissue for surgical replacement (Vacanti and Langer, 1999). Initial attempts to create vascular tissue engineered blood vessels resulted in the manufacturing of grafts with poor mechanical properties (Weinberg and Bell, 1986; L'Heureux et al., 1993). The next generation of tissue engineered vascular grafts had dramatically improved mechanical properties. However, it took several months to develop tissue engineered vascular grafts with desired optimal mechanical properties (L'Heureux et al., 1998). Elastin and collagen are the main two components of extracellular matrix that determine the biomechanical properties of the vascular wall. Dynamic mechanical conditioning of the cardiovascular grafts accelerates the production of these two proteins (Kim et al., 1999; Seliktar et al., 2000; Nerem and Selictar, 2001). For example, the periodical radial distention resulted in accelerated maturation of tissue engineered vascular grafts (Niklason et al., 1999). In an in vitro experiment with a rabbit pulmonary artery subjected to longitudinal stretch, the rate of protein synthesis in smooth muscle cells, percentage of procollagen type I-positive cells, and the rate of cell replication was related to the magnitude of the longitudinal stretch. After four days of longitudinal stretch, the rate of elastin and collagen synthesis and the contents of actin and elastin in the pulmonary artery segment increased dramatically (Kolpakov et al., 1995).

It has been recently reported that endothelial cell and smooth muscle cell replication rates in rabbit carotid arteries increased > 50-fold and > 15-fold, respectively, after three days of longitudinal stretch in vivo (Jackson et al., 2002). This data indicates that longitudinal stretch of arteries is an important mechanical force for accelerated cell proliferation, the production rate of the extracellular matrix, and vascular remodeling.

We hypothesized that longitudinal stretch will accelerate tissue engineered vascular wall maturation and result in a tissue engineered vascular graft with optimal mechanical properties. In order to accomplish this goal, we developed a perfusion bioreactor with the capacity for biomechanical conditioning using longitudinal strain. The capacity of the bioreactor to perform both circumferential and longitudinal strain was tested on silicone tubes and natural arteries.

2.2 Bioreactor Design

The newly designed bioreactor was developed with a closed loop perfusion system that can be used either for non-pulsatile or pulsatile flow in the CETC. The bioreactor consists of a chamber, two peristaltic pumps, and tubing (Figures 1 - 2).

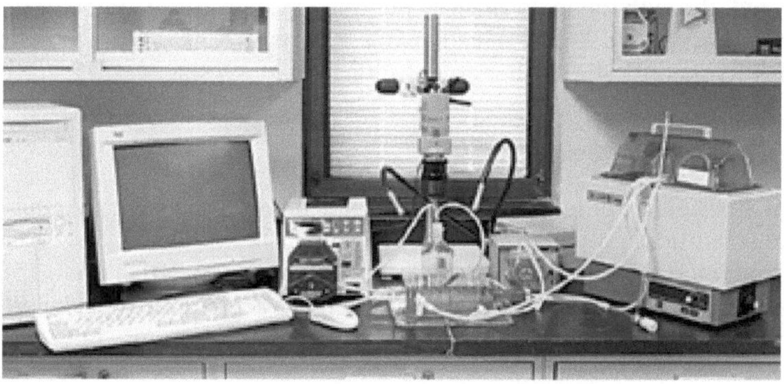

Figure 1. The perfusion bioreactor system.

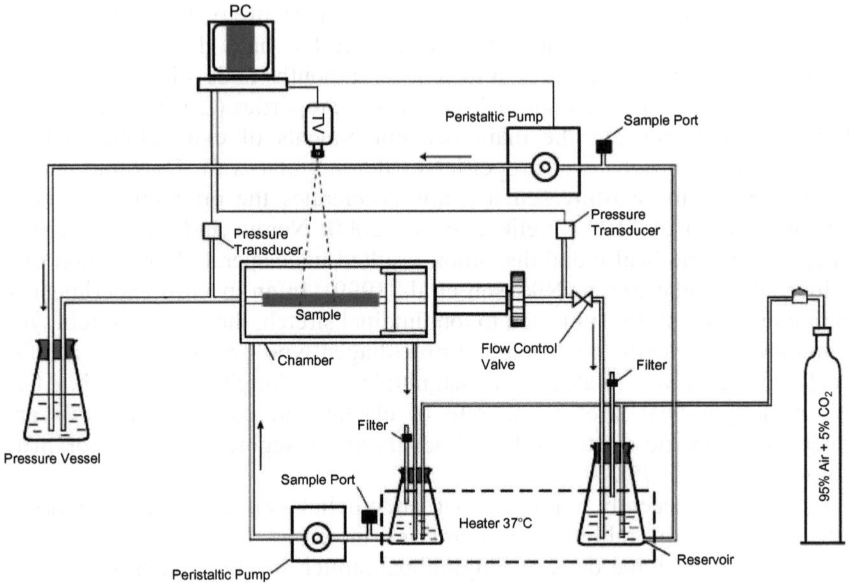

Figure 2. Experimental setup of the perfusion bioreactor.

There are two systems of media perfusion: one is for inside and another one is for outside perfusion of the CETC. A medium continuously circulates through the closed-loop system. A third system includes a digital camera and pressure transducers which control of biomechanical properties of the CETC. The whole system is a highly isolated cell culture setting, which provides a high level of sterility, gas supply, and can fit into a standard humidified incubator. The bioreactor can be sterilized by ethylene oxide and assembled with a standard screwdriver.

The CETC (inner diameter from 2 mm up to 6 mm) is cannulated on both ends with removable male luer locks (Cole Parmer Instrument Co., Vernon Hills, IL). One cannula connects from one side of the silicon tubing and another cannula from the other side to the stainless steel tube with a thread, which provides longitudinal stretch to the construct by means of special knurled nut. Nalgen 50 platinum – cured silicon tubing (Fisher Scientific, Atlanta, GA) is used to transmit the perfusate from the construct, through the main (inside) pumping system (L/S Computer Compatible Drive with Easy Load Pump Head, Cole Parmer), back to the pressure vessel (PyrexPlus Reagent Bottle, Fisher Scientific).

2.3 Systems of Media Perfusion

The main peristaltic computer-operated digital pump allows a changing flow rate inside the construct in a wide range: from 28 to 1700 ml/min. A flow control valve allows for an increase in mean static pressure inside the construct up to 150 mm Hg. The pressure inside the construct is controlled by a pressure transducer, which is connected to the computer. Pressure transducers (EPB-C02-5P; Entran Sensors &

Electronics Co., Fairfield, NJ) are connected through the 5 Digit Meter & Power supply (Entran Sensors & Electronics Co.) to a computer. These transducers are used to monitor intraluminal pressure, which can be increased by a manual pinch flow control valve (H. P. Rotaflo Stopcock, Fisher Scientific) attached just distal to the chamber. A compact, variable - speed peristaltic pump (L/S Compact, Low – flow Drive with Easy Load Pump Head, Cole – Parmer) with flow rate from 2.1 to 560 ml/min, and silicon tubing are used for the outer perfusion system.

A sample infusion port (Thermogreen LB-2 Septa, Supelco Co., Bellefonte, PA) is also used to allow sterile injection of liquids into the perfusate. For temperature control, reservoirs (Kimax Bottles with Tubulation, Fisher Scientific) are placed into a standard 5-liter temperature bath (Fisher Scientific), and the CO_2 level is titrated to achieve desired pH depending on the culture medium used in the system. Sterile gas exchange in the bottles is accomplished with a 0.2 μm filter. Physiologic pH, pO_2 and pCO_2 are maintained by infusing a 5% CO_2–air mixture into the gas exchange regions and allowing outflow some distance away. A standard water–jacketed incubator may also be used for this purpose. A gas analyzer (CDI 500) is used for verification of proper pH, pO_2 and pCO_2.

2.4 Flow Chamber

The culture chamber is constructed completely from transparent polycarbonate. The chamber is covered with a gasketed lid. The gasket is made from polymer "Sylgar-184" (Dow Corning Co., MI). The mechanism of the watertight chamber (Figure 3) allows periodical changing of the longitudinal strain of the construct during the mechanical conditioning.

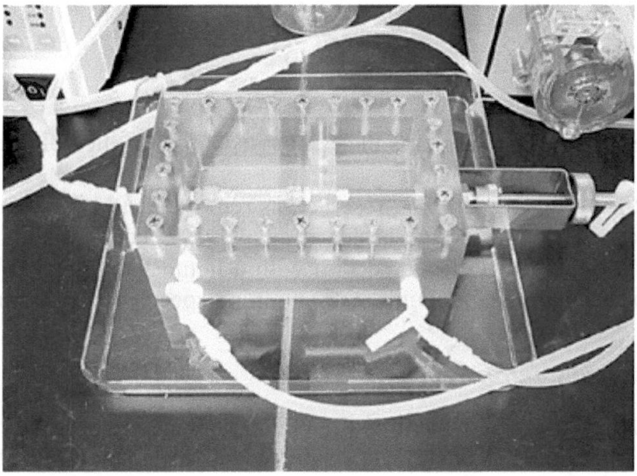

Figure 3. The watertight chamber of the bioreactor.

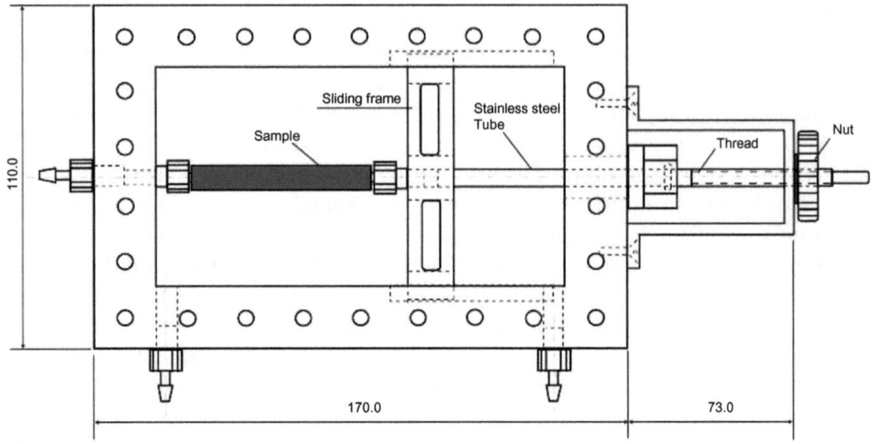

Figure 4. Watertight chamber scheme (a chamber lid is not shown: all dimensions are in mm).

The range of the longitudinal strain is 0% to 200%. This mechanism includes a sliding frame, a stainless steel tube with a thread, and a knurled nut (Figure 4). Turning the knurled nut moves the frame to the right and increases the length of the tubular construct.

In steady flow, the shear stress in the CETC is determined by the flow rate and viscosity of the medium and the diameter of the vessel (Ku, 1997):

$$\tau = \frac{32 \mu Q}{\pi D^3}, \tag{1}$$

where τ is the wall shear stress, μ is the fluid viscosity, Q is the volumetric flow rate and D is the CETC inner diameter.

For puslatile flow, the damping reservoir is removed and a T-connector with an extra 60 cm of tubing is added (Chesler et al., 1998). The oscillation amplitude is adjusted by shortening this elastic section with a hemostat clamp. Pulse frequency depends on the main peristaltic pump with pumping speeds of 1-600 rpm.

2.5 Biomechanical Control System

The biomechanical control system of the CETC includes two pressure transducers and a digital TV camera (Kodak, MDS 100), which are connected to the computer. The pressure inside the tubular construct is calculated as a mean value of the two transducers. This system allows the recording of a relationship between the radius of

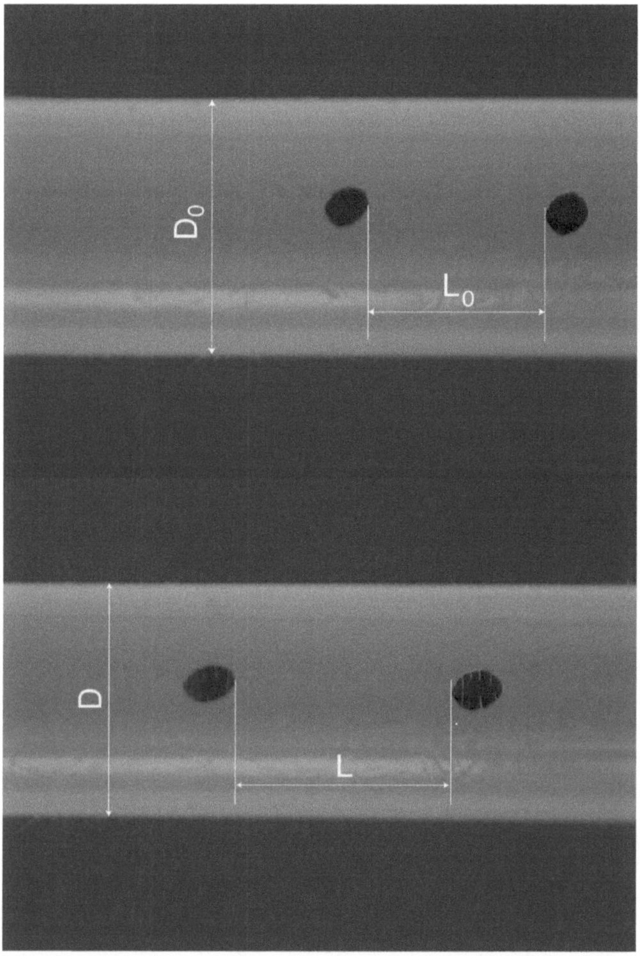

Figure 5. Non-deformed state of the CETC (upper): L_0 initial length, D_0 initial diameter; Deformed state of the CETC (lower): L length and D diameter of the CETC after longitudinal stretch, respectively.

CETC and pressure, as in static and in dynamic regimes. The chamber and control system are also used for the investigation of burst strength of the CETC. The diameter of the CETC and distance between two marks on the specimen are measured using digital TV camera (Figure 5).

The longitudinal (ε_1) and circumferential (ε_2) strains of the CETC are calculated as:

$$\varepsilon_1 = (L - L_0)/L_0; \quad \varepsilon_2 = (D - D_0)/D_0 \qquad (2)$$

where L_0 and D_0 are the initial (non-deformed) length between two marks and the diameter of the CETC, respectively; L and D are the length and diameter after deformation, respectively.

2.6 Functional Capacities

The capability of the bioreactor was estimated using a native bovine carotid artery as a tubular construct. Two ink marks were placed in the middle of the sample to determine the longitudinal strain during stretch of the construct. The experimental results show that pressure–circumferential strain has a non-linear character (Figure 6). The largest deformation of the artery is observed until internal pressure is 40 mmHg. Also, it was shown that increasing the longitudinal strain leads to a decreased diameter of the construct (Figure 7). Longitudinal strain and circumferential strain have a non-linear relationship.

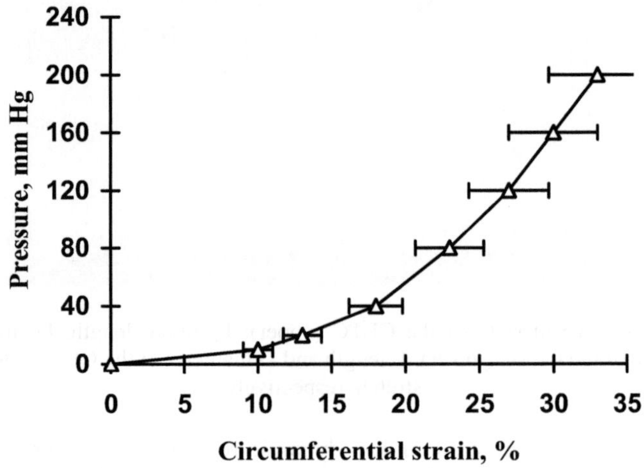

Figure 6. The pressure – circumferential strain relationship for a native bovine carotid artery (n=5).

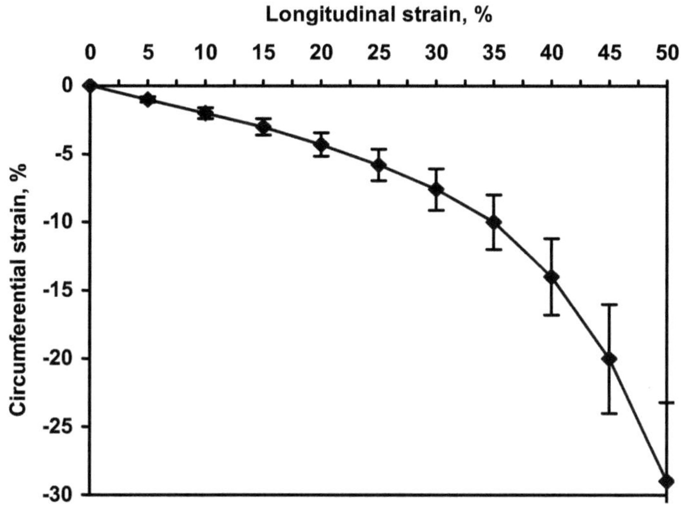

Figure 7. The longitudinal strain – circumferential strain relationship for a native bovine carotid artery (n=5).

Thus, the bioreactor can provide a new critical component of biomechanical conditioning (longitudinal strain), which is essential to mimic the embryonic mechanical vascular environment. A unique feature of this bioreactor is a functional capacity to control flow rate, pressure, and longitudinal strain of the vascular tissue engineered graft with high precision.

The best reported results in vascular tissue engineering are achieved when classical seeding or embedding techniques are combined with dynamic biomechanical conditioning (Nerem & Selitar, 2001). Several perfusion apparatuses and bioreactors have already been developed for these purposes. For example, Mangiarua et al. (1992) developed a perfusion apparatus for short-term studies of related vascular changes associated with hypertention. Chesler et al. (1998) presented perfusion techniques that enable the use of physiologic stretch of porcine arteries in an *ex vivo* system. Another *in vitro* flow system was developed to investigate the effect of laminar flow on extracellular matrix deposition and tissue development (Jockenhoevel et al., 2002). A perfusion bioreactor was also used for the pre-conditioning of endothelial progenitor cells (EPCs) seeded on an acellular vascular scaffold (Kaushal et al, 2001). Surowiec et al. (2000) and Conklin et al (2000) developed a vascular perfusion organ culture system that generates pulsatile flow by the combination of a cam-driven syringe, a peristaltic pump, and a compliance chamber. Hoerstrup et al. (2001) reported an *in vitro* pulsatile perfusion system to grow seeded vascular constructs under "biomimetic" flow conditions. In contrast to previous studies, the pulsed flow of nutrient media was directed immediately through the vascular lumen, thereby generating direct shear stress to the luminal surface, as well as periodical radial distension of the vessel wall. All reported perfusion apparatuses can only provide mechanical conditioning in the circumferential direction using internal pressure during the maturation of the construct. The novel distinguishing feature of our bioreactor is the capacity to provide longitudinal strain to the tissue engineered constructs.

Normal arteries and veins in the human body are under a 3D stress-strain state that includes circumferential, radial and longitudinal strain (Kasyanov, 1974). Almost all arteries in vivo are under substantial longitudinal stretch (axial strain). An *in situ* longitudinal strain of arteries is 40% to 65% (Learoyd and Taylor, 1966). The magnitude of longitudinal stretch in vivo can be altered by the growth or atrophy of the surrounding perivascular tissues. Due to close attachment of blood vessels to the surrounding tissues, any changes in tissue volume results in alterations in the level of longitudinal strain applied to these vessels. Therefore, local tissue growth or atrophy during development, weight gain/loss, pregnancy, parturition, exercise regimens, or a host of pathophysiological changes will alter the mechanical forces applied to the vascular wall.

In a recent review (Salacinski et al., 2001) it was clearly shown that compliance mismatch (in circumferential and longitudinal directions) has an important role in vascular graft failure. Kolpakov et al (1995) demonstrated that longitudinal stretch can increase the rate of total protein synthesis in medial smooth muscle cells (SMC) and the total actin content (normalized to DNA) of the pulmonary artery segment in vitro. The increase in the rate of proliferation of medial SMC and adventitial fibroblasts induced by longitudinal stretch in vitro suggests that this mechanical force can also act as a hyperplastic stimulus in the intact pulmonary artery. Also, longitudinal stretch can increase the rate of matrix protein synthesis and accumulation in the in vitro pulmonary artery. It was recently reported that endothelial cell and smooth muscle cell replication rates, and extracellular matrix deposition in rabbit carotid arteries were increased dramatically after three days of longitudinal stretch *in vivo* (Jackson et al., 2002). Thus, it is logical to assume that longitudinal strain can accelerate wall maturation in tissue engineered vascular graft. It remains to be demonstrated what specific stretching regime (one-step elongation or gradual multi-step elongation) will provide the best outcome. The design of perfusion chamber of our bioreactor allows us to employ both regimes. Because longitudinal strain is an important element of the embryonic vascular biomechanical environment, we designed a bioreactor with the capacities for both circumferential and longitudinal strains. Longitudinal strain in combination with the capacity for circumferential strain can maximally imitate the embryonic vascular biomechanical environment and perform the acceleration of vascular wall maturation. Application of defined longitudinal strain can be used for the production of a vascular graft that is mechanically strong. This device is also suitable to study the role of dynamic mechanical conditioning in vessel wall maturation and remodeling.

3. PERFUSION BIOREACTORS FOR VASCULARIZED TISSUE ENGINEERED 3D MYOCARDIAL TISSUE.

3.1. Minibioreactor For 3D Vascularised Myocardial Tissue Constructs

The behavior of myocardial cells in a two-dimensional vs. a three-dimensional cell culture system is dramatically different. The latter conditions more closely simulate in vivo realities and give a more in vivo phenotype. Thus, future bioassays for testing the potential direct or indirect (side) effects of drugs on myocardial structure and function (including genomics and proteomics) should be based on developing three-dimensionally reconstructed tissue. We have shown recently that closely placed cell aggregates fuse when placed into three-dimensional collagen gels can be manipulated into tissue constructs of any desired geometrical form. We have also shown that endothelial cell aggregates placed in three-dimensional collagen gels form sprouts and eventually fuse and form a network-like structure. Thus, it is logical to assume that co-mixtures of endothelial cell and cardiomyocyte aggregates could be manipulated to form three-dimensional vascularized "cardiac" tissue. However, the vascularization of a three-dimensional tissue construct per se does not automatically guarantee adequate perfusion. In order to get adequate perfusion the microvascular network of this assembled three-dimensional tissue construct must be perfused or seamlessly integrated with a perfusion minibioreactor. Although several groups are already involved in the development of tissue engineered myocardial tissue (Eschenhagen et al., 2002; Carrier et al., 1999; 2002) to our knowledge this task has not been effectively solved. The absence of an effective system of perfusion that is seamlessly integrated with the assembly of the three-dimensional tissue is the main obstacle in the development of a three-dimensional myocardial tissue-based assay. We have shown that fragments of quail brachio-cephalic arteries placed in collagen gel culture can form endothelial sprouts from the both cut ends. If additional cuts or holes are made along the line of explanted vessel, endothelial spouts appear in these sites as well. Based on this observation, we hypothesize that porous hollow fibers (synthetic microtubes) pre-seeded with endothelial cells and then placed in a three-dimensional collagen gel will result in the formation of endothelial sprouts invading the collagen gel at each of the holes in the microtube. Lumenization of these spouts and their integration with the microvasculature of fabricated three-dimensional myocardial tissue will establish a perfused vascularized three-dimensional myocardial tissue construct. Realization of this project needs expertise both in engineering and vascular biology. It could be accomplished in 5 principal steps:

Step 1. Fabrication of a three-dimensional myocardial tissue construct from self-assembling myocyte aggregates in a gel of collagen type 1.

Myocytes from chicken embryos will be allowed to aggregate then placed close to each other to form a metastable cube-like porous tissue construct. The dynamics of the tissue fusion process will be monitored using confocal microscopy. The angle between adjacent cell aggregates will be used as a criterion of completeness of the tissue fusion process.

Our preliminary experiments clearly indicate that closely placed in 3D collagen type 1 gel homotypic cell aggregates are fused into 3D tissue construct of desired geometrical form (Mironov et al., 2003).

Step 2. Fabrication of a vascularized three-dimensional myocardial tissue construct from self-assembling aggregates of myocyte and endothelial cells in a three-dimensional collagen type 1 gel.

Mixtures of myocyte aggregates and endothelial cell aggregates will be placed proximal to each other in a mosaic-like pattern in a three-dimensional collagen type 1 gel. These aggregates will fuse to form a metastabile cube-like "endothelialized" myocardial tissue construct. The dynamics of the tissue fusion process will be monitored using confocal microscopy. The angle between adjacent cell aggregates will be used as a criterion of completeness of the tissue fusion process. Formation of a network of lumenized vessels will be assessed by histological analysis. The fluorescently labeled cell culture media will allow monitoring of the dynamics of lumen formation in vivo (Davis and Camarillo, 1986).

Step 3. Evaluate synthetic porous hollow fiber tubes as a permissive architecture for endothelial sprouting.

Hollow fiber synthetic tubes from standard hollow fiber bioreactors will be used. By using laser-based evaporation (laser ablation) pores of different diameters from 10 to 100 microns will be created. Endothelial cells will be introduced into the lumen of the hollow fiber tubes to establish an internal endothelial monolayer. These endothelialized tubes will be placed within a three-dimensional collagen gel and cultured for various times. The relationship of diameter of the laser-ablated hole in

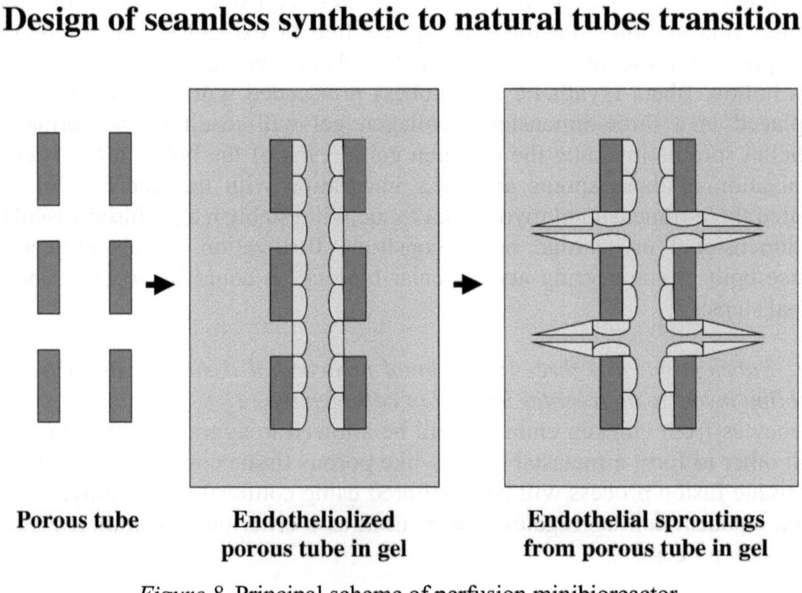

Figure 8. Principal scheme of perfusion minibioreactor.

the tube vs. the extent of sprout formation will be determined. The optimal diameter for sprout formation will be use for manufacturing tubes to be employed in minibioreactor design (Figure 8).

Step 4. Designing and manufacturing of perfused minibioreactor.

The minibioreactor consists of a roller pump and a system of hollow fiber microtubes. The overall design for this minibioreactor is typical for perfusion type of bioreactors (Figure 9).

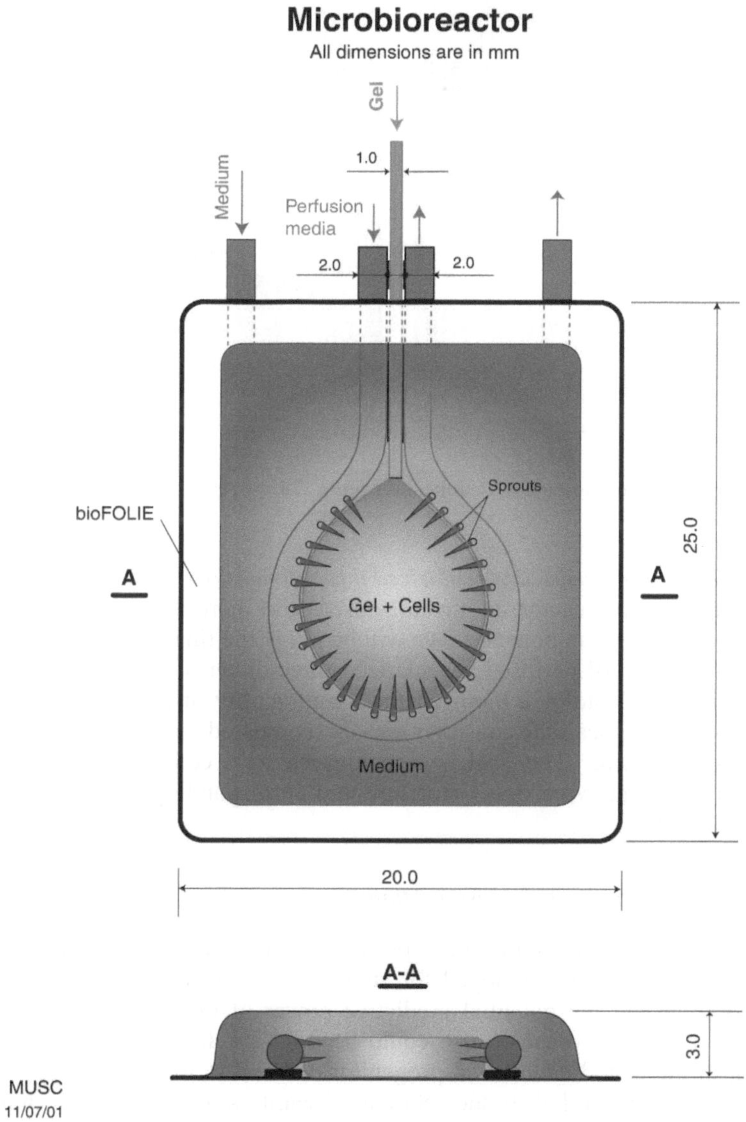

Figure 9. Schematic of the mechanism of endothelial sprouting from a hollow fiber tube.

Initially we were able to create a very simple chamber with loop like hollow fiber tube (Figure 10-11).

However, the ultimate goal is to create a perfusion chamber the size and shape of a standard histological slide. It will be very compatible with automatic reader and image analysis systems. The manufacture of the first prototype of perfusion minibioreactor demonstrates that proposed design is technically achievable, but need further optimization. Our goal was to make it simple, cheap, user-friendly, easy operated and disposable.

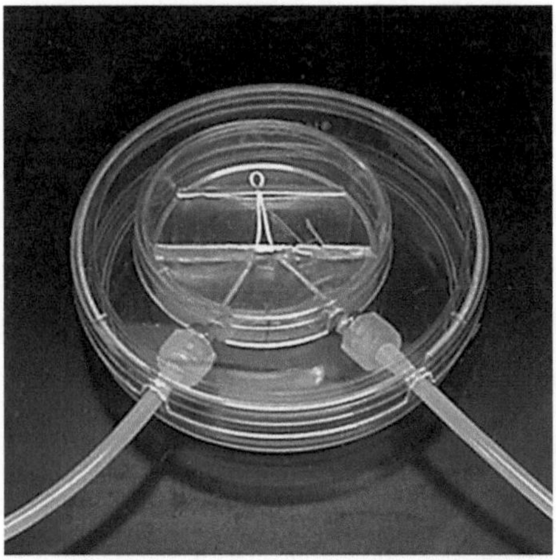

Figure 10. Perfusion minibioreactor chamber

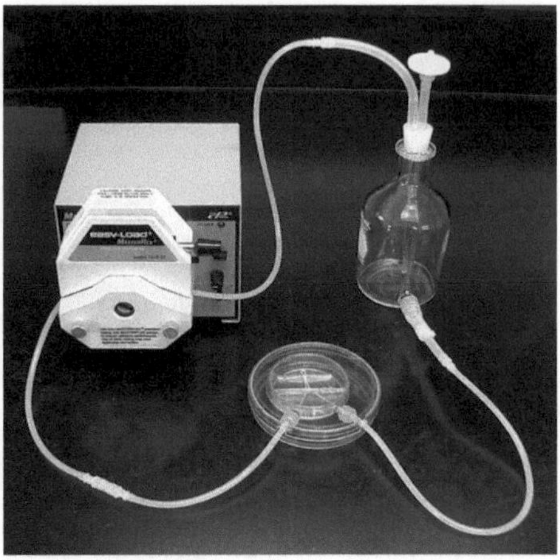

Figure 11. Overall view of minibioreactor system

Step 5. Fabrication of perfused vascularized three-dimensional myocardial tissue.

Once the microbioreactor is manufactured and porous hollow fiber tubes optimized for sprouting angiogenesis we will be ready the final step of our proposed research - fabrication of perfused vascularized three-dimensional myocardial tissue construct. Viability, histology, biomechanics, contractility of perfused vascularized three-dimensional myocardial construct will be compared to embryonic, newborn and adult myocardium. The goal is to achieve a 95% cell viability rate, and a myocardium with histotypic tissue structure and corresponding biomechanical and contractile properties.

3.2 Possible Applications Of Perfusion Minibioreactors

Bioassays using tissue-engineered, perfused vascularized three-dimensional myocardial tissues will be employed for studies on the mechanism of accelerated tissue maturation and the potential predictive power of this tissue-based assay in drug testing. The minibioreactor also could be used for direct visualization of the recruitment of circulated endothelial progenitor cell and stem cells into myocardial vasculature and myocardial tissue. Similar "seamless polymer-tissue junction" design principles could be employed for upscaled type of hollow fiber bioreactor, which will allow to growth larger mass of vascularized myocardial, skeletal tissue or any other tissue. However, the design must take into account the fact that living myocardial tissue is contractile tissue. It will create additional challenge for designers. Utilization of special elastic polymers can be used to overcome this technical barrier.

3.3. Perfusion Bioreactor For Tissue Engineered Tubular Myocardial Constructs

Early in the evolution and ontogenesis heart has a geometrical form of tube. Thus, it is logical to assume that cardiovascular tissue engineers will eventually try to recapitulate this form. Recently developed cell sheet technology, centrifugal casting and organ printing allows at least theoretically to achieve this goal. The Auger group from Quebec has demonstrated that rolling of cohesive smooth muscle cell sheet resulted in the formation of functional vascular tube (L'Heureux et al., 1998). It is already shown that several sheet of myocardial monolayers can fused and behave as functional unit (Shimizu et al., 2002). Okano's group (personal communication) has already been able to fabricate a myocardial tube by rolling several preliminary fused myocardial sheets. (Figure 12). However, without adequate vascularization and perfusion, relatively thick myocardial tissue will not survive. Even the presence of a non-perfused microvascular network will not guarantee survival. In order to fabricate a sustainable living myocardial tube, a system of intraconstruct perfusion must be designed and fabricated. Theoretical and mathematical modeling already allows the design of a vascular tree for myocardial tissue construct with any desired geometrical form. Combining the rolling of a myocardial sheet and a smooth muscle sheet will be eventually possible to create a vascularized myocardial tube suitable for intraconstruct perfusion. The nest step will be to create a bioreactor with a triple perfusion system. One perfusion circuit is used to necessary for the perfused tube, second one is used to maintain wet environment and the third one for perfusing the intraconstruct vasculature. The principal scheme of the triple perfusion bioreactor is presented in Figure 12. Another challenge is how to fix this fragile myocardial tube to perfusion inlet and outlet tube. We suggest the use of an intravascular balloon as mechanism for initial fixation, and maintaning intratubular perfusion.

The first prototype of a triple perfusion bioreactor is under development. The rapid development of tissue engineering technology does not provide time for delay. Moreover we strongly believe that development of the bioreactor technology and tissue-engineering technology must be parallel. What is the point of creating a tissue engineered myocardial tubular construct that has no chance of long-term survival. Additionally, a special design requirement for the triple perfusion bioreactor is its ability to produce longitudinal strain on the tissue engineered myocardial tubular construct .The perfused vascularized contractile tissue engineered myocardial tubular construct will be an important milestone in the development of a tissue engineered ventricular assist device. The demand for such devices in pediatric cardiothoracic surgery is very strong. The absence of a device forces pediatric cardiovascular surgeons to perform several very invasive and costly procedures on their growing patients. This an urgent priority in the emerging field of cardiovascular tissue engineering.

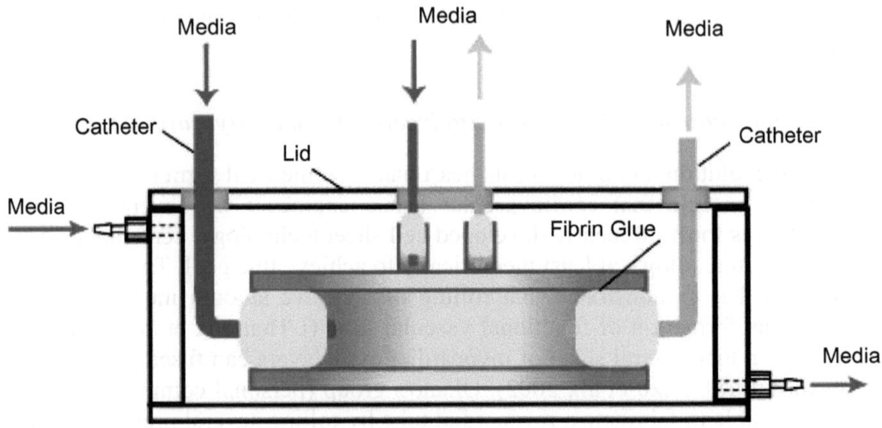

Figure 12. Scheme of triple perfusion bioreactor for tissue engineered ventricular assist device.

3.4. Perfusion Bioreactor For Printed Heart

The ultimate goal of cardiovascular tissue engineering is fabrication a complete human heart, or as leading Canadian tissue engineer Michael Sefton once put it "a heart in the box". The race for accomplishment this ambitious goal is already started. According our knowledge already several well-equipped and funded groups in USA, Europe, Israel and Japan are already involved in this competition. In some aspects the recent situation resembles the competition for designing a total artificial heart. It is obvious that only a multidisciplinary and long-term funded and well-managed group can accomplish this goal. We believe that recently proposed concept of organ printing (Mironov et al., 2003) offers the most realistic approach for engineering a living human heart. Our preliminary data strongly suggest that printing of the branching vascular tree is feasible, therefore a printed heart can be is viable and could be a subject of further growth remodeling and conditioning. Only a carefully designed post processing based on sophisticated chemical and mechanical conditioning and accelerated tissue and organ maturation will allow the creation of a mature beating functional human "heart in the box". The fact that printed heart is not a completely mature organ provides a special requirement for design of perfusion bioreactor. First of all it must provide maximal flexibility for applying different regime of mechanical and chemical conditioning. Second and most important perfusion bioreactor must have "built in " capacities for noninvasive biomonitoring of tissue and organ structural and functional maturation with the constant collection and monitoring of circulated biochemical markers of tissue maturation. Finally, capacity for controlled injection circulated stem and progenitor cells must be also carefully considered. One of easiest ways to incorporate the bioreactor biomonitoring system is by making the perfused printed tissue engineered heart suitable for ultrasound analysis. A draft scheme of a bioreactor with ultrasound based biomonitoring system is shown in Figure 13.

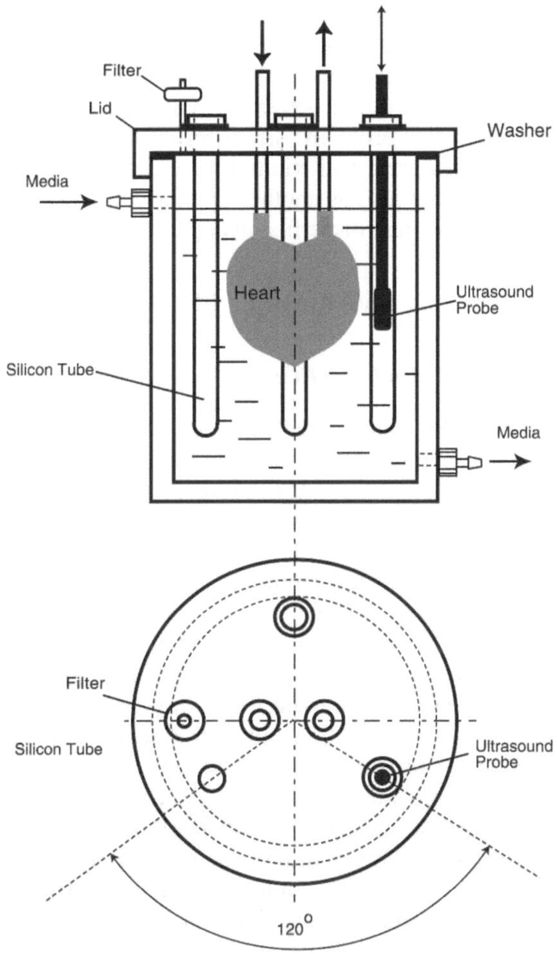

Figure 13. Scheme of pefusion bioreactor for printed tissue engineered heart with capacities for ultrasound biomonitoring.

Although functional capacities of modern ultrasound technologies for precise estimation level of maturity of printed and engineered myocardial tissue and heart are still limited, they are constantly being improved. A recently proposed process was to use contrast-enhanced ultrasound (CEU) with microbubbles targeted to alpha (v)-integrins, to allow visualization of high-resolution intraorgan angiogenesis

(Leong-Poi et al., 2003). Another example is ultrasound based elasticity imaging to allow estimation of collagen accumulation and cross-linking (Kanai et al., 2003). Although today ultrasound technology is characterized by lower resolution, it is safe to predict that similar new technologies will be developed in the future. Other powerful clinical bioimaging technologies such as MRI are also subjects of constant improvement and innovations. In any case, whatever tissue engineering technology will be used, the optimal design of perfusion bioreactor is one the prerequisite for the effective engineering of living heart. Although the engineered heart is considered by some not to be in the near future, several groups around the world are already systematically working on this problem. Engineering living "heart in the box" will be considered not only as a great and triumph of emerging tissue engineering and organ printing technology, but also as another great accomplishment in the long and successful history of development of perfusion apparatus and bioreactors.

4. COMMERCIALLY AVAILABLE PERFUSION BIOREACTORS

4.1 TEMS™ BioReactor™ (EnduraTEC)

The EnduraTEC ITEMS (Intelligent Tissue Engineering via Mechanical Stimulation) BioReactor (Figure 14) is a scalable biodynamic environmental system that provides a complete closed-loop nutrient-flow subsystem capable of introducing mechanical stresses and strains to a tissue engineered construct during its growth cycle. In addition to mechanical stimulation, the researcher can query the tissue construct to evaluate its mechanical condition. By using EnduraTEC's WinTest® software, the tissue's complex modulus can be estimated. This procedure allows adjustment of the stimulation profile to best meet the desired growth plan for the chosen tissue engineered construct. Using the ITEMS BioReactor, a researcher can seed the construct and then apply pressure/flow waveforms of varying shapes and frequencies to stimulate cell proliferation and matrix deposition leading to constructs with properties suitable for use in humans. The waveforms are generated using a computer-controlled dynamic pumping system and allows experimentation with different pressure/flow conditions. The ITEMS BioReactor is available in a single chamber design or in a six-chamber design for development of tissue engineered blood vessels, grafts, or valves. The EnduraTEC ITEMS BioReactor provides a nondestructive means for stimulating the tissue and evaluating the strength of the developing tissue construct during its growth cycle. The six-station ITEMS BioReactor represents an advancement in bioreactor technology in that several tissues can be developed simultaneously, allowing for more thorough and quantitative analysis of how a bioreactor environment enhances tissue growth. In addition, the researcher can address important bioprocess engineering issues such as scale-up and tissue uniformity. These criteria will become important as tissue engineered medical products move from the research laboratory to a production environment.

Figure 14. Commercial perfusion bioreactor ITEMS™ BioReactor™

4.2 LIFEPORT™ (Organ Recovery Systems, Inc.)

LifePort™ (Figure 15) has introduced a new class of devices designed to establish a continuum of organ care, spanning the critical time between donation and transplantation. The first in the new class is the LifePort™ Kidney Transporter. Innovative, durable, and exceptionally portable, the LifePort™ Kidney Transporter offers a new tool for assessing and treating traditional, expanded criteria, and nonheartbeating donor kidneys. LifePort™ represents the new state-of-the-art in hypothermic organ perfusion, and a new standard of care for transplantable kidneys. LifePort™ Liver, Pancreas and Heart Perfusion Transporters in development will share the same proprietary technology platform—an integrated, precision-engineered system supported by organ-specific accessories and solutions.

Pre-clinical trials required by the FDA as a prerequisite for submission were successfully completed at the University of Wisconsin. Canine kidneys were preserved successfully for 48 hours in the LifePort™ Kidney Transporter, and then transplanted with no complications. Adapted and modified LifePort™ could be used in cardiovascular tissue engineering or at least be a good prototype for designing research perfusion bioreactor.

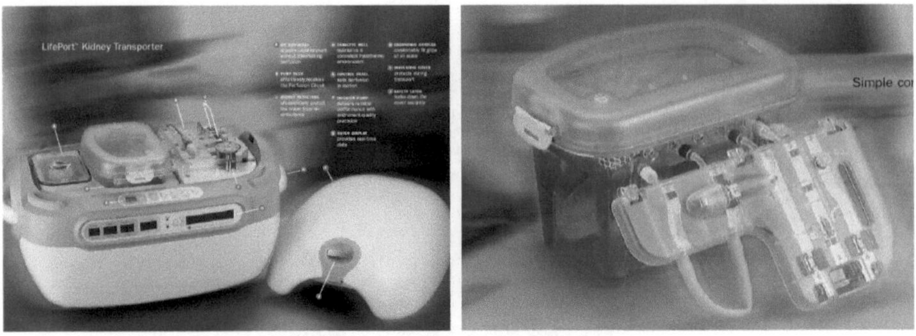

Figure 15. Chamber of the commercial perfusion bioreactor LifePort™

4.3 POPS™ (TransMedics, Inc)

Portable Organ Preservation System (TM) (Figure 16) is an experimental organ preservation and transportation system under development by TransMedics, Inc. of Woburn, The POPS maintains the human heart in its normal functioning physiologic state, with continuous blood flow. The system weigh is approximately 70 pounds and is shaped like a box that can fit in an ambulance or private jet. The system has a four-hour battery, built-in handles, and detachable wheels. It consists of a portable electro-perfusion device, biocompatible organ-specific disposable components, and proprietary chemical solutions that bathe the organ. A smaller unit, weighing about 55 pounds, could fit on the seat of a commercial airplane. It is shaped like a box that can fit on an airline seat. It has a four-hour battery, built-in handles and detachable wheels. Using current methods -- whereby organs are packed in coolers filled with ice and special solutions -- organs can only be preserved safely for limited periods of

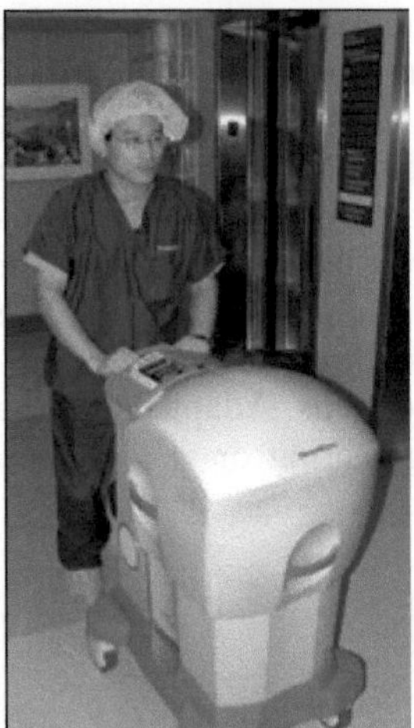

Figure 16. Commercial perfusion bioreactor POPS™

time. The cold ischemic time for the heart is the shortest -- once it is removed from the donor and has no blood supply, it has a "shelf life" of approximately six hours. Current technology provides only a small window of opportunity to transport and transplant an organ, thereby greatly limiting the availability of organs to those in need. Transplantation organ experts believe that a longer preservation time would allow the sharing of organs across greater distances, and more patients would benefit from life-saving transplants. POPS could provide transplant teams with sufficient time to perform a complete range of tests on organs that we currently do not consider suitable for transplantation, such as those from older donors or those with questionable function. Furthermore, perfusion with POPS may result in improved organ function thus 'resuscitating' previously unusable organs and might significantly increase the number of organs that can be used for transplantation. The new technique might also reduce or even eliminate a reperfusion injury to the donor organ that sometimes results when oxygenated blood is reintroduced into the organ during transplant surgery. The reoxygenation injury initiates a process of cell self-destruction compromising the function of the graft. POPS could serve as a prototype system for a perfusion bioreactor for a tissue engineered heart.

5. CONCLUSIONS

Cardiovascular tissue engineering is a fast growing field of tissue engineering. Further successful development of this field will require both the optimization of existing perfusion bioreactors and development of a new generation. Existing perfusion apparatus and systems must be adapted to specific requirement of tissue engineered tissue constructs. The broad spectrum of perfusion bioreactors from perfusion minibioreactors for 3D vascularized tissue constructs to whole organ perfusion apparatus with incorporated system of biomonitoring of tissue and organ maturation needs to be developed. A multidisciplinary approach based on closed collaboration of tissue engineers, cardiovascular biologists, physiologists, cardiologists, biomechanical engineers and software specialists will be necessary to designing high performing perfusion bioreactors for cardiovascular tissue engineering. We strongly believe that new beautiful Lindbergh-Carrel Prize for outstanding contribution for development perfusion technology for organ preservation and growth designed by famous Italian sculptor Carlo Zoli as result of a joint initiative between the Charles and Anne Lindbergh Foundation (USA), the Alexis Carrel Foundation (Italy) and the Medical University of South Carolina is an appropriate way to celebrate the contribution to this important field of biomedical engineering by the new generation of inventors and innovators (Figure 17).

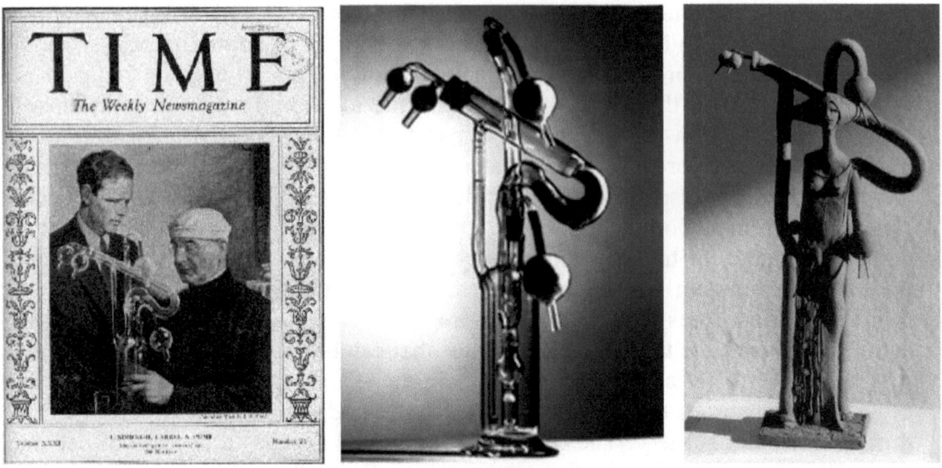

Figure 17. (left) Lindbergh and Carrel with their Oxigenator (picture from front page of journal "Time"); (middle) Lindbergh-Carrel Oxigenator; (right) Lindbergh-Carrel Prize designed by Italian postmodernist surrealist sculptor Carlo Zoli (some experts call Carlo Zoli - "Salvador Dali in third dimension"). Woman depicted in sculpture is Charles Lindbergh's sister-in-law Elizabeth Monroe, who died from an inoperable congenital heart disease. This death was the motivation for Lindbergh to develop the perfusion apparatus and oxygenator, seen behind Elizabeth, which would allow open-heart surgery.

6. ACKNOWLEDGEMENTS

This research is supported by the NASA/EPSCOR grant "Cardiovascular Tissue Engineering". and COBRE grant 1P20 RR 16434 from the National Research Resource Center (NIH)

7. REFERENCES

Carrier RL, Rupnick M, Langer R, Schoen FJ, Freed LE, Vunjak-Novakovic G. 2002. Perfusion improves tissue architecture of engineered cardiac muscle. Tissue Eng 8:175-188.

Carrier RL, Papadaki M, Rupnick M, Schoen FJ, Bursac N, Langer R, Freed LE, Vunjak-Novakovic G. 1999. Cardiac tissue engineering: cell seeding, cultivation parameters, and tissue construct characterization. Biotechnol Bioeng 64:580-589.

Chesler NC, Conklin BS, Han H-C, Ku D: 1998. Simplified ex vivo artery culture techniques for porcine arteries. J Vascular Invest 4: 123–127.

Conklin BS, Surowiec SM, Lin PH, Chen C. 2000. A simple physiologic pulsatile perfusion system for the study of intact vascular tissue. Med Engineering and Physics 22: 441-449.

Davis GE, Camarillo CW. 1996. An alpha 2 beta 1 integrin-dependent pinocytic mechanism involving intracellular vacuole formation and coalescence regulates capillary lumen and tube formation in three-dimensional collagen matrix. Exp Cell Res. 224:39-51.

Eschenhagen T, Didie M, Munzel F, Schubert P, Schneiderbanger K, Zimmermann WH. 2002. 3D engineered heart tissue for replacement therapy. Basic Res Cardiol. 97 Suppl 1:I146-52.

Folkman J. 1971. Tumor angiogenesis: therapeutic implications. N Engl J Med. 285: 1182-1186.

Hibino N, Imai Y, Shin-oka T, Aoki M, Watanabe M, Kosaka Y, Matsumura G, Konuma T, Toyama S, Murata A, Naito Y, Miyake T. 2002. First successful clinical application of tissue engineered blood vessel. Kyobu Geka. 55: 368-373. (Japanese)

Hoerstrup SP, Zund G, Sodian R, Schnell AM, Grunenfelder J, Turina MI. 2001. Tissue engineering of small caliber vascular grafts. Europ J Cardio-thoracic Surg 20: 164-169.

Jackson ZS, Gotlieb AI, Langille BL. 2002. Wall tissue remodeling regulates longitudinal tension in arteries. Circ Res 90: 918–925.

Jockenhoevel S, Zundo G, Hoerstrup SP, Schnell A, Turina M: 2002. Cardiovascular tissue engineering: a new laminar flow chamber for in vitro improvement of mechanical tissue properties. ASAIO Journal 48: 8-11.

Kasyanov V. 1974. Anisotropic nonlinear-elastic model of the large human blood vessels. Polymer Mechanics 10: 756-764.

Kaushal S, Amiel GE, Guleserian K J, Shapira OM, Perry T, Sutherland FW, Rabkin E, Moran AM, Schoen FJ, Atala A, Soker S, Bischoff J, Mayer JE. 2001. Functional small-diameter neovessels created using endothelial progenitor cells expanded ex vivo. Nature Medicine 7: 1035-1040.

Kim B-S, Nikolovski J, Banadio J, Mooney DJ. 1999. Cyclic mechanical strain regulates the development of engineered smooth muscle tissue. Nature Biotechnology 17: 979-983.

Kolpakov V, Rekhter MD, Gordon D, Wang WH, Kulik TJ. 1995. Effect of mechanical forces on growth and matrix protein synthesis in the in vitro pulmonary artery analysis of the role of individual cell types. Circulation Research 77: 823-831.

Ku DN. 1997. Blood flow in arteries. Ann Rev Fluid Mech 29: 399-434.

Learoyd BM, Taylor MG. 1966. Alterations with age in the viscoelastic properties of human arterial walls. Circ Res 18: 278–292.

L'Heureux N, Germain L, Labbe R, Auger FA. 1993. In vivo construction of a human blood vessel from cultured vascular cells: a morphologic study. J.Vasc.Surg 17: 499-509.

L'Heureux N, Paquet S, Labbe R, Germain L, Auger FA. 1998. A completely biological tissue-engineered human blood vessel. FASEB J. 12: 47-56.

Mangiarua EI, Moss N, Lemke SM, McCumbee WD, Szarek JL, Gruetter CA. 1992. Morphological and contractile characteristics of rat aortae perfused for 3 or 6 days in vitro. Artery 19: 14-38.

Mironov V, Boland T, Trusk T, Forgacs G, Markwald RR. 2003. Organ printing: computer-aided jet-based 3D tissue engineering. Trends Biotechnol. 21:157-61.

Mironov V, Kasyanov V, McAllister K, Oliver S, Sistino J, Markwald R. 2003. Perfusion bioreactor for vascular tissue engineering with capacities for longitudinal stretch. J Craniofac Surg.14: 340-347.

Naito Y, Imai Y, Shin'oka T, Kashiwagi J, Aoki M, Watanabe M, Matsumura G, Kosaka Y, Konuma T,Hibino N, Murata A, Miyake T, Kurosawa H. 2003. Successful clinical application of tissue-engineered graft for extracardiac Fontan operation. J Thorac Cardiovasc Surg. 125:419-20.

Nerem RM, Seliktar D. 2001. Vascular tissue engineering. Annu Rev Biomed Eng 3: 225-243.Niklason LE, Gao J, Abbott WM, Hirschi KK, Houses S, Marini R, Langer R. 1999. Functional arteries grown in vitro. Science 284: 489-493.

Salacinski HJ, Goldner S, Giudiceandrea A, Hamilton G, Seifalian AM. 2001. The mechanical behavior of vascular graft: a review. J Biomat Applic 15: 241-278.

Seliktar D, Black RA, Vito RP, Nerem RM. 2000. Dynamic mechanical conditioning of collagen-gel blood vessel constructs including remodeling in vitro. Ann. Biomed. Engineering 28: 351-362.

Surowiec SM, Conklin BS, Li JS, Lin PH, Weiss VJ, Lumsden AB, Chen C. 2000. A new perfusion culture system used to study human vein. J Surg Res 88: 34–41.

Vacanti JP, Langer R. 1999. Tissue engineering: the design and fabrication of living replacement devices for surgical reconstruction and transplantation. Lancet 345: 32-34.

Weinberg C, Bell E. 1986. A blood vessel model constructed from collagen and cultured vascular cells. Science 231: 397-400.

CHAPTER 14

HAEMATOPOIETIC CULTURE SYSTEMS

L. SAFINIA[1], N. PANOSKALTSIS[2] AND A. MANTALARIS[1]

[1]*Department of Chemical Engineering & Chemical Technology, Imperial College London, London, UK*

[2]*Department of Haematology, Imperial College London, London, UK*

1. HAEMATOPOIESIS AND BONE MARROW

Bone marrow (BM), located in the medullary cavity of bone, is the haematopoietic tissue and a primary lymphoid organ. In healthy adults, it produces about 2.5 billion red blood cells, 2.5 billion platelets, and 1 billion granulocytes per kilogram of body weight per day, but the production rate may fluctuate from nearly zero to many times normal according to the physiological need (Abboud and Lichtman 2001). The bone marrow has an elaborate three-dimensional architecture composed of the haematopoietic cells, the supporting stroma (reticular cells, osteocytes, adipocytes, vascular endothelium) and the extracellular matrix (ECM) (Charbord 2001). This architecture supports haematopoiesis, a continual process that is commonly referred to in three-stages: proliferation, commitment, and differentiation. Within these three stages, extensive expansion and maturation of the haematopoietic cells occurs, and a single HSC may be capable of more than 50 cell divisions and generate up to 10^{15} cells (McNiece and Briddell 2001). The haematopoietic inductive microenvironment (HIM) regulates these processes (Trentin 1970) through the unique 3-D structural and chemical microenvironment created by the stromal and haematopoietic cells and the ECM, and the positive and negative growth factors these cells secrete, that regulate the survival, proliferation, and differentiation of HSCs (Figure 1).

The fate of HSCs may be determined by specific microenvironments (Weissman 1994). According to this model, rare niches within the HIM promote the self-renewal rather than the differentiation of HSCs and it is the space within each niche

J.B. Chaudhuri & M. Al-Rubeai (eds.), Bioreactors for Tissue Engineering, 309 - 334.

that limits further self-renewal. When the stem cell niche is full, the excess is pushed into an adjacent niche with a microenvironment that promotes commitment,

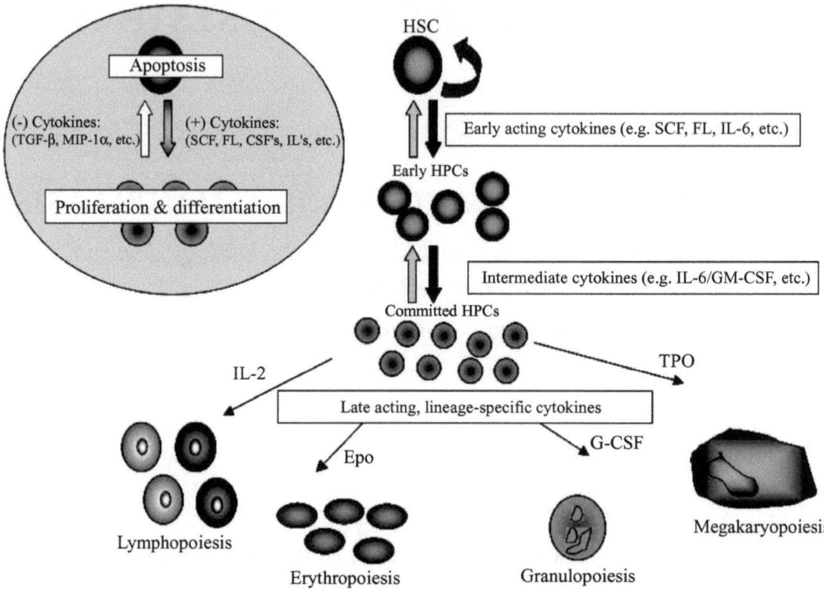

Figure 1. The haematopoietic system hierarchy. Haematopoiesis is a three-stage process that includes proliferation, commitment, and differentiation. It is modulated by the haematopoietic inductive microenvironment, which includes the cytokine networks.

differentiation and maturation of blood cells that can egress from the marrow into the peripheral blood via the marrow sinusoids. Microenvironmental niches that selectively regulate cell populations may be due to the presence and/or absence of specific positive or negative growth factors, as well as concentration gradients of membrane-bound cytokines that are able to instruct HSCs to differentiate or remain quiescent (Aiuti et al. 1998; Quesenberry et al. 1994). Unfortunately, these interactions and the proteins responsible for them are difficult to define in commonly used 2-D culture systems since the 3-D microenvironment found *in vivo* is difficult to mimic *ex vivo* (Mantalaris et al. 1998).

2. HAEMATOPOIETIC STEM/PROGENITOR CELL ASSAYS

The largest impediment to the successful *ex vivo* expansion of HSCs has been the difficulty in accurately identifying and enumerating true HSCs. Although the "gold standard" assay that defines the HSC is the ability of a single such cell to reconstitute long-term haematopoiesis in the recipient host (Quesenberry and Colvin

2001), a variety of *in vitro* and *in vivo* assays have been developed to help identify stem cells in other ways.

2.1. In vitro Assays

Flow cytometry may sort and further characterise HSCs from an enriched population based on the surface and intracellular phenotype. This procedure is both rapid and relatively inexpensive for defining HSCs both prior to and following expansion cultures (Madlambayan et al. 2001). The cell-surface glycoprotein cluster designation (CD)34 was one of the first surface markers that could easily define human cells with haematopoietic potential (Krause et al. 1996). Haematopoietic stem cells may be identified within $CD34^+$ subsets (Shih et al. 2002), such as those expressing $CD34^+/CD38^-$ or $CD34^+$/Thy-1^+ phenotypes (Craig et al. 1993; Sutherland et al. 1990). However, HSCs can also be identified within the $CD34^-$ cell subset. Thus, characterisation based on surface phenotype alone is not sufficient to define the HSC (Osawa et al. 1996). Fluorescent dyes that indicate cell proliferative status have become a method of choice for identifying and isolating HSCs, in particular when combined with cell-surface markers. One of these dyes, Hoechst-33342, can define a small subset of murine and human BM cells (Side Population or SP), which are enriched for HSCs capable of long-term haematopoietic reconstitution (Goodell et al. 1996). Rhodamine-123, a mitochondrial dye, can also identify these primitive stem cells by flow cytometry (Spangrude and Johnson 1990). However, the identification of HSC phenotype and function or genotype is based on studies of cells prior to culture. This relationship is not dependably preserved *in vitro* (Zandstra et al. 1997) and therefore the usefulness of these markers as predictors of haematopoietic potential following *in vitro* culture is unclear (Shih et al. 2002).

Functional assays for HSCs *in vitro* can detect and quantify lineage-restricted as well as multi-, bi-, and mono-potential subpopulations or colonies derived from haematopoietic progenitor cells. Specific factors are required in the supportive medium, depending on the type of colony-forming potential being evaluated. Committed haematopoietic progenitor cells form colonies of differentiated cells on semi-solid culture media and the number and type of progenitors can be enumerated by these colony-forming unit (CFU), burst-forming unit (BFU), and high-proliferative potential colony-forming cell (HPP-CFC) assays (Eaves and Eaves 1997; Williams et al. 1987) with the suffix indicating the lineage commitment of the colony. For example, colony-forming unit-granulocte/macrophage (CFU-GM) is a progenitor committed to the granulocytic and macrophage lineages. Despite the potential for multi-lineage commitment of some of these progenitors, they have a limited capacity for self-renewal *in vitro* and therefore cannot be defined as HSCs based on these characteristics alone. By contrast, the long-term culture-initiating cell (LTC-IC) assay fulfils the HSC criterion of being able to initiate and sustain myelopoiesis, but not the two other criteria of self-renewal and long-term lymphopoiesis (Verfaillie 2001). Although the LTC-IC is better at quantifying haematopoietic cells that appear to be more primitive than most colony forming cells

(CFCs), it is also more complicated in that it requires co-culture of haematopoietic cells with a supportive stroma feeder layer, with or without exogenous cytokines (Sutherland et al. 1990) .

2.2. In vivo Assays

Similar to the *in vitro* assays, the *in vivo* HSC functional assays can also be classified by the ability of the cells to generate colonies (colony-forming unit-spleen or CFU-S) and the ability to reconstitute long-term lympho-haematopoiesis (long-term repopulating assay or LTRA) in a recipient. The CFU-S assay evaluates the capability of bone marrow or spleen cells from a donor to form visible colonies on the surface of the spleen of lethally irradiated recipients. The spleens of the recipients are harvested and analysed 8 ($CFU\text{-}S_8$) and 12 ($CFU\text{-}S_{12}$) days following the bone marrow or splenic stem cell transplantation (Till and McCulloch 1961). The colonies appearing on the spleen surface after 8 days ($CFU\text{-}S_8$) do not arise from HSCs and represent predominantly uni-potential precursor cells, whereas CFU-S_{12} are derived predominantly from multi-potential haematopoietic progenitors and HSCs. By contrast, the LTRA is the best method by which to measure HSC activity since donor cells must show the capacity to self-renew, differentiate, and reconstitute haematopoiesis in the transplanted host. Although in mice, syngeneic donor/recipient pairs may be used in order to avoid rejection of HSCs, the assay of human HSCs necessitates the use of animals that are incapable of rejecting human cells (Civin et al. 1996; Larochelle et al. 1996; McCune et al. 1998). These transplants that use xenogeneic microenvironments in order to assay human HSCs may not enumerate all human HSCs since the xenogeneic host marrow may not be as effective as the human microenvironment at homing, growth and differentiation of human HSC (Verfaillie 2001).

3. PROCESS DESIGN PARAMETERS FOR HAEMATOPOIETIC CULTURES

Haematopoiesis is a dynamic process regulated *in vivo* by the haematopoietic microenvironment and controlled by local parameters such as pH, oxygen concentration, nutrient supply and demand, concentration and type of growth factors, and the cellular and molecular environment surrounding the cells (Noll 2002). Although defining the dynamic interactions between all of these components can be quite challenging, it is only by doing so that optimal culture parameters may be identified for the *ex vivo* expansion of haematopoietic stem and progenitor cells for human clinical applications (Nielsen 1999).

3.1. Cell Source and Inoculum Density

Human haematopoietic stem and progenitor cells can be derived for clinical use from bone marrow, mobilised peripheral blood (MPB), and umbilical cord blood (CB) from either autologous (from the patient) or allogeneic (from a related or unrelated donor) sources (for the various collection procedures refer to (Collins 1996; McAdams 1995; Nielsen 1999)). The best source and starting population of cells is unknown (McAdams 1996). Although most experience in clinical haematopoietic stem cell transplantation has been with BM and MPB, recently, CB has also been used as an alternative source of HSCs for the allogeneic transplantation of patients who require but lack an HLA-matched stem cell donor (approximately 30% of patients who need a transplant actually have a compatible donor) (Cairo and Wagner 1997). In contrast to BM and MPB sources, CB is readily available, easily collected and banked and potentially available to groups underrepresented in transplant registries. CB may also permit a greater degree of mismatch between donor and recipient due to the naiveté of the immune cells within the graft (De La Selle et al. 1996) and has the least amount of viral contamination with common pathogens compared with other stem cell sources (McAdams 1996). The main disadvantage of CB as a HSC source is the small volume collected (~100 ml) and, although the product is enriched in stem and progenitor cells and enough to engraft children (average weight of 20kg), the total number of stem cells is insufficient for adult engraftment. The successful expansion of HSCs from CB *ex vivo* without changing the phenotype of the cells would provide a donor source for haematopoietic stem cells and other haematopoietic components for regular transfusion as well as a stem cell source for the potential renewal of other tissues.

Since the starting population in any raw stem cell source is heterogeneous, haematopoietic stem and precursor cells may be enriched prior to *in vivo* and *in vitro* use (McAdams 1995). However, enriched $CD34^+$ cell subsets derived from different cell sources may not have equivalent capacities for expansion (Kim 1999; Mayani 1998). For example, although there are fewer $CD34^+$ cells in a CB graft than in a BM graft, CB $CD34^+$ subsets produce more CFCs that have a greater capacity to proliferate when compared with those derived from BM (Kim 1999). Most stem cell transplant centres use whole mononuclear cells (MNCs) from BM, MPB and CB and some centres do enrich for $CD34^+$ cells prior to transplant in pursuit of a "cleaner" product with fewer T cells in an allograft and possibly fewer tumour cells in the case of an autograft (Koller 1993; McAdams 1995; Poloni et al. 1997). For *ex vivo* expansion of HSCs, a bioreactor culture will support equal numbers of total cells and CFU-GMs with similar volumetric culture requirements, regardless of the extent of purification of the input cells (Koller et al. 1995b; McAdams 1995). Compared with an enriched product or one that has been manipulated *ex vivo*, transplantation of BM MNCs may be better in that there is rapid neutrophil and platelet engraftment as well as long-term stem cell engraftment, possibly due to the presence of "accessory cells" in the MNCs (Bachier et al. 1999; Koller et al. 1999). Although there was a 10-fold expansion of CFU-GM in this study, it was countered by a reduction of LTC-IC and $CD34^+/Lin^-$ cells to half or less of the input amount which may have accounted for the inferior engraftment kinetics (Bachier et al. 1999; Koller et al. 1999).

Seeding density, growth factor supplements, and the rate of medium exchange are the most important factors that influence progenitor cell output in long-term haematopoietic cultures (Zandstra et al. 1994). In addition to these factors, knowledge of the composition of stem, progenitor, and accessory cells in the culture is required since consumption of nutrients and growth factors varies according to the type of responding cell or the specific lineage of the progenitors (Zandstra et al. 1997). These components are especially pertinent in defining conditions for transport-limited (static cultures) and homogeneous environments (stirred suspension cultures). In human BM MNC static cultures, a high seeding density (e.g. 5×10^5 cells/well) produced higher total numbers of cells and CFCs, whereas a low seeding density (e.g. 5×10^4 cells/well) increased the expansion capacity of the culture but also resulted in greater depletion of CFCs (Koller 1996). By decreasing the seeding density from 3×10^4 to 1.5×10^3/ml, there was enhanced expansion of total cells and progenitor cells by up to 20-fold and 3-fold respectively, yet LTC-ICs were maintained (Poloni et al. 1997). Furthermore, when BM MNCs were cultured at sufficiently high cell densities (e.g. 5×10^5 cells/well) the culture could maintain and expand LTC-IC independent of the effects of pre-formed stroma (Koller 1996). When a lower seeding density of MNCs was used (e.g. 5×10^4 cells/well), there was variability in culture output and the cells depended on support provided by accessory stromal cells, similar to that seen in cultures of $CD34^+$-enriched cells. Homogenous environments provided by stirred suspension bioreactor cultures are also dependent on seeding density, with the highest cell density ($0.1\text{-}1 \times 10^6$ cells/ml) of BM MNCs resulting in a significant net expansion of primitive haematopoietic progenitor cells (Zandstra et al. 1994). The range of seeding density can vary depending on the culture system, the use of preformed stromal 'feeder' layer, or the use of scaffolds with typical ranges for inocula in the order of $1\text{-}3 \times 10^6$ MNCs per ml and $0.5\text{-}1 \times 10^4$ $CD34^+$ cells per ml (Koller 1996; Koller et al. 1995b; McAdams 1995; Zandstra et al. 1994). Given this variability in the setting of different culture parameters, it is vital that individual laboratories determine optimal seeding density from a given starting population under specific experimental conditions.

3.2. Culture Media

Haematopoietic cultures are usually supplemented with foetal calf and horse sera (about 25% v/v) or bovine serum albumin as well as with a cocktail of cytokines (Alley 1983; Bohmer 1989; Douay 2001). These xenogenic sera enhance culture performance since they are a source of essential nutrients, hormones and growth factors, induce or enhance the action of cytokines (Guba 1992; Tsukada 1992), protect the cell against damage (Collins 1996), and influence the growth and survival of haematopoietic progenitor cells (Bohmer 1989). However, the proteins in serum may also negatively affect the physiochemical properties and kinetics of the culture by interfering with components of cell signalling (Table 1). For example, transforming growth factor-β (TGF-ß) found in serum facilitates the expansion of granulocytic and monocytic lineages (Lill et al. 1994) yet inhibits that of erythroid and megakaryocytic lineages (Dybedal and Jacobsen 1995; Fortunel 2000; Krystal et al. 1994) from cultures of $CD34^+$ BM cells. For clinical applications, the addition of xenogenic serum poses a risk of contamination by foreign and/or infectious proteins and lacks the definition and standardisation required of media components for human use . Autologous serum or plasma used at low concentrations (1-2%) may be a viable alternative for animal serum since it can support the expansion of haematopoietic cells (Brugger et al. 1993; Rosenzwajg et al. 1996).

Specially formulated serum-free media may also support the *ex vivo* expansion of HSCs and haematopoietic progenitor cells (HPCs) when supplemented with the appropriate cytokines, (Brown 1997). Although serum-containing media are better for total cell expansion, serum-free media enhance the clonogenic potential of human BM $CD34^+$ cells (Table 1) (Almeida-Porada 2000). Since serum aides the growth and maintenance of the stromal layer (McAdams 1996), serum-free media without other additives are unable to support the adherent cell layer, even with pre-treatment of the surface area of the culture with poly-D-lysin, collagen or fibronectin (Brown 1997). Hence, serum-free media are more suited for stroma-free cultures.

Table 1. Summary of the features of serum-containing and serum-free media for the cultivation of haematopoietic stem and progenitor cells.

Medium	Comments	Ref.
Serum	• Essential source of nutrients, hormones and growth factors • Suitable for stroma-dependent and stroma-independent cultures • Enhanced total cell number expansion • Ill-defined and poorly characterized; uncontrolled lot to lot variability • Restricted cell lineage differentiation due to the presence of TGF-β • Regulatory problems for clinical applications • Use of autologous plasma as replacement for animal sera	(Fortunel 2000; Guba 1992; Krystal et al. 1994; Noll 2002)
Serum-Free	• Chemically defined and simple composition • Enhanced culture uniformity and experimental reproducibility • Enhanced total clonogenic potential • Requires the addition of cytokines as well as serum replacement supplements • Universal serum-free formulation does not exist (limited number of complete formulation are available) • Suitable only for stroma-independent cultures	(Almeida-Porada 2000; Brown 1997; Collins 1996; McAdams 1996; Noll 2002)

One of the most salient features of any haematopoietic culture system is the combination, concentration and maintenance of cytokines it delivers to cells that express the relevant receptors within the culture over time, since these are the components that influence the cumulative response(s) obtained (Audet 1998). Cytokine receptors belong to three main groups: the tyrosine kinase receptors, the haematopoietic growth factor receptors, and the gp130 receptors (Audet 1998). c-kit, the receptor for stem cell factor (SCF) (Broudy 1997) and flk-2/flt3, the receptor for flt3 ligand (Lyman and Williams 1995), are examples of the tyrosine kinase receptors that affect the early stages of haematopoiesis since they are expressed on primitive haematopoietic cells. When combined with other stimulatory cytokines, SCF promotes the survival and growth of primitive haematopoietic progenitors (Brandt 1994); little expansion is observed in cultures supplemented with SCF alone (Ohmizono et al. 1997). Flt3 ligand may also mediate the expansion and maintenance of HSC (Lyman and Jacobsen 1998) by recruiting them into the cell cycle (Shah et al. 1996) as well as by promoting survival, likely by inhibiting apoptosis (Meyer and Drexler 1999).

Ligands for the cytokine receptor family include interleukin-3 (IL-3) and thrombopoietin (TPO), both of which are active on primitive progenitors. IL-3 is required, along with SCF and flt3 ligand, for optimal expansion of human LTC-IC from BM $CD34^+/CD38^-$ cells (Petzer et al. 1996). However, IL-3 may be detrimental to the pluripotentiality of murine progenitors (Matsunaga et al. 1998), and when added in inappropriately high concentration to human progenitors it induces terminal differentiation of HPCs (Zandstra et al. 1997). Similar to the actions of SCF, TPO supports the proliferation and long-term maintenance of primitive haematopoietic progenitors and repopulating HSCs when used in combination with other stimulatory cytokines or in the presence of stroma (Kobayashi et al. 1996; Won et al. 2000; Yagi et al. 1999) but has little effect on the expansion of HSCs in suspension cultures when used alone (Kobayashi et al. 1997).

The group of gp130 receptors includes receptors for IL-6, IL-11, oncostatin-M, and leukaemia inhibitory factor (LIF). IL-6 directly enhances the proliferation of haematopoietic cells and stimulates myeloid proliferation in synergy with other growth factors (Ogawa and Clark 1988). Stromal cells make IL-11 which, with SCF and flt3 ligand, can expand adult murine stem cells *ex vivo* (Holyoake et al. 1996). *Ex vivo* expansion of purified murine$CD34^+/Thy\text{-}1^+$ cells can be further enhanced with the actions of LIF on co-cultivated stromal cell lines which, with IL-3, IL-6, SCF, and granulocyte/macrophage colony stimulating factor (GM-CSF) resulted in a 150 fold expansion of cells retaining the $CD34^+/Thy\text{-}1^+$ phenotype (Shih et al. 2002).

Novel factors, not originally identified as a result of their effects on haematopoietic cells, may also have profound effects on HSC growth and differentiation. These factors include the bone morphogenetic proteins, which affect $CD34^+/CD38^-/Lin^-$ HSC (Bhatia et al. 1999), and the Notch ligand, Jagged-1 (Karanu et al. 2000). Although the role of inhibitory cytokines, such as TGF-β, macrophage inflammatory protein-1α (MIP-1α), and tumour necrosis factor-α (TNF-α) *in vivo* is unclear, *in vitro* these cytokines may cause a transient expansion of cells (Madlambayan et al. 2001; Verfaillie 2001). TGF-β was responsible for the progenitor cell quiescence observed 5 to 7 days after the last medium exchange in long-term cultures (Eaves et al. 1991) and MIP-1α added to cultures prevented cells from cycling (Broxmeyer et al. 1993). The addition of both MIP-1α and platelet factor-4 to human haematopoietic cultures containing IL-3 preserved LTC-ICs for at least 8 weeks (Verfaillie et al. 1994). TNF-α added to cultures stimulated with cytokines such as SCF and flt3 ligand could inhibit the proliferation of progenitor cells, likely by promoting apoptosis (Rusten et al. 1994; Veiby et al. 1996).

3.3. Culture Systems

A stroma-dependent long-term murine bone marrow culture system was first developed by Dexter and colleagues (Dexter et al. 1977) and later adapted for the human system (Gartner 1980). These conventional bone marrow cultures use flat, two-dimensional surfaces of tissue culture dishes. This culture system poses the major challenge for the *ex vivo* expansion of haematopoietic stem and progenitor cells: the provision of a culture environment that could adequately support the expansion of clinically relevant cell numbers of the desired cell type(s) from relatively small starting populations (Cabral 2001).

Two main approaches may be considered for *in vitro* haematopoiesis, namely stroma-free and stroma-dependent cultures. Although stroma-free cultures are easier to standardise and maintain (Verfaillie 2001), stroma-based cultures provide a more physiologic microenvironment to the haematopoietic stem and progenitor cells and can facilitate expansion of human CB $CD34^+$ cells (by 600-fold) and CFUs as well as cobblestone area-forming cells (CAFCs) (Kanai 2000; Koller 1992a). For clinical applications, stromal feeders have been used to support the retroviral transduction of autologous HSCs (Emmons et al. 1997) and to purge autologous stem cell products of cancer cells (leukaemia) (Barnett et al. 1994; Coutinho et al. 1990) *ex vivo*. Autologous feeders may be advantageous when used in cultured products intended for clinical use because the feeder cells don't contaminate the cultivated cells intended for human transplantation (Verfaillie 2001). However, defects in the stromal feeder cells of the donor due to *in vivo* exposure to chemotherapy and radiation (Mayani 1996) or to the disease itself (Bhatia et al. 1995) may limit the haematopoietic support that these cells can offer *in vitro*. Hence, murine or human stromal feeders have been cloned from a variety of sources to try to standardise the co-cultivation system (Verfaillie 2001).

Different devices for culture have been developed (Table 2). The most widely used of these devices for the expansion of HSCs are the static culture systems, which include well-plates, T-flasks and gas-permeable blood bags (Chabannon et al. 2000; Collins et al. 1998). Gas-permeable polypropylene bags are preferred for clinical applications since they provide a closed and easily manipulated system, although, in contrast to flask cultures, they do not support the development of stromal cells. Optimal seeding density for expansion of total cells, $CD34^+$ cells, clonal progenitors, CFU-GM, BFU-E and LTC-ICs in this system has been noted at concentrations of 5×10^4 $CD34^+$ BM cells/ml or 1×10^4 CB $CD34^+$ cells/ml using serum-free media in a 14-day culture (Douay 2001). When the seeding density was increased, the production of terminally differentiated granulocytes also increased whilst overall cell expansion decreased (Giarratana 1998). Despite their widespread use, static cultures are limited to relatively low-density cultures with low total cell output. These limitations are imposed on the system because of the inability of the user to readily monitor and control the culture microenvironment with parameters

Table 2. Summary of the culture systems used for haematopoietic and progenitor cells.

Culture System	Comments	Ref.
Static	• Spatial inhomogeneities • Lack of on-line monitoring and control • Lack of automation	(Chabannon et al. 2000; Collins et al. 1998)
Suspension (Stirred bioreactor)	• Homogeneous environment • Controlled, closed and reproducible • Shear stress limitations • Scaleable	(Sardonini and Wu 1993; Zandstra et al. 1994)
Perfusion		
Hollow-Fibre	• Shear-free • Spatial inhomogeneities • Difficulty in monitoring and control • Scale-up difficulties	(Sardonini and Wu 1993)
Perfusion chambers	• Relatively homogeneous • Problem with over-dilution • Medium-intensive culture system • Scale-up not straightforward • Problems with monitoring and control • Clinical application (Aastrom Biosciences)	(Collins 1996; Koller et al. 1993; Oh 1994; Palsson et al. 1993; Sandstrom et al. 1995)
Airlift	• High cell density growth • Bubble shear and breakage caused by gas sparging	(Highfill 1996; Sardonini and Wu 1993)

such as concentration gradients (pH, metabolites, etc.), the requirement for repeated handling during feeding, and the limited surface area available for expansion of HSCs (Chabannon et al. 2000; Collins et al. 1998). Bioreactors were developed to overcome these limitations by enabling better control of culture parameters, automated "hands off" feeding strategies with reduced risk of contamination (Cabral 2001; Koller 1993), and the ability to support high cell densities as well as accessory cells. At the point of writing this chapter, only one bioreactor system (Aastrom Biosciences; Ann Arbor, Michigan) - an automated perfusion system - has FDA approval in the USA and has obtained a CE mark in Europe (Koller 1993; Koller 1998).

Perfusion bioreactors remove harmful metabolites and maintain a constant supply of growth factors and nutrients by continually replacing culture medium. These bioreactors have also been used to analyse the consumption and production of specific growth factors in relation to the cells present in the culture (Koller et al. 1995a) and have demonstrated that changes in the perfusion schedule of the culture significantly influence the secretion of growth factors by stromal cells (Caldwell et al. 1991b). This effect of perfusion on stromal function suggests that instead of adding large amounts of growth factors to the cultures, haematopoietic support may be improved by optimal maintenance of the stroma. Perfusion bioreactors are good at expanding and maintaining LTC-ICs and CFCs (Koller 1993; Koller et al. 1993; Palsson et al. 1993). Too much perfusion may, however, over-dilute soluble factors that are produced endogenously (Collins 1996) or mechanically damage the cells in culture. Three types of perfusion bioreactor systems have been reported, namely the flat-bed perfusion system (Koller 1993), the grooved-bed perfusion system (Horner et al. 1998; Sandstrom et al. 1996), and the hollow-fibre system (Sardonini and Wu 1993).

Flat-bed perfusion systems contain culture chambers which are designed to maintain low laminar flow (0.5ml/min per chamber) (Koller et al. 1993). The chambers are inclined with the outlet at a higher point to reduce the removal of non-adherent cells during perfusion. The first 24 hours of the culture passes without perfusion in order to facilitate stromal cell attachment (Koller 1993). Thereafter, medium flow is used and after a specified period of culture, non-adherent cells are harvested by aspirating the medium, whereas adherent cells are removed after trypsinisation (Koller 1993). This system has been used to expand BM MNCs (20-25-fold) and CFU-GM (10-30 fold) in two-week cultures that developed a stromal layer (Koller 1993), or CD34-enriched cultures on pre-formed stroma (Koller et al. 1995b) and may facilitate the expansion of all types of human haematopoietic progenitors and stem cells from BM, CB, and MPB (Koller et al. 1993). LTC-IC expansion (7.5-fold) and 3 billion cells from a 10-15 ml BM aspirate may be obtained (Koller 1993). Using this system, expansion of CFU-GM (11-fold), BFU-E (2.5-fold), and colony-forming unit-multipotential progenitors (CFU-mix; 2.4-fold) in a 7-day culture of CB MNCs on preformed stromal cells has been reported (Koller et al. 1993). Thus, with cytokine support and under the best conditions tested, the flat-bed perfused bioreactors could produce adult-sized stem cell grafts from CB and MPB required for clinical applications and are currently being evaluated in clinical trials. Initial results suggest that human repopulating HSCs can be preserved (Verfaillie 2001).

Grooved-bed perfusion bioreactors facilitate stroma-free haematopoietic perfusion cultures and are designed with chambers that contain multiple microgrooves, 200 μm across and 200 μm deep, separated every 200 μm. In contrast to the flat-bed reactor, the microgrooves of the grooved-bed perfusion reactor are perpendicular to the direction of flow and can therefore accommodate a higher perfusion rate (2.5 ml/min) without disruption of the cells. However, this same conformation also presents difficulties in monitoring and controlling the microenvironment near the cells since it creates inhomogeneities in the concentration of soluble factors due to the isolation of the cells from the main convective flow of the medium (Horner et al. 1998). These reactors may supported the expansion of CFU-GM (19.4-fold) and maintain LTC-IC (64% of input after a 15-day culture) (Horner et al. 1998; Sandstrom et al. 1996) from MPB MNC cultures and have also been studied in the cultivation of CB cells.

Hollow-fibre perfusion bioreactors offer a shear-free environment for cell proliferation but they also present inherent difficulties in scale-up since the spaces between the fibre modules are not kept constant (Sardonini and Wu 1993). This inconsistency results in fluctuations in oxygen transfer and added complexity when trying to monitor culture parameters. Sardonini *et al.* employed a hollow-fibre bioreactor with 1.0 ft^2 fibre surface area and an extracapillary space of 12 ml and innoculated it with 4.4 x 10^7 low density (LD) MNCs. Medium was perfused into the hollow-fibre cartridge at a flow rate of 50 ml/min. However, when compared with the control (static) cultures, the hollow-fibre bioreactor did not support the expansion of haematopoietic cells (Sardonini and Wu 1993). Although the factors that restrict cell expansion in perfusion bioreactors are unclear, some limitations that may apply include cytokine concentrations, oxygen delivery and the surface area available. Harvesting half the cell population every 3 to 4 days, starting on day 11 of culture has been noted to improve the capability of adult human bone marrow MNCs to expand in perfusion bioreactors (Oh 1994).

Since static cultures have intrinsic limitations to cell expansion and perfusion systems are medium-intensive and scale-up can be complex due to the geometry of the systems (Collins et al. 1998), stirred culture systems may be preferred for clinical applications. Stirred bioreactors have been well characterised for the culture of both microbial and mammalian cells and they provide a homogeneous, well-controlled, closed, and reproducible culture environment. They are superior to most static and perfusion systems in terms of sampling, data collection, and control of culture parameters and can be scaled-up relatively easily for clinical use (Madlambayan et al. 2001). The stirred culture has been used successfully for the cultivation of BM MNCs in serum-containing medium (Sardonini and Wu 1993; Zandstra et al. 1994), for MPB and CB MNCs and $CD34^+$ cells in both serum and serum-free media (Collins et al. 1998) and for the culture of natural killer cells isolated from peripheral blood mononuclear cells (Pierson et al. 1996). Since haematopoietic cells are sensitive to shear stresses (McDowell and Papoutsakis 1998), low shear rates are necessary in this culture system in order to avoid physical damage (agitation rates of 30-60 rpm) (Collins et al. 1998; Sardonini and Wu 1993; Zandstra et al. 1994). Agitation could also have a profound effect on the type of cells that will expand and the extent of the expansion; macrophages tolerate suspension whereas fibroblasts and endothelial cells are unlikely to do well in stirred cultures (Nielsen 1999). It is unclear if extensive HSC self-renewal can be sustained in the stirred culture since suspension culture systems do not mimic the complex three-dimensional (3-D) microenvironment found in BM. In an attempt to mimic the marrow's microenvironment and provide a scaffold for the haematopoietic stem cells, the use of microspheres within the system has resulted in expansion of cell numbers (Sardonini and Wu 1993).

Based on normal bone marrow physiology, novel 3-D culture systems have been developed in order to mimic the *in vivo* bone marrow HIM by enhancing cell-cell and cell-matrix interactions and thereby promote self-renewal and differentiation of stem and progenitor cells. This contrasts with most *ex vivo* culture systems that have focused on the addition of cytokines alone and is not optimised for blood cell development or for the retention of long-term generative and regenerative capacity (Chute et al. 1999; Dao et al. 1998). Collagen-coated nylon mesh, used as a template to grow BM cells, was capable of maintaining multilineage differentiation in 3-D, but CFU numbers declined (Naughton et al. 1990; Naughton and Naughton 1989; Naughton et al. 1991). Recently, a packed bed bioreactor using porous microspheres supported multilineal differentiation of both murine and human BM cells in the absence of exogenous growth factors (Mantalaris et al. 1998; Wang et al. 1995). This bioreactor culture had a 3-D growth configuration that mimicked that of BM *in vivo*, and in the presence of cytokines, including physiological concentrations erythropoietin (Epo), this system could support extensive erythropoiesis when compared with a traditional flask culture (Mantalaris et al. 1998). Another 3-D culture reactor used a tantalum-coated porous biomaterial that could enhance the maintenance and retroviral transduction of HPCs, improve the cell pluripotency (Rosenzweig et al. 1997), and support BM $CD34^+$ cell expansion (1.5-fold) in the absence of exogenous cytokines (Bagley 1999). In addition, 3-D non-woven fibrous matrices have been used for the culture of human CB cells and could prolong the proliferation process and support more total $CD34^+$ cells when compared with a 2-D film (Abboud and Lichtman 2001).

Finally, the airlift packed bed bioreactor system has also been used for the expansion of HSCs and HPCs (Highfill 1996; Sardonini and Wu 1993). This bioreactor system was designed with a central draft tube and an annular region packed with fibreglass disks that harboured a mean diameter of 13.5 μm and a surface area of 109 cm^2. These packed disks were used to establish pre-formed stromal layers to mimic the *in vivo* environment. Medium was sent upwards through the draft tube by the gas sparging and then downwards through the stacked disks. Over 11-weeks of culture, a 0.5 L system resulted in the production of 3.6×10^8 BM cells. However, this packed bed did not simulate the BM microenvironment and the gas sparging caused bubble shear and breakage that further inhibited cell expansion (Highfill 1996).

3.4. Feeding Schedule and Culture Duration

In vivo, BM is perfused at a rate of 0.1 ml/cm^3/min (Martiat 1987) which is sufficient to provide nutrients to and remove waste from haematopoietic cells under static conditions. Hence, various feeding strategies have been developed in the cultivation of haematopoietic cells to try and recapitulate the needs of the HIM (Caldwell et al. 1991a; Koller 1993; Martinson 1998; McAdams 1995; Schwartz et al. 1991). Medium exchange is required in order to overcome growth restrictions imposed by mass transfer limitations of oxygen and nutrients (Koller 1996; Oh 1994; Zandstra et al. 1994) and the depletion of substrates and the accumulation of the metabolic by-products such as lactate that result in a drop in culture pH (Caldwell et al. 1991a). Frequent medium exchange enhanced the cell output and CFU-GM in static cultures, especially at high cell densities of BM MNCs in the presence of preformed stroma (Koller 1996). As noted above, perfusion systems offer automated, continuous, and controlled medium exchange that require less handling and do not depopulate the nonadherent cell fractions during feeding (McAdams 1995), all advantages over static cultures. Perfusion cultures of PB MNCs or CD34$^+$ cells have more LTC-ICs in comparison to static cultures (McAdams 1995) and allow for more expansion of total and colony-forming cells (Schwartz et al. 1991).

The optimal duration of haematopoietic cultures is unclear. When CD34$^+$ cells from MPB were expanded *ex vivo*, two-phases of culture growth were observed. In the first 7-10 days, a pre-CFU phase was noted and then, past day 10, a second phase was observed wherein CFU-GM decreased and mature myeloid cell numbers increased (Haylock et al. 1992). This decrease in the production of haematopoietic progenitors during days 12-15 of culture is not uncommon and may be due to inadequate feeding protocols and high cell concentrations (Bregni et al. 1992). Although several pre-clinical studies have reported continuously increasing levels of expansion of all haematopoietic lineages for several months (Piacibello et al. 1998), the results of the aforementioned clinical studies suggest that HSC expansion in static liquid cultures should be limited to 7-14 days.

3.5. pH

Gradients of pH affect the *ex vivo* expansion of HSCs (Akatov 1985), cell physiology and receptor expression (Carswell 2000; Hevehan 2000) as well as cell differentiation (Carswell 2000; Endo 1994; Fischkoff 1984; Hevehan 2000; McAdams et al. 1997; McAdams et al. 1998; Tennant 2000; Yang 2002). The optimum pH is dependent on the haematopoietic cell lineage (Table 3) (Carswell 2000; Endo 1994; Fischkoff 1984; Hevehan 2000; McAdams et al. 1997; McAdams et al. 1998; Tennant 2000; Yang 2002). For example, T cells proliferate better at pH 7.0-7.2 (Carswell 2000) whereas erythroid differentiation is best at pH 7.6 (Endo 1994; McAdams et al. 1997; McAdams et al. 1998). An increase in the population of benzidine positive cells, suggesting erythroid lineage, was also noted by examining the synergistic effect of pH 7.54 and compounds such as hemin and BSA that induce the differentiation of erythroid cells (Endo 1994). The differentiation, maturation

Table 3. Summary of the effects of pH onhaematopoietic stem and progenitor cell cultures.

Target cell lineage	**Optimum pH range**	**Experimental Conditions**	**Comments**	**Ref.**
T-Lymphocytes	7.0-7.2	Culture Duration: 14-21 days Feeding schedule: 1-3 days Oxygen Tension: 21%	3-fold increase in proliferating capacity pH 7.4 enhances apoptosis	(Carswell 2000)2}
Erythropoiesis	7.6	Culture Duration: 14-21 days Feeding schedule: Every 6 days Oxygen Tension: 5%	Promotion of erythroid differentiation Promotion of erythroid maturation towards later stages (increased numbers	(Endo 1994; McAdams et al. 1997; McAdams et al. 1998)5}
Granulopoieisis	7.07-7.21	Culture Duration: 13 days Feeding schedule: Dilution feeding Oxygen Tension: 5% & 20%	Enhanced granulocytic cell proliferation and differentiation Synergistic effect of low oxygen tension	(Hevehan 2000; McAdams et al. 1997)3}
Eosinophilopoiesis	7.6-7.8	Culture Duration: 7 days Feeding schedule: None Oxygen Tension: 21%	Promoted eosinophilic differentiation Development of granules similar to eosinophilic precursors	(Fischkoff 1984)6}
Megakaryopoiesis	7.4-7.6	Culture Duration: 21 days Feeding schedule: Dilution feeding Oxygen Tension: 20%	Promoted differentiation, maturation (polyploidisation), and apoptosis of megakaryocyte cells	(Yang 2002)7}

(polyploidisation), and apoptosis of megakaryocytes ($CD41^+$ cells) was enhanced in the presence of pH of 7.4-7.6 (Yang 2002), whereas the differentiation of CFU-GM was optimal at pH 7.2-7.4 (McAdams et al. 1997). Thus, in these methylcellulose cultures, macrophage progenitors (CFU-M) were more sensitive to acidic pH than granulocytic progenitors (CFU-G) such that at pH 6.7, the only detectable colonies remaining were CFU-G. Furthermore, the effects of pO_2 and pH on granulopoiesis were synergistic such that at low pO_2 (5%) and pH 7.07-7.21 the production of total cells and neutrophil precursors were much more that that observed at 20% pO_2 and pH 7.38 (Hevehan 2000). Interestingly, CB-derived progenitor cells were more resistant to pH variation (pH 6.7 and pH 7.6) with respect to CFU-GM growth in methylcellulose cultures than were PB-derived progenitor cells (McAdams et al. 1997).

Expression of surface receptors, which mediate cell adhesion, catalysis of surface associated reactions, and signal transduction, is pH-dependent. For example, G-CSF is one of the main regulators of granulopoiesis and, when it interacts with its receptor, it stimulates granulocyte differentiation and proliferation. The expression of G-CSF receptor was enhanced at pH 7.22 and decreased with a concomitant increase in pH (Hevehan 2000). Similarly, expression of the IL-2 receptor, critical for T-cell proliferation, was optimal at pH 7.0 and decreased in a more alkaline environment (Carswell 2000).

Cellular metabolites in haematopoietic cultures result in acidification of the medium and growth inhibition (Akatov 1985). Within each culture, pH variations of 0.5 pH units can occur unless controlled. Control of pH has been attempted through the use of buffers and by modulating cellular oxidative activity by decreasing the concentration of glucose in the culture medium (Tennant 2000). More importantly, the spatial in-homogeneities inherent to each culture system and within each separate culture have yet to be addressed and make pH monitoring and control essential for consistent and reproducible culture results (Jeevarajan 2002). Hence, the control of pH should be a critical factor in the design and operation of bioreactors.

3.6. Oxygen Tension

The oxygen (O_2) tension of BM *in vivo* is between 10 and 50 mmHg (Ishikawa 1988; Koller 1993). Oxygen tension may determine the performance of haematopoietic cell cultures by affecting the growth of HPCs (Cipolleschi 1997; Ishikawa 1988; Ivanovic 2002), increasing the responsiveness of cells to cytokines (Ivanovic 2002; Rich 1982), and influencing cytokine production (Rich 1986) (Table 4). At low O_2 concentrations (2-7%), the size and number of haematopoietic colonies in semisolid cultures is significantly enhanced (Cipolleschi 1997; Cipolleschi et al. 1993; Ishikawa 1988; Ivanovic 2000; Koller 1992a; Rich 1982). Specifically, the clonal expansion of human BM CFU-Cs, BFU-Es and CFU-mix is better at 7% O_2 when compared with that at 19% O_2 (Ishikawa 1988). The production of progenitor as well as mature cells is dependent upon oxygen tension (Cipolleschi 1997; Ishikawa 1988; Ivanovic 2002; Koller 1992a; Mostafa 2000;

Rich 1982). Under hypoxic conditions (1%), the production of erythroid bursts and the size and number of colony-froming unit-megakaryocyte (CFU-MKs) was enhanced (1% and 5% O_2 respectively) whilst mature erythrocyte and $CD41^+$ megakaryocyte generation with high ploidy, proplatelet formation and apoptosis was favoured at 20% oxygen (Cipolleschi 1997; Mostafa 2000). This *in vitro* situation may resemble the *in vivo* situation, where megakaryocyte maturation and platelet production occurs adjacent to the BM sinus, exposed to a higher venous pO_2 (approximately 8%). By contrast, HSCs and HPCs reside in the core of the bone marrow haematopoietic compartment, where the average pO_2 is 1-5% (Pennathur-Das 1987). Cells are more responsive to cytokines under low oxygen tension and may account for the cell growth kinetics of HSCs and HPCs in hypoxic conditions (Ivanovic 2002; Koller 1992a; Mostafa 2000; Naldini 1997). Specifically, colony-forming unit-erythroid (CFU-E) and BFU-E formation has been associated with an enhanced sensitivity to Epo under low oxygen tension (Rich 1982). Oxygen tension has also been associated with the production of haematopoietic growth factors, such as the production of EPO by macrophages (Rich 1986). Thus, oxygen tension for haematopoietic cell culture may be optimised and may be co-dependent on the cytokines used and on the developmental and maturational state or potential of the cells in culture.

In optimising *in vitro* cultures, low oxygen tension is appropriate for the production and maintenance of progenitor cells, but as the number of cells increase in the culture, O_2, in addition to other factors, limits this system. Oxygen availability depends on culture configuration, cell density, and gas-phase oxygen tension (McAdams 1996) and by increasing this gas-phase O_2 tension, agitation rates or perfusion rates, oxygen can be calibrated and maintained within the system. In order to overcome the increased oxygen requirements due to increased cell density, increased feeding, or use of perfusion are preferred to simply increasing the gas-phase oxygen tension since the latter method can lead to oxygen toxicity (Koller 1992b).

High oxygen levels can drastically change the phenotype of haematopoietic cultures (Erslev 1983; Ishikawa 1988; Rich 1986) due to increased cellular metabolism and the generation of reactive oxygen intermediates. The resultant accumulation of mature myeloid cells, such as neutrophils and macrophages, leads to the progressive decline of progenitor cell production (Rich 1982). Hence, the use of antioxidants such as reduced glutathione, vitamin E, and 2-mercaptoethanol or the use of scavengers of hydroxyl radicals and hydrogen peroxide, such as mannitol and catalase, may be able to reduce this oxygen toxicity (Ishikawa 1988; Rich 1982). The local concentration of oxygen in the cellular microenvironment is, however, more important than the gas phase oxygen level (Koller 1993) and slight deviation in molecular oxygen concentration could have a detrimental effect on cellular function and cellular physiology. Hence, the optimisation of local oxygen concentration as well as O_2 tension gradients is required for the controlled *ex vivo* expansion of haematopoietic cells.

Table 4. Summary of the effects of oxygen tension on haematopoietic stem and progenitor cell cultures.

Oxygen Tension	Comments	Ref.
Low (1% - 7%)	• Enhanced size and number of haematopoietic colonies in semi solid cultures • Optimal O_2 tension for BFU-E & CFU-E: 3.5% • Optimal O_2 tension for CFU-GM: 2% & 3.5% • Enhanced responsiveness of the cells to cytokines (e.g. Epo) • Enhanced production of growth factors (Epo, IL-3, etc.) • Decreased oxygen toxicity to cells • Inhibition of terminal expansion and maturation of the erythroid and megakaryocyte lineages • Addition of antioxidants is required	(Ishikawa 1988; Rich 1986; Rich 1982)5}
High (19% - 20%)	• Progressive decline in progenitor cell production due to increase in reactive oxygen intermediates • Enhanced production of mature erythrocytes • Enhanced megakaryocyte differentiation, maturation and apoptosis • Increased oxygen toxicity to cells • Addition of antioxidants is recommended	(Cipolleschi 1997; Rich 1982)4}

4. CONCLUSIONS

Haematopoietic stem cells have been applied successfully in the clinic for over 30 years. This experience, the relative ease with which HSCs can be identified and obtained from a variety of sources and the potential plasticity of these cells makes them ideal for use in haematologic and non-haematologic conditions. Despite these advantages and the significant progress that has been made in the characterisation of factors that govern haematopoiesis, enrichment and *ex vivo* expansion of repopulating (and possibly plastic) HSCs remains elusive. Conventional 2-D cultures are insufficient to meet the complex demands required and small deviations in the culture parameters can profoundly affect the final cell output. The application of factorial and composite designs to HSC cultures is required in order to fully appreciate the effects and interdependence of stimulatory and inhibitory factors as well as the culture parameters on haematopoietic culture systems. Furthermore, *ex vivo* expanded HSCs must be safe to use in humans and meet regulations guided by good manufacturing practice (GMP) requirements for clinical therapeutics which includes the development of suitable, closed culture systems that can be easily controlled and monitored. The engineering of optimal haematopoietic cell culture systems requires the design of new expansion systems that mimic the *in vivo* bone marrow environment that is able to self-regulate and operate under reliable and reproducible conditions. Such a system would offer a broad spectrum of possibilities for different culture strategies in the cultivation of various cell types – from stem cells to differentiated cells for gene, cellular, and tissue therapies.

5. REFERENCES

Abboud CN, Lichtman MA. 2001. Structure of the Marrow and the Hemopoietic Microenvironment. In: Seligsohn U, editor. Hematology. New York: McGraw-Hill. p 29-58.

Aiuti A, Friedrich C, Sieff C, Gutierrez-Ramos J. 1998. Identification of distinct elements of the stromal microenvironment that control human hematopoietic stem/progenitor cell growth and differentiation. Experimental Hematology 26:143-157.

Akatov V, Lezhnev, EI., Vexler, AM., Kublik, LN. 1985. Low pH value of pericellular medium as a factor limiting cell proliferation in dense cultures. Experimental Cell Research 160:412-418.

Alley C, MacDermott, RP., Stewart, CC. 1983. The Effect of Serum on Human Marrow Mononuclear Cell Proliferation and Maturation. Journal of the Reticuloendothelial Society 34:271-278.

Almeida-Porada G, Brown, RL., MacKintosh, FR., Zanjani, ED. 2000. Evaluation of Serum-Free Culture Conditions Able to Support the Ex Vivo Expansion and Engraftment of Human Hematopoietic Stem CElls in Human-to Sheep Xenograft Model. Journal of Hematotherapy & Stem Cell Research 9:683-693.

Audet J, Zanstra, PW., Eaves, CJ., Piret, JM. 1998. Advances in hematopoietic stem cell culture. Current Opinion in Biotechnology 9:146-151.

Bachier CR, Gokmen E, Teale J, Lanzkron S, Childs C, Franklin W, Shpall E, Douville J, Weber S, Muller T and others. 1999. *Ex-vivo* expansion of bone marrow progenitor cells for hematopoietic reconstitution following high-dose chemotherapy for breast cancer. Experimental Hematology 27:615-623.

Bagley J, Rosenzweig, M., Mark, DF., Pykett, MJ. 1999. Extended culture of multipotent hematopoietic progenitors without cytokine augmentation in a novel three-dimensional device. Experimental Hematology 27:496-504.

Barnett MJ, Eaves CJ, Phillips GL, Gascoyne RD, Hogge DE, Horsman DE, Humphries RK, Klingemann HG, Lansdorp PM, Nantel SH and others. 1994. Autografting with cultured marrow in chronic myeloid leukemia: results of a pilot study. Blood 84:724-732.

Bhatia M, Bonnet D, Wu D, Murdoch B, Wrana J, Gallacher L, Dick JE. 1999. Bone morphogenetic proteins regulate the developmental program of human hematopoietic stem cells. Journal of Experimental Medicine 189:1139-1148.

Bhatia M, McGlave PB, Dewald GW, Blazar BR, Verfaillie CM. 1995. Abnormal function of the bone marrow microenvironment in chronic myelogenous leukemia: role of malignant stromal macrophages. Blood 85:3636-3645.

Bohmer R. 1989. Interaction of Serum and Colony-Stimulating Factor for Survival of a Factor-Dependent Hemopoietic Progenitor Cell Line. Journal of Cellular Physiology 139:531-537.

Brandt J, Bhalla, K., Hoffman, R. 1994. Effects of interleukin-3 and c-kit ligand on the survival of various classes of human hematopoietic progenitor cells. Blood 83(6):1507-1514.

Bregni M, Magni M, Siena S, Di Nicola M, Bonadonna G, Gianni AM. 1992. Human peripheral blood hematopoietic progenitors are optimal targets of retroviral-mediated gene transfer. Blood 80:1418-1422.

Broudy VC. 1997. Stem cell factor and hematopoiesis. Blood 90:1354-1364.

Brown R, Xu, FS., Dusing, SK., Li, Q., Fischer, R., Patchen, M. 1997. Serum-Free Conditions for Cells Capable of Producing Long-Term Survival in Lethally Irradiated Mice. Stem Cells 15:237-245.

Broxmeyer HE, Sherry B, Cooper S, Lu L, Maze R, Beckmann MP, Cerami A, Ralph P. 1993. Comparative analysis of the human macrophage inflammatory protein family of cytokines (chemokines) on proliferation of human myeloid progenitor cells. Interacting effects involving suppression, synergistic suppression, and blocking of suppression. Journal of Immunology 150:3448-3458.

Brugger W, Mocklin W, Heimfeld S, Berenson RJ, Mertelsmann R, Kanz L. 1993. *Ex vivo* expansion of enriched peripheral blood $CD34^+$ progenitor cells by stem cell factor, interleukin-1 (IL-1), IL-6, IL-3, interferon g and erythropoietin. Blood 81:2579-2584.

Cabral JMS. 2001. *Ex vivo* expansion of hematopoietic stem cells in bioreactors. Biotechnology Letters 23:741-751.

Cairo MS, Wagner JE. 1997. Placental and/or umbilical cord blood: an alternative source of hematopoietic stem cells for transplantation. Blood 90:4664-4671.

Caldwell J, Locey B, Clarke MF, Emerson SG, Palsson BO. 1991a. Influence of medium exchange schedules on metabolic, growth, and GM-CSF secretion rates of genetically engineered NIH-3T3 cells. Biotechnology Progress 7:1-8.

Caldwell J, Palsson BO, Locey B, Emerson SG. 1991b. Culture perfusion schedules influence the metabolic activity and granulocyte-macrophage colony-stimulating factor production rates of human bone marrow stromal cells. Journal of Cellural Physiology 147:344-353.

Carswell K, Papoutsakis, ET. 2000. Extracellular pH affects the proliferation of cultured human T-cells and their expression of the interleukin-2 receptor. Journal of Immunology 23(6):669-674.

Chabannon C, Olivero S, Blaise D, Maraninchi D, Viens P. 2000. *Ex vivo* expansion of human hematopoietic progenitors and cells to support high-dose chemoradiation therapy: Five years of clinical experience. Cytokines, Cellular & Molecular Therapy 6:97-108.

Charbord P. 2001. Microenvironmental Cell Populations Essential for the Support of Hematopoietic Stem Cells. In: Zon LI, editor. Hematopoiesis: A Developmental Approach. New York: Oxford University Press. p 691-701.

Chute JP, Saini AA, Kampen RL, Wells MR, Davis TA. 1999. A comparative study of the cell cycle status and primitive cell adhesion molecule profile of human $CD34^+$ cells cultured in stroma-free versus porcine microvascular endothelial cell cultures. Experimental Hematology 27:370-379.

Cipolleschi M, D'Ippolito, G., Bernabei, PA., Caporale, R., Nannini, R., Mariani, M., Fabbiani, M., Rossi-Ferrini, P., Olivotto, M., Sbarba, PD. 1997. Severe hypoxia enhances the formation of erythroid bursts from human cord blood cells and the maintenance of BFU-E in vitro. Experimental Hematology 25:1187-1194.

Cipolleschi MG, Dello Sbarba P, Olivotto M. 1993. The role of hypoxia in the mainetance of hematopoietic stem cells. Blood 82:2031-2037.

Civin CI, Almeida-Porada G, Lee MJ, Olweus J, Terstappen LW, Zanjani ED. 1996. Sustained, retransplantable, multilineage engraftment of highly purified adult human bone marrow stem cells *in vivo*. Blood 88:4102-4109.

Collins P, Miller, WM., Papoutsakis, ET. 1996. Ex vivo culture systems for hematopoietic cells. Current Opinion in Biotechnology 7(2):223-230.

Collins PC, Miller ME, Papoutsakis ET. 1998. Stirred culture of peripheral and cord blood hematopoietic cells offers advantages over traditional static systems for clinically relevant applications. Biotechnology & Bioengineering 59:534-543.

Coutinho LH, Testa NG, Chang J, Morgenstern G, Harrison C, Dexter TM. 1990. The use of cultured bone marrow cells in autologous transplantation. Progress in Clinical and Biological Research 333:415-432.

Craig W, Kay R, Cutler RL, Lansdorp PM. 1993. Expression of Thy-1 on human hematopoietic progenitor cells. Journal of Experimental Medicine 177:1331-1342.

Dao MA, Hashimo K, Kato I, Nolta JA. 1998. Adhesion to fibronectin maintains regenarative capacity during *ex vivo* culture and transduction of human hematopoietic stem and progenitor cells. Blood 92:4612-4621.

De La Selle V, Gluckman E, Bruley-Rosset M. 1996. Newborn blood can engraft adult mice without inducing graft versus host disease across non H-2 antigens. Blood 87:3977-3984.

Dexter TM, Allen TD, Lajtha LG. 1977. Conditions controlling the proliferation of haematopoietic stem cells *in vivo*. Journal of Cellular Physiology 91:335-344.

Douay L. 2001. Experimental Culture Conditions Are Critical for *Ex Vivo* Expansion of Hematopoietic Cells. Journal of Hematotherapy & Stem Cell Research 10:341-346.

Dybedal I, Jacobsen SE. 1995. Transforming growth factor beta (TGF-beta), a potent inhibitor of erythropoiesis: neutralizing TGF-beta antibodies show erythropoietin as a potent stimulator of murine burst-forming unit erythroid colony formation in the absence of a burst-promoting activity. Blood 86:949-957.

Eaves CJ, Cashman JD, Kay RJ, Dougherty GJ, Otsuka T, Gaboury LA, Hogge DE, Lansdorp PM, Eaves AC, Humphries RK. 1991. Mechanisms that regulate the cell cycle status of very primitive hematopoietic cells in long-term human marrow cultures. II. Analysis of positive and negative regulators produced by stromal cells within the adherent layer. Blood 78:110-117.

Eaves CJ, Eaves AC. 1997. Stem cell kinetics. Baillieres Clinical Haematology 10:233-257.

Emmons RV, Doren S, Zujewski J, Cottler-Fox M, Carter CS, Hines K, O'Shaughnessy JA, Leitman SF, Greenblatt JJ, Cowan K and others. 1997. Retroviral gene transduction of adult peripheral blood or marrow-derived $CD34^+$ cells for six hours without growth factors or on autologous stroma does not improve marking efficiency assessed *in vivo*. Blood 89:4040-4046.

Endo T, Ishibashi, Y., Okana, H., Fukumaki, Y. 1994. Sgnificance of pH on differentiation of human erythroid cell lines. Leukemia Research 18:49-54.

Erslev A, et al. 1983. Structure and function of the marrow. al. WWe, editor: New York: McGraw-Hill. 75-83 p.

Fischkoff S, Pollak, A., Gleich, GJ., Testa, JR., Misawa, S., Reber, TJ. 1984. Eosinophilic differentiation of the human promyelocytic leukemia cell line. Journal of Experimental medicine 160:179-196.

Fortunel N, Hatzfeld, A. and Hatzfeld, JA. 2000. Transforming growth factor-B: pleiotropic role in the regulation of hematopoiesis. Blood 96(6):2022-2036.

Gartner S, Kaplan, HS. 1980. Long-term culture of human bone marrow cells. Proc. Natl. Acad. Sci. (USA) 77:4756-4759.

Giarratana M-C, Kobari, L., Neildez Nguyen, TMA., Firat, H., Bouchet, S., Lopez, M., Gorin, N-C., Thierry, D., Douay, L. 1998. Cell culture bags allow a large extent of ex vivo expansion of LTC-IC and functional mature cells which can subsequently be frozen: interest for large-scale clinical appplication. Bone Marrow Transplantation 22(7):707-715.

Goodell M, Brose K, Paradis G, Conner A, Mulligan A. 1996. Isolation and functional properties of murine hematopoietic stem cells that are replicating *in vivo*. Journal of Experimental Medicine 183:1797-1806.

Guba S, Gottschalk, LR., Jing, YH., Mulligan, T., Emerson, SG. 1992. Bone marrow stromal fibroblasts secrete interleukin-6 and granulocyte- macrophage colony-stimulating factor in the absence of inflammatory stimulation: demonstration by serum-free bioassay, enzyme-linked immunosorbent assay, and reverse transcriptase polymerase chain reaction. Blood 80(5):1190-1198.

Haylock DN, To LB, Dowse TL, Juttner CA, Simmons PJ. 1992. *Ex vivo* expansion and maturation of peripheral blood $CD34^+$ cells into the myeloid lineage. Blood 80:1405-1412.

Hevehan D, Papoutsakis, ET., Miller, WM. 2000. Physiologically significant effect of pH and oxygen tension on granulopoiesis. Experimental Hematology 28:267-275.

Highfill J, Haley, SD., Kompala, DS. 1996. Large-scale production of murine bone marrow cells in an airlift packed bed bioreactor. Biotechnology and Bioengineering 50:514-520.

Holyoake TL, Freshney MG, McNair L, Parker AN, McKay PJ, Steward WP, Fitzsimons E, Graham GJ, Pragnell IB. 1996. *Ex vivo* expansion with stem cell factor and interleukin-11 augments both short-term recovery posttransplant and the ability to serially transplant marrow. Blood 87:4589-4595.

Horner M, Miller ME, Ottino JM, Papoutsakis ET. 1998. Transport in a grooved perfusion flat-bed bioreactor for cell therapy applications. Biotechnology Progress 14:689-698.

Ishikawa Y, Ito, T. 1988. Kinetics of hematopietic stem cells in a hypoxic culture. European Journal of Haematology 40:126-129.

Ivanovic Z, Bartolozzi, B., Bernabei, PA., et al. 2000. Incubation of murine bone marrow cells in hypoxia ensures the maintenance of marrow-repopulating ability together with the expansion of commited progenitors. British Journal of Haematology 108:424-429.

Ivanovic Z, Belloc, F., Faucher,J., et al. 2002. Hypoxia maintains and interleukin-3 reduces the pre-colony-forming cell potential of dividing CD34+ murine bone marrow. Experimental Hematology 30:67-73.

Jeevarajan A. 2002. Continuous pH Monitoring in a Perfused Bioreactor System Using an Optical pH Sensor. Biotechnology AND Bioengineering 78(4):467-472.

Kanai M, Hirayama, F., Yamaguchi, M., Ohkawara, J., Sato, N., Fukazawa, K., Yamashita, K., Kuwabara, M., Ikeda, H. & Ikebuchi, K. 2000. Stromal cell-dependent ex vivo expansion of human cord blood and augmentation of transplantable stem cell activity. Bone Marrow Transplantation 26:837-844.

Karanu FN, Murdoch B, Gallacher L, Wu DM, Koremoto M, Sakano S, Bhatia M. 2000. The notch ligand jagged-1 represents a novel factor of human hematopoietic stem cells. Journal of Experimental Medicine 192:1365-1372.

Kim D, Fujiki, Y., Fukushima, T., Ema, H., Shibuya, A., Nakauchi, H. 1999. Comparison of Hematopoietic Activities of Human Bone Marrow and Umbilical Cord Blood CD34 Positive and Negative Cells. Stem Cells 17:286-294.

Kobayashi M, Laver JH, D LS, Kato T, Miyazaki H, Ogawa M. 1996. Thrombopoietin supports proliferation of human primitive hematopoietic cells in synergy with steel factor and/or interleukin-3. Blood 88:429-436.

Kobayashi M, Laver JH, Lyman SD, Kato T, Miyazaki H, Ogawa M. 1997. Thrombopoietin, steel factor and the ligand for flt3/flk2 interact to stimulate the proliferation of human hematopoietic progenitors in culture. International. Journal of Hematology 66:423-434.

Koller M, Bender, JG., Papoutsakis, ET. & Miller, WM. 1992a. Effects of synergistic cytokine combinations, low oxygen, and irradiated stroma on the expansion of human cord blood progenitors. Blood 80(2):403-411.

Koller M, Emerson, SG. & Palsson, BO. 1993. Large-Scale Expansion of Human Stem and Progenitor Cells From Bone Marrow Mononuclear Cells in Continuous Perfusion Cultures. Blood 82(2):378-384.

Koller M, et al. 1992b. Reduced Oxygen Tension Increases Hematopoiesis in Long-term Culture of Human Stem and Progenitor Cells from Cord Blood and Bone Marrow. Experimental Hematology 20:264-270.

Koller M, Manchel, I., Maher, RJ., Goltry, KL., Armostrong, D., Smith, AK. 1998. Clinical-Scale Human Umbilical Cord Blood Cell Expansion in a Novel Automated Perfusion Culture System. Bone Marrow Transplantation 21:653-663.

Koller M, Manchel, I., Palsson, MA., Maher, RJ. 1996. Different Measures of Human Hematopoietic Cell Culture Performance Are Optimized under Vastly Different Conditions. Biotechnology and Bioengineering 50:505-513.

Koller MR, Bender JG, Miller ME, Papoutsakis ET. 1993. Expansion of primitive human hematopoietic progenitors in a perfusion bioreactor system with IL-3, IL-6, and stem cell factor. Bio/Technology 11:358-63.

Koller MR, Bradley TR, Palsson BO. 1995a. Growth factor consumption and production in perfusion cultures of human bone marrow correlate with specific cell production. Experimental Hematology 23:1275-1283.

Koller MR, Manchel I, Newsom BS, Palsson MA, Palsson BO. 1995b. Bioreactor expansion of human bone marrow: comparison of unprocessed, density-separated and CD34-enriched cells. Journal of Hematotherapy 4:159-169.

Koller MR, Oxender M, Jensen TC, Goltry KL, Smith AK. 1999. Direct contact between CD34+/Lin- cells and stroma induces a soluble activity that specifically increases primitive hematopoietic cell production. Experimental Hematology 27:734-741.

Krause DS, Fackler MJ, Civin CI, May WS. 1996. CD34: structure, biology, and clinical utility. Blood 87:1-13.

Krystal G, Lanm V, Dragowska W, Takahashi C, Appel J. 1994. Transforming growth factor beta 1 is an inducer of erythroid differentiation. Journal of Experimental Medicine 180:851-860.

Lansdorp PD, W. 1992. Long-Term Erythropoiesis from Constant Numbers of CD34+ Cells in Serum-free Cultures initiated with Highly Purified Progenitor Cells from Human Bone Marrow. Journal of Experimental Medicine 175:1501-1509.

Larochelle A, Vormoor J, Hanenberg H, Wang JC, Bhatia M, Lapidot T, Moritz T, Murdoch B, Xiao XL, Kato I and others. 1996. Identification of primitive human hematopoietic cells capable of repopulating NOD/SCID mouse bone marrow: implications for gene therapy. Nature Medicine 2:1329-1337.

Lill MC, Lynch M, Fraser JK, Chung GY, Schiller G, Glaspy JA, Souza L, Baldwin GC, Gasson JC. 1994. Production of functional myeloid cells from CD34-selected hematopoietic progenitor cells using a clinically relevant *ex vivo* expansion system. Stem Cells 12:626-637.

Lyman SD, Jacobsen SE. 1998. c-kit ligand and Flt3 ligand: stem/progenitor cell factors with overlapping yet distinct activities. Blood 91:1101-1134.

Lyman SD, Williams DE. 1995. Biology and potential clinical applications for flt3 ligand. Current Opinion in Hematology 2:177-181.

Madlambayan GJ, Rogers I, Casper RF, Zandstra PW. 2001. Controlling Culture Dynamics for the Expansion of Hematopoietic Stem Cells. Journal of Hematotherapy & Stem Cell Research 10:481-492.

Mantalaris A, Keng P, Bourne P, Chang AYC, Wu JHD. 1998. Engineering a human bone marrow model: A case study on *ex vivo* erythropoiesis. Biotechnology Progress 14:126-133.

Martiat P, Ferrant, A., Cogneau, M., Bol, A., Rodhain, J., Michaux, JL., Sokal, G. 1987. Assessment of bone marrow blood flow using positron emmission tomography: No relationship with bone cellularity. Brithish Journal of Haematology 66:307-310.

Martinson J, Unverzagt, K., Schaeffer, A., Smith, SL., Loudovaris, M., Schneidkraut, MJ., Bender, JG. and Van Epps, DE. 1998. Neutrophil Precursor Generation: Effect of Culture Conditions. Journal of Hematotherapy 7:463-471.

Matsunaga T, Hirayama F, Yonemura Y, Murray R, Ogawa M. 1998. Negative regulation by interleukin-3 (IL-3) of mouse early B-cell progenitors and stem cells in culture: transduction of the negative signals by betac and betaIL-3 proteins of IL-3 receptor and absence of negative regulation by granulocyte-macrophage colony-stimulating factor. Blood 92:901-907.

Mayani H. 1996. Composition and function of the hematopoietic microenvironment in human myeloid leukemia. Leukemia 10:1041-1047.

Mayani HaL, PM. 1998. Biology of Human Umbilical Cord Blood-Derived Hematopoietic Stem/Progenitor Cells. Srem Cells 16:153-165.

McAdams T, Miller, WM. & Papoutsakis, ET. 1996. Hematopoietic cell culture therapies (part I): cell culture considerations. Trends in Biotechnology 14:341-349.

McAdams T, Sandstrom, CE., Miller, WM., Bender, JG. and Paputsakis, ET. 1995. Ex vivo expansion of primitive hematopoietic cells for cellular therapies: An overview. Cytotechnology 18:133-146.

McAdams TA, Miller WM, Papoutsakis ET. 1997. Variations in culture pH affect the cloning eficiency and differentiation of progenitor cells in *ex vivo* haematopoiesis. British Journal of Haematology 97:889-895.

McAdams TA, Miller WM, Papoutsakis ET. 1998. pH is a potent modulator of erythroid differentiation. British Journal of Haematology 103:317-325.

McCune JM, Namikawa R, Kaneshima H, Shultz LD, Lieberman M, Weissman IL. 1998. The SCID-hu mouse: murine model for the analysis of human hematolymphoid differentiation and function. Science 24:1632-1639.

McDowell CL, Papoutsakis ET. 1998. Increased agitation intensity increases CD13 receptor surface content and mRNA levels, and alters the metabolism of HL60 cells cultures in stirred tank bioreactors. Biotechnology & Bioengineering 60:239-250.

McNiece I, Briddell R. 2001. *Ex vivo* expansion of hematopoietic progenitor cells and mature cells. Experimental Hematology 29:3-11.

Meyer C, Drexler HG. 1999. FLT3 ligand inhibits apoptosis and promotes survival of myeloid leukemia cell lines. Leukemia Lymphoma 32:577-581.

Mobest D, Goan, S., Junghahn, I., Winkler, J., Fichtner, I., Becker, M., De Lima-Hahn, E., Mertelsmann, R. & Henschler, R. 1999. Differential Kinetics of Primitive Hematopoietic Cells Assayed In Vitro and In Vivo During Serum-Free Suspension Culture of CD34+ Blood Progenitor Cells. Stem Cells 17:152-161.

Mobest D, Mertelsmann, R. & Henschler, R. 1998. Serum-Free ex vivo Expansion of CD34+ Hematopoietic Progenitor Cells. Biotechnology and Bioengineering 60(3):341-347.

Mostafa S, Miller, WM & Papoutsakis, ET. 2000. Oxygen tension influences the differentiation, maturation and apoptosis of human megakaryocytes. British Journal of Haematology 111:879-889.

Naldini A, Carraro, F., Silvestri, S., et al. 1997. Hypoxia Affects Cytokine Production and Proliferation Responses by Human Peripheral Mononuclear Cells. Journal of Cellular Physiology 173:335-342.

Naughton BA, Jacob L, Naughton GK. 1990. A three-dimensional culture system for the growth of hematopoietic cells. Progress in Clinical and Biological Research 333:435-445.

Naughton BA, Naughton GK. 1989. Hematopoiesis on nylon mesh templates - comparative long-term bone-marrow culture and the influence of stromal support cells. Annals of the New York Academy of Sciences 554:125-140.

Naughton BA, Tjota A, Sibanda B, Naughton GK. 1991. Hematopoiesis on suspended nylon screen-stromal cell microenvironments. Journal of Biomechanical Engineering - Transactions of the ASME 113:171-177.

Nielsen LK. 1999. Bioreactors for Hematopoietic Cell Culture. Annual Reviews in Biomedical Engineering 1:129-152.

Noll T, Jelinek, N., Schmidt, S., Biselli, M. & Wandrey, C. 2002. Cultivation of Hematopoietic Stem and Progenitor cells: Biochemical Engineering Aspects. Advances in Biochemical Engineering 74:111-128.

Ogawa M, Clark SC. 1988. Synergistic interaction between interleukin-6 and interleukin-3 in support of stem cell proliferation in culture. Blood Cells 14:329-335.

Oh D, Koller, MR. & Palsson, BO. 1994. Frequent Harvesting from Perfused Bone Marrow Cultures Results in in Increased Overall Cell and Progenitor Expansion. Biotechnology and Bioengineering 44:609-616.

Ohmizono Y, Sakaba H, Kimura T, Tanimukai S, Matsumura T, Miyazaki H, Lyman SD, Sonoda Y. 1997. Thrombopoietin augments *ex vivo* expansion of human cord blood-derived hematopoietic progenitors in combination with stem cell factor and flt3 ligand. Leukemia 11:524-530.

Osawa M, Hanada K, Hamada H, Nakauchi H. 1996. Long-term lymphohematopoietic reconstitution by CD34-low/negative hematopoietic stem cells. Science 273:242-245.

Palsson BO, Paek S-H, Schwartz RM, Palsson MA, Lee G-M, Silver S, Emerson SG. 1993. Expansion of human bone marrow progenitor cells in a high cell density continuous perfusion system. Bio/Technology 11:368-372.

Pennathur-Das RL, L. 1987. Augmentation of In Vitro Human Marrow Erythropoiesis Under Physiological Oxygen Tensions Is Mediated by Monocytes and T Lymphocytes. Blood 69(3):899-907.

Petzer AL, Zandstra PW, Piret JM, Eaves CJ. 1996. Differential cytokine effects on primitive ($CD34^+CD38^-$) human hematopoietic cells: novel responses to flt3-ligand and thrombopoietin. Journal of Experimental Medicine 183:2551-2558.

Piacibello W, Sanavio F, Garetto L, Severino A, Dane A, Gammaitoni I, Agglietta M. 1998. Differential growth factor requirement of primitive cord blood hematopoietic stem cells for self-renewal and amplification versus proliferation and differentiation. Leukemia 12:718-727.

Pierson BA, Europa AF, Hu WS, Miller JS. 1996. Production of human natural killer cells for adoptive immunotherapy using a computer-controlled stirred-tank bioreactor. Journal of Hematotherapy 5:474-483.

Poloni A, Giarratana MC, Kobari L, Firat H, Bouchet S, Gorin NC, Douay L. 1997. The *ex vivo* expansion capacity of normal human bone marrow cells is dependent on experimental conditions: role of cell concentration, serum and $CD34^+$ cell selection in stroma-free cultures. Hematological Cellular Therapy 39:49-58.

Quesenberry P, Colvin GA. 2001. Hematopoietic stem cells, progenitor cells, and cytokines. In: Seligsohn U, editor. Hematology. New York: McGraw-Hill. p 153-174.

Quesenberry P, Crittenden R, Lowry P, Kittler E, Rao S, Peters S, Ramshaw H, Stewart F. 1994. *In vitro* and *in vivo* studies of stromal niches. Blood Cells 20:97-106.

Rich I. 1986. A Role for the Macrophage in Normal Hemopoiesis. II. Effect of Varying Physiology Oxygen Tensions on the Release of Hemopoietic Growth Factors from Bone-marrow-derived Macrophage in vitro. Experimental Hematology 14:746-751.

Rich I, & Kubanek, B. 1982. The effect of reduced oxygen tension on colony formation of erythropoietic cells in vitro. British Journal of Haematology 52:579-588.

Rosenzwajg M, Canque B, Gluckman JC. 1996. Human dendritic cell differentiation pathway from $CD34^+$ hematopoietic precursor cells. Blood 87:535-544.
Rosenzweig M, Pykett MJ, Marks DF, Johnson RP. 1997. Enhanced maintenance and retroviral transduction of primitive hematopoietic progenitor cells using a novel three-dimensional culture system. Gene Therapy 4:928-936.
Rusten LS, Smeland EB, Jacobsen FW, Lien E, Lesslauer W, Loetscher H, Dubois CM, Jacobsen SE. 1994. Tumor necrosis factor-alpha inhibits stem cell factor-induced proliferation of human bone marrow progenitor cells *in vitro*. Role of p55 and p75 tumor necrosis factor receptors. Journal of Clinical Investigation 94:165-172.
Sandstrom CE, Bender JG, Miller ME, Papoutsakis ET. 1996. Development of a novel perfusion chamber to retain nonadherent cells and its use for comparison of human "mobilized" peripheral blood mononuclear cell cultures with and without irradiated bone marrow stroma. Biotechnology & Bioengineering 50:493-504.
Sandstrom CE, Bender JG, Papoutsakis ET, Miller WM. 1995. Effects of $CD34^+$ cell selection and perfusion on *ex vivo* expansion of peripheral blood mononuclear cells. Blood 86:958-970.
Sardonini CA, Wu YJ. 1993. Expansion and differentiation of human hematopoietic cells from static cultures through small-scale bioreactors. Biotechnology Progress 9:131-137.
Schwartz R, Palsson B, Emerson S. 1991. Rapid medium perfusion rate significantly increases the productivity and longevity of human bone marrow cultures. Proc. Natl. Acad. Sci. (USA) 88:6760-6764.
Shah AJ, Smogorzewska EM, Hannum C, Crooks GM. 1996. Flt3 ligand induces proliferation of quiescent human bone marrow $CD34^+CD38^-$ cells and maintains progenitor cells *in vitro*. Blood 87:3563-3570.
Shih C-C, DiGiusto D, Forman SJ. 2002. *Ex Vivo* Expansion of Transplantable Human Hematopoietic Stem Cells: Where Do We Stand in the Year 2000? Journal of Hematotherapy & Stem Cell Research 9:621-628.
Spangrude GJ, Johnson GR. 1990. Resting and activated subsets of mouse multipotent hematopoietic stem cells. Proceedings of the National Academy of Sciences USA 87:7433-7477.
Sutherland HJ, Lansdorp PM, Henkelman DH, Eaves AC, Eaves CJ. 1990. Functional characterization of individual human hematopoietic stem cells cultured at limiting dilution on supportive marrow stromal layers. Proceedings of the National Academy of Sciences USA 87:3584-3588.
Tennant G. 2000. Control of pH in human long-term bone marrow cultures with low-glucose medium containing zwitterion buffer lengthens the period of haematopoiesis activity. Brithish Journal of Haematology 109:785-787.
Till JE, McCulloch EA. 1961. A direct measurement of the radiation sensitivity of normal mouse bone marrow cells. Radiation Research 14:213-222.
Trentin JJ. 1970. Influence of hematopoietic organ stroma (hematopoietic inductive microenvironments) on stem cell differentiation. In: Gordon AS, editor. Regulation of hemopoiesis. New York: Appleton-Century-crofts. p 161-186.
Tsukada J, Misago, M., Kikuchi, M., Sato, T., Ogawa, R., Ota, T., Oda, S., Morimoto, I., Chiba, S. & Eto, S. 1992. Interaction between recombinant human erythropoietin and serum factor(s) on murine megakaryocyte colony formation. Blood 80(1):37-45.
Veiby OP, Jacobsen FW, Cui L, Lyman SD, Jacobsen SE. 1996. The flt3 ligand promotes the survival of primitive hemopoietic progenitor cells with myeloid as well as B lymphoid potential. Suppression of apoptosis and counteraction by TNF-alpha and TGF-beta. Journal of Immunology 157:2953-2960.
Verfaillie CM. 2001. *Ex Vivo* Expansion of Hematopoietic Stem Cells. In: Zon LI, editor. Hematopoiesis: A Developmental Approach. New York: Oxford University Press. p 119-129.
Verfaillie CM, Hurley R, Bhatia R, McCarthy JB. 1994. MIP-1a combined with IL-3 conserves primitive human LTBMC-IC for at least 8 weeks in *ex vivo* "stroma-non-contact" cultures. Journal of Experimental Medicine 179:643-649.
Wang T-Y, Brennan JK, Wu JHD. 1995. Multilineal hematopoiesis in a three-dimensional murine long-term bone marrow culture. Experimental Hematology 23:26-32.
Weissman I. 1994. Developmental switches in the immune system. Cell 76:207-218.
Williams DE, Hangoc G, Cooper S, Boswell HS, Shadduck RK, Gillis S, Waheed A, Urdal D, Broxmeyer HE. 1987. The effects of purified recombinant murine interleukin-3 and/or purified natural murine CSF-1 *in vivo* on the proliferation of murine hifg- and low-proliferative potential colony-forming cells: demonstration of *in vivo* synergism. Blood 70:401-403.

Won JH, Cho SD, Park SK, Lee GT, Baick SH, Suh WS, Hong DS, Park HS. 2000. Thrombopoietin in synergy with other cytokines for expansion of cord blood progenitor cells. Journal of Hematotherapy and Stem Cell Research 9:465-473.

Yagi M, Ritchie KA, Sitnicka E, Storey C, Roth GI, Bartelmez S. 1999. Sustained ex vivo expansion of hematopoietic cells mediated by thrombopoietin. Proceedings of the National Academy of Sciences USA 96:8126-8131.

Yamaguchi M, Hiramaya, F., Kanai, M., Sato, N., Fukazawa, K., Yamashita, K., Sawada, K., Koike, T., Kuwabara, M., Ikeda, H. & Ikebuchi, K. 2001. Serum-free coculture system for ex vivo expansion of human cord blood primitive progenitos and SCID mouse-reconstituting cells using human bone marrow primary stromal cells. Experimental Hematology 29:174-182.

Yang H, Miller, WM., Papoutsakis, ET. 2002. High pH promotes megakaryocytic maturation and apoptosis. Stem Cells 20:320-328.

Zandstra PW, Conneally E, Petzer AL, Piret JM, Eaves CJ. 1997. Cytokine manipulation of primitive human hematopoietic cell self-renewal. Proceedings of the National Academy of Sciences USA 94:4698-4703.

Zandstra PW, Eaves CJ, Piret JM. 1994. Expansion of Hematopoietic Progenitor Cell Populations in Stirred Suspension Bioreactors of Normal Human bone Marrow Cells. Bio/Technology 12:909-914.

CHAPTER 15

MONITORING THE PERFORMANCE OF TISSUE ENGINEERING BIOREACTORS USING MAGNETIC RESONANCE IMAGING AND SPECTROSCOPY

A.A. NEVES AND K.M. BRINDLE

Department of Biochemistry, University of Cambridge, Cambridge, UK

1. INTRODUCTION

The use of intensive bioreactors, that can maintain high cell densities, has become common practice in industry, in particular for the production of biopharmaceuticals. These include fixed- and fluidized-beds (Racher and Griffiths, 1993; Wang et al., 2002) spin filters (Deo et al., 1996), acoustic filters (Gorenflo et al., 2002) and hollow-fibre (Gorter et al., 1993) bioreactors. As culture systems, bioreactors offer close control of the physical properties of the medium, as well as its chemical composition. Intensive systems, such as the ones described above, have the additional advantage of allowing the recovery of the final product without cell contamination and typically with high productivities. These systems have also been used in the last few years in the field of tissue engineering (Pei et al., 2002; Ratcliffe and Niklason, 2002). The major applications of intensive bioreactors in this area include the cultivation *ex vivo* of bioartificial tissues for implantation and the culture of large quantities of cells which can be used directly in cell-based therapies or incorporated in extra-corporeal devices to be used as surrogate organs, e.g. the bioartificial liver (Allen and Bhatia, 2002; Samuel, 2002; van de Kerkhove et al., 2002). Optimizing the performance of intensive bioreactors for tissue engineering applications requires investigation of the effects of alternative designs and operating

J.B. Chaudhuri & M. Al-Rubeai (eds.), Bioreactors for Tissue Engineering, 335 - 352.

strategies on the performance of such systems, in terms of cell viability, tissue formation and histological properties.

Nuclear magnetic resonance (NMR) imaging (MRI) and spectroscopy (MRS) are amongst the few analytical techniques able to obtain information non-invasively from living systems. NMR is a tremendously versatile technique that is able to detect a variety of atomic nuclei found in biological systems. These include ^{1}H, ^{31}P, ^{23}Na, ^{39}K and also non-abundant nuclei, such as ^{13}C and ^{15}N, which, through the use of ^{13}C- and ^{15}N-labelled substrates, can be used to study metabolic fluxes. Measurements can be made from a variety of biological entities, including isolated cells (Franks et al., 1996), perfused organs (Desmoulin et al., 1990), animals (Koretsky and Williams, 1992) and human beings (Gruetter et al., 1992).

Clinical MRI, in particular, has evolved in the past decade from an imaging technique that reported purely on anatomical properties to one that is now able to investigate dynamically a variety of tissue and organ functions (Koretsky, 2002; Shu et al., 2002). MRS, on the other hand, has played historically an important role in the analysis of metabolic pathways in a wide range of systems, from isolated cells to humans. Both of these techniques have been successfully applied in the past to investigate the performance of intensive bioreactor systems (Donoghue et al., 1992; Brindle, 1998). More recently, NMR imaging and spectroscopy have been used to study the development of engineered tissues both *in vitro* and *in vivo* (Constantinidis et al., 2002).

The purpose of this chapter is to describe the range of NMR-based techniques currently available to the tissue engineer and the type of information that can be obtained with them to guide the design and operation of tissue engineering bioreactors.

2. CELLULAR ENERGETICS ON-LINE USING ^{31}P MRS

^{31}P MRS can be used to measure the cellular concentrations of ATP and inorganic phosphate (P_i). The concentration of free ADP in solution is normally below the detection limit. However, in cells containing the enzyme creatine kinase, the free cytosolic ADP concentration can be estimated from the observed concentrations of phosphocreatine and ATP, and the pH (Brindle et al., 1989; Brindle, 1995). This technique therefore provides a unique insight into cellular energy status.

The intra- and extracellular pHs can be estimated from the chemical shifts of the P_i resonances, if these are resolved. The observation that the chemical shift of the P_i resonance in the ^{31}P spectrum could be used to determine the intracellular pH was originally made in the 1970s (Moon and Richards, 1973). The later realization that these non-invasive measurements could also be made on skeletal muscle tissue (Hoult et al., 1974) led to the birth of "*in vivo* NMR". More recently, the technique has been used to monitor cell concentration and stability in various other systems, including hybridoma cells in a membrane bioreactor (Fernandez et al., 1988) and in a hollow-fibre system (Gillies et al., 1991), yeast in a NMR airlift bioreactor (Melvin and Shanks, 1996) and perfused glioma cells following hyperosmotic shock in culture, as a model for brain injury (Lien et al., 1992). Hesse and co-workers employed ^{31}P MRS to measure the intracellular pH of perfused fungi (Hesse et al., 2000).

^{31}P MRS has also found application in the field of tissue engineering. ^{31}P spectroscopy was used to monitor cell viability and to control the pH in hepatocyte-loaded hollow-fibre bioreactors (Macdonald et al., 1998). Similar systems have been used as bioartificial livers. Papas and co-workers used ^{31}P MRS to study the effect of short-term hypoxia on the bioenergetic status, metabolism, and insulin secretion capability of perfused pancreatic constructs (Papas et al., 2000). The concentrations and spin-lattice relaxation times of phosphorus-containing metabolites in neocartilage, growing in an NMR-compatible hollow-fibre bioreactor, have been also investigated (Petersen et al., 2000).

We have used ^{31}P MRS to evaluate the effect of medium flow rate on the development of bioartificial meniscal cartilage in a perfusion bioreactor. The system was operated with a range of global flow rates between 20 and 60 ml.min^{-1} for a period of up to two weeks (Neves et al., 2003). ^{31}P spectra were acquired every 3-4 days in order to follow the energetics of the cell population within a single construct in the bioreactor. One of these spectra, acquired after 8 days of cultivation at 60 ml.min^{-1}, is shown in Figure 1. A splitting of the inorganic phosphate peak (P_i) was assumed to be due to its partition between two different compartments of

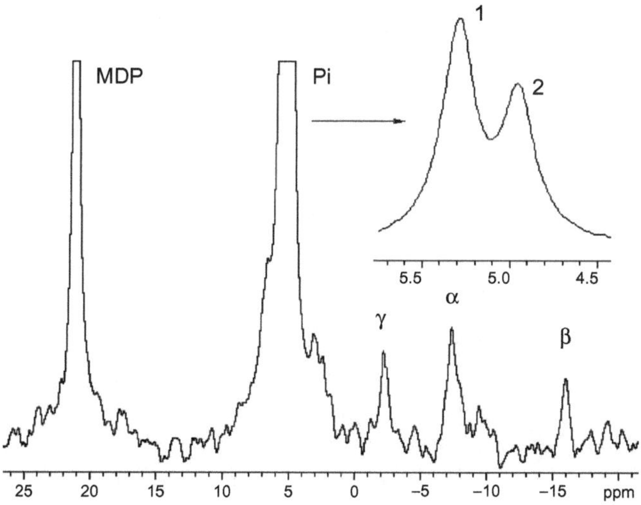

Figure 1. ^{31}P spectrum of meniscal fibrochondrocyte cells growing in a construct produced at a flow rate of 60 ml.min^{-1} in a perfusion bioreactor. The spectrum was acquired after 8 days of cultivation and is the sum of 3000 transients which took 180 minutes to acquire. At this stage, the cell density in the construct (volume of 1.13 cm^3) was approximately 1.5 x 10^8 cells. The inset is a magnification of the P_i peak which shows that it has two components (labelled 1 and 2), corresponding to two compartments in the bioreactor with different pHs (see section 2 in the text). Abbreviations: P_i– inorganic phosphate; and γ-, α- and β-phosphates, respectively, of nucleoside triphosphates; MDP– methylene diphosphonate, used as an internal standard in a capillary tube inside the bioreactor; The nucleoside phosphates are predominantly ATP and ADP.

different pHs. One of these compartments is the bulk medium surrounding the construct, of which the pH is automatically regulated by the system bio-controller. The other compartment corresponds to the microenvironment of the cells, i.e. the inner volume of the construct which becomes more acidic throughout cultivation, due to mass transport limitations, caused by cell growth and extracellular matrix (ECM) deposition. The first peak of the P_i doublet, at lower field (see inset in Figure 1), corresponding to the bulk medium (pH 7.40 in this example), can be used regularly throughout the run to validate, and correct if necessary, the bio-controller readings. The second peak, at higher field, corresponds to the inner volume of the construct which has a lower pH (7.06). The medium flow rate had a significant effect on the energetics of the cell population, as shown in Figure 2. At the lowest flow rate tested of 20 $ml.min^{-1}$ the levels of NTP declined for the first four days of cultivation, which was probably due to limitations in the mass transport of nutrients, in particular oxygen. At 30 and 40 $ml.min^{-1}$ there was an increase of the NTP levels indicating cell proliferation and higher metabolic activity. The use of higher flow rates (50 and 60 $ml.min^{-1}$) induced a rapid increase of the NTP levels of the construct, during the first 7 days of cultivation, followed by a dramatic decline until the end of the cultivation period, possibly due to the destructive effect of flow-induced shear stress at this range of flow rates. An optimal flow rate of 40 $ml.min^{-1}$ was therefore identified for this particular system using ^{31}P MRS. These results are in accordance with evidence from other NMR-based methods, as we shall discuss later in the text.

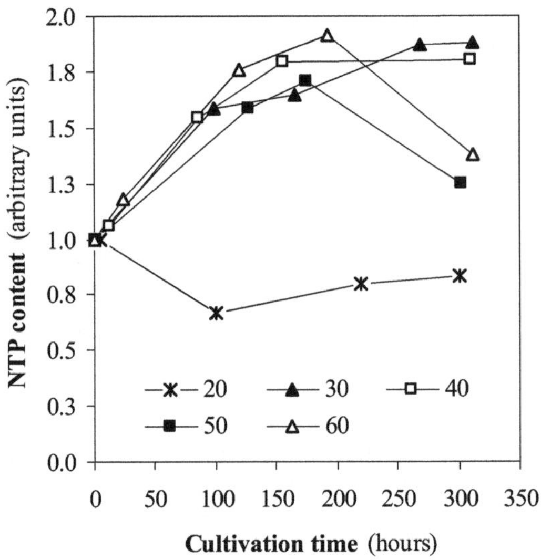

Figure 2. Time courses of NTP content of meniscal cartilage constructs produced in a perfusion bioreactor, as a function of the global medium flow rate. NTP content is expressed in arbitrary units. The flow rates are in $ml.min^{-1}$.

The major applications of ^{31}P MRS to optimize the performance of tissue engineering bioreactors are essentially two-fold. The technique can be used to follow cell growth and death, through the associated increase or reduction of the NTP signals. Cell death, in particular, results normally in a reduction of the cellular energy status (increased P_i concentrations) and in a subsequent loss of high-energy phosphate compounds. ^{31}P MRS can also be applied to monitor the efficiency of nutrient supply to a particular developing construct in the bioreactor, as this will affect directly the energy status of the cell population within that construct.

3. CELL GROWTH AND DISTRIBUTION

Viable cell concentration is perhaps the single most important parameter to determine in intensive bioreactor operation. Quantitative metabolic measurements and modelling in such systems often require an estimate of cell concentration. The use of nutrient uptake rates is a traditional non-invasive approach to estimate this parameter. However this assumes constant specific consumption rates throughout the cultivation period, an assumption which is often not justified (Mancuso et al., 1990).

van Zijl and co-workers, exploited the low apparent diffusion coefficient (ADC) of intracellular water molecules to estimate their fraction in immobilised breast cancer cells (van Zijl et al., 1991). A similar method was used to monitor cell volume and water exchange time in mammalian cells that had been exposed to osmotic stress and to conditions that induce apoptosis (Pfeuffer et al., 1998).

In a diffusion-weighted spectroscopy experiment, the signal intensity is a function of the magnitude and duration of the magnetic field gradients applied in the experiment and the diffusion time, according to Equation 1.

$$\ln(S/S_0) = -\gamma^2 G^2 \delta^2 (\Delta-\delta/3) D = -b D \quad (1)$$

where D is the apparent diffusion coefficient of water, γ is the gyromagnetic ratio of the observed nucleus, in this case proton, δ and G are the duration and strength of the pulsed magnetic field gradients respectively, and Δ (the diffusion time) is the time between the leading edges of these field gradient pulses. S and S_0 are the measured signal intensities in the presence and absence of the gradient pulses, respectively. b is a measure of the diffusion weighting in units of $m^{-2}.s$.

A plot of the natural logarithm of the signal intensity of water protons in a cell-filled bioreactor versus b value gave a bi-exponential plot, representing two water fractions with different ADCs. The slopes of these two components (see Figure. 3) can be used to estimate the ADC values of the water molecules within the cell mass and exterior to it. The intercepts of their asymptotes can be used to estimate the water fractions in these two compartments, according to Equation 2. This estimate assumes that the relaxation times of the water molecules in the two compartments are similar.

$$CVF = \exp(A_1)/\exp(A_2) \quad (2)$$

where CVF is the volume of the cell mass in the bioreactor and A_1 and A_2 are, respectively, the intercepts of the asymptotes for the water molecules in the two compartments in the bioreactor (see Figure 3).

The diffusion-weighting principle can also be incorporated into imaging experiments to produce images that can distinguish cells in various environments. By using a variety of gradient amplitudes, these imaging experiments can also be used to produce maps of the ADC of water. Neeman and co-workers used one of these methods to distinguish a rim of viable cells in multicellular tumour spheroids, from a necrotic core region (Neeman et al., 1991). The ADC of the water molecules inside the cells was much lower than the ADC of the surrounding interstitial water and the water in the necrotic core. The changes in cell volume, monitored by diffusion-weighted MRI, were measured in a rat model of brain injury (Verheul et al., 1994). The results showed that the temporal course of the intensity changes in the diffusion-weighted images paralleled the progressive shrinkage of the extracellular space, as determined from electrical impedance measurements.

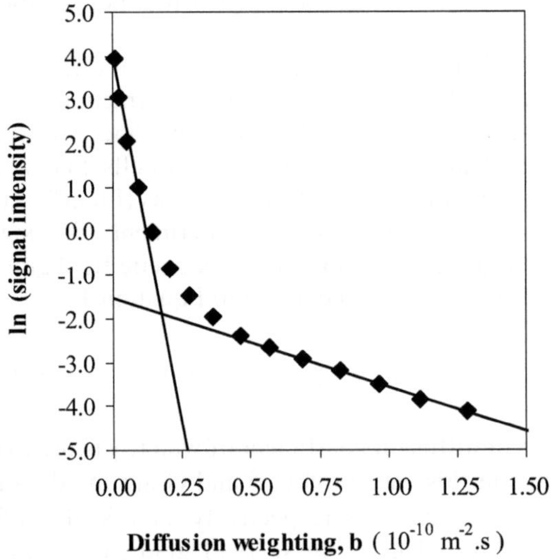

Figure 3. Bi-exponential plot of the natural logarithm of the water proton signal intensity in a diffusion-weighted MRS experiment on a cell and ECM-filled bioreactor, as a function of the diffusion weighting (b). The intercepts of the asymptotes of the two exponential components of the curve can be used to estimate the volume fraction of cells and ECM in the bioreactor (see Equations 1 and 2, section 2 in the text). The b value was incremented by increasing the amplitude of the gradient pulses in a STEAM experiment (van Zijl et al., 1991).

We have used diffusion-weighted MRI methods in the past to map the growth and distribution of CHO cells in a hollow-fibre bioreactor (Callies et al., 1994). A conventional, non-diffusion-weighted image, which maps only the water distribution, showed the distribution of the fibres in the reactor but not where the cells were located. However, the cells were clearly visible in a diffusion-weighted image. This was also the case in more recent studies in our lab using CHO cells immobilized in macroporous cubic carriers in a perfusion bioreactor (Thelwall et al., 2001). A conventional image, in this case, was only able to show the distribution of the carriers inside the bioreactor. Using diffusion-weighted imaging, a thin layer of cells (approximately 0.8-mm thick) was identified at the periphery of the carriers, indicating cell growth restrictions, due to limitations in the mass transport of oxygen and other nutrients into the carrier.

These methods were used more recently to monitor the effect of flow rate in a perfusion bioreactor on meniscal cartilage cell growth and ECM deposition (Neves et al., 2003). Diffusion-weighted MR images acquired from a meniscal cartilage construct cultivated at 60 ml.min^{-1} are shown in Figure 4.

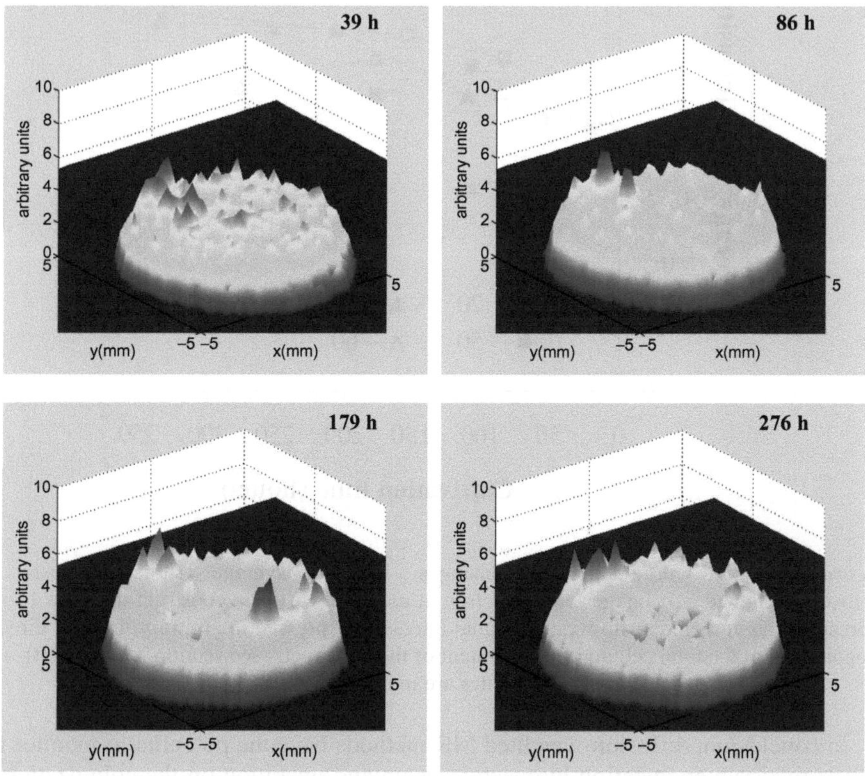

Figure 4. Diffusion-weighted MR images acquired from a meniscal cartilage construct cultivated at a flow rate of 60 ml.min^{-1} in a perfusion bioreactor. The image plane was within the construct and perpendicular to the direction of flow. The cultivation time is expressed in hours. In these diffusion-weighted images, signal intensity, expressed in arbitrary units, is related to cell and ECM density (see section 3 in the text).

An initial increase in cell content was observed followed by a subsequent decline in the second half of the culture, identified in these images as a loss of signal intensity at the center of the construct. Figure 5 shows the time-dependent changes in signal intensities in diffusion-weighted images of constructs perfused at different flow rates. The changes observed paralleled, to some extent, those observed in nucleoside triphosphate (NTP) content of the constructs determined by ^{31}P MRS (Figure 2). This is consistent with signal intensity in diffusion-weighted MR images being predominantly an indication of cell content.

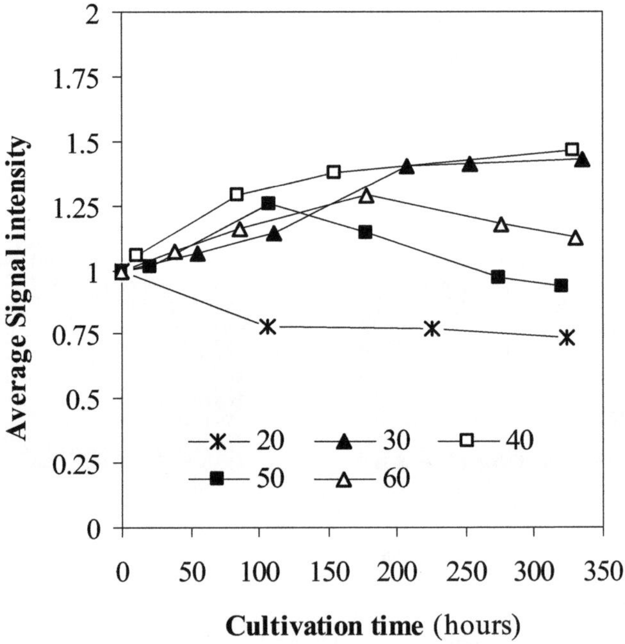

Figure 5. Relative time-dependent changes in the average signal intensity in diffusion-weighted MR images acquired from a meniscal cartilage construct during a 14-day bioreactor run at different flow rates. Signal intensity, expressed in arbitrary units, is directly proportional to the total cell and ECM content of the constructs (see section 3 in the text). The flow rates are in $ml.min^{-1}$.

In conclusion, diffusion-weighted MR methods have the potential to monitor the effects of tissue engineering bioreactor design and operation on the colonization of particular regions in a developing construct. Furthermore, they are able to provide quantitative estimates of the volume of the cell mass and ECM deposition.

4. DIFFUSIVITY OF SMALL MOLECULES IN BIOARTIFICIAL TISSUES

In theory, a diffusion-weighted experiment can be used to map the diffusion coefficient of any molecule in a bioreactor for which a detectable MR signal is available. However, in practice, for reasons of sensitivity, this will most of the time be water, although by using larger voxel sizes, it may be possible to detect the diffusivity of other species, albeit with lower spatial resolution. An alternative way to measure the diffusion of a small molecule, other than water, in these systems, is to measure the diffusion of a molecule or ion able to affect the intensity of the water signal in an MR image. This method provides much better sensitivity and spatial resolution and allows the diffusion of the molecule or ion to be monitored over much longer time scales. Such molecules, which can enhance the intensity of the water signal in an MR image, are known as contrast agents (CAs). These are used routinely in clinical MRI to enhance tissue contrast and to assess tissue perfusion. Gadolinium diethylenetriaminepentaacetate (Gd-DTPA) is one of the most commonly used agents (Gadian, 1996).

In a contrast agent-based MRI experiment, the location and concentration of the reagent is detected through its effect on the spin-lattice relaxation rate ($1/T_1$) of exchanging water protons. The main advantage of using Gd-DTPA is that, at low concentrations (typically less than 2 mM), the spin-lattice relaxation rate is a linear function of its concentration (Thelwall et al., 2001). As a consequence, it is possible to obtain a full concentration map of Gd-DTPA based on a T_1 image (i.e., an image in which pixel intensity is equal to the spin-lattice relaxation rate). The acquisition of a series of these maps, following the injection of the CA into the system, allows the diffusivity of the reagent to be estimated. We used this method to measure the diffusion coefficient of Gd-DTPA in CHO cell aggregates growing in macroporous cubic carriers in a perfusion bioreactor (Thelwall et al., 2001). These studies showed that cell proliferation occurred predominantly at the periphery of the carriers, where the lowest diffusion coefficients were estimated. Fast appearance of the CA in the centre of the carriers indicated that the flow of medium in these regions was driven by convection. These findings agreed with determinations of cell distribution using diffusion-weighted MRI.

We recently used contrast agent-enhanced MRI to estimate the diffusivity of Gd-DTPA in bioartificial meniscal cartilage constructs produced in a perfusion bioreactor (Neves et al., 2003). Perfusion of meniscal constructs was measured by adding the CA to the culture medium. Analysis of the changes in signal intensity, which were associated with a reduction in the T_1-relaxation rate of water protons, was performed by acquiring a series of images of a single construct in the bioreactor, after the injection of the CA into the system. This is illustrated in Figure 6, which shows a series of T_1-weighted images acquired from a construct after injection of the CA on day 14 of cultivation. An increase in signal intensity, approximately 7 minutes after CA injection into the system, indicated its arrival in the spaces surrounding the construct. A slow increase in signal intensity within the construct was later observed, as it became infiltrated by the CA. Figure 7 shows the time-dependent changes in CA concentration in two regions of interest (ROIs) located within and outside the construct, in a saggital cross section. The diffusion

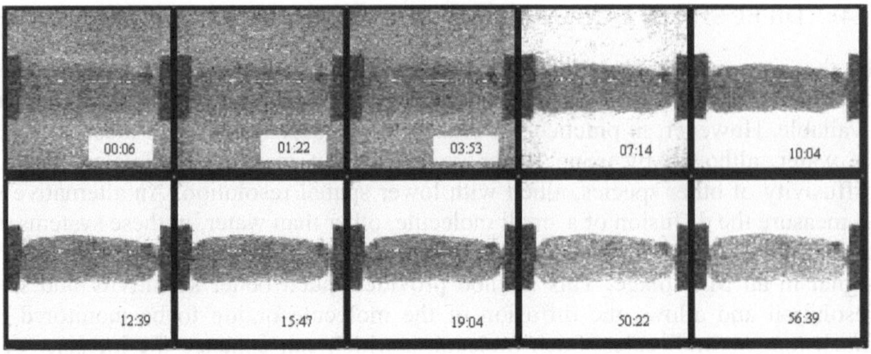

Figure 6. Time course of Gd-DTPA inflow into a meniscal cartilage construct on day 14 of cultivation at a perfusion flow rate of 30 ml.min^{-1}. Images are maps of T_1 relaxation rate, which is directly proportional to GdDTPA concentration (see section 4 in the text). The image plane was perpendicular to the plane of the construct and parallel to the direction of flow. The numbers indicate the time (minutes:seconds) after the injection of the contrast agent into the system. At 7 minutes the contrast agent appears in the spaces around the construct and in the subsequent 50 minutes penetrates the inner regions of the construct.

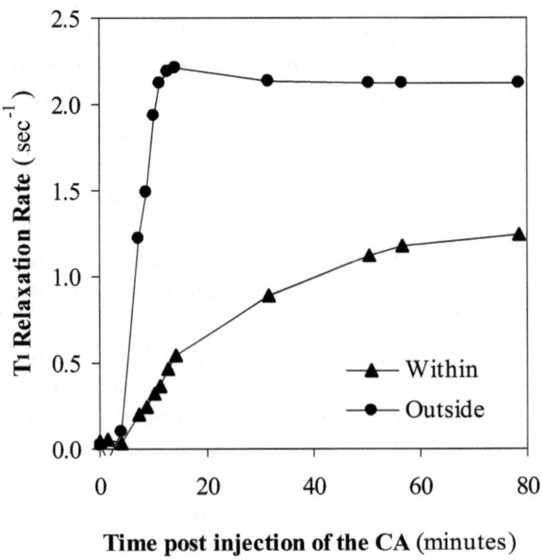

Figure 7. Time-dependent changes in contrast agent concentration (expressed as T_1-relaxation rate) in a specific region of interest (ROI) within the construct. The perfusion profiles correspond to a construct perfused at 30 ml.min^{-1}. Two ROIs were selected; outside and within the construct in the region exposed directly to flow. These ROIs measured $3 \times 9.5 \times 0.85$ mm.

coefficient of the CA within the internal ROI was determined using an infinite slab model, a simple one-dimensional, non-steady-state diffusion model (Crank, 1964). A value of $(7.0 \pm 0.5)\times10^{-11}$ $m^2.s^{-1}$ was obtained. This is similar to a value of 9.2×10^{-11} $m^2.s^{-1}$ reported by Foy and Blake for the diffusion of Gd-DTPA in native human articular cartilage (Foy and Blake, 2001), indicating that this region of the construct has similar properties to those of the native tissue, in terms of permeability to a relatively low molecular weight compound (the molecular weight of Gd-DTPA is 547 Da).

This experiment can also be used to estimate the volume fraction of the construct from which the agent is excluded, thus providing an indication of the volume filled with cells and their ECM. By day 14, at a flow rate of 30 $ml.min^{-1}$, the excluded volume fraction of the construct was estimated at 55%. Removal of the construct at this stage and subsequent analysis of the cell content, through DNA quantification (Kim et al., 1988), gave a cell volume fraction of 53% of the final wet weight of the construct, assuming a volume of 10^{-8} cm^3 for a fibrochondrocyte cell (Neves et al., 2003). The total collagen and GAG contents were measured using specific chemical assays (Woessner, 1961; Farndale et al., 1986) at 0.6 and 0.11 % of the wet weight thus making only a small contribution to the excluded volume fraction.

In conclusion, contrast-agent enhanced MRI is a valuable method for the determination of useful parameters in bioreactor design. These include the diffusivity of a small molecule (CA) in specific regions of the bioartificial tissue and the porosity of the tissue. This information is particularly useful for the screening of different elements of tissue design, including alternative scaffold geometries.

5. NON-INVASIVE FLOW IMAGING INSIDE A DEVELOPING CONSTRUCT

NMR-based methods have been used over the last four decades for measuring flow in various systems. The principle of *time-of-flight* (TOF) is one of the most commonly used. In a TOF experiment, a slice-selective excitation pulse is used to perturb the magnetisation within a voxel, normally enclosed in a plane perpendicular to the direction of flow (Axel et al., 1986; Wehrli et al., 1986). The signal is acquired after a very short period of time from the same slice. The movement of unlabelled spins into the voxel of interest produces an increase in the apparent spin-lattice relaxation rate ($R_1=1/T_1$). The T_1 map produced by this imaging sequence, when appropriately calibrated, can be used to produce a map of the axial flow velocities within a blood vessel or in a bioreactor.

The principle of TOF has been applied in clinical MRI to monitor blood flow within specific arteries in the body. The technique is known as Magnetic Resonance Angiography (MRA), and its major applications include the early diagnosis of brain death (Karantanas et al., 2002), coronary disease (Hardy et al., 2000) and thoracic vascular lesions (Ko et al., 1999). These applications were reviewed by Lamminen and Keto (Lamminen and Keto, 1996).

NMR flow imaging techniques, including TOF, have also been used in the past to measure fluid flow in bioreactors. Using ^1H NMR, the motion of proton spins was tracked to obtain velocity distributions in the extracapillary space of a cell-free hollow-fibre bioreactor (HFBR), as a function of position and time (Hammer et al., 1990). Flow in a solid-wall tube (phantom) was first used to calibrate pixel intensities with axial velocities in these systems. Flow-sensitive MRI techniques, modified with diffusion damping pulse sequences, were used to indirectly image cell growth patterns in commercial HFBRs (Donoghue et al., 1992). The application of NMR-based methods to visualize and unravel complex, heterogeneous transport processes in porous biological systems, has been reviewed recently (As and Lens, 2001). This includes applications in biofilm modelling, bioreactors, microbial mats and sediments.

We have implemented a TOF-based method for characterising the axial flow velocity profiles for different cross sections of a tissue-engineering perfusion bioreactor. The T_1 maps produced using this imaging sequence, were calibrated and used to produce a map of axial flow velocities inside a developing meniscal cartilage construct. The time courses of the mean axial velocities across one of the meniscal constructs in bioreactors operated at 30 and 60 $ml.min^{-1}$ are shown in Figure 8.

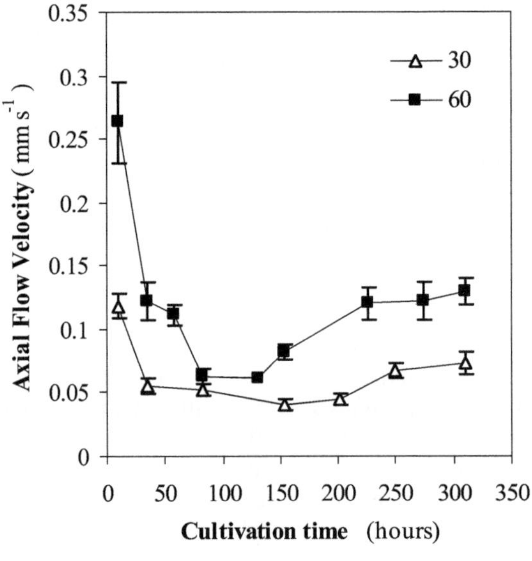

Figure 8. Time-dependent changes in the axial flow velocity across the centre of a meniscal cartilage construct produced at two different perfusion flow rates. Flow rates are expressed in $ml.min^{-1}$. Values are means of the axial flow velocity (n=3). Error bars represent standard deviations.

A marked decrease of the mean flow velocities was observed at both flow rates between day 1 and day 7 of cultivation. These were consistent with increased cell densities determined using ^{31}P MRS measurements of NTP content, diffusion-weighted MRI experiments and from contrast agent-enhanced MRI measurements of the excluded volume fraction. The higher flow rate clearly produced better perfusion of the constructs, albeit being destructive after 7 days of cultivation, reducing cell and matrix content at the centre of the constructs. This effect is illustrated in Figure 9 which compares the axial flow velocity profile across a construct (A), with the axial diffusion-weighted MR image (B), both acquired after 14 days of cultivation at 60 ml.min^{-1}.

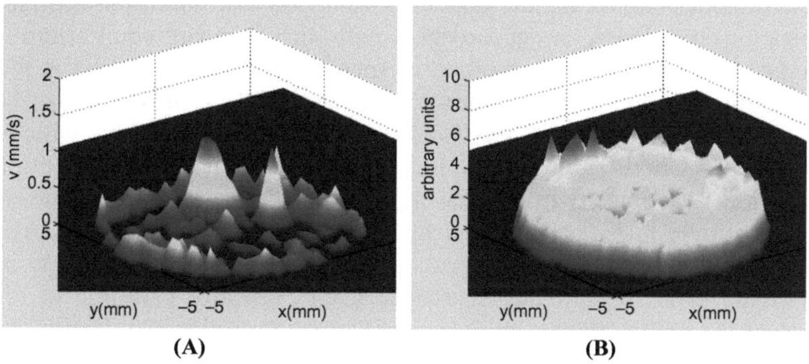

Figure 9. MRI characterisation of a meniscal cartilage construct produced at a perfusion flow rate of 60 ml.min^{-1}, at day 14 of cultivation. (A) Axial flow velocity profile, in mm.s^{-1} (B) Axial diffusion-weighted MR image. Signal intensity, which is in arbitrary units, is proportional to the total cell and ECM content of the construct (see section 3 in the text).

In summary, NMR-based methods designed to measure flow, including the ones based on the principle of TOF described here, provide a valuable insight into tissue permeability throughout the cultivation period. They enable the tissue engineer to analyze the convective (macroscopic) flow of medium through the tissue and are therefore complementary to the contrast agent-based methods which are able to describe the diffusive (microscopic) flow of medium within a functional construct. Other techniques, such as those based on Laser Doppler (Begley and Kleis, 2000) can also be used to measure flow. However, their use requires the introduction of sensors into the bioreactor which may affect its performance, as could the introduction of tracer particles, normally required by this technique.

6. NON-INVASIVE ESTIMATE OF ECM COMPOSITION

The use of contrast agents to detect degenerative changes in orthopaedic tissues, predominantly articular and meniscal cartilage, has been tested recently in the clinic. Imaging methods that allow visualization of cartilage integrity should revolutionize our understanding of cartilage degenerative diseases and the surgeon's ability to diagnose and establish efficacious treatment strategies (Burstein et al., 2000). MRI has proven to be well suited for examining lesions of both articular and meniscal cartilage and bone tissues. Furthermore, it can detect lesions before they become apparent on radiographs (Adams et al., 1991; McAlindon et al., 1991). Various MRI sequences, with and without the injection of intra-articular contrast agents (e.g.

Gd-DTPA), allow the visualization of degenerative changes in cartilage disease. In osteoarthritis (OA), in particular, there is a characteristic loss of proteoglycans in the superficial zones (Poole, 1993), with nearly complete loss of these macromolecules in the later stages of this degenerative disease (Buckwalter and Martin, 1995). MR methods for measuring glucosaminoglycan (GAG) depletion are therefore desirable to follow the progression of OA, as they are the main constituents of proteoglycans. All these methods exploit the fact that the mobile ion concentrations change with the concentrations of GAG, in accordance with the Donnan equilibrium principle (Maroudas, 1968). A possible method of measuring cartilage GAGs is to measure the distribution of the charged contrast agent $Gd(DTPA)^{2-}$ in the tissue, because it will distribute in relation to the GAG concentrations. In this method, which has been validated *ex-vivo*, both in bovine and human cartilage (Bashir et al., 1999), the GAG concentration is estimated from measurements of the T_1-relaxation rate, which, as mentioned above, is linearly related to $Gd(DTPA)^{2-}$ concentration. GAG content, estimated using this technique, was validated against a biochemical assay in excised human tissue. The method has also been used clinically (Trattnig et al., 1999). Delivery of the contrast agent by intravenous or intra-articular injection permits its entry into cartilage. Several *in vivo* studies have validated this method to map GAG distribution in cartilage. Histology results obtained after excision matched the distribution predicted with Gd-DTPA-enhanced MRI (Bashir et al., 1997). In summary, the technique provides a validated, quantitative, sensitive and specific measure for GAG content of cartilage *in vitro*, where the equilibration of the tissue with the contrast agent can be ensured. The potential of the method *in vivo* has also been demonstrated, where it is able to map spatial variations in GAG distribution.

Collagen is the major solid macromolecular constituent of articular and meniscal cartilage, making up approximately 20 % and 23 %, respectively of the tissue wet weights (Mow et al., 1990). In the early stages of OA, collagen concentration does not change markedly (Buckwalter and Martin, 1995). However, it is believed that there are molecular changes as a consequence of disease progression, namely partial changes in the type of collagen in the tissue. Several MRI methods have been developed to evaluate collagen concentration in cartilage. These include T_2-weighted MRI (Dardzinski et al., 1997) and magnetisation transfer (MT) techniques (Gray et al., 1995). None, however, is specific for collagen alone. T_2 contrast in cartilage was one of the earliest contrast mechanisms to be investigated (Lehner et al., 1989). Both T_1 and T_2 relaxation times depend on tissue hydration. T_2 however depends on many other factors. Differences in T_2 across the cartilage may also reflect changes in collagen orientation (Mlynarik et al., 1996; Goodwin et al., 1998). Magnetisation transfer is related to the interactions of water molecules and large macromolecules. It has been suggested that collagen is the main source of MT in articular cartilage (Kim et al., 1993). Other studies suggest though that GAG also contributes to MT, although to a lesser extent (Wachsmuth et al., 1997). The MT effect depends on collagen structure and concentration. It is therefore different in articular cartilage, in which the predominant form is type II collagen, when compared to meniscal cartilage, where type I collagen is the main constituent. More work needs to be done to relate MR parameters with other measurement modalities, in order to develop a more quantitative modelling of the MT effect due to the various constituents of tissues (Burstein et al., 2000).

Parameters derived from MR images, including T_1, T_2, MT and ADC (apparent diffusion coefficient) have also been used *in vitro* to quantitatively assess the composition of cartilage tissue in bioreactors. Potter and co-workers have correlated the biochemical composition of bioartificial cartilage produced in a HFBR with these MRI parameters. These parameters were also used to evaluate the effects of different growth factors on cell physiology and tissue composition (Potter et al., 2000).

In summary, MRI parameters can be related to cartilage tissue composition both *in vitro* in bioreactors and *in vivo*. The application of these techniques to study the composition and physiology of other bioartificial tissues in bioreactors may also be possible, provided that the appropriate calibrations are established between the parameters of interest.

7. CONCLUSIONS

The NMR-based methods described here are able to evaluate local changes in cell density and metabolic activity, perfusion and flow velocities within a functional engineered tissue. The application of NMR techniques is particularly attractive in the context of bioartificial organs. The application of such techniques over the past 25 years to the study of intact biological systems, including man, has generated a wealth of data, including spectroscopic and imaging profiles of healthy and diseased tissues. This database is a valuable tool that can be used to assess the performance of cells from a given tissue, when incorporated into a bioartificial organ, and/or the physical and morphological properties of an engineered tissue as compared to its native counterpart. This approach could be useful, for example, for monitoring the progress of skin grafts, or to assess GAG and collagen contents locally in a repair site within articular or meniscal cartilage. The possibility of using the same suite of techniques both *in vitro*, during the production stage of the tissue, and *in vivo*, post implantation, is probably unique to NMR and represents a tremendous advantage of this technique.

8. ACKNOWLEDGEMENTS

KMB would like to thank the BBSRC and Smith & Nephew Plc for supporting the tissue engineering research that has been performed in his laboratory. AAN acknowledges FCT and the Portuguese Ministry for Science and Higher Education for his PhD scholarship.

9. REFERENCES

Adams, M.E., Li, D.K., McConkey, J.P., Davidson, R.G., Day, B., Duncan, C.P. and Tron, V. 1991. Evaluation of cartilage lesions by magnetic resonance imaging at 0.15 T: comparison with anatomy and concordance with arthroscopy. J Rheumatol. 18: 1573-1580.

Allen, J.W. and Bhatia, S.N. 2002. Improving the next generation of bioartificial liver devices. Semin Cell Dev Biol. 13: 447-54.

As, H.V. and Lens, P. 2001. Use of 1H NMR to study transport processes in porous biosystems. J Ind Microbiol Biotechnol. 26: 43-52.

Axel, L., Shimakawa, A., MacFall, J. 1986. A time-of-flight method of measuring flow velocity by magnetic resonance imaging. Magn Reson Imaging. 4: 199-205.

Bashir, A., Gray, M.L., Hartke, J. and Burstein, D. 1999. Nondestructive imaging of human cartilage glycosaminoglycan concentration by MRI. Magn Reson Med. 41: 857-65.

Bashir, A., Gray, M.L., Boutin, R.D. and Burstein, D. 1997. Glycosaminoglycan in articular cartilage: in vivo assessment with delayed Gd(DTPA)(2-)-enhanced MR imaging. Radiology. 205: 551-558.

Begley, C.M. and Kleis, S.J. 2000. The fluid dynamic and shear environment in the NASA/JSC rotating-wall perfused-vessel bioreactor. Biotechnol Bioeng. 70: 32-40.

Brindle, K.M. 1998. Investigating the performance of intensive mammalian cell bioreactor systems using magnetic resonance imaging and spectroscopy. Biotech Genetic Eng Rev. 15: 499-520.

Brindle, K.M. 1995. Enzyme-catalyzed exchange: magnetisation transfer measurements. In: D. M. G. a. R.K.Harris editor. Encyclopedia of Nuclear Magnetic Resonance. Chichester: John Wiley and Sons Ltd. p 1902-1909.

Brindle, K.M., Blackledge, M.J., Challiss, R.A. and Radda, G.K. 1989. 31P NMR magnetization-transfer measurements of ATP turnover during steady-state isometric muscle contraction in the rat hind limb in vivo. Biochemistry. 28: 4887-93.

Buckwalter, J.A. and Martin, J. 1995. Degenerative joint disease. Clin Symp. 47: 1-32.

Burstein, D., Bashir, A. and Gray, M.L. 2000. MRI techniques in early stages of cartilage disease. Invest Radiol. 35: 622-38.

Callies, R., Jackson, M.E. and Brindle, K.M. 1994. Measurements of the growth and distribution of mammalian cells in a hollow-fiber bioreactor using nuclear magnetic resonance imaging. Biotechnology (N Y). 12: 75-78.

Constantinidis, I., Stabler, C.L., Long, R., Jr. and Sambanis, A. 2002. Noninvasive monitoring of a retrievable bioartificial pancreas in vivo. Ann N Y Acad Sci. 961: 298-301.

Crank, J. 1964. The Mathematics of Diffusion. London: Oxford University Press.

Dardzinski, B.J., Mosher, T.J., Li, S., Van Slyke, M.A. and Smith, M.B. 1997. Spatial variation of T2 in human articular cartilage. Radiology. 205: 546-50.

Deo, Y.M., Mahadevan, M.D. and Fuchs, R. 1996. Practical considerations in operation and scale-up of spin-filter based bioreactors for monoclonal antibody production. Biotechnol Prog. 12: 57-64.

Desmoulin, F., Confort-Gouny, S., Masson, S., Bernard, M., Doddrell, D.M. and Cozzone, P.J. 1990. Application of reverse-DEPT polarization transfer pulse sequence to study the metabolism of carbon-13-labeled substrates in perfused organs by 1H NMR spectroscopy. Magn Reson Med. 15: 456-61.

Donoghue, C., Brideau, M., Newcomer, P., Pangrle, B., DiBiasio, D., Walsh, E. and Moore, S. 1992. Use of magnetic resonance imaging to analyze the performance of hollow-fiber bioreactors. Ann N Y Acad Sci. 665: 285-300.

Farndale, R.W., Buttle, D.J. and Barrett, A.J. 1986. Improved quantitation and discrimination of sulphated glycosaminoglycans by use of dimethylmethylene blue. Biochim Biophys Acta. 883: 173-7.

Fernandez, E.J., Mancuso, A. and Clark, D.S. 1988. NMR spectroscopy studies of hybridoma metabolism in a simple membrane bioreactor. Biotechnol Progress. 4: 173-83.

Foy, B.D. and Blake, J. 2001. Diffusion of paramagnetically labeled proteins in cartilage: Enhancement of the 1-D NMR imaging technique. J Magn Reson. 148: 126-134.

Franks, S.E., Kuesel, A.C., Lutz, N.W. and Hull, W.E. 1996. 31P MRS of human tumor cells: effects of culture media and conditions on phospholipid metabolite concentrations. Anticancer Res. 16: 1365-74.

Gadian, D.G. 1996. NMR And Its Applications To Living Systems. London: Oxford University Press.

Gillies, R.J., Scherer, P.G., Raghunand, N., Okerlund, L.S., Martinez-Zaguilan, R., Hesterberg, L. and Dale, B.E. 1991. Iteration of hybridoma growth and productivity in hollow fiber bioreactors using 31P NMR. Magn Reson Med. 18: 181-192.

Goodwin, D.W., Wadghiri, Y.Z. and Dunn, J.F. 1998. Micro-imaging of articular cartilage: T2, proton density, and the magic angle effect. Acad Radiol. 5: 790-798.

Gorenflo, V.M., Smith, L., Dedinsky, B., Persson, B. and Piret, J.M. 2002. Scale-up and optimization of an acoustic filter for 200 L/day perfusion of a CHO cell culture. Biotechnol Bioeng. 80: 438-44.

Gorter, A., van de Griend, R.J., van Eendenburg, J.D., Haasnoot, W.H. and Fleuren, G.J. 1993. Production of bi-specific monoclonal antibodies in a hollow-fibre bioreactor. J Immunol Methods. 161: 145-150.

Gray, M.L., Burstein, D., Lesperance, L.M. and Gehrke, L. 1995. Magnetization transfer in cartilage and its constituent macromolecules. Magn Reson Med. 34: 319-325.

Gruetter, R., Novotny, E.J., Boulware, S.D., Rothman, D.L., Mason, G.F., Shulman, G.I., Shulman, R.G. and Tamborlane, W.V. 1992. Direct measurement of brain glucose concentrations in humans by 13C NMR spectroscopy. Proc Natl Acad Sci U S A. 89: 1109-1112.

Hammer, B.E., Heath, C.A., Mirer, S.D. and Belfort, G. 1990. Quantitative flow measurements in bioreactors by nuclear magnetic resonance imaging. Bio/Technology. 8: 327-330.

Hardy, C.J., Saranathan, M., Zhu, Y. and Darrow, R.D. 2000. Coronary angiography by real-time MRI with adaptive averaging. Magn Reson Med. 44: 940-946.

Hesse, S.J., Ruijter, G.J., Dijkema, C. and Visser, J. 2000. Measurement of intracellular (compartmental. pH by 31P NMR in Aspergillus niger. J Biotechnol. 77: 5-15.

Hoult, D.I., Busby, S.J., Gadian, D.G., Radda, G.K., Richards, R.E. and Seeley, P.J. 1974. Observation of tissue metabolites using 31P nuclear magnetic resonance. Nature. 252: 285-287.

Karantanas, A.H., Hadjigeorgiou, G.M., Paterakis, K., Sfiras, D. and Komnos, A. 2002. Contribution of MRI and MR angiography in early diagnosis of brain death. Eur Radiol. 12: 2710-2716.

Kim, D.K., Ceckler, T.L., Hascall, V.C., Calabro, A. and Balaban, R.S. 1993. Analysis of water-macromolecule proton magnetization transfer in articular cartilage. Magn Reson Med. 29:, 211-215.

Kim, Y.J., Sah, R.L., Doong, J.Y. and Grodzinsky, A.J. 1988. Fluorometric assay of DNA in cartilage explants using Hoechst 33258. Anal Biochem. 174: 168-176.

Ko, S.F., Wan, Y.L., Ng, S.H., Lee, T.Y., Cheng, Y.F., Wong, H.F. and Hsieh, M.J. 1999. MRI of thoracic vascular lesions with emphasis on two-dimensional time-of-flight MR angiography. Br J Radiol. 72: 613-620.

Koretsky, A.P. 2002. Functional assessment of tissues with magnetic resonance imaging. Ann N Y Acad Sci. 961: 203-205.

Koretsky, A.P. and Williams, D.S. 1992. Application of localized in vivo NMR to whole organ physiology in the animal. Ann Rev Physiol. 54: 799-826.

Lamminen, A. and Keto, P. 1996. MRI angiography. Duodecim. 112: 1533-1542.

Lehner, K.B., Rechl, H.P., Gmeinwieser, J.K., Heuck, A.F., Lukas, H.P. and Kohl, H.P. 1989. Structure, function, and degeneration of bovine hyaline cartilage: assessment with MR imaging in vitro. Radiology. 170: 495-499.

Lien, Y.H., Zhou, H.Z., Job, C., Barry, J.A. and Gillies, R.J. 1992. In vivo 31P NMR study of early cellular responses to hyperosmotic shock in cultured glioma cells. Biochimie. 74: 931-9.

Macdonald, J.M., Grillo, M., Schmidlin, O., Tajiri, D.T. and James, T.L. 1998. NMR spectroscopy and MRI investigation of a potential bioartificial liver. NMR Biomed. 11: 55-66.

Mancuso, A., Fernandez, E.J., Blanch, H.W. and Clark, D.S. 1990. A nuclear magnetic resonance technique for determining hybridoma cell concentration in hollow fiber bioreactors. Bio/Technology. 8: 1282-1285.

Maroudas, A. 1968. Physicochemical properties of cartilage in the light of ion exchange theory. Biophys J. 8: 575-595.

McAlindon, T.E., Watt, I., McCrae, F., Goddard, P. and Dieppe, P.A. 1991. Magnetic resonance imaging in osteoarthritis of the knee: correlation with radiographic and scintigraphic findings. Ann Rheum Dis. 50: 14-19.

Melvin, B.K. and Shanks, J.V. 1996. Influence of aeration on cytoplasmic pH of yeast in an NMR airlift bioreactor. Biotechnol Prog. 12: 257-265.

Mlynarik, V., Degrassi, A., Toffanin, R., Vittur, F., Cova, M. and Pozzi-Mucelli, R.S. 1996. Investigation of laminar appearance of articular cartilage by means of magnetic resonance microscopy. Magn Reson Imaging. 14: 435-442.

Moon, R.B. and Richards, J.H. 1973. Determination of intracellular pH by 31P magnetic resonance. J Biol Chem. 248: 7276-7278.

Mow, V.C., Fithian, D.C. and Kelly, M.A. 1990. Fundamentals of articular cartilage and meniscus biomechanics. In: J. W. Ewing editor. Articular Cartilage and Knee Joint Function. New York: Raven Press Ltd. p 1-18.

Neeman, M., Jarret, K.A., Sillerud, L.O. and Freyer, J.P. 1991. Self-diffusion of water in multicellular spheroids measured by magnetic resonance microimaging. Cancer Res. 51: 4072-4079.

Neves, A.A., Medcalf, N. and Brindle, K.M. 2003. Functional assessment of tissue engineered meniscal cartilage using magnetic resonance imaging and spectroscopy. Tissue Eng. 9: 49-57.

Papas, K.K., Long, R.C., Jr., Constantinidis, I. and Sambanis, A. 2000. Effects of short-term hypoxia on a transformed cell-based bioartificial pancreatic construct. Cell Transplant. 9: 415-422.

Pei, M., Solchaga, L.A., Seidel, J., Zeng, L., Vunjak-Novakovic, G., Caplan, A.I. and Freed, L.E. 2002. Bioreactors mediate the effectiveness of tissue engineering scaffolds. FASEB J. 16: 1691-1694.

Petersen, E.F., Fishbein, K.W., McFarland, E.W. and Spencer, R.G. 2000. 31P NMR spectroscopy of developing cartilage produced from chick chondrocytes in a hollow-fiber bioreactor. Magn Reson Med. 44: 367-372.

Pfeuffer, J., Flogel, U. and Leibfritz, D. 1998. Monitoring of cell volume and water exchange time in perfused cells by diffusion-weighted 1H NMR spectroscopy. NMR Biomed. 11: 11-18.

Poole, A.R. 1993. Cartilage in health and disease. In: McCarthy D.J. Arthritis and Allied Conditions: A Textbook of Rheumatology. Philadelphia: Lea & Febinger. p 279-333.

Potter, K., Butler, J.J., Horton, W.E. and Spencer, R.G. 2000. Response of engineered cartilage tissue to biochemical agents as studied by proton magnetic resonance microscopy. Arthritis Rheum. 43: 1580-1590.

Racher, A.J. and Griffiths, J.B. 1993. Investigation of parameters affecting a fixed bed bioreactor process for recombinant cell lines. Cytotechnology. 13: 125-131.

Ratcliffe, A. and Niklason, L.E. 2002. Bioreactors and bioprocessing for tissue engineering. Ann N Y Acad Sci. 961: 210-215.

Samuel, D. 2002. The bioartificial liver: hopes and realities. Gastroenterol Clin Biol. 26: 721-723.

Shu, S.Y., Wu, Y.M., Bao, X.M., Wen, Z.B., Huang, F.H., Li, S.X., Fu, Q.Z. and Ning, Q. 2002. A new area in the human brain associated with learning and memory: immunohistochemical and functional MRI analysis. Mol Psychiatry. 7: 1018-1022.

Thelwall, P.E., Neves, A.A. and Brindle, K.M. 2001. Measurement of bioreactor perfusion using dynamic contrast agent-enhanced magnetic resonance imaging. Biotechnol Bioeng. 75: 682-690.

Trattnig, S., Mlynarik, V., Breitenseher, M., Huber, M., Zembsch, A., Rand, T. and Imhof, H. 1999. MRI visualization of proteoglycan depletion in articular cartilage via intravenous administration of Gd-DTPA. Magn Reson Imaging. 17: 577-583.

van de Kerkhove, M.P., Di Florio, E., Scuderi, V., Mancini, A., Belli, A., Bracco, A., Dauri, M., Tisone, G., Di Nicuolo, G., Amoroso, P., Spadari, A., Lombardi, G., Hoekstra, R., Calise, F. and Chamuleau, R.A. 2002. Phase I clinical trial with the AMC-bioartificial liver. Academic Medical Center. Int J Artif Organs. 25: 950-959.

van Zijl, P.C., Mooney, C.T., Faustino, P., Pekar, J., Kaplan, O. and Cohen, J.S. 1991. Complete separation of intracellular and extracellular information in NMR spectra of perfused cells by diffusion-weighted spectroscopy. Proc Nat Acad Sciences. 88: 3228-3232.

Verheul, H.B., Balazs, R., Berkelbach van der Sprenkel, J.W., Tulleken, C.A., Nicolay, K., Tamminga, K.S. and van Lookeren Campagne, M. 1994. Comparison of diffusion-weighted MRI with changes in cell volume in a rat model of brain injury. NMR Biomed. 7: 96-100.

Wachsmuth, L., Juretschke, H.P. and Raiss, R.X. 1997. Can magnetization transfer magnetic resonance imaging follow proteoglycan depletion in articular cartilage? Magma. 5: 71-78.

Wang, M.D., Yang, M., Huzel, N. and Butler, M. 2002. Erythropoietin production from CHO cells grown by continuous culture in a fluidized-bed bioreactor. Biotechnol Bioeng. 77: 194-203.

Wehrli, F.W., Shimakawa, A., Gullberg, G.T. and MacFall, J.R. 1986. Time-of-flight MR flow imaging: selective saturation recovery with gradient refocusing. Radiology. 160: 781-785.

Woessner, J.F. 1961. The determination of hydroxyproline in tissue and protein samples containing small proportions of this imino acid. Arch Bioch Bioph. 93: 440-447.

CHAPTER 16

CRYOPRESERVATION OF HEPATOCYTES FOR BIOARTIFICIAL LIVER DEVICES

M.H. GRANT AND D. STEVENSON

Bioengineering Unit, University of Strathclyde, Glasgow, UK

1. INTRODUCTION

Supply of sufficient numbers of functional viable cells is one of the most controversial issues surrounding tissue and organ engineering. While use of autologous cells derived from the recipient patient is a feasible option for replacement of tissues such as cartilage, skin and bone, this is not a practical possibility for replacement of the function of organs such as liver. The ideal scenario would be that bioartificial liver (BAL) systems used to treat humans would contain human hepatocytes, but that is, at present, not possible. The metabolic function of the liver declines rapidly post mortem, and most available human donated liver tissue is used in orthotic liver transplantation programmes. Due to the risk of inter-species transfer of viruses, xenotransplantation is generally unacceptable, particularly where the engineered tissue is to be transplanted into the body (Blusch et al., 2002). However, replacement of organ function with extracorporeal devices has lead to the development of BAL devices containing primary porcine cells (Demetriou, 1998; Flendrig et al., 1997; Gerlach et al., 1996; Watanabe et al., 1997). Cryopreservation is the only practical option for maintaining a stock or bank of hepatocytes for clinical use, either in a BAL or for transplantation. Primary cultures of these cells do not proliferate, and their function is unstable in culture with many liver specific functions being lost in conventional monolayer cultures within 24-72 hours (Grant et al., 1985; LeCluyse et al.,1996; Watts et al., 1995). Thus, storage as cultures is not an option.

The first publication on the successful freezing of cells for storage was the revival of suspensions of spermatozoa following cryopreservation reported in Nature in 1949 by Polge and co-workers. Since then cryopreservation of suspensions of cells has become a matter of routine for preserving stocks of cell lines derived from

J.B. Chaudhuri & M. Al-Rubeai (eds.), Bioreactors for Tissue Engineering, 353 - 372.

virtually every tissue in the body, and also for more complex structures such as embryos (Gerber et al., 2002). Hepatocytes are not readily cryopreserved and do not survive freezing in suspension, and despite the efforts of many scientists over the years this area is still contentious. The reasons why hepatocytes are so fragile and sensitive to damage during the freezing process are probably many-fold. The cells are isolated by perfusion of liver tissue with collagenase, and this results in membrane damage which may not have the opportunity to repair before the cells are treated with a cryoprotectant and frozen. This makes the cells particularly sensitive to membrane damage during the cryopreservation process itself. Furthermore, the cells are biochemically very complex, and retention of their liver-specific functions in culture post-thaw is a difficult task. Finally, the large size and slow water transport characteristics (Yano et al., 1996) of the hepatocyte may be another reason why they are difficult to cryopreserve compared with other cell types.

In this chapter we describe the different systems that have been used to cryopreserve hepatocytes, discuss their practicality in terms of providing cells for BAL and transplantation, and describe our experiences with cryopreservation of monolayers of hepatocytes.

2. MECHANISMS OF CRYOINJURY

What mechanisms lead to cell death during cryopreservation? As a cell and its surrounding environment are cooled, extracellular ice forms. The rate of solute incorporation into ice is slower than the rate of ice formation; therefore the extracellular ice excludes dissolved solutes during its formation, and this causes a supercooled extracellular hypertonic liquid environment to form in front of the propagating ice barrier. These solutes form a concentration gradient proportional to the distance from the ice-liquid interface (Körber et al., 1986). During this time, solutes can also reach their eutectic point and precipitate out of solution, causing a change in pH (Mazur, 1965). Furthermore, as an ice barrier propagates, anions are selectively incorporated into the ice crystal and cations are selectively excluded. This differential inclusion of cations and anions into a propagating ice barrier causes a charge separation and has been dubbed the Workman-Reynolds effect after its discoverers (Workman and Reynolds, 1950). The charge separation can be of the order of a few volts (Körber et al., 1986). Collectively, these phenomena, which can be injurious to a cell depending on the length of exposure and the severity of the change, are referred to as "solution effects" (Mazur, 1965). Hence, extracellular ice formation leads to hepatocellular injury due to solution effects.

An extracellular propagating ice barrier has the capability of excluding not only solutes, but also the cells themselves. If a cell is floating freely in suspension without some form of anchorage, a propagating ice boundary can exclude a cell and physically move it along its axis of propagation. This exclusion is caused by van Der Waal's forces within the ice barrier. Whether a cell is excluded or encapsulated by a propagating ice boundary is dependent on the velocity of the ice propagation and the viscosity of the solution (Körber et al., 1986). The result of this exclusion on a population of cells in suspension is that cells are compacted into solution channels formed between ice crystals, leading to deleterious mechanical stress (Ishiguro and Rubinsky, 1998). Hence, extracellular ice leads to hepatocellular injury due to mechanical stress.

Since the extracellular environment becomes hypertonic during cooling (and since cryoprotectants themselves typically create a hypertonic environment), the cell responds osmotically by expelling water and dramatically decreasing in volume. This leads to an increase in intracellular solute concentration and dehydration. According to theoretical simulations, a typical hepatocyte treated with 1.33 M dimethylsulphoxide (DMSO), the most commonly used cryoprotectant, at 22°C and cooled at 10°C/min to –12°C (the approximate nucleation temperature) loses 87.4% of its water volume because of the osmotic stress created by the DMSO (Karlsson et al., 1993). Mammalian cells find it difficult to cope with volume changes greater than 40% (Pegg, 1994). Intracellular dehydration can also seriously affect cellular lipid bilayers such as the plasma membrane or organelle membranes, causing geometric strain (lateral compression), topological strain (transition of the lipid bilayer into a hexagonal II phase, which is essentially tubes of water surrounded by lipids), and spontaneous demixing (separation of the membrane into two fluid phases with different compositions, such as a separate protein and lipid phase for example) (Wolfe and Bryant, 1999).

If a cell is unfortunate enough to develop a large quantity of intracellular ice, it will certainly die. As intracellular ice develops, plasma membranes rupture, mitochondrial membrane potentials decouple, proteins denature, and essential cofactors are released into the extracellular medium. The result is blebbing, lysis and death. Ironically, the more dehydrated a cell the less likely it is to form intracellular ice, so a balance is at play between too much and too little dehydration during cryopreservation.

On top of these cold induced injuries, the cryoprotectant itself can lead to hepatocellular injury. Hepatocytes are always exposed to a cryoprotectant prior to freezing; hepatocytes frozen without a cryoprotectant are not viable (Koebe et al., 1990; Loretz et al., 1989), and it is therefore a necessary evil of current cryopreservation protocols. By far the most common cryoprotectant is DMSO. Its use is, unfortunately, a double-edged sword, being both cryoprotective in some respects and toxic in others. DMSO protects cells during the freeze-thaw process by preventing intracellular ice formation (IIF) (McGann and Walterson, 1987), interacting electrostatically with the plasma membrane (Anchordoguy et al., 1991), and by preventing protein denaturation (Arakawa et al., 1990). Paradoxically, DMSO can also promote protein denaturation, depending on the conditions of exposure (Arakawa et al., 1990). Low concentrations of DMSO are cytotoxic, and healthy hepatocytes exposed to it will rapidly lose viability (Beecherl et al., 1987; Watts and Grant, 1996), and the ability to synthesise albumin (Koebe et al., 1990), urea, and protein (Fuller et al., 1980). DMSO has an osmolarity of 1.7 osmol (Diener et al., 1993), which is about five times higher than physiological solutions (Hengstler et al., 2000). Hence, DMSO can lead to hepatocellular injury by rapid water loss, osmotic stress (Karlsson et al., 1993) and protein denaturation (Arakawa et al., 1990).

Changing the freezing rate of a cell determines the type of cryoinjury it will undergo. The two-factor hypothesis of cell death due to freezing was proposed by Peter Mazur in the late 1960s. If a freezing rate is too slow, or if the water permeability of the cell membrane is very high, the increased length of time the cell is exposed to deleterious solution effects kills the cell. If the freezing rate is too fast, or if the water permeability of the cell membrane is very low, deleterious IIF occurs because the cell has not had enough time to dehydrate sufficiently. The optimum cooling rate is slow enough to prevent IIF but fast enough to minimize the time the cell is exposed to solution effects (Mazur, 1965; 1970).

3. CRYOPRESERVATION IN SUSPENSIONS

Suspension cryopreservation remains the most widely used method of preserving cell lines and also stocks of most primary cells. Typical suspension cryopreservation protocols call for hepatocytes to be frozen immediately upon isolation, although cells are sometimes preincubated with carbogen prior to freezing (Silva et al., 1999). For freezing protocols, cells are suspended in one of many media, such as Modified Eagle's medium, Dulbecco's Modified Eagle's medium, Leibovitz L-15 medium, Williams' E medium, Eagle's basal medium, Waymouth's 752/1 medium, or foetal calf serum (FCS), (Beecherl et al., 1987; Chesné and Guillouzo, 1988; Coundouris et al., 1990; Diener et al., 1993; Koebe et al., 1990; Le Cam et al., 1976; Loretz et al., 1989), though none offer particular advantage in terms of viability, membrane integrity, recovery, or function post thaw. FCS is often included in the cryopreservation media. As early as 1953, FCS (plus glycerol) was known for its cryoprotective properties in rat oocyte cryopreservation (Parkes and Smith, 1953). FCS (plus DMSO) has also been used successfully as a cryoprotectant during the cryopreservation of dog hepatocyte slices (Fisher et al., 1996), monkey hepatocytes in suspension (Sun et al., 1990), rat hepatocytes in suspension (Madan et al., 1999; Gomez et al., 1984; Desai and Hawksworth, 1990; Loretz et al., 1989; Lawrence and Benford, 1991) and foetal human liver derived haematopoietic progenitor cells (FHL $CD34^{+}$) in suspension (Zhao et al., 2001).

The cryopreservation media invariably contains a cryoprotectant such as DMSO, which is by far the most effective in maintaining cell function and viability post-thaw compared with other cryoprotectants such as propanediol, glycerol (Chesné and Guillouzo, 1988), polyvinylpyrrolidone, or dextran (Loretz et al., 1989). Typically a concentration of 10% DMSO is used; although higher or lower concentrations will also work, they are not as effective (Diener et al., 1993). To reduce the osmotic shock experienced by cells exposed to DMSO, a "stepwise addition" of DMSO is sometimes performed, i.e. cells are gradually exposed to increasing concentrations of DMSO (Diener et al., 1993; Koebe et al., 1990). We have found this practice to be quite beneficial to cells cryopreserved as monolayers.

The density of the cell suspension for cryopreservation is typically between 1 and 20 million cells/ml. Densities higher than 7 million cells/ml have been found to result in significantly reduced viability (Diener et al, 1993). This suspension is then frozen at a freezing rate of the order of –0.3°C/min to –20°C/min. Faster cooling rates, ≤–50°C/min, invariably lead to IIF damage (Harris et al., 1991; Hubel et al., 1991), which causes reduced viability and increased morphological abnormalities such as blebbing (Griffiths et al., 1979). So, plunging suspensions directly in liquid nitrogen is not an option.

As cells freeze, they either undergo spontaneous ice nucleation, or nucleation can be induced either by a rate controlled freezer (Diener et al., 1993), or by something as simple as touching the side of the cryotube with metal clamps dipped in liquid nitrogen (Beecherl et al., 1987). During nucleation, cells release latent heat of fusion, which can be deleterious to the cell.

Thawing is always "rapid," and performed by submerging the suspension cryotube in a 37°C water bath. Dilution of cryoprotectant can be either by direct addition of media, or stepwise removal to reduce osmotic shock. Often, cells are washed once immediately upon thawing with a 0.6-0.8M glucose solution (Chesne and Guillouzo, 1988) in order to reduce osmotic shock and renew the energy status of the cell. This practice has been associated with decreased viability and GSH levels in some studies (Coundouris et al., 1990); failure to remove such a high concentration of glucose will also result in decreased post-thaw attachment in suspension cryopreserved rat hepatocytes (Lawrence and Benford, 1991).

Before culturing thawed cells post-cryopreservation, some researchers choose to purify cell suspensions with iso-density Percoll centrifugation. Viable rat liver parenchymal cells have densities between 1.06 g/ml and 1.11 g/ml in an iso-osmotic Percoll medium, and these viable cells can be readily separated from dead cells and other hepatic cell types (Kreamer et al., 1986). Iso-density Percoll centrifugation results in an increase in viability and function, but also a corresponding decrease in attachment efficiency and recovery (Coundouris et al., 1990; 1993; de Sousa et al., 1996; Diener et al., 1993; Hengstler et al., 2000; Innes et al., 1988; Kreamer et al., 1986; Powis et al., 1987; Santone et al., 1989; Utesch et al., 1992).

In general, there are huge losses of hepatocytes with suspension cryopreservation protocols. These losses are usually discussed using two terms: recovery, and attachment efficiency. Recovery refers to the % of the hepatocytes that were originally cryopreserved that are recovered post-thaw after undergoing the cryopreservation procedure. It varies greatly from laboratory to laboratory, from as low as 33% (Powis et al., 1987) to as high as 78% (Diener et al., 1993; Utesch et al., 1992) in suspension cryopreserved rat hepatocytes without Percoll purification, and from as low as 22% (Diener et al., 1993; Utesch et al., 1992) to as high as 64% (Coundouris et al., 1990; 1993) with Percoll purification.

Attachment refers to the % of post-thaw hepatocytes able to attach to the bottom of a tissue culture dish. This is a very energy intensive process requiring a great deal of oxygen (Rotem et al., 1994); non-viable hepatocytes cannot attach and are washed away with subsequent media changes. Attachment varies greatly from laboratory to laboratory, from as low as 4% (Novicki et al., 1982), to as high as 73% (Swales et al., 1996) without Percoll purification, and from as low as 17% (Innes et al., 1988) to as high as 75% (Santone et al., 1989) with Percoll purification in rat suspension cryopreservation protocols.

So, the actual % of post-thaw cultured hepatocytes available for experiments after undergoing suspension cryopreservation is [(% Recovery)•(% Attachment)]/100. For example, if a hypothetical suspension cryopreservation protocol had, say, 78% recovery, and 75% attachment, then the % of hepatocytes on that culture dish after cryopreservation would be 78•75/100=58.5%. But the term "78% recovery and 75% attachment" sounds much better than "58.5% true recovery," so the latter is never used. Aspiring cryobiologists beware: a suspension cryopreservation researcher may proudly report "100% function compared with non-cryopreserved controls," in the knowledge that only 10% of the original population of cells was assayed, the other 90% were lost during cryopreservation and washing!

Thus, the major limitation to cryopreserving hepatocyte suspensions as a means of providing cells for tissue engineering and transplantation is the fact that on thawing there is typically a low recovery of cells, and those cells that are recovered do not attach efficiently in culture. In contrast, when cell lines are cryopreserved in suspension, even if the recovery post-thaw is low, the surviving cells will divide and regenerate a sustainable population. There is no possibility of generating a sustainable population of hepatocytes using this approach because the cells are non-proliferating.

4. CRYOPRESERVATION IN 3-DIMENSIONAL SYSTEMS

In vivo hepatocytes exist in close association with each other, and with the extracellular matrix in the form of the basal lamina. There is little doubt that cell-to cell contact is important in vitro for maintaining cell viability during culture and also during cryopreservation. In this context, hepatocytes have been cryopreserved in many different configurations (Table 1) which preserve, to different extents, the cell-to cell contact, either in the form of aggregates or whole pieces of tissue, and/or the cell to matrix contacts, by the use of artificial polymers or collagen based substrates.

Table 1. Configurations for cryopreservation of hepatocytes.

Suspensions
Cell monolayers
Liver slices
Spheroids
Encapsulated cells
Microcarriers
Collagen sandwiches

Several protocols exist for cryopreserving liver slices (Glockner et al., 2001; de Graaf et al., 2002; Maas et al., 2000; Vanhulle et al., 2001) by immersion in liquid nitrogen. These studies show that, upon thawing, viability and function decline rapidly, and are retained for only a period of up to 24 hours post-thaw. Generally, phase I cytochrome P450 (CYP) dependent monooxygenase activities are better maintained during this period than the phase II conjugation reactions of glucuronidation and sulphation (Glockner et al., 2001; Vanhulle et al., 2001). These thawed slices can only be used for short-term studies and the difficulty with their application lies in the delivery of nutrients and the penetration of substrates through to the interior of the slice.

The importance of maintaining 3-D configuration for the retention of differentiated status of cultured hepatocytes was realised in the late 1980s. For hepatocytes to function effectively in vitro a cellular polarity similar to the situation in vivo should be established. In the liver in vivo plasma proteins such as albumin are secreted across the basolateral (sinusoidal) surfaces of the cells to enter the circulation, whereas bile salts are secreted across the apical (bile canalicular) surface to enter the biliary system (Figure1).

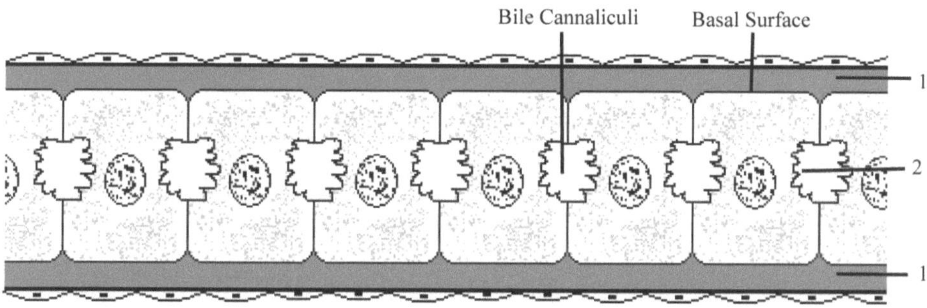

Figure 1. Hepatocyte polarity; apical and basal surfaces. (1, Space of Disse; 2 apicalsurface.)

It is now well established that the relationship between epithelial cells and the extracellular matrix controls the maintenance of cell polarity. This was first demonstrated when a monolayer of mammary duct epithelial cells cultured on collagen was overlaid with a second layer of collagen, and the monolayer of cells re-organised and differentiated into tubular structures (Hall et al., 1982). The pioneering work on hepatocytes cultured between two layers of collagen gel (sandwich system) was carried out by Dunn and co-workers (Dunn et al., 1989; 1991). Their work showed that the collagen gel sandwich culture system supported the secretion of albumin, transferrin, fibrinogen, bile acids and urea for at least six weeks, and maintained a distribution of actin filaments similar to that observed in vivo. When this configuration was used to cryopreserve hepatocytes, the presence of the extracellular matrix surrounding the cells significantly altered the IIF characteristics of the hepatocytes resulting in the preservation of more viable cells compared with suspension freezing protocols (Borel Rinkes et al., 1992a; 1992b;

Hubel et al., 1991; Koebe et al., 1990). The method was developed for freezing rat hepatocytes and showed a survival rate (in terms of morphology) on thawing of only 30%, signs of severe cytoplasmic damage in all cells, and a decrease in albumin secretion down to 20% of control cultures. However, this has been applied with more success to porcine hepatocytes (Koebe et al., 1996a; 1996b; 1999). The method uses a programmable freezer unit to control rates of freezing to –80°C, and the authors demonstrate that the conditions outlined in Table 2 support viability, morphology and differentiated function of cryopreserved cells post-thaw to an optimal extent.

Table 2. Freezing rates used for optimal cryopreservation of collagen sandwich cultures of hepatocytes (Koebe et al 1996b)

Time (minute)	*Temperature (°C)*	*Protocol*
0	37	37°C for 15 minutes
15	4	4°C for 10 minutes
25	4	4°C to –4°C @–1°C/min
33	–4	–4°C to –10°C @ –7.2°C/min
33.8	–10	–10°C for 20 minutes
53.8	–10	–10°C to –30°C @ –1°C/min
73.8	–30	–30°C to –80°C @ –10°C/min
78.8	–80	Long term storage at –80°C

The parameters they measured included cytochrome P450 (CYP1A1) dependent metabolism of ethoxycoumarin (ECOD), albumin secretion and cell viability (by morphology and transmission electron microscopy). After an initial rapid decline to 20-30% of control (non-frozen) values, the activity of ECOD recovered in cultured cells between 7-9 days post-thaw.

The scientists working on this model system for cryopreservation concede that it is unlikely to improve the situation of cell supply for bioreactors for BAL (Koebe et al., 1996a). It uses a low cell density (3 x 10^6 cells are frozen in 25cm^2 flasks), which necessarily involves the storage of a large volume, making it an expensive option. The system traps dead cells and debris between the two layers of collagen, and access to the cell layer by nutrients, substrates and metabolites is restricted. This latter aspect would be of vital importance in the design of a BAL device based on this configuration of cells and matrix. The oxygen demand of the cells is very high, particularly when they are actively synthesising proteins, detoxifying compounds from plasma and generally carrying out their liver specific functions (Smith et al., 1996). It would be technically difficult to supply sufficient oxygen to the cells embedded in a collagen gel sandwich configuration, as discussed by Foy and co-workers (1994). The use of a programmable freezer unit appears to be necessary to freeze the collagen sandwiches successfully. In our experience, freezing collagen gels without controlled freezing rates results in an unstable, mushy gel on thawing which does not retain its physical strength, and is difficult to handle.

The presence of the stabilising influence of contact with the matrix during cryopreservation processes has been mimicked successfully by freezing porcine hepatocytes attached to microcarriers. The methodology was developed by Demitriou and co-workers (Demitriou et al., 1986a; 1986b; Naik et al., 1997) and is used to provide the cell source for the HepatAssist® device marketed by Circe Biomedical (http://www.circebio.com/) currently undergoing clinical trials in America and Europe. The efficacy of the cells in this device has yet to be proven, and the improvements in the clinical conditions of patients may be due more to the charcoal column incorporated as part of the device than to the presence of the liver cells. Attachment of primary cultures of hepatocytes to microcarriers is dependent on high partial pressures of oxygen, and as a result the density of hepatocytes attaching is low (Foy et al., 1993) and can limit the number of cells available to seed bioreactors.

Numerous studies have shown the therapeutic potential of encapsulated hepatocytes and pancreatic islet cells as bioartificial organs (Dixit et al., 1990; 1992a; 1992b; Bruni and Chang, 1989; Rajotte 1999; De Vos et al., 1993). The application of this technology in transplantation and tissue engineering has been limited by low mechanical strength, and by degradation of the capsules and induction of inflammatory responses. However, the encapsulation of cells affords protection from the cryopreservation processes, and several authors have advocated this as a convenient means of establishing a bank of stored hepatocytes (Guyomard et al., 1996; Muraca et al., 2000). Most encapsulation membranes are based on alginates, and recently Canaple et al. (2001) used a composite of sodium alginate, cellulose sulphate and poly(methylene-6-guanidine) to store mouse hepatocytes for up to 4 months in liquid nitrogen without detrimental effect on albumin secretion levels. The capsule membrane should also provide immunoisolation on transplantation. However, this is a difficult task to achieve as the cells must exchange a variety of proteins, cytokines and metabolites with the plasma. The molecular cut off required for the capsular membrane is believed to be around 100kD, allowing exchange of albumin and albumin-bound compounds, but excluding immunoglobulins and complement factors (Morris, 1996).

5. EXPERIENCES WITH CRYOPRESERVATION OF MONOLAYERS

5.1 Introduction

Most of the data available in the literature on the stability of function in primary cultures of hepatocytes are from experiments where the cells are attached as monolayers to collagen coated polystyrene culture dishes. If such a simple system could be utilised for a BAL, there would be no barrier to nutrient supply or substrate access, and the cells would be readily oxygenated. BAL systems based on flat plate cultures where the cells are present as monolayers on membranes have been designed (Bader et al., 1995; 1997; 2000; Smith et al., 1997; 2000), but because of manufacturing difficulties (there are presently no commercially available modules) they have not advanced as far as the cartridge designs. A monolayer cryopreservation protocol first introduced by Ohno, in 1991 was further developed in our laboratory (Watts and Grant, 1996). In this protocol, primary rat hepatocytes are cultured on collagen coated tissue culture dishes before being cryopreserved, thereby circumventing the problem of low post-thaw recovery and attachment efficiency exhibited by hepatocytes frozen in suspension. We have recently modified this method enabling us to cryopreserve intact monolayers of cells which retain many aspects of function and relate our experience with it to date.

5.2 Method

Hepatocytes were isolated from male Sprague Dawley rats (180-260 g), which had *ad libitum* access to food and water, by collagenase (from Gibco BRL Life Technologies, Paisley, Scotland) perfusion (Moldeus et al., 1978). Viability of the hepatocyte suspensions was typically 80-95%, as determined by Trypan Blue exclusion. 60mm Falcon tissue culture dishes coated with type 1 collagen (0.0435 mg per cm^2) isolated from rat tail tendons as described by (Elsdale and Bard, 1972) were seeded with 3×10^6 viable cells in 2 ml of medium. The cultures were incubated in 5% CO_2/95% air at 37°C for 4 hours in Chee's medium, supplemented with 2.5 μg ml^{-1} Fungizone, 50 IU ml^{-1} penicillin, 50 μg ml^{-1} streptomycin (all from Gibco BRL Life Technologies Ltd, Paisley, Scotland), and 5% (v/v) FCS (Sigma Chemical Co., Poole, Dorset, UK). After 4 hours, the cells were washed twice with 2 ml medium and 3 ml fresh medium was added. Medium was replaced again at 20 hours, and every 24 hours thereafter.

At 24 hours, cultures were cryopreserved by the protocol of Watts and Grant (1996) and McKay et al. (2001), with minor alterations. Cultures were incubated with cryopreservation media consisting of 90%FCS + 10%DMSO (CM90) for 30 minutes at room temperature. CM90 was then removed completely, leaving the cultures covered by a thin film of approximately 100μl (McKay et al., 2001) and immediately placed in a polystyrene box with a lid (Biorad, 18cm internal height, 21cm x 21cm internal width, 4cm thick lid and base, 3.5cm thick side walls). The box was placed in a standard (non-rate controlled) –78°C freezer. 24 hours later, cultures were thawed by placing them immediately on the aluminium shelf of a 37°C incubator until ice disappeared. Chee's medium was added to each culture, and cultures were returned to the incubator.

At 24 hours, non-cryopreserved controls were also maintained, and were pretreated with CM90 in the same manner as cultures to be cryopreserved were, but instead of being placed in the cryopreservation box and frozen, the cryopreservation medium was aspirated, Chee's medium was added, and cultures were returned to the incubator.

Freezing and thawing rate measurements were taken using a picolog TC-08 datalogger with type T thermocouple. The following assays were then measured in non-cryopreserved and cryopreserved 96 hour post-isolation cultures (i.e. 48 hours post-thaw). Viability was assessed by the ability of cultures to deacetylate carboxyfluorescein diacetate (CFDA) into its fluorescent product, carboxyfluorescein (CF); stained cells were imaged by confocal laser scanning microscopy (CLSM) (McKay et al., 2001). Membrane integrity was assessed by ethidium bromide (EtBr) exclusion, as visualised by CLSM. For each culture dish, 2 images were examined with an average of 27 cells per image. The area of an image was 45500μm^2. The function of the cells was measured by the activities of two of the major pathways of xenobiotic metabolism in the liver, cytochrome P450 dependent monooxygenation and the activity of UDP-glucuronyltransferase. The latter was measured by the ability of cultures to metabolise the flavonoid, kaempherol, into two glucuronide metabolites detected and quantified by High Performance Liquid Chromatography (HPLC) (Oliveira and Watson, 2000). 96 hour cultures were incubated with 1.5ml 100μM kaempferol in Krebs-Henseleit buffer for 1 hour at 37ºC. The activity of the isoenzymes of cytochrome P-450 was measured by the ability of cultures to metabolise testosterone into multiple hydroxylated metabolites detected and quantified by HPLC (Grant et al., 2000). 96 hour cultures were incubated for 1 hour with 6ml 100μM testosterone in Hank's Balanced Salt Solution HBSS at 37ºC.

5.3 Results and Discussion

The morphology of the cells is shown in Figure 2 after staining the cells with CFDA and EtBr. The de-acetylation of CFDA in viable cells produces CF which accumulates inside the cells staining them with bright green fluorescence. A typical flattened hepatocyte morphology can be seen in the monolayer of cells which have been cryopreserved (B) and in the control non-cryopreserved cultures (A). Both cultures show intact cells with prominent nuclei. There are traces of red EtBr staining in figure B indicating some nuclear staining in dead cells.

a) NCP

b) CP

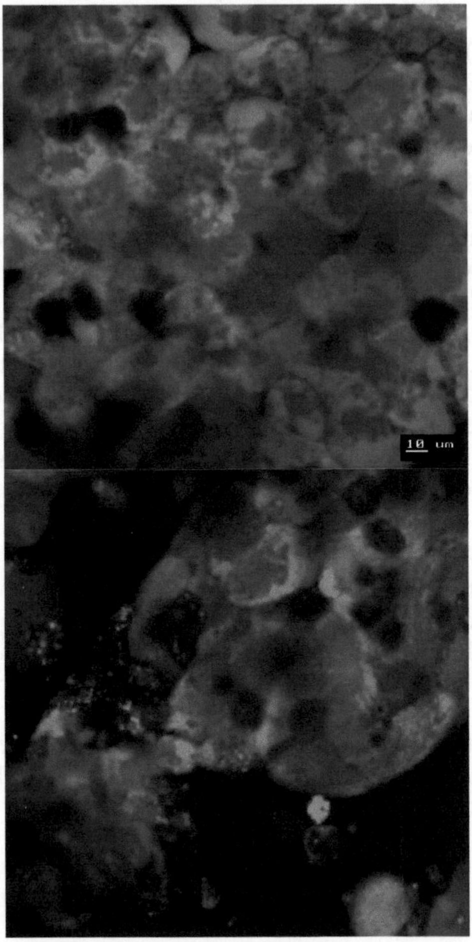

Figure 2. CLSM images of CFDA and EtBr stained 96h non-cryopreserved (a) and cryopreserved (b) rat hepatocyte monolayer cultures treated with CM90. NCP:non-cryopreserved. CP:cryopreserved. The voltage of the laser was the same for each image.

Table 3 shows that the intensity of the fluorescence generated from CFDA is significantly less after cryopreservation, and there are a greater number of EtBr stained nuclei. However, the area of colonies attached in the cultures is not significantly reduced post-thaw. These data have been collected 48 hours after the cells were thawed, and re-established in culture, so the cells have remained attached securely to the collagen-coated dish after the cryopreservation process. The area of the colonies of the cryopreserved cultures within the images measured was 80% of that of the colonies of non-cryopreserved cells, indicating that 80% of the cells had survived the cryopreservation process.

Table 3. CFDA fluorescent intensity, hepatocyte colony area, and average number of EtBr nuclei counted per image of 96h non-cryopreserved and cryopreserved rat hepatocyte monolayer cultures treated with CM90. NCP:non-cryopreserved. CP:cryopreserved. Values are mean±SEM, n=3 experiments. Two replicate plates per experiment were performed, and 2 images per plate were taken. An average of 27 hepatocytes per image were analysed.

Treatment	Average EtBr Nuclei Count	CFDA Colony Intensity (arbitrary units)	Colony Area (μm^2)
NCP	0.3±0.2	73.5±10.2	39533±2706
CP	13±9	40.7±4.7*	31522±4169

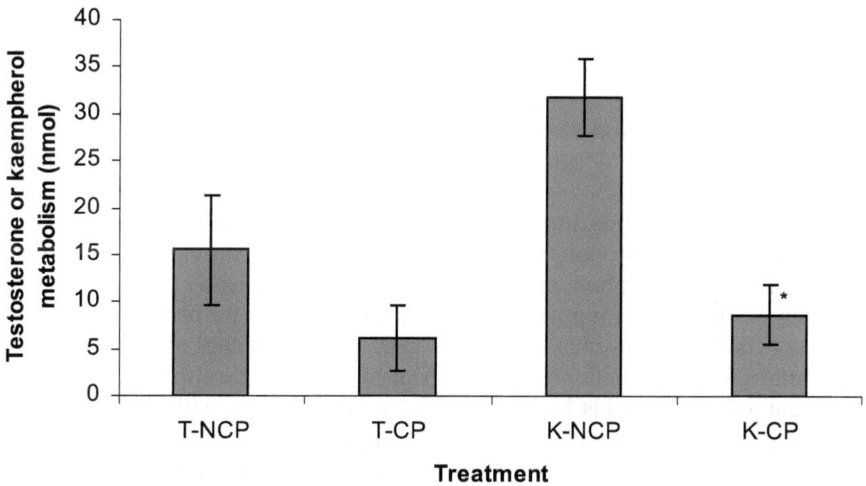

Figure 3. Total testosterone (T) and kaempherol (K) metabolism of 96h non-cryopreserved and cryopreserved rat hepatocyte monolayer cultures treated with CM90. NCP:non-cryopreserved. CP:cryopreserved. Values are mean±SEM, n=3 experiments. *Cryopreserved values which are significantly different from non-cryopreserved controls, ANOVA followed by Dunnett's test.

The total metabolism of testosterone and kaempherol was decreased in cryopreserved cells compared with non-cryopreserved controls (Figure 3). Total testosterone metabolism was decreased to 40% of the control level, and the alteration in the profile of metabolites formed after freezing and thawing the cells is detailed in Table 4. The major metabolite formed was androstenedione both in cryopreserved and non-cryopreserved hepatocytes, and this is consistent with what we and others have previously reported for cultured hepatocytes (Grant et al., 2000; Kern et al., 1997). The formation of the 6-α-hydroxy and 7-α-hydroxy- testosterone metabolites was the most labile after freezing the cells, and they represent the activity of the CYP 2A1 family of P450 isoenzymes (Grant et al., 2000; Wortelboer et al., 1990; Hengstler et al., 2000). CYP2C11 is the major form of P450 in the male rat liver (Schenkman et al., 1989), and it is solely responsible for the formation of 2-□-hydroxytestosterone, and along with CYP2B1 catalyses the 16-□-hydroxylation and the formation of androstenedione. It appears to be much more stable than CYP 2A1, and is retained at levels of between 20 and 56% of those in the control cells.

Kaempherol metabolism is more labile than that of testosterone after freezing, and the total production of glucuronide was maintained at only 27% of the level in the control non-cryopreserved cultures. Two glucuronide metabolites are produced from kaempherol by hepatocytes (Oliveira et al., 2002), and their relative formation was not changed in the cells post-cryopreservation. Recently, nuclear magnetic resonance studies have been applied to measure low-molecular weight substances present in culture supernatants from cryopreserved cells (Dabos et al., 2002). These studies have indicated that there is a defect in glycolysis and in the Krebs cycle in cryopreserved primary porcine hepatocytes. These are the main ATP producing pathways in hepatocytes, and lack of energy to activate the UDP-glucuronic acid donor for glucuronidation reactions may explain why this conjugation reaction is more labile than phase I cytochrome P450 reactions in post-thaw cryopreserved cells.

This protocol for freezing cells attached to a collagen-coated surface allows survival of a greater proportion of the cell population than can be achieved with suspension cryopreservation, and is much simpler than freezing cells in collagen gel sandwiches. It holds great potential for the supply of cells for plate bioreactors. The bioreactor designed in our laboratory uses polycarbonate plates with integral oxygenation for attachment of monolayers of hepatocytes (Smith et al., 1997; 2000). At present our design holds 13 stacked polycarbonate plates, with an area of 442 cm^2 per plate. It is seeded with $1.5\text{-}1.8\times10^5$ hepatocytes per cm^2, and is currently undergoing trials with animal models of liver failure. We propose to freeze collagen-coated plates seeded with monolayers of hepatocytes, and establish a bank of plates ready to use in the bioreactor module.

Table 4. Testosterone hydroxylation (nmol/culture dish/60min) by 96h non-cryopreserved and cryopreserved rat hepatocyte monolayer cultures treated with CM90. NCP:non-cryopreserved. CP:cryopreserved. Values are mean±SEM, n=3. *Cryopreserved values which are significantly different from non-cryopreserved controls, ANOVA followed by Dunnett's test. The major CYP isoenzyme(s) responsible for formation of each metabolite is indicated (Hengstler et al., 2000; Wortelboer et al., 1990).

	6-α-OHT	7-α-OHT	6-β-OHT	16-α-OHT	16-β-OHT	2-α-OHT	Androstenedione
CYP	2A1	2A1	3A1, 2C13	2B1, 2C11	2B1, 2B2	2C11	2B1, 2C11
NCP	0.087±0.006	0.622±0.428	2.161±0.365	2.844±1.902	0.285±0.254	1.603±0.42	7.928±3.205
CP	0.008±0.001*	0.039±0.037	0.325±0.098*	0.982±0.863	0.056±0.054	0.276±0.1*	4.421±2.392
% CP to NCP	9.3±1.7	6.3±2.1	15±2.3	34.5±9.5	19.6±5.3	17.2±8.3	55.7±7.8

6. THE WAY FORWARD FOR CRYOPRESERVING LIVER CELLS FOR TISSUE ENGINEERING

From the experiments conducted on the cryopreservation of suspensions, encapsulated cells, collagen gel sandwiches, and monolayer cultures of hepatocytes, several general facts emerge. Regardless of the system used, the cells lose viability and function to some extent, they require a cryoprotectant to survive at all, and the period of cryogenic storage time does not appear to influence the viability and function on thawing. Each method has its problems, and the method to recommend depends to a large extent on the design of bioreactor being considered. Cartridge-based designs, such as the HepatAssist® device, are conveniently seeded with microcarrier cryopreserved cells, but the contribution that the cells make to detoxification of the patients' plasma has been questioned. Encapsulated cryopreserved cells appear to be very suitable for transplantation of hepatocytes in vivo, but the mechanical strength of the capsules may not withstand the shear stress of fluid flow in a bioreactor environment. This method of cryopreservation has been widely applied to the storage of several different cell types (Karlsson and Toner, 1996; Grout et al., 1990), including bone marrow cells (Stockschlader et al., 1996) and pancreatic islets (Rajotte, 1999). Cryopreservation of monolayers of hepatocytes provides a convenient supply of cells for plate bioreactors and we envisage establishing a bank of stored frozen seeded plates for the plate bioreactor designed in our laboratory. The protocol described in this chapter has been successfully applied to porcine hepatocytes for this purpose.

7. ACKNOWLEDGEMENTS

Cryopreservation research in the authors' laboratory is supported by BBSRC grant number 78/E15110. The authors are grateful to Hazel Thomson for critical reading of the manuscript.

8. REFERENCES

Anchordoguy TJ, Cecchini CA, Crowe JH, Crowe LM. 1991. Insights into the cryoprotective mechanism of dimethyl sulfoxide for phospholipid bilayers. Cryobiology 28: 467-473.

Arawaka T, Carpenter JF, Kita YA, Crowe JH. 1990. The basis for toxicity of certain cryoprotectants: a hypothesis. Cryobiology 27: 401-415.

Bader A, Knop E, Fruhauf N, Crome O, Boker K, Christians U, Oldhafer K, Ringe B, Pichlmayr R, Sewing KF . 1995. Reconstruction of liver tissue in vitro: geometry of characteristic flat bed, hollow fiber, and spouted bed bioreactors with reference to the in vivo liver. Artif.Organs 19: 941-950.

Bader A, Fruehauf N, Steinhoff G, Haverich A. 1997. Development of a bioartificial liver by reconstructing 3-D cell plates and sinusoidal analogues. Artificial Organs 21: 526.

Bader A, De Bartolo L, Haverich A. 2000. High level benzodiazepine and ammonia clearance by flat membrane bioreactors with porcine liver cells. J. Biotechnol. 81: 95-105.

Beecherl EE, Sorensen EMB, Aggarwal SJ, Diller KR. 1987. Effect of cooling rate on the cryopreservation of neonatal Sprague-Dawley primary cultured hepatocytes. Cryo-Letters 8: 276-283.

Blusch JH, Patience C, Martin U. 2002. Pig endogenous retroviruses and xenotransplantation. Xenotransplantation 9: 242-251.

Borel Rinkes I, Toner M, Ezzeu R, Tompkins R, Yarmush M. 1992a. Effects of dimethylsulfoxide on cultured rat hepatocytes in sandwich configuration. Cryobiology 29: 442-453.
Borel Rinkes I, Toner M, Sheeha SJ, Tompkins RG, Yarmush ML. 1992b. Long-term functional recovery of hepatocytes after cryopreservation in a three-dimensional culture configuration. Cell Transplant. 1: 281-292.
Bruni S and Chang TM. 1989. Hepatocytes immobilised by microencapsulation in artificial cells: effects on hyperbilirubinemia in Gunn rats. Biomater. Artif. Organs 17: 403-411.
Canaple L, Nurdin N, Angelova N, Saugy D, Hunkeler D, Desvergne B. 2001. Maintenance of primary murine hepatocyte functions in multicomponent polymer capsules--in vitro cryopreservation studies. J.Hepatol. 34: 11-18.
Chesné C, Guillouzo C. 1988. Cryopreservation of isolated rat hepatocytes: a critical evaluation of freezing and thawing conditions. Cryobiology 25: 323-330.
Coundouris JA, Grant MH, Simpson JG, Hawksworth GM. 1990. Drug metabolism and viability studies in cryopreserved rat hepatocytes. Cryobiology 27: 288-300.
Coundouris JA, Grant MH, Engeset J, Petrie JC, Hawksworth GM. 1993. Cryopreservation of human adult hepatocytes for use in drug metabolism and toxicity studies. Xenobiotica 23: 1399-1409.
Dabos KJ, Parkinson JA, Hewage C, Nelson LJ, Sadler IH, Hayes PC, Plevris JN. 2002. ^{1}H NMR spectroscopy as a tool to evaluate key metabolic functions of primary porcine hepatocyte after cryopreservation. NMR Biomed. 15, 241-250.
de Graaf, IAM, Geerlinks A, Koster HJ. 2002. Incubation at 37°C prior to cryopreservation decreases viability of liver slices after cryopreservation by rapid freezing. Cryobiology 45: 1-9.
de Sousa G, Guillouzo A, Nicolas F, Lorenzon G, Placidi M, Rahmani R, Guillouzo A. 1996. Increase of cytochrome P-450 1A and glutathione transferase transcripts in cultured hepatocytes from dogs, monkeys, and humans after cryopreservation. Cell Biol. and Toxicol. 12: 351-358.
Demetriou AA, Whiting J, Levenson SM, Chowdhury NR, Schechner R, Michalski S, Feldman D, Chowdhury JR. 1986a. New method of hepatocyte transplantation and extracorporeal liver support. Ann.Surg. 204: 259-271.
Demetriou AA, Levenson PM, Novikoff PM, Novikoff AB, Chowdhury NR, Whiting J, Reisner A, Chowdhury JR. 1986b. Survival, organization and function of microcarrier-attached hepatocytes transplanted in rats. Proc. Nat. Acad. Sci. USA 83: 7475-7479.
Demetriou AA. 1998. Support of the acutely failing liver: state of the art. Ann. Surg. 228: 14-15.
Desai ADJ, Hawksworth GM. 1990. Cryopreservation of rat hepatocytes with high attachment efficiency and mixed function oxidase activity post thawing. Biochemical Society Transactions 18: 1214.
De Vos P, Wolters GH, Fritschy WM, Van Schilfgarade R. 1993. Obstacles in the application of microencapsultation in islet tranplantation. Int. J. Artif. Organs 16: 205-212.
Diener B, Utesch D, Beer N, Oesch F. 1993. A method for the cryopreservation of liver parenchymal cells for studies of xenobiotics. Cryobiology 30: 116-127.
Dixit V, Davasi R, Arthur M, Brezina M, Lewin K, Gitnick G. 1990. Restoration of liver function in Gunn rats without immunosuppression using transplanted microencapsulated hepatocytes. Hepatology 12: 1342-1349.
Dixit V, Arthur M, Gitnick G. 1992a. Repeated transplantation of microencapsulated hepatocytes for sustained correction of hyperbilirubinemia in Gunn rats. Cell Transplant. 1: 725-759.
Dixit V, Arthur M, Reinhardt R, Gitnick G. 1992b. Improved function of microencapsulated hepatocytes in a hybrid bioartificial liver support system. Artif. Organs 16: 336-341.
Dunn JCY, Tomkins RG, Yarmush ML. 1991. Long-term in vitro function of adult hepatocytes in a collagen sandwich configuration. Biotechnology Progress 7: 237-245.
Dunn JCY, Yarmush ML, Koebe HG, Tomkins RG. 1989. Hepatocyte function and extracellular matrix geometry: long-term culture in a sandwich configuration. FASEB Journal 3: 174-177.
Elsdale T, Bard J. 1972. Collagen substrata for studies on cell behaviour. J.Cell Biol. 54: 626-637.
Fisher RL, Hasal SJ, Sanuik JT, Hasal KS, Gandolfi AJ, Brendel K. 1996. Cold- and cryopreservation of dog liver and kidney slices. Cryobiology 33: 163-171.
Flendrig LM, la Soe JW, Jorning GG, Steenbeek A, Karlsen OT, Bovee WM, Ladiges NC, te Velde AA, Chamuleau RA. 1997. In vitro evaluation of a novel bioreactor based on an integral oxygenator and a spirally wound nonwoven polyester matrix for hepatocyte culture as small aggregates. J.Hepatol. 26: 1379-1392.
Foy BD, Lee J, Morgan J, Toner M, Tompkins RG, Yarmush ML. 1993. Optimization of hepatocyte attachment to microcarriers: Importance of oxygen. Biotechnology and Bioengineering 42: 579-588.

Foy BD, Rotem A, Toner M, Tomkins RG, Yarmush ML. 1994. A device to measure the oxygen uptake rate of attached cells: importance in bioartificial organ design. Cell Transplantation 3: 515-527.

Fuller BJ, Morris GJ, Nutt LH, Attenburrow VD. 1980. Functional recovery of isolated rat hepatocytes upon thawing from -196° C. Cryo-Letters 1: 139-146.

Geber S, Sales L, Sampaio MA. 2002. Laboratory techniques for human embryos. Reprod Biomed Online 5: 211-218.

Gerlach T, Zeimer R, Neuhaus P. 1996. Fulminant liver failure: relevance of extracorporeal hybrid liver support systems. Int.J.Artif.Org. 19: 7-13.

Glockner R, Rost M, Pissowotzki K, Muller D. 2001. Monooxygenation, conjugation and other functions in cryopreserved rat liver slices until 24 h after thawing. Toxicology 161: 103-109.

Gomez LJM, Lopez P, Castell JV. 1984. Biochemical functionality and recovery of hepatocyte after deep freezing storage. In Vitro 20: 826-832.

Grant MH, Melvin MAL, Shaw P, Melvin WT, Burke MD. 1985. Studies on the maintenance of cytochromes P-450 and b_5 monooxygenases and cytochrome reductases in primary cultures of rat hepatocytes. FEBS 190: 99-103.

Grant MH, Anderson K, McKay G, Wills M, Henderson C, Macdonald C. 2000. Manipulation of the phenotype of immortalised rat hepatocytes by different culture configurations and by dimethyl sulphoxide. Hum.Exp.Toxicol. 19: 309-317.

Griffiths JB, Cox CS, Beadle DJ, Hunt CJ, Reid DS. 1979. Changes in cell size during the cooling, warming and post-thawing periods of the freeze-thaw cycle. Cryobiology 16: 141-151.

Grout B, Morris J, Mchellan M. 1990. Cryopreservation and the maintainance of cell lines. Trends. Biotechnol. 8: 293-297.

Guyomard C, Rialland L, Fremond B, Chesne C, Guillouzo A. 1996. Influence of alginate gel entrapment and cryopreservation on survival and xenobiotic metabolism capacity of rat hepatocytes. Toxicol.Appl.Pharmacol. 141: 349-356.

Hall HG, Farson, Bissell MJ. 1982. Lumen formation by epithelial cell lines in response to collagen overlay: a morphogenetic model in culture. Proc. Natl. Acad. Sci. 79: 4672-4676.

Harris CL, Toner M, Hubel A, Cravalho EG, Yarmush ML, Tomkins RG. 1991. Cryopreservation of isolated hepatocytes: intracellular ice formation under various chemical and physical conditions. Cryobiology 28: 436-444.

Hengstler JG, Utesch D, Steinberg P, Platt KL, Diener B, Ringel M, Swales N, Fischer T, Biefang K, Gerl M, Bottger T, Oesch F. 2000. Cryopreserved primary hepatocytes as a constantly available in vitro model for the evaluation of human and animal drug metabolism and enzyme induction. Drug Metab Rev. 32: 81-118.

Hubel A, Toner M, Cravalho EG, Yarmush ML, Tomkins RG. 1991. Intracellular ice formation during the freezing of hepatocytes cultured in a double collagen gel. Biotechnology Progress 7: 554-559.

Innes GK, Fuller BJ, Hobbs KE. 1988. Functional testing of hepatocytes following their recovery from cryopreservation. Cryobiology 25: 23-30.

Ishiguro H and Rubinsky B. 1998. Influence of fish antifreeze proteins on the freezing of cell suspension with cryoprotectant penetrating cells. In. J. Heat Mass Transfer 41(13): 1907-1915.

Karlsson JOM, Cravalho EG, Rinkes IHMB, Tompkins RG, Yarmush ML, Toner M. 1993. Nucleation and growth of ice crystals inside cultured hepatocytes during freezing in the presence of dimethyl sulfoxide. Biophysical Journal 65: 2524-2536.

Karlsson JO and Toner M. 1996. Long-term storage of tissues by cryopreservation: critical issues. Biomaterials 17: 243-256.

Kern A, Bader A, Pichlmayr R, Sewing KF. 1997. Drug metabolism in hepatocyte sandwich cultures of rats and humans. Biochem.Pharmacol. 54: 761-772.

Koebe HG, Dunn JCY, Toner M, Sterling LM, Hubel A, Cravalho EG, Yarmush ML, Tomkins RG. 1990. A new approach to the cryopreservation of hepatocytes in sandwich culture configuration. Cryobiology 27: 576-584.

Koebe HG, Dähnhardt C, Müller-Höcker J, Wagner H, Schildberg FW. 1996a. Cryopreservation of porcine hepatocyte cultures. Cryobiology 33: 127-141.

Koebe HG, Pahernik SA, Thasler WE, Schildberg FW. 1996b. Porcine hepatocytes for biohybrid artificial liver devices: a comparison of hypothermic storage techniques. Artif.Organs 20: 1181-1190.

Koebe HG, Muhling B, Deglmann CJ, Schildberg FW. 1999. Cryopreserved porcine hepatocyte cultures. Chem.Biol.Interact. 121: 99-115.

Körber Ch, Englich S, Schwindlke P, Scheiwe MW, Rau G, Hubel A, Cravalho EG. 1986. Low temperature light microscopy and its application to study freezing in aqueous solutions and biological cell suspensions. Journal of Microscopy 141: 263-276.
Kreamer BL, Staecker JL, Sawada N, Sattler GL, Hsia MT, Pitot HC. 1986. Use of a low-speed, iso-density percoll centrifugation method to increase the viability of isolated rat hepatocyte preparations. In Vitro Cellular Developmental Biology 22: 201-211.
Lawrence JN, Benford DJ. 1991. Development of an optimal method for the cryopreservation of hepatocytes and their subsequent monolayer culture. Toxicology in Vitro 5: 39-50.
Le Cam A, Guillouzo A, Freychet P. 1976. Ultrastructural and biochemical studies of isolated adult rat hepatocytes prepared under hypoxic conditions. Cryopreservation of hepatocytes. Exp.Cell Res. 98: 382-395.
LeCluyse EL, Bullock PL, Parkinson A, Hochman JH. 1996. Cultured rat hepatocytes. Pharm.Biotechnol. 8: 121-159.
Loretz LJ, Li AP, Flye MW, Wilson AGE. 1989. Optimization of cryopreservation procedures for rat and human hepatocytes. Xenobiotica 19: 489-498.
Maas WJ, Leeman WR, Groten JP, van de Sandt JJ. 2000. Cryopreservation of precision-cut rat liver slices using a computer- controlled freezer. Toxicol. In Vitro 14: 523-530.
Madan A, Dehaan R, Carroll K, Lecluyse E, Parkinson A. 1999. Effect of cryopreservation on cytochrome P-450 enzyme induction in cultured rat hepatocytes. Drug Metabolism and Disposition 27: 327-335.
Mazur P. 1965. Causes of injury in frozen and thawed cells. Federation Proceedings 24: S-175 - S-182.
Mazur P. 1970. Cryobiology: the freezing of biological systems. Science 168: 939-949.
McGann LE, Walterson ML. 1987. Cryoprotection by dimethyl sulfoxide and dimethyl sulfone. Cryobiology 24: 11-16.
McKay GC, Henderson C, Goldie E, Connel G, Westmoreland C, Grant MH. 2002. Cryopreservation of rat hepatocyte monolayers: cell viability and cytochrome P450 content in post-thaw cultures. Toxicology in Vitro 16: 71-79.
Moldeus P, Hogberg J, Orrenius S. 1978. Isolation and use of liver cells. Methods Enzymol. 52: 60-71.
Morris PJ. 1996. Immunoprotection of therapeutic cell transplants by encapsulation. Trends Biotechnol. 14: 163-167.
Muraca M, Vilei MT, Zanusso E, Ferraresso C, Granato A, Doninsegna S, Dal Monte R, Carraro P, Carturan G. 2000. Encapsulation of hepatocytes by SiO_2. Transplant.Proc. 32: 2713-2714.
Naik S, Santangini HA, Trenkler DM, Mullon CJP, Solomon BA, Pan J, Jauregiu HD. 1997. Functional recovery of porcine hepatocytes after hypothermic or cryogenic preservation for liver support systems. Cell Transplantation 6: 447-454. .
Novicki D, Irons GP, Strom SC, Jirtle R, Michalopoulos G. 1982. Cryopreservation of isolated rat hepatocytes. In Vitro 18: 393-399.
Ohno T, Saijo-Kurita K, Miyamoto-Eimori N, Kurose T, Aoki Y, Yosimura S. 1991. A simple method for in situ freezing of anchorage-dependent cells including rat liver parenchymal cells. Cytotechnology 5: 273-277.
Oliveira EJ, Watson DG. 2000. In vitro glucuronidation of kaempferol and quercetin by human UGT-1A9 microsomes. FEBS Lett. 471: 1-6.
Oliveira EJ, Watson DG, Grant MH. 2002. Metabolism of quercetin and kaempferol by rat hepatocytes and the identification of flavonoid glycosides in human plasma. Xenobiotica 32: 279-287.
Parkes AS, Smith AU. 1953. Regeneration of rat ovarian tissue grafted after exposure to low temperatures. Proc.R.Soc.Lond.B Biol.Sci 140: 455-470.
Pegg DE. 1994. Cryobiology: life in the deep freeze. Biologist 41: 53-56.
Polge C, Smith AU, Parkes AS. 1949. Revival of spermatozoa after vitrification and dehydration at low temperatures. Nature 164: 666.
Powis G, Santone KS, Melder DC, Thomas L, Moore DJ, Wilke TJ. 1987. Cryopreservation of rat and dog hepatocytes for studies of xenobiotic metabolism and activation. Drug Metabolism and Disposition 15: 826-832.
Rajotte RV. 1999. Islet cryopreservation protocols. Ann. NY Acad. Sci. 875: 200-207.
Rotem A, Toner M, Bhatia S, Foy BD, Tompkins RG, Yarmush ML. 1994. Oxygen is a factor determining in vitro tissue assembly: effect on attachment and spreading of heaptocytes. Biotech. Bioeng. 43: 654-660.

Santone KS, Melder DC, Powis G. 1989. Studies of chemical toxicity to fresh and cryopreserved rat hepatocytes. Toxicology and Applied Pharmacology 97: 370-376.

Schenkman JB, Thummel KE, Favreau LV. 1989. Physiological and pathophysiological alterations in rat hepatic cytochrome P-450. Drug Metab Rev. 20: 557-584.

Silva JM, Day SH, Nicoll-Griffith DA. 1999. Induction of cytochrome-P450 in cryopreserved rat and human hepatocytes. Chemico-Biological Interactions 121: 49-63.

Smith MD, Smirthwaite AD, Cairns DE, Cousins RB, Gaylor JD,. 1996. Techniques for measurement of oxygen consumption rates of hepatocytes during attachment and post-attachment. Int. J. Artif. Organs 19: 36-44.

Smith MD, Airdrie I, Courtney JM, Cousins RB, Ekeval E, Grant MH, Gaylor JDS. 1997. Development and characterisation of a hybrid artificial liver bioreactor with integral membrane oxygenation. In: Crepaldi G, Demetriou AA, Muraca M, editors. Bioartifical Liver Support: The Critical Issues. Italy: C.I.C. Edizioni International. p 27-35.

Smith MD, Grant MH, Cousins RB, Gaylor JDS. 2000. A review of bioartificial liver development from an artificial organ engineering perspective. In: Berry MN, Edwards AM editors. The Hepatocyte Review. Dordrecht: Kluwer Academic Publishers. p 521-541.

Stockschlader M, Kruger W, tomDieck A, Horstmann M, Aetnoder M, Loliger C. 1996. Use of cryopreserved bone marrow in unrelated allogenic transplantation. Bone Marrow Transplantation 17: 197-199.

Sun EL, Aspar DG, Ulrich RG, Melchior GW. 1990. Cryopreservation of cynomolgus monkey (*macaca fascicularis*) hepatocytes for subsequent culture and protein synthesis studies. In Vitro Cellular Developmental Biology 25: 147-150.

Swales NJ, Luong C, Caldwell J. 1996. Cryopreservation of rat and mouse hepatocytes. I. Comparative viability studies. Drug Metabolism and Disposition 24: 1218-1223.

Utesch D, Diener B, Molitor E, Oesch F, Platt K-L. 1992. Characterization of cryopreserved rat liver parenchymal cells by metabolism of diagnostic substrates and activities of related enzymes. Biochemical Pharmacology 44: 309-315.

Vanhulle VP, Martiat GA, Verbeeck RK, Horsmans Y, Calderon PB, Eeckhoudt SL, Taper HS, Delzenne N. 2001. Cryopreservation of rat precision-cut liver slices by ultrarapid freezing: influence on phase I and II metabolism and on cell viability upon incubation for 24 hours. Life Sci. 68: 2391-2403.

Watanabe FD, Mullon CJ, Hewitt WR, Arkadopoulos N, Kahaku E, Eguchi S, Khalili T, Arnaout W, Shackleton CR, Rozga J, Solomon B, Demetriou AA. 1997. Clinical experience with a bioartificial liver in the treatment of severe liver failure. A phase I clinical trial. Ann.Surg. 225: 484-491.

Watts P, Grant MH. 1996. Cryopreservation of rat hepatocyte monolayer cultures. Human & Experimental Toxicology 15: 30-37.

Watts P, Smith MD, Edwards I, Zammit V, Brown V, Grant H. 1995. The influence of medium composition on the maintenance of cytochrome P- 450, glutathione content and urea synthesis: a comparison of rat and sheep primary hepatocyte cultures. J.Hepatol. 23: 605-612.

Wolfe J, Bryant G. 1999. Freezing, drying, and/or vitrification of membranes. Cryobiology 39: 103-129.

Workman EJ, Reynolds SE. 1950. Electrical phenomena occuring during the freezing of dilute aqueous solutions and their possible relationship to thunderstorm electricity. Physical Review 79: 254-259.

Wortelboer HM, de Kruif CA, van Iersal AAJ, Falke HE, Noordhoek J, Blaauboer BJ. 1990. The isoenzyme pattern of cytochrome P450 in rat hepatocytes in primary culture, comparing different enzyme activities in microsomal incubation and in intact monolayers. Biochem. Pharmacol. 40: 2525-2534.

Yano M, Marinelli RA, Roberts SK, Balan V, Pham L, Tarara JE, de Groen PC, LaRusso NF. 1996. Rat hepatocytes transport water mainly via a non-channel-mediated pathway. J.Biol.Chem. 271: 6702-6707.

Zhao J, Hao HN, Thomas RL, Lyman WD. 2001. An efficient method for the cryopreservation of fetal human liver hematopoeitic progenitor cells. Stem Cells 19: 212-218.

INDEX

MIX
Papier aus verantwortungsvollen Quellen
Paper from responsible sources
FSC® C105338

If you have any concerns about our products,
you can contact us on
ProductSafety@springernature.com

In case Publisher is established outside the EU,
the EU authorized representative is:
Springer Nature Customer Service Center GmbH
Europaplatz 3, 69115 Heidelberg, Germany

Printed by Libri Plureos GmbH
in Hamburg, Germany